Advances in
Heterocyclic Chemistry

Volume 27

Editorial Advisory Board

R. A. Abramovitch
A. Albert
A. T. Balaban
S. Gronowitz
T. Kametani
C. W. Rees
Yu. N. Sheinker
H. A. Staab
M. Tišler

Advances in
HETEROCYCLIC CHEMISTRY

Edited by

A. R. KATRITZKY

Department of Chemistry
University of Florida
Gainesville, Florida

A. J. BOULTON

School of Chemical Sciences
University of East Anglia
Norwich, England

1980

ACADEMIC PRESS
A Subsidiary of Harcourt Brace Jovanovich, Publishers
New York   London   Toronto   Sydney   San Francisco

Volume 27

COPYRIGHT © 1980, BY ACADEMIC PRESS, INC.
ALL RIGHTS RESERVED.
NO PART OF THIS PUBLICATION MAY BE REPRODUCED OR
TRANSMITTED IN ANY FORM OR BY ANY MEANS, ELECTRONIC
OR MECHANICAL, INCLUDING PHOTOCOPY, RECORDING, OR ANY
INFORMATION STORAGE AND RETRIEVAL SYSTEM, WITHOUT
PERMISSION IN WRITING FROM THE PUBLISHER.

ACADEMIC PRESS, INC.
111 Fifth Avenue, New York, New York 10003

United Kingdom Edition published by
ACADEMIC PRESS, INC. (LONDON) LTD.
24/28 Oval Road, London NW1 7DX

LIBRARY OF CONGRESS CATALOG CARD NUMBER: 62–13037

ISBN 0–12–020627–7

PRINTED IN THE UNITED STATES OF AMERICA

80 81 82 83   9 8 7 6 5 4 3 2 1

Contents

CONTRIBUTORS vii

PREFACE ix

1-Azabicyclo[3.1.0]hexanes and Analogs with Further Heteroatom Substitution
DAVID ST. C. BLACK AND JAMES E. DOYLE

I. Introduction	1
II. Systems with One Heteroatom	2
III. Systems with Two Heteroatoms	10
IV. Systems with Three Heteroatoms	24
V. NMR Spectroscopic Data	29

Heteroaromatic Radicals, Part II: Radicals with Group VI and Groups V and VI Ring Heteroatoms
PETER HANSON

I. Introduction	32
II. Radicals from Oxygen Heterocycles	33
III. Radicals from Sulfur Heterocycles	68
IV. Radicals from Selenium Heterocycles	107
V. Radicals from Heterocycles with Mixed Group VI Heteroatoms	110
VI. Radicals from Heterocycles with Group V and Group VI Heteroatoms	112
VII. Conclusion	148

The 1,2- and 1,3-Dithiolium Ions
NOËL LOZAC'H AND MADELEINE STAVAUX

I. Introduction	152
II. The 1,2-Dithiolium Ion	153
III. The 1,3-Dithiolium Ion	199

Advances in Imidazole Chemistry
M. R. Grimmett

I. Introduction	242
II. Formation of the Imidazole Ring	242
III. Physical Properties	270
IV. Chemical Properties	282
V. Appendix	323

Cumulative Index of Titles 327

Contributors

Numbers in parentheses indicate the pages on which the authors' contributions begin.

DAVID ST. C. BLACK, *Department of Chemistry, Monash University, Clayton, Victoria 3168, Australia* (1)

JAMES E. DOYLE, *Department of Chemistry, Monash University, Clayton, Victoria 3168. Australia* (1)

M. R. GRIMMETT, *Chemistry Department, University of Otago, P.O. Box 56, Dunedin, New Zealand* (241)

PETER HANSON, *Department of Chemistry, University of York, Heslington, York YO1 5DD, England* (31)

NOËL LOZAC'H, *Institut des Sciences de la Matière et du Rayonnement, Université de Caen, 14032 Caen Cedex, France* (151)

MADELEINE STAVAUX, *Institut des Sciences de la Matière et du Rayonnement, Université de Caen, 14032 Caen Cedex, France* (151)

Preface

Volume 27 contains four contributions. Black and Doyle discuss "1-Azabicyclo[3.1.0]hexanes and Analogs with Further Heteroatom Substitution." Part II of the treatment of "Heteroaromatic Radicals" by Hanson deals with radicals with group VI heteroatoms and completes the review of the heteroaromatic free radicals begun in Volume 25. In "The 1,2- and 1,3-Dithiolium Ions," Lozac'h and Stavaux bring up to date the review by Prinzbach and Futterer which appeared in 1966 in Volume 7 of this publication. Finally, Grimmett has updated his own review of imidazole chemistry which appeared in 1970 in Volume 12.

A. R. KATRITZKY
A. J. BOULTON

1-Azabicyclo[3.1.0]hexanes and Analogs with Further Heteroatom Substitution

DAVID ST. C. BLACK AND JAMES E. DOYLE

Department of Chemistry, Monash University, Clayton, Australia

I. Introduction	1
II. Systems with One Heteroatom	2
1-Azabicyclo[3.1.0]hexanes	2
1. Preparations of 1-Azabicyclo[3.1.0]hexanes	2
2. Reactions of 1-Azabicyclo[3.1.0]hexanes	8
III. Systems with Two Heteroatoms	10
A. 1,2-Diazabicyclo[3.1.0]hexanes	10
B. 1,3-Diazabicyclo[3.1.0]hexanes	10
1. Preparations of 1,3-Diazabicyclo[3.1.0]hexanes	10
2. Preparations of 1,3-Diazabicyclo[3.1.0]hex-3-enes	11
3. Reactions of 1,3-Diazabicyclo[3.1.0]hex-3-enes	13
C. 1,4-Diazabicyclo[3.1.0]hexanes	15
D. 1,5-Diazabicyclo[3.1.0]hexanes	16
E. 3-Oxa-1-azabicyclo[3.1.0]hexanes	17
F. 6-Oxa-1-azabicyclo[3.1.0]hexanes	18
1. Preparations of 6-Oxa-1-azabicyclo[3.1.0]hexanes	18
2. Reactions of 6-Oxa-1-azabicyclo[3.1.0]hexanes	19
G. 3-Thia-1-azabicyclo[3.1.0]hexanes	23
IV. Systems with Three Heteroatoms	24
A. 1,3,5-Triazabicyclo[3.1.0]hexanes	24
1. Preparations of 1,3,5-Triazabicyclo[3.1.0]hexanes	24
2. Reactions of 1,3,5-Triazabicyclo[3.1.0]hexanes	25
B. Oxadiazabicyclo[3.1.0]hexanes	26
C. 4,6-Dioxa-1-azabicyclo[3.1.0]hexanes	27
D. 2-Thia-3-oxa-1-azabicyclo[3.1.0]hexanes	27
V. NMR Spectroscopic Data	29

I. Introduction

The azabicyclo[3.1.0]hexane ring system (**1**) may be considered as arising from the corresponding carbobicyclic ring system by the replacement of one of the carbon atoms with nitrogen. Position in the ring system is then identified, beginning from a bridgehead atom, with preference being given to the heteroatom, and numbering sequentially through the larger ring to the three-membered ring. The parent 1-azabicyclo[3.1.0]hexane system (**1**) is numbered as shown.

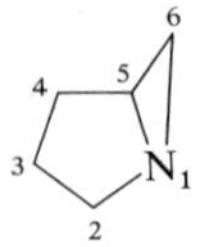

(**1**)

Several reports of this type of compound in the early literature have been reviewed elsewhere.[1,2] However, most of the work in this area has been undertaken since these reviews, especially with respect to analogs with further heteroatom substitution.

II. Systems with One Heteroatom

1-AZABICYCLO[3.1.0]HEXANES

1. *Preparations of 1-Azabicyclo[3.1.0]hexanes*

In 1948, 1-ethyl-1-azoniabicyclo[3.1.0]hexane chloride (**3**), postulated by Fuson and Zirkle[3] to be an intermediate in the conversion of 2-chloromethyl-

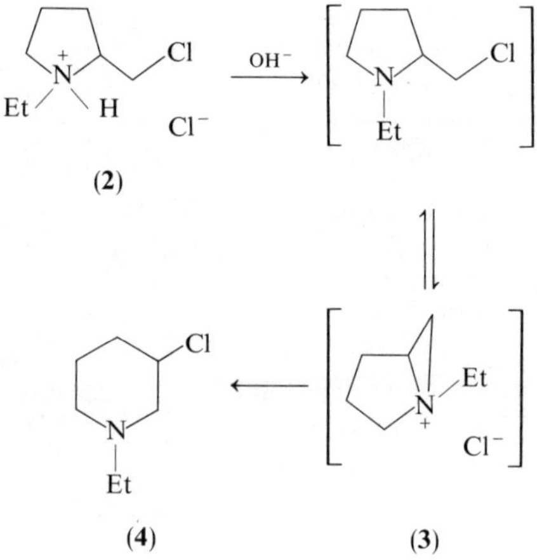

---

[1] W. L. Mosby, *in* "The Chemistry of Heterocyclic Compounds" (W. L. Mosby, ed.), Vol. XV, pp. 15–25. Wiley (Interscience), New York, 1961.

[2] L. C. Behr, *in* "The Chemistry of Heterocyclic Compounds" (R. H. Wiley, ed.), Vol. XVII, p. 295. Wiley (Interscience), New York, 1962.

[3] R. C. Fuson and C. L. Zirkle, *J. Am. Chem. Soc.* **70**, 2760 (1948).

1-ethylpyrrolidinium chloride (2) to its piperidine analog (4) by aqueous sodium hydroxide at 20°C, has since been isolated as its perchlorate salt,[4] as have other examples of this ring system.[5]

Eleven years later, a German patent described[6] the preparation of ethyl 1-azabicyclo[3.1.0]hexane-5-carboxylate (6) by the reaction of ethyl piperidine-3-carboxylate (5) with *tert*-butyl hypochlorite, followed by treatment with base.

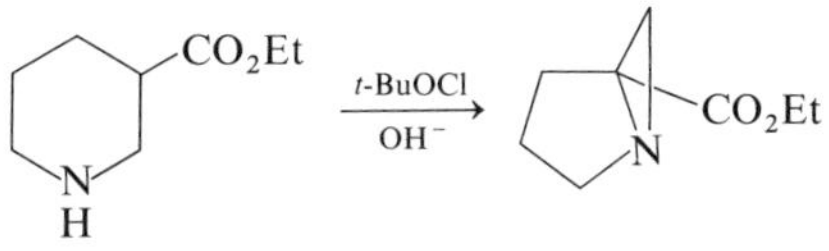

(5) → (6)

a. *Thermal Decomposition of Olefinic Azides.* The thermal decomposition of δ,ε-unsaturated azides in hydrocarbon solvents has been shown by Logothetis[7] to lead to at least two compounds, the major product being an imine and the minor product a 1-azabicyclohexane (see Table I).

The ratio of products formed from 5-azidohex-1-ene (10) shows a marked dependence upon reaction conditions. Thus, the ratio of imine 11 to bicyclic compound 12 in toluene (111°C) after 4 hours is 2.5:1, whereas in xylene (138°C), for 0.5 hour, the ratio increases to 8:1.

The products of the room-temperature reactions of 5-azido-5-methylhex-1-ene (7) and 5-azidohex-1-ene (10) were assigned as the triazolines 18 on the basis of their ^{1}H NMR and infrared spectra. Decomposition of these substances under the reaction conditions gave rise to the same mixture of products previously formed. The high-temperature reaction was therefore assumed to proceed via this intermediate in all cases. The azabicyclohexane can form from this intermediate by ring opening to 19, followed by

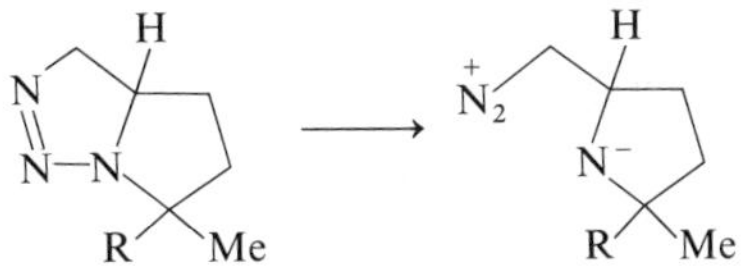

(18)   (19)
R = H or Me

---

[4] C. F. Hammer, S. R. Heller, and J. H. Craig, *Tetrahedron* **28**, 239 (1972).
[5] E. M. Fry, *J. Org. Chem.* **30**, 2058 (1965), and references cited therein.
[6] E. Merck A G, German Patent 1,054,088 (1959) [*CA* **55**, 8439 (1961)].
[7] A. L. Logothetis, *J. Am. Chem. Soc.* **87**, 749 (1965).

TABLE I
1-AZABICYCLO[3.1.0]HEXANES FROM AZIDES

Azide	Decomposition conditions	Products (percentage of distilled product)	
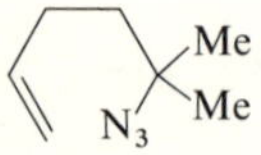 (7)	Cyclohexane 80°C, 8 h	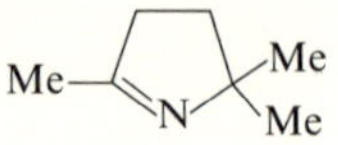 (8) 48%	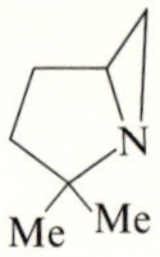 (9) 22%
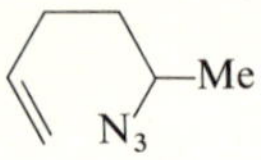 (10)	Toluene 111°C, 4 h	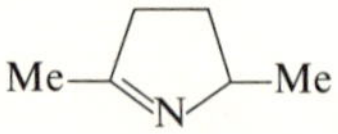 (11) 36%	 (12) 14%
(10)	Xylene 138°C, 18 h	44%	6%
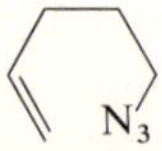 (13)	o-Dichlorobenzene 120–130°C, 2 h	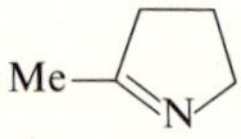 (14) 63%	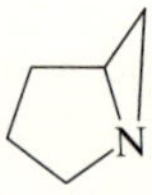 (1) 9%
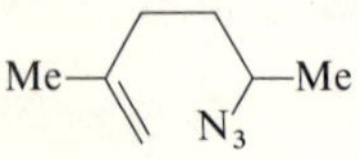 (15)	Xylene 138°C, 18 h	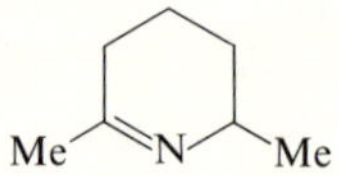 (16) 28%	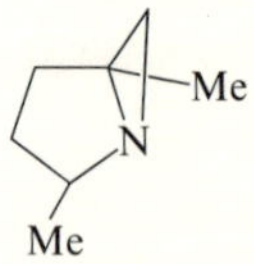 (17) 22%

attack of the ring nitrogen on carbon and loss of nitrogen (N_2). The imines **8**, **11**, and **14** can also be formed by extrusion of nitrogen and a 1,2-hydride shift, followed by double-bond formation. The case of imine **16** requires a methylene shift from **20** in a similar manner.

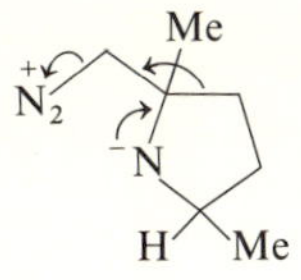

(20)

b. *Preparation Involving Substitution Reactions.* The parent 1-azabicyclo[3.1.0]hexane **1** was prepared by Buyle,[8] in 25% yield, by the conversion of prolinol (**21**) into 2-(chloromethyl)pyrrolidinium chloride (**22**) by the action of thionyl chloride and subsequent treatment of this salt with aqueous sodium hydroxide. Similarly, Gassman and Fentiman[9] treated L-proline with hot sulfuric acid, to form the sulfate ester **23** which was converted into **1** in 50% yield. An attempt[9] to form the bicyclic compound **1** from the sulfate ester (**24**) of 3-hydroxypiperidine led only to trace formation of the desired product.

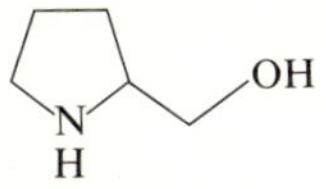

(21) 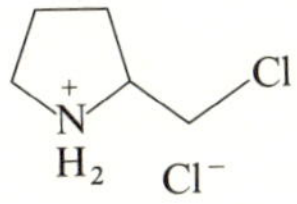 (22) 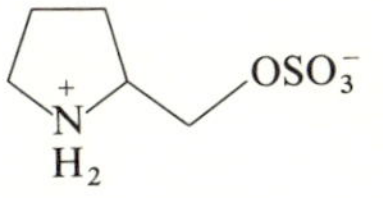 (23) 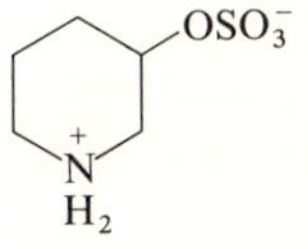 (24)

Recently, Laurent[10,11] prepared 6-methyl- and 6,6-dimethyl-5-phenyl-1-azabicyclo[3.1.0]hexane (**27**) by hydroboration of the 3-allylaziridines **25** to the alcohols **26**, followed by bromination and cyclization.

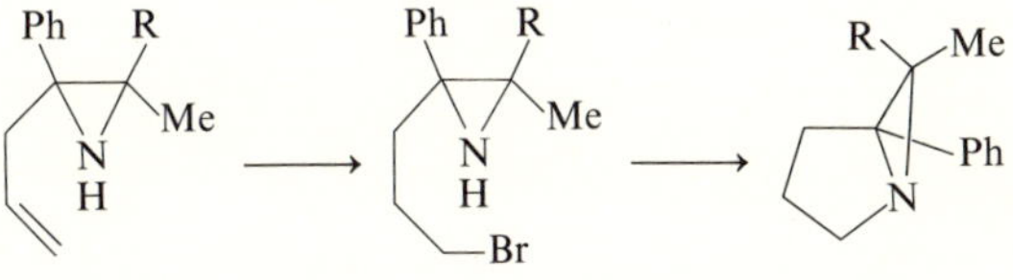

(25) (26) (27)

Addition of bromine to the *gem*-diphenyl substituted δ,ε-unsaturated amines **28** was shown[12,13] to afford mixtures of the pyrrolidinium and piperidinium salts, **29** and **30**, which were converted into the 1-azabicyclohexanes **31** by treatment with sodium hydride in dimethylformamide.

[8] R. Buyle, *Chem. Ind.* (*London*), 195 (1966).
[9] P. G. Gassman and A. Fentiman, *J. Org. Chem.* **32**, 2388 (1967).
[10] D. A. Laurent, *Chimia* **30**, 274 (1976).
[11] R. Chaabouni, A. Laurent, and B. Marquet, *Tetrahedron Lett.*, 3149 (1976).
[12] D. E. Horning and J. M. Muchowski, *Can. J. Chem.* **52**, 1321 (1974).
[13] D. St.C. Black and J. E. Doyle, *Aust. J. Chem.* **31**, 2247 (1978).

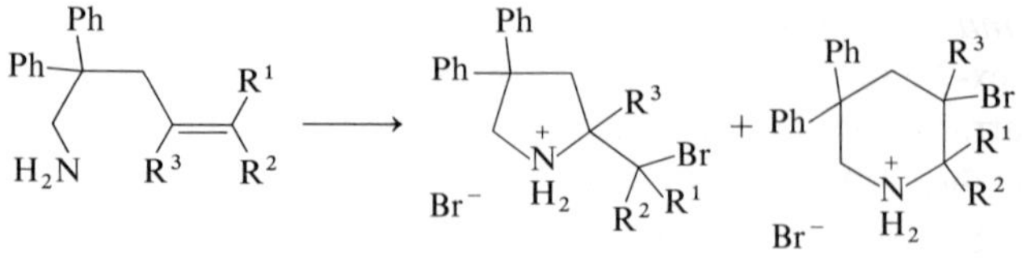

(28) (29) (30)

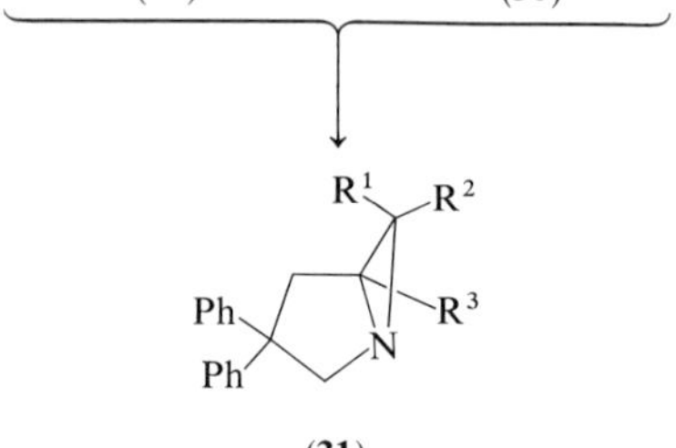

	R¹	R²	R³
a:	H	H	H
b:	H	H	Me
c:	H	Ph	H
d:	Me	Me	H

(31)

However, similar treatment of alkyl-substituted δ,ε-unsaturated alkylamines with bromine and sodium hydride fails to yield azabicyclohexanes, but unsaturated acyclic products are formed instead. Suitable bulky substituents are thus apparently essential for the satisfactory synthesis of azabicyclohexanes from δ,ε-unsaturated amines.[13]

Azabicyclohexenes can also be prepared by substitution reactions, but again the method is not general. Treatment of the 2H-pyrroles (32a,b) with sodium ethoxide in ethanol gives the azabicyclohexenes (33a,b).[14,15] In the case of 32b, the pyridine compound 34b is also formed and the 2H-pyrrole 32c affords only the pyridine compound 34c and no azabicyclohexene 33c.[15]

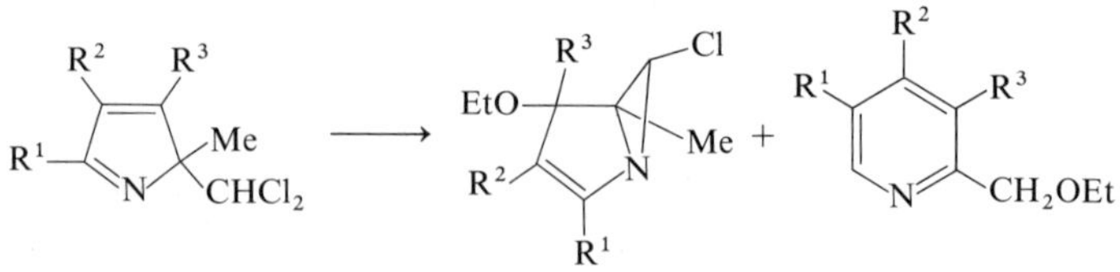

(32) (33) (34)

	R¹	R²	R³
a:	Me	H	H
b:	CMe₃	H	H
c:	Me	Me	Me

Compound 33a can be reconverted to the 2H-pyrrole 32a by treatment with hydrochloric acid.[14]

---

[14] R. Nicoletti and M. L. Forcellese, *Gazz. Chim. Ital.* **97**, 148 (1967).
[15] R. Nicoletti, M. L. Forcellese, and C. Germani, *Gazz. Chim. Ital.* **97**, 685 (1967).

c. *Thermal Rearrangement of 2-Allyl-2H-azirines.* The 1-azabicyclo[3.1.0]hex-2-ene (**37**) has been postulated as an intermediate in the thermal rearrangement of the 2H-azirine **35** at 180°C to the pyrrole **38** and the pyridine **40**,[16] the latter formed via the intermediate **39**. The formation of **37** is thought to arise via intramolecular addition of the vinyl nitrene **36**.

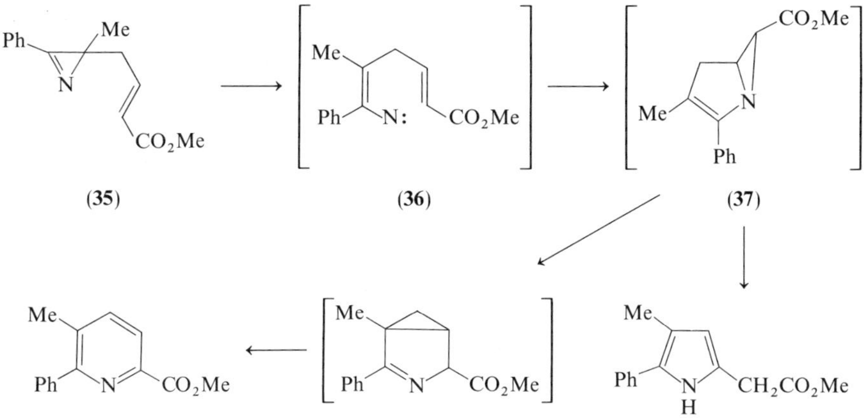

d. *Reaction of 1-Pyrroline 1-Oxides with Ylids.* Reaction of triethylphosphonosodioacetate with the aldonitrones **41a–c** forms the ethyl 1-azabicyclohexane-6-carboxylates **42**, sometimes with the enamines **43**.[17–20]

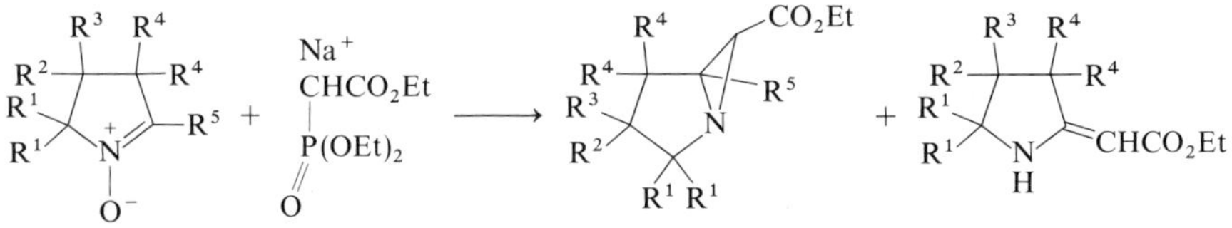

	R^1	R^2	R^3	R^4	R^5
a:	Me	H	H	H	H
b:	Me	H	H	Me	H
c:	Me	Ph	H	Me	H
d:	H	Me	Me	H	Me

[16] A. Padwa and P. H. J. Carlsen, *Tetrahedron Lett.*, 433 (1978).
[17] D. St.C. Black and V. C. Davis, *J. C. S., Chem. Commun.*, 416 (1975).
[18] D. St.C. Black and V. C. Davis, *Aust. J. Chem.* **29**, 1735 (1976).
[19] E. Breuer, S. Zbaida, J. Pesso, and S. Levi, *Tetrahedron Lett.*, 3103 (1975).
[20] S. Zbaida and E. Breuer, *Tetrahedron* **34**, 1241 (1978).

Reaction of the same phosphorus ylid with the nitrone **41d**,[17,18] or of the ylid derived from diethylcyanomethylphosphonate,[19,20] with the aldo-nitrone **41a** gives rise only to the corresponding azabicyclohexane. However, the reaction is not general for other phosphonates.

e. *Reaction of 1-Pyrroline 1-Oxides with Dimethyl Acetylenedicarboxylate.* Ylids **47** were postulated[21,22] to form from the 1,3-dipolar cycloadducts **45** of 2-alkyl-1-pyrroline 1-oxides **44** and dimethylacetylene dicarboxylate via the azabicyclohexanes **46**. However, these compounds (**46**) have neither been isolated nor detected, possibly because of instability conferred on the aziridine ring by the presence of two electron-withdrawing substituents.

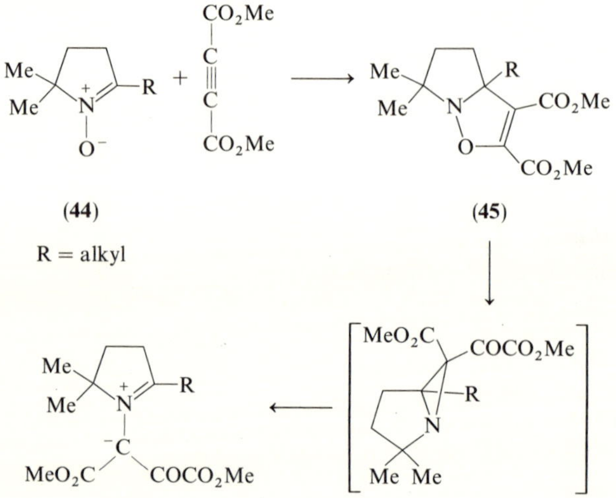

2. Reactions of 1-Azabicyclo[3.1.0]hexanes

a. *Ring-Destroying Reactions.* Treatment of the bicyclic compounds **1** or **9** with methyl iodide[7] caused cleavage of the N—C5 bond, to give quaternary salts of the corresponding 3-iodopiperidines **48** and **49**.

---

[21] R. Grigg, *Chem. Commun.*, 607 (1966).
[22] D. St.C. Black, R. F. Crozier, and V. C. Davis, *Synthesis*, 205 (1975).

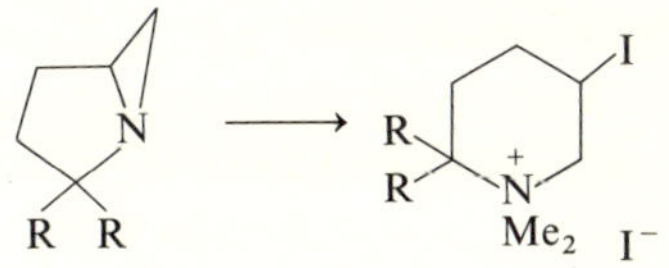

Sec. II]   1-AZABICYCLO[3.1.0]HEXANES AND ANALOGS

(1) R = H
(9) R = Me

(48) R = H
(49) R = Me

Similarly, treatment of the parent compound **1** with acetic anhydride in hexane at low temperature has been shown[23] to form the intermediate **50**. Subsequent attack by acetate ion at C5 and rupture of the N—C5 bond gives rise to the piperidine **51** whereas attack at C6 with cleavage of the N—C6 bond affords **52**.

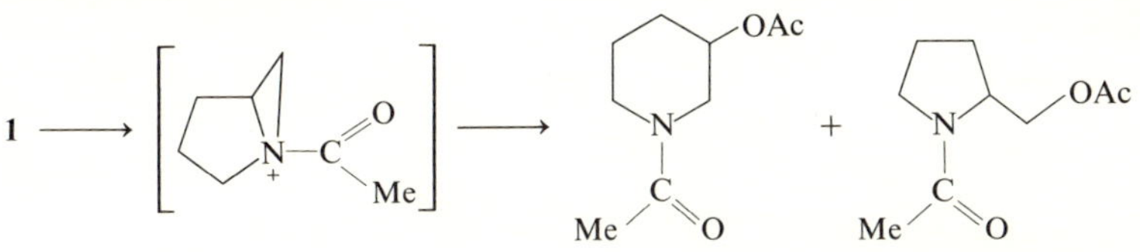

(50)   (51)   (52)

Ethyl 1-azabicyclo[3.1.0]hexane-5-carboxylate **6** with hydrochloric acid gives 3-chloropiperidine-3-carboxylic acid hydrochloride **53** by ring opening with concomitant hydrolysis. Catalytic reduction with hydrogen over palladium on charcoal causes cleavage of both aziridine N—C bonds of the parent azabicyclohexane **1**, to give 2-methylpyrrolidine (**54**) and piperidine (**55**) in the ratio 2:1.[9]

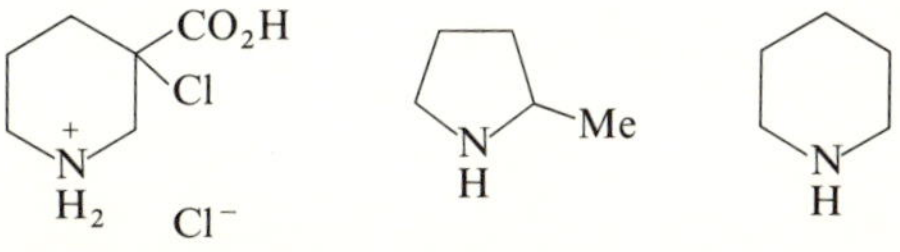

(53)   (54)   (55)

b. *Ring-Preserving Reactions.* Under mild conditions **6** can be hydrolyzed to the corresponding free acid.[6] The esters **42** have been converted to the corresponding benzylamides by reaction with benzylamine.[17,18]

c. *Polymerization.* Several authors[7-9] have reported that this class of compounds undergoes ready polymerization which is catalyzed by ammonium salts and a boron trifluoride complex.[24]

[23] J. L. Wong and D. O. Helton, *J. C. S., Chem. Commun.*, 352 (1973).
[24] E. F. Razvodovskii, A. V. Nekrasov, L. M. Pushchaeva, I. S. Morozova, M. A. Markevich, A. A. Berlin, and N. S. Enikolopyan, *J. Macromol. Sci., Chem.* **8**, 241 (1974)[*CA* **80**, 133922a (1974)].

III. Systems with Two Heteroatoms

A. 1,2-DIAZABICYCLO[3.1.0]HEXANES

Photorearrangement of the 2,3-diazabicyclo[3.1.0]hexene dicarboxylic ester **56** gave the 1,2-diazabicyclohexene **57**, which at 20°C decomposes to give the alkene **58**.[25]

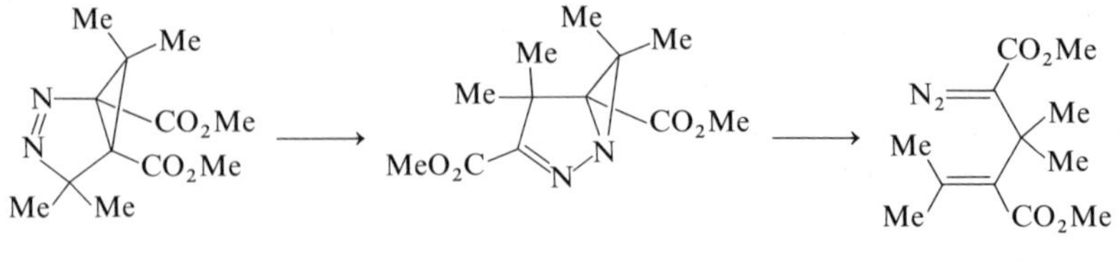

B. 1,3-DIAZABICYCLO[3.1.0]HEXANES

1. *Preparations of 1,3-Diazabicyclo[3.1.0]hexanes*

Treatment of 2-aminomethylaziridine (**59**) with a variety of aromatic or aliphatic aldehydes or ketones in refluxing ethanol gives rise to several examples of 2-substituted 1,3-diazabicyclo[3.1.0]hexanes **60**.[26,27] The ^{1}H NMR spectra of these diazabicyclohexanes have been reported[28] and the crystal structure of 2-*p*-bromophenyl-1,3-diazabicyclo[3.1.0]hexane has been determined.[29]

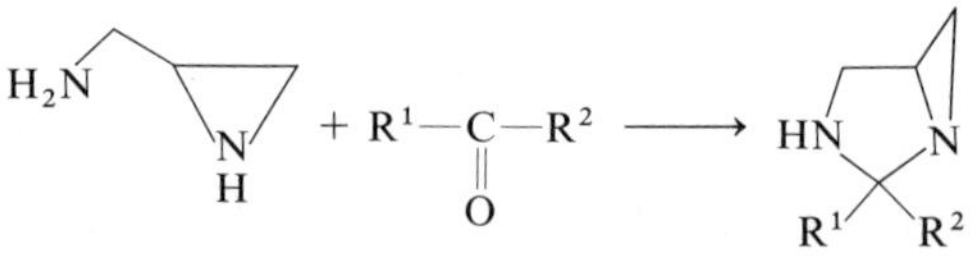

R^1 = H or Me; R^2 = Me, aryl, HOCH$_2$CH(OH), MeCH(OH), COMe, CH$_2$COMe

---

[25] M. Franck-Neumann, D. Martina, and C. Dietrich-Buchecker, *Tetrahedron Lett.*, 1763 (1975).
[26] S. Hillers, M. Lidaks, A Eremeev, and V. A. Kholodnikov, *Khim. Geterotsikl. Soedin.*, 483 (1972) [*CA* **77**, 61880 (1972)].
[27] S. Hillers, A. V. Eremeev, and V. A. Kholodnikov, *Khim. Geterotsikl. Soedin.*, 425 (1975) [*CA* **83**, 28155 (1975)].
[28] S. Hillers, E. Liepins, A. V. Eremeev, and V. A. Kholodnikov, *Khim. Geterotsikl. Soedin.*, 940 (1975) [*CA* **83**, 155201 (1975)].
[29] S. A. Hiller, Y. Y. Bleidelis, A. A. Kemme, and A. V. Eremeyev, *J. C. S., Chem. Commun.*, 130 (1975).

Matsumoto and Maruyama[30] have prepared more highly substituted examples of this system by the reaction of several aziridines **61** with a large excess of 2-phenylazirine **62a** in benzene or toluene, to give, in most cases, the isomeric diazabicyclohexanes **63** and **64** as crystalline solids (77–81%). This condensation is not restricted to N-cyclohexyl- or benzoyl-substituted aziridines.[30]

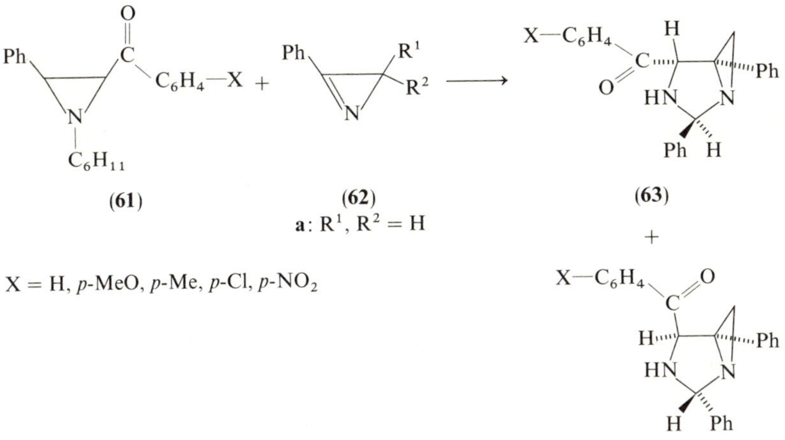

X = H, p-MeO, p-Me, p-Cl, p-NO$_2$

2. Preparations of 1,3-Diazabicyclo[3.1.0]hex-3-enes

Heine prepared[31,32] several 1,3-diazabicyclo[3.1.0]hex-3-enes **66** by the reaction of trans-2-aryl-3-aroylaziridines **65** with aldehydes or ketones in ethanolic ammonia containing ammonium bromide, at 20°C.

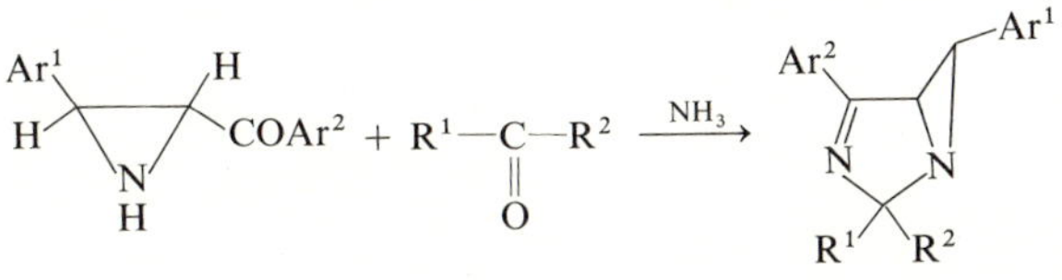

Ar^1 mainly p-NO$_2$C$_6$H$_4$; Ar^2 mainly Ph
R^1 = H, Me, or alicyclic; R^2 = alkyl or aryl

Since the initial trans-aziridine could be regenerated by allowing the bicyclic compound **66** to stand in moist acetic acid, no epimerization took

---

[30] K. Matsumoto and K. Maruyama, *Chem. Lett.*, 759 (1973).
[31] H. W. Heine, R. H. Weese, R. A. Cooper, and A. J. Durbetaki, *J. Org. Chem.* **32**, 2708 (1967).
[32] H. W. Heine, R. H. Weese, and R. A. Cooper, U.S. Patent 3,609,165 (1971)[*CA* **75**, P151794 (1971)].

place during the reaction.[31,32] Padwa[33,34] similarly converted 1,2-dibenzoyl-2-phenylaziridine (67), into a mixture of *exo* and *endo* isomers of the diazabicyclohexene 68.

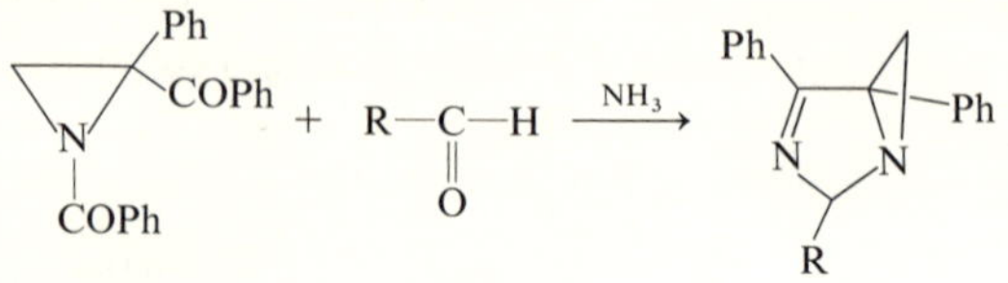

(67) R = H or Ph (68)

Benzonitrile-*p*-nitrobenzylide (69) reacts *in situ* with 2*H*-azirines 62 to form 1,3-diazabicyclo[3.1.0]hex-3-enes 70 in low to moderate yields.[35] Other

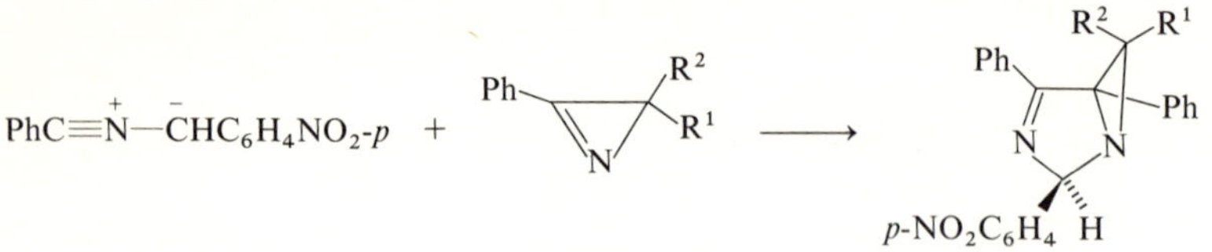

(69) (62) (70)

a: $R^1, R^2 = H$
b: $R^1 = Ph; R^2 = H$

products isolated from this reaction mixture were shown to arise from degradation of the initially formed bicyclic products. Padwa[36] has shown that 2-phenylazirine (62a) undergoes various photochemical reactions which exhibit the characteristics of a 1,3-dipole, the structure of which has been represented as 71 ($R^1 = R^2 = H$). In the absence of a dipolarophile, azirines 62 dimerize[33,37] to form isomeric diazabicyclohex-3-enes 72 (35–85%).

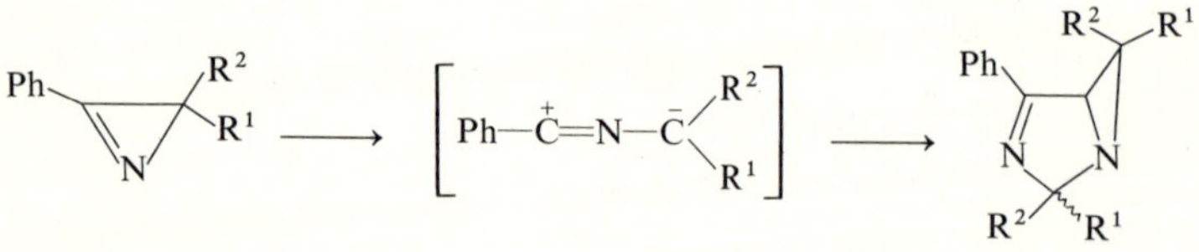

(62) (71) (72)

R^1 = H, Me, Ph, or *β*-naphthyl
R^2 = H

[33] A. Padwa, M. Dharan, J. Smolanoff, and S. I. Wetmore, *J. Am. Chem. Soc.* **95**, 1954 (1973).
[34] A. Padwa, J. Smolanoff, and S. I. Wetmore, *J. Org. Chem.* **38**, 1333 (1973).
[35] N. S. Narasimhan, H. Heimgartner, H. J. Hansen, and H. Schmid, *Helv. Chim. Acta* **56**, 1351 (1973).
[36] A. Padwa, M. Dharan, J. Smolanoff, and S. I. Wetmore, *J. Am. Chem. Soc.* **95**, 1945 (1973).
[37] N. Gakis, M. Märky, H. J. Hansen, and H. Schmid, *Helv. Chim. Acta* **55**, 748 (1972).

Sec. III.B] 1-AZABICYCLO[3.1.0]HEXANES AND ANALOGS

Further photorearrangement products of **71** and **72** were also isolated. If two azirines are selected so that one absorbs substantially more radiation than the other, it is possible to obtain mixed condensation products. Thus, when 2-phenylazirine (**62a**) and 2,3-diphenylazirine (**62b**) were irradiated[34] together, the products consisted of a small quantity of tetraphenylpyrazine and the two diazabicyclohexanes, **73** and **74**, which were formed by combination of 2-phenylazirine with the ylid from 2,3-diphenylazirine. No other bicyclic products were reported.

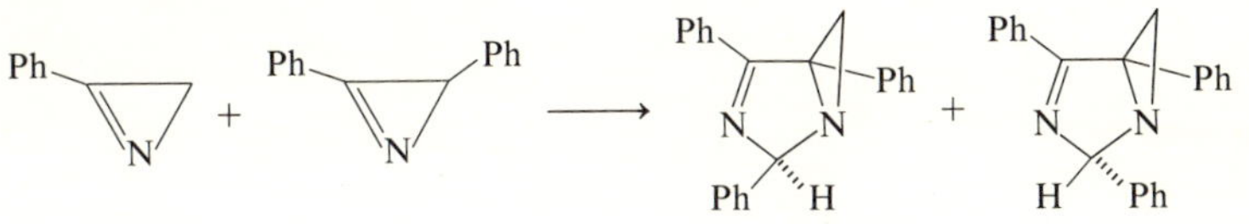

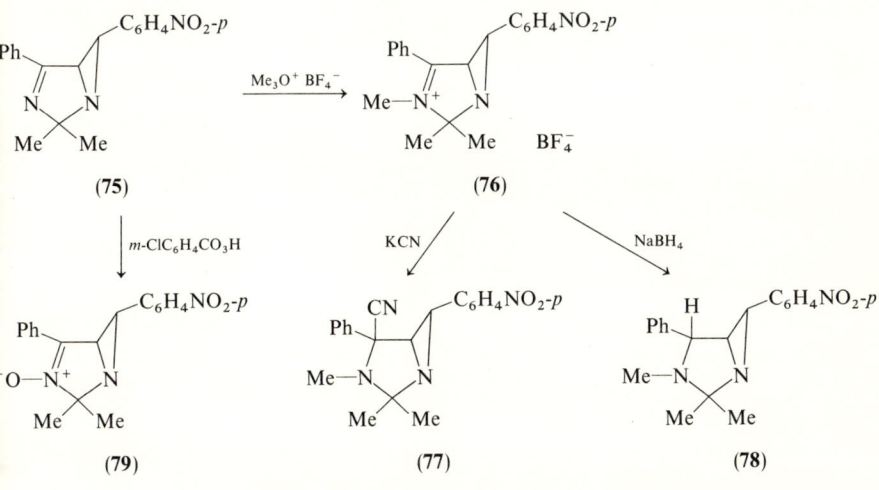

(62a)   (62b)   (73)   (74)

3. Reactions of 1,3-Diazabicyclo[3.1.0]hex-3-enes

a. *Ring-Preserving Reactions.* Trimethyloxonium tetrafluoroborate adds to the 1,3-diazabicyclo[3.1.0]hex-3-ene **75** to give the fluoroborate salt **76**.[38] This activates the imine bond toward attack by cyanide ion to give the nitrile **77**. Interestingly, the same salt (**76**) can be reduced by excess of sodium borohydride[38] to give the corresponding *N*-methylated and saturated 1,3-diazabicyclohexane **78** in good yield (87%). Ring opening was not observed. The diazabicyclohexene **75** has also been oxidized with *m*-chloroperbenzoic acid to the corresponding bicyclic nitrone **79**.[38]

---

[38] H. W. Heine, T. A. Newton, G. J. Blosick, K. C. Irving, C. Meyer, and G. B. Corcoran, *J. Org. Chem.* **38**, 651 (1973).

b. *Ring-Destroying Reactions.* Removal of the acidic proton attached to C2 of the bicyclic aziridine **80** with sodium methoxide in methanol induces cleavage of the N—C5 bond and generation of the trisubstituted pyrimidine **81**.[31,32,39]

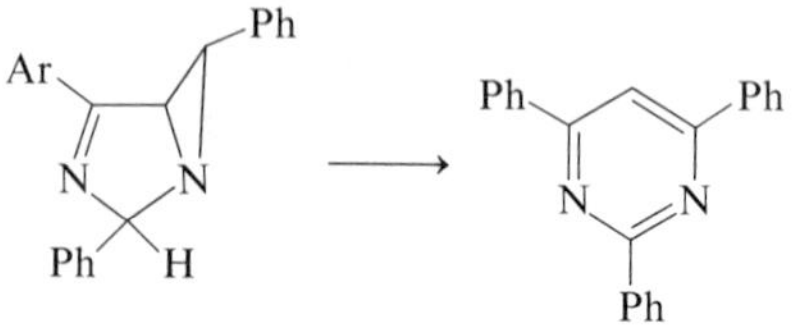

Ar = Ph, *p*-NO$_2$C$_6$H$_4$

Irradiation of a benzene solution of the bicyclic compounds **82** at 15°C leads to the formation of a labile intermediate assigned the enediimine structure **83**. At 50°C, this forms the *cis*-2,3-dihydropyrazine **84**.[39,40] It has been suggested that the ring opening of the diazabicyclohexene ring occurs in a stepwise fashion via an azomethine ylid intermediate such as **85**, which would arise from cleavage of the aziridine C—C bond.[39-41] The major product from the irradiation of the compound **82a** in methanol was the imidazoline **86** which is possibly formed by the addition of methanol across the ylid **85**; no dihydropyrazine **84** is formed under these conditions.[39]

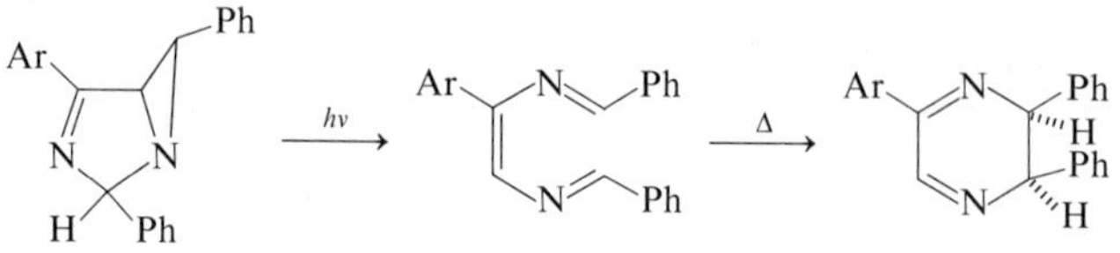

a: Ar = Ph
b: Ar = *p*-ClC$_6$H$_4$
c: Ar = biphenylyl

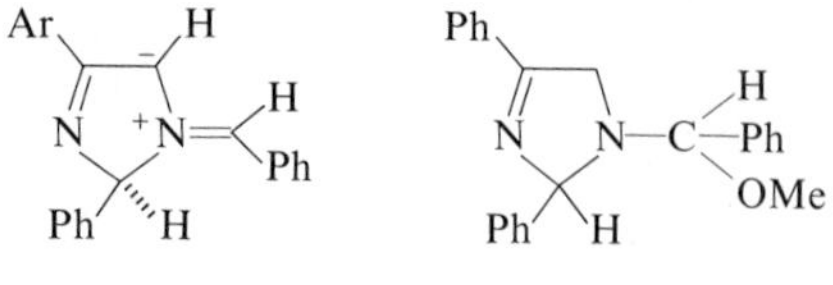

Heine[42] has shown that in refluxing xylene the diazabicyclohexene ring cycloadds a dipolarophile to form a 1,3-diazabicyclo[3.3.0]octane. Thus, **75** with diethyl acetylenedicarboxylate, diethyl fumarate, or dibenzoyl-

---

[39] A. Padwa and E. Glazer, *J. Am. Chem. Soc.* **94**, 7788 (1972).
[40] T. DoMinh and A. M. Trozzolo, *J. Am. Chem. Soc.* **92**, 6997 (1970).
[41] T. DoMinh and A. M. Trozzolo, *J. Am. Chem. Soc.* **94**, 4046 (1972).
[42] H. W. Heine, A. B. Smith, and J. D. Bower, *J. Org. Chem.* **33**, 1097 (1968).

ethylene forms the bicyclic compounds **87**, **88**, and **89**, respectively. More

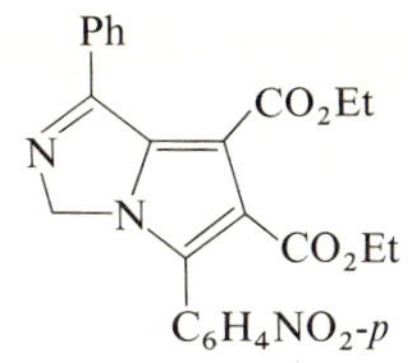

(87) 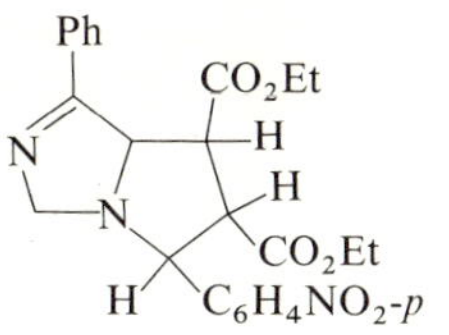   (88) 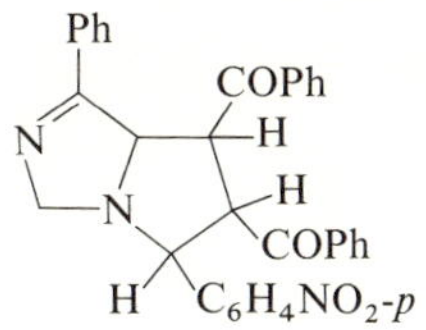   (89)

recently Padwa[43] has investigated the reaction of the triphenyl compound **82a** with dimethyl maleate and dimethyl fumarate. The same cycloadducts formed under both thermal and photochemical conditions. Those from reaction with dimethyl fumarate are outlined in Eq. (1). The mechanism of this cycloaddition is also believed to involve an azomethine ylid similar to **85**.[43]

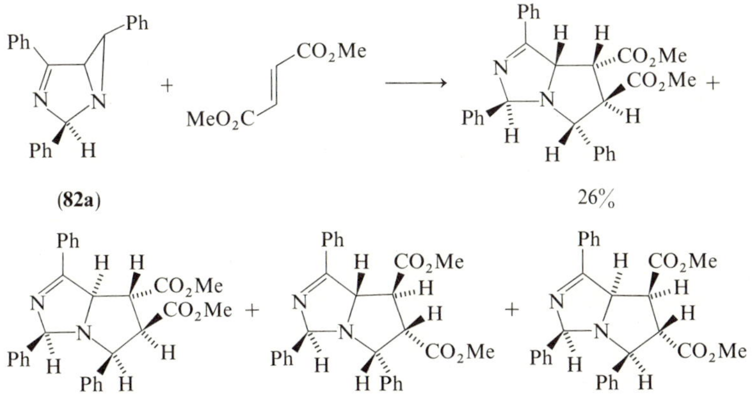

(82a)   26%

27%   13%   13%   (1)

C. 1,4-Diazabicyclo[3.1.0]hexanes

2-Phenylazirine **62a** is converted by copper(II) bromide in carbon tetrachloride at low temperature into structure **90**, assigned on the basis of the following reactions.[44] The bicyclic product decomposed at ambient temperature to afford 2-bromomethyl-2,4-diphenyl-2H-imidazole (**91**), while reaction with methanolic potassium hydroxide gave 2,6-diphenylpyrazine (**92**). These reactions were consistent with the proposed structure **90** but are

[43] A. Padwa and E. Glazer, *J. Org. Chem.* **38**, 284 (1973).
[44] K. Hayashi, K. Isomura, and H. Taniguchi, *Chem. Lett.*, 1011 (1975).

equally appropriate for structure **93** in which the bromine is a substituent of the aziridine ring. The AB quartet at $\delta = 3.9$ and 4.4 ppm ($J = 15$ Hz) for the original product indicated the presence of a methylene group having nonequivalent protons. As aziridine protons with phenyl bridgehead substitution typically absorb in the range 1.4–3.2 ppm (see Section V), the aziridine resonances of **90** appear to occur at particularly low field.

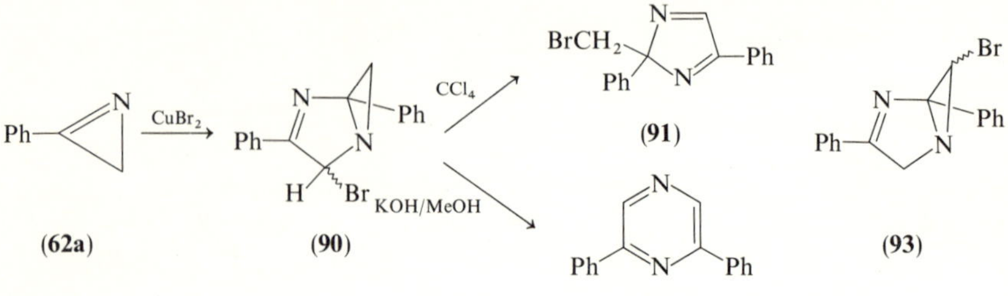

D. 1,5-DIAZABICYCLO[3.1.0]HEXANES

2-Methyl-1,5-diazabicyclo[3.1.0]hexane (**94**)[45] is formed in 90% yield by reaction of 1,3-diaminobutane with formalin and sodium hypochlorite. The reaction presumably proceeds via initial formation of 4-methyl hexahydropyrimidine which subsequently undergoes oxidation to **94** (Eq. 2).

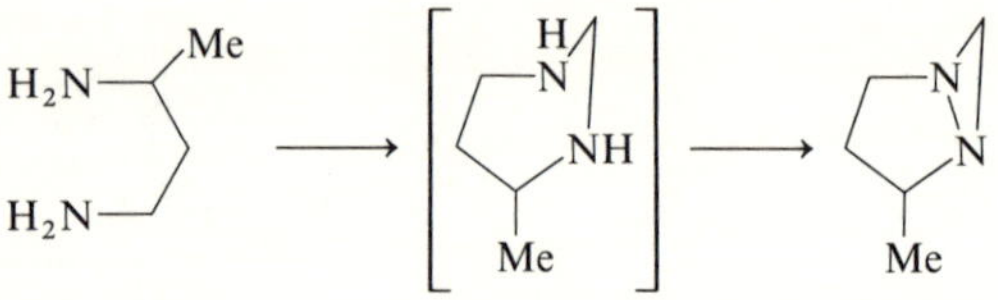

$$\tag{2}$$

5,5-Dimethylpyrazolidin-3-one (**95a**) combines with benzaldehyde to give the betaine **96a** which undergoes photoisomerization in quantitative yield to 4,4-dimethyl-6-phenyl-1,5-diazabicyclo[3.1.0]hexan-2-one (**97a**).[46] Similarly, the betaine **96b** leads to the corresponding bicyclic compound **97b**. Solutions of the bicyclic compound **97b** revert to the betaine on standing at room temperature[47] while compound **97a** forms the betaine **96a** on exposure to moist air[46]; both compounds also revert to the betaine in the presence of acid.

---

[45] R. Ohme, E. Schmitz, and P. Dolge, *Chem. Ber.* **99**, 2104 (1966).
[46] M. Schulz and G. West, *J. Prakt. Chem.* **315**, 711 (1973).
[47] M. Schulz and G. West, *J. Prakt. Chem.* **312**, 161 (1970).

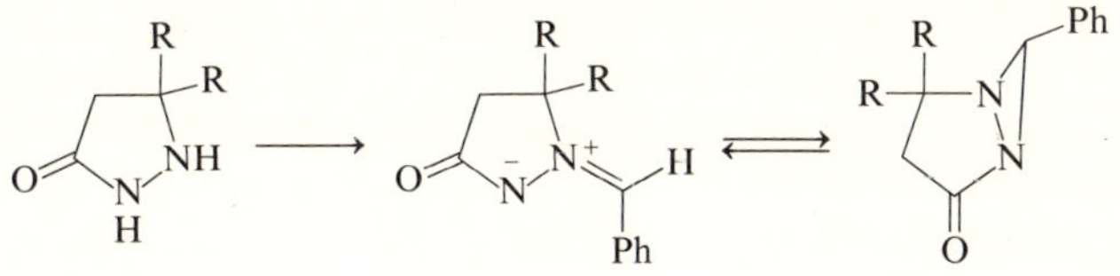

(95) (96) (97)

a: R = Me
b: R = H

Reaction of the diazabicyclohexane **97b** with methanolic sodium methoxide[47] results in nucleophilic attack at the amide function, to cause ring opening to the aziridine ester **98**. Potassium hydroxide[46] also attacks the diazabicyclohexane **97a** at the amide, but to cleave both rings to the hydrazone **99**. The less nucleophilic potassium *tert*-butoxide converts **97b** into the pyrimidine **100** (26%).[48]

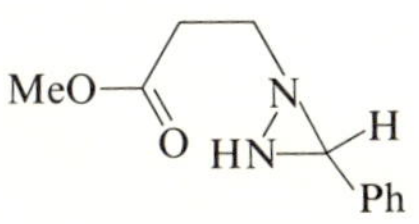

(98) 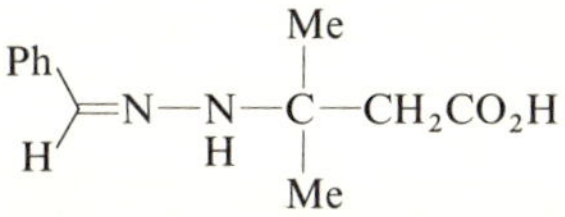 (99) 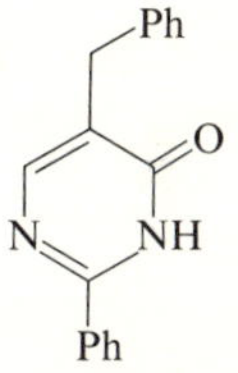 (100)

E. 3-Oxa-1-azabicyclo[3.1.0]hexanes

The oxazolidone tosylate **101** with sodium hydride in benzene at 55°C gives the aziridine **102** (48%).[49] This aziridine reacts with sodium *p*-cresolate in hexamethylphosphoramide to yield two products. Attack of the anion at C5 of the aziridine **102** forms the oxazinone **103**, while attack at C6 leads to the oxazolidinone **104**.

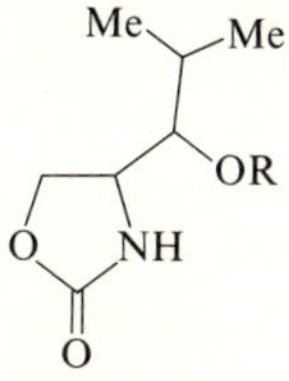

(101) R = Ts
(104) R = C_6H_4Me-*p*

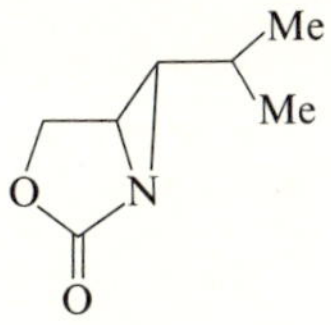

(102)

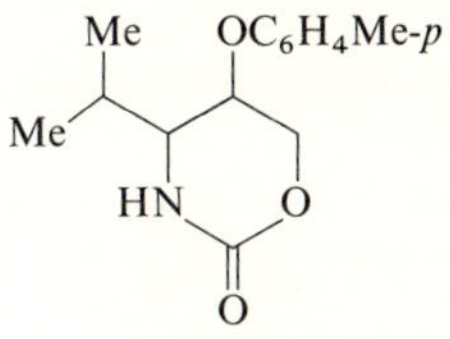

(103)

[48] M. Schulz and G. West, *J. Prakt. Chem.* **316**, 999 (1974).
[49] J. Marchand, F. Rocchiccioli, M. Païs, and F. X. Jarreau, *Bull. Soc. Chim. Fr.*, 4699 (1972).

F. 6-OXA-1-AZABICYCLO[3.1.0]HEXANES

1. *Preparations of 6-Oxa-1-azabicyclo[3.1.0]hexanes*

Irradiation of many pyrroline-1-oxides **105** leads to 6-oxa-1-azabicyclo[3.1.0]hexanes **106** in fair to good yields.[50-58]

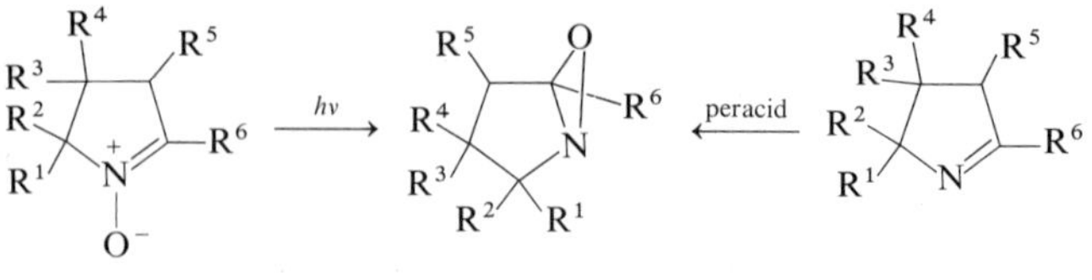

(105)   (106)   (107)

R^1, R^2 = H or Me; R^3 = H, Me or Ph; R^4 = H or Me; R^5 = H or Me; R^6 = H, Me, aryl or CN

1-Pyrrolines **107** can also afford the oxazabicyclohexanes **106** on oxidation with hydrogen peroxide,[50] or more commonly with peracids.[51-56] Both the photorearrangement and the oxidation reactions are widely applicable.

In the case of 2-aryl-1-pyrroline 1-oxides (e.g., **108**) ultraviolet irradiation has been shown[53,54] to form only the isomer (e.g., **109**) in which the oxaziridine ring is located on the same side of the five-membered ring as the substituent at C3. However, when the 2-substituent is alkyl or hydrogen, there is no stereoselectivity and the two isomers are formed in equal proportions.[54-56]

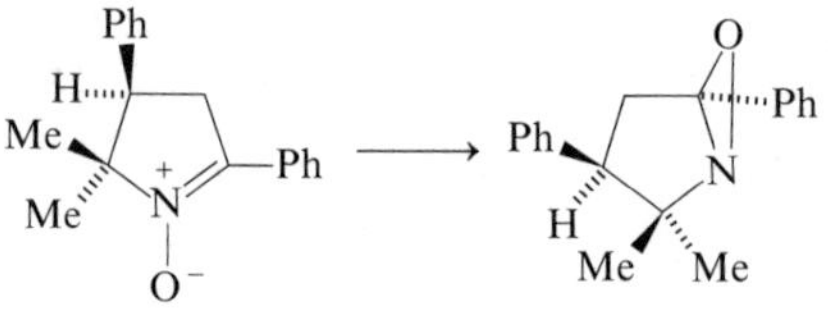

(108)   (109)

The peracid oxidation of 1-pyrrolines (e.g., **110**) is stereoselective and

---

[50] R. Bonnett, V. M. Clark, and Sir A. Todd, *J. Chem. Soc.*, 2102 (1959).
[51] L. S. Kaminsky and M. Lamchen, *J. Chem. Soc. C*, 2295 (1966).
[52] L. Kaminsky and M. Lamchen, *J. C. S., Chem. Commun.*, 130 (1965).
[53] J. B. Bapat and D. St.C. Black, *J. C. S., Chem. Commun.*, 73 (1967).
[54] J. B. Bapat and D. St.C. Black, *Aust. J. Chem.* **21**, 2507 (1968).
[55] D. St.C. Black and K. G. Watson, *Aust. J. Chem.* **26**, 2159 (1973).
[56] D. St.C. Black and K. G. Watson, *Aust. J. Chem.* **26**, 2505 (1973).
[57] J. B. Bapat, D. St.C. Black, and G. Newland, *Aust. J. Chem.* **27**, 1591 (1974).
[58] D. St.C. Black, N. A. Blackman, and A. B. Boscacci, *Tetrahedron Lett.*, 175 (1978).

produces the oxazabicyclohexanes (e.g., **111**) in which the oxaziridine ring is located trans to the substituent at C3.[53,54]

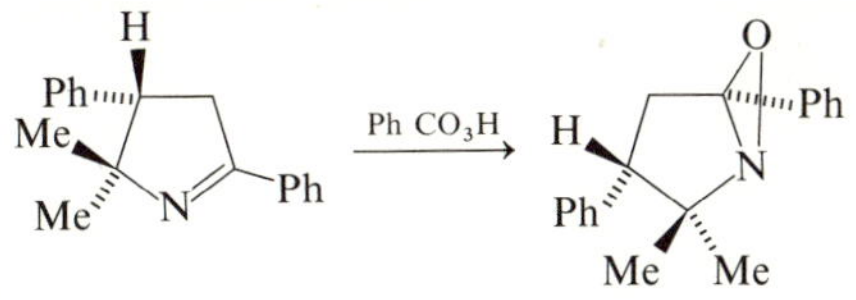

(110) → Ph CO₃H → (111)

2. Reactions of 6-Oxa-1-azabicyclo[3.1.0]hexanes

Treatment of C-aryl oxaziridines (e.g., **109**) with dilute acid causes[54] C—O bond fission to the corresponding nitrone **108**. However, similar treatment of C-alkyl oxaziridines (e.g., **112**) leads to the 1,4-dicarbonyl compound (e.g., **113**) by N—O bond fission.[50,51,54] The mode of ring opening is influenced by the stabilization of the positive charge resulting from initial protonation on oxygen.

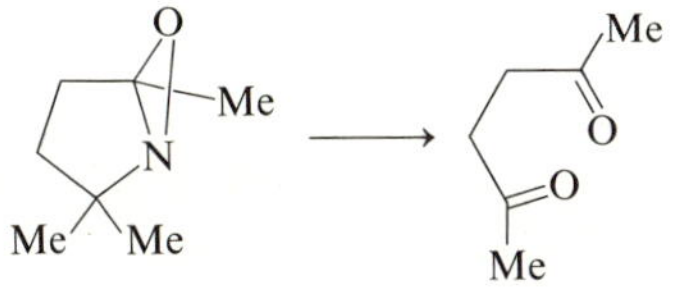

(112) → (113)

Acid hydrolysis of the oxazabicyclohexane **114a** yielded the hydroxamic acid **115** and methyl 3-hydroxyaminobutanoate (**116**).[59,60] Compound **114b**

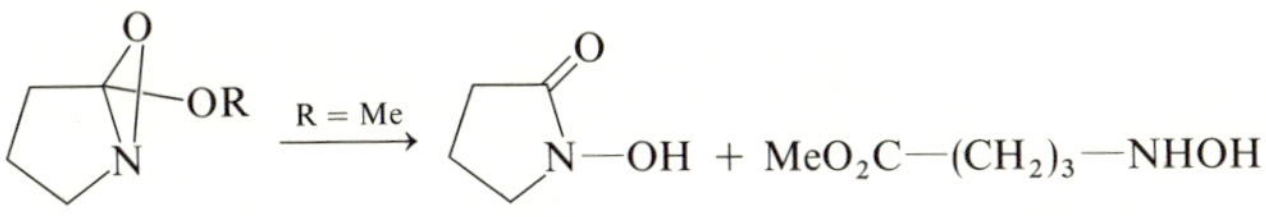

(114) —R = Me→ (115) N—OH + MeO₂C—(CH₂)₃—NHOH (116)

a: R = Me
b: R = Et

underwent thermal decomposition to ethyl 4-iminobutanoate (**117**), which was isolated as its trimer **118**.[60] The pyrolysis of more stable oxazabicyclohexanes (e.g., **119**) under various conditions gives acylazetidines (e.g., **120**), pyrrolines (e.g., **121**), and pyrrolidinones (e.g., **122**).[61,62]

[59] D. Thomas and D. H. Aue, *Tetrahedron Lett.*, 1807 (1973).
[60] D. H. Aue and D. Thomas, *J. Org. Chem.* **39**, 3855 (1974).
[61] L. S. Kaminsky and M. Lamchen, *J. Chem. Soc. C*, 2128 (1967).
[62] J. B. Bapat and D. St.C. Black, *Aust. J. Chem.* **21**, 2521 (1968).

(117) (118)

(119) (120) (121) (122)

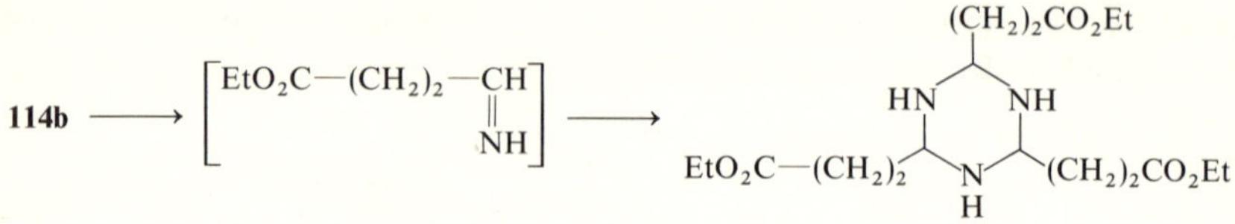

The type of breakdown which occurs on ultraviolet irradiation of an oxazabicyclohexane is very dependent on the substituent at the 5-position.[51,54,56] The products include pyrrolidinones (e.g., **124** from **123**,[51] and **126** from **125**[56]) and *N*-acylazetidines (e.g., **120** from **119**,[51] **127** from **125**,[56] and **129**

(123) (124)

(125) (126) (127)

(128) (129)

from **128**[58]) as a result of rearrangement, and occasionally pyrrolines (e.g., **121** from **119**[51]) as a result of deoxygenation. Irradiation of the 4-oxo-6-

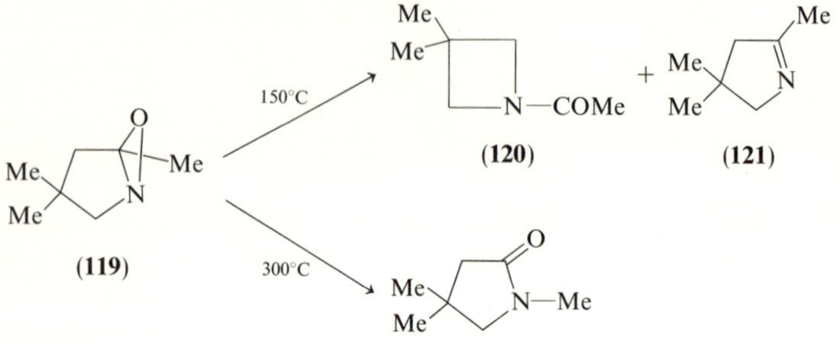

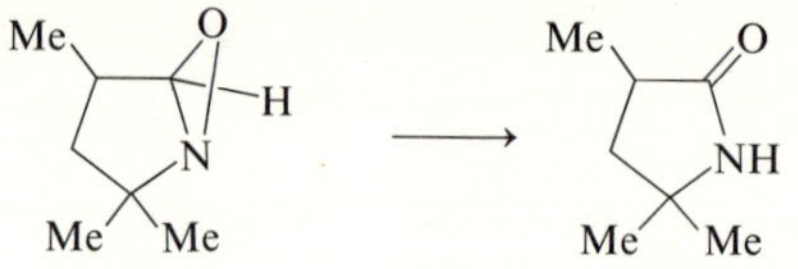

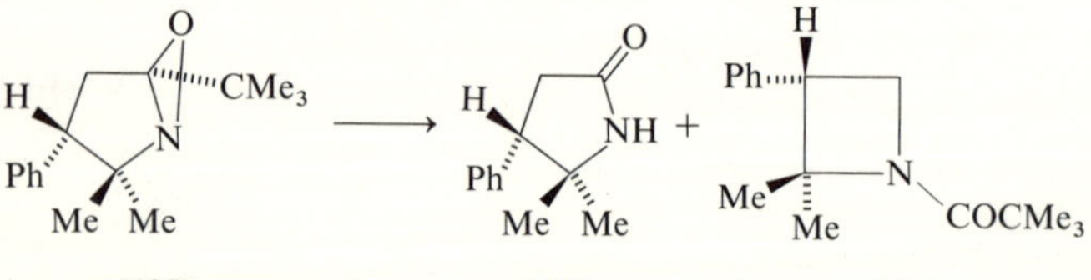

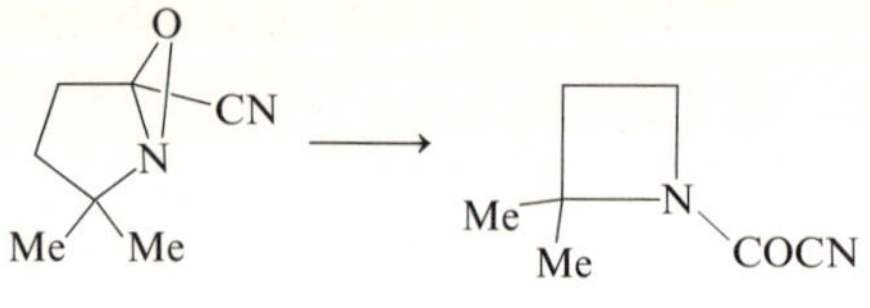

oxa-1-azabicyclohexanes **130a,b** gives the azetidinones **131a,b** and the oxazinones **132a,b**, but compound **130c** gives[63,64] only the related oxazinone **132c**.

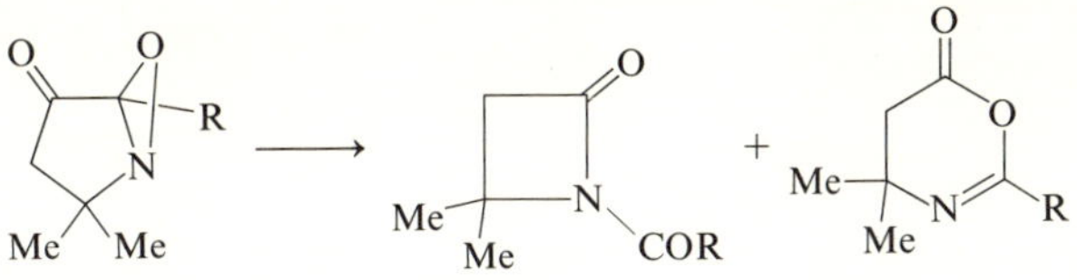

(130)   (131)   (132)

a: R = Me
b: R = Ph
c: R = CMe$_3$

In contrast, the 6-oxa-1-azabicyclo[3.1.0]hex-3-enes **133** are stable to further irradiation.[65]

Oxazabicyclohexanes **134** have been postulated[66] as undetected intermediates in the photolytic conversion of 2-acyl-1-pyrroline 1-oxides into pyrrolidinones and acyl pyrrolines.

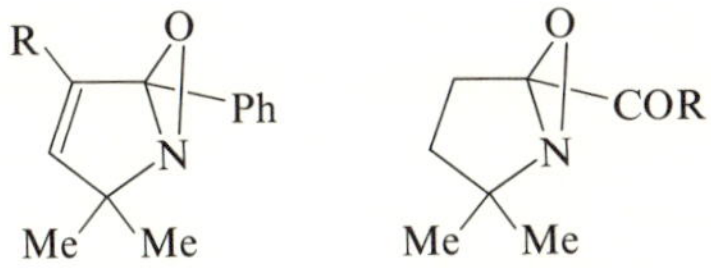

(133)   (134)

R = alkyl   R = H or Ph

Oxazabicyclohexanes (e.g., **106**) can be deoxygenated readily under mild conditions to afford the corresponding pyrrolines (e.g., **107**). Deoxygenation reagents include[55] thiourea, triphenylphosphine sulfide, potassium thiocyanate, potassium ethylxanthate, and potassium selenocyanate. The 6-thia-1-azabicyclo[3.1.0]hexanes **135** have been suggested[55] as possible intermediates in some of these reactions.

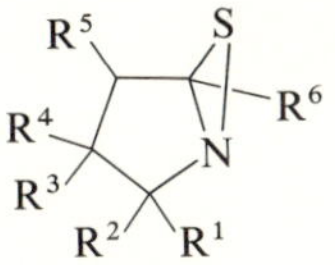

R^1, R^2 = H or Me; R^3 = H, Me, or Ph;
R^4 = H or Me; R^5 = H or Me; R^6 = H, Me, or Ph

(135)

---

[63] D. St.C. Black and A. B. Boscacci, *J. C. S., Chem. Commun.*, 129 (1974).
[64] D. St.C. Black and A. B. Boscacci, *Aust. J. Chem.* **30**, 1109 (1977).
[65] D. St.C. Black, N. A. Blackman, and L. M. Johnstone, *Aust. J. Chem.* **32**, 2025 (1979).
[66] D. St.C. Black and A. B. Boscacci, *Aust. J. Chem.* **30**, 1353 (1977).

Iron(II) sulfate also causes deoxygenation[67] of 5-substituted oxazabicyclohexanes (e.g., **136**) to give the pyrrolines (e.g., **137**) except when the 5-substituent is hydrogen,[67] *tert*-butyl,[67] or cyano,[68] in all of which cases the related pyrrolidinones (e.g., **126**) are formed.

(136) **(137)** **(126)**

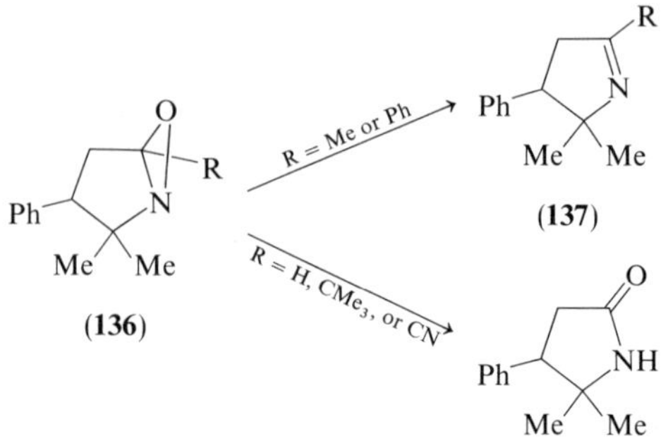

Treatment of the oxazabicyclohexane **138** with *m*-chloroperbenzoic acid leads to the oxidation of the pyridine nitrogen, to give the *N*-oxide **139** in good yield.[54] However, the oxaziridine ring itself can be oxidized in the absence of another easily oxidizable function: the ethoxy oxaziridine **114b** is converted by *m*-chloroperbenzoic acid into the oximino ester **141**, via the unstable *N*-oxide **140**.[60]

(138) **(139)**

(140) **(141)**

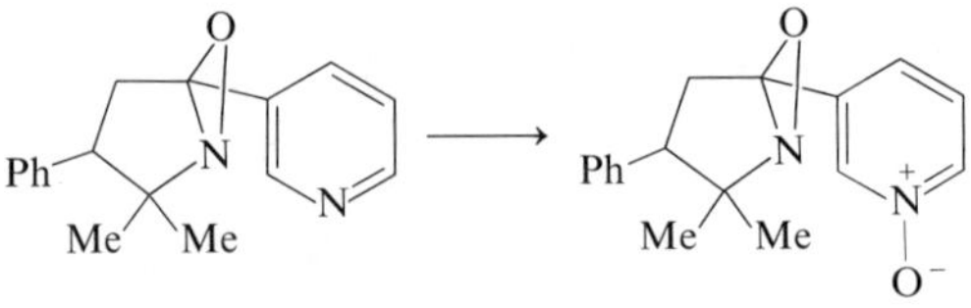

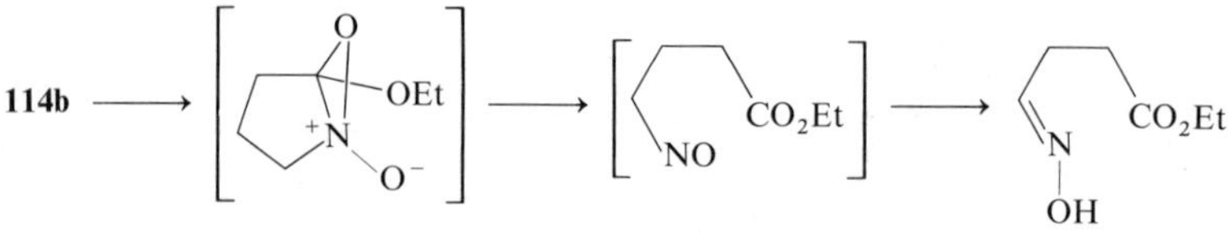

[67] D. St.C. Black and K. G. Watson, *Aust. J. Chem.* **26**, 2515 (1973).
[68] D. St.C. Black and N. A. Blackman, *Aust. J. Chem.* **32**, 2035 (1979).

G. 3-Thia-1-azabicyclo[3.1.0]hexanes

N-Substituted 6-aminopenicillanic acids can be readily reduced to the corresponding alcohols from which the tosylates **142a,b** are prepared. Treatment of **142b** with sodium bicarbonate in methanol affords the 3-thia-1-azabicyclohexane **143b**, whereas **143a** can be obtained from the tosylate **142a** with methanolic diethylamine.[69]

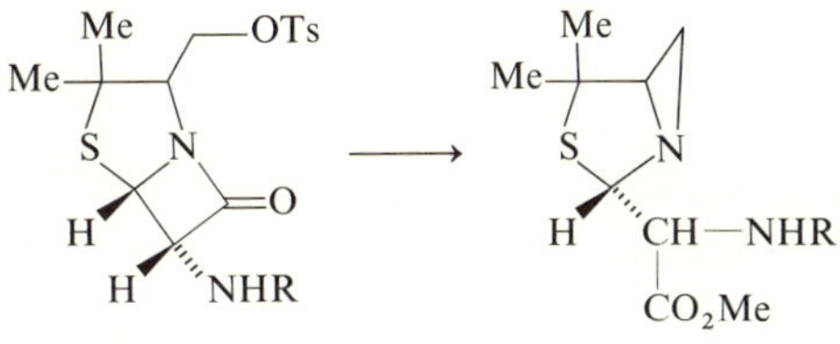

(142)   (143)

a: R = PhCH$_2$CO
b: R = Ph$_3$C

Treatment of the bicyclic compound **143a** with sodium bicarbonate in methanol yields the episulfide **145**,[69] presumably by removal of the proton α to the ester group, to give the intermediate **144** which then rearranges to the observed product.

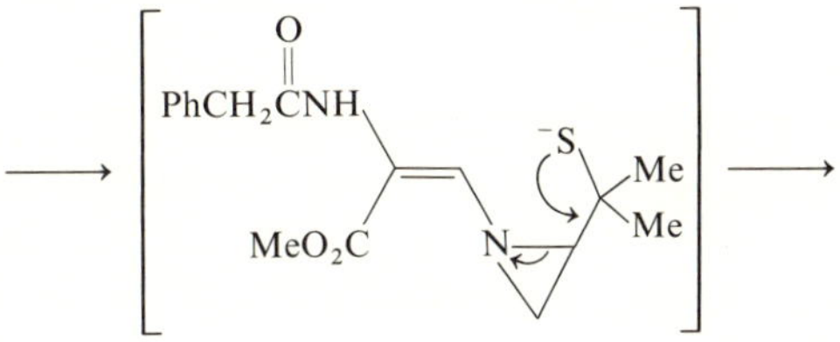

(144)

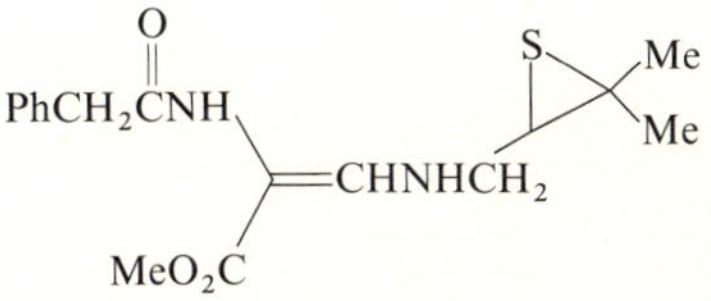

(145)

The addition of aqueous hydrochloric acid to the azabicyclohexanes **143a,b** causes breakdown of the aziridine ring. In contrast to this, the *p*-methoxybenzyl ester **146** can be hydrolyzed to the unstable carboxylic

[69] M. R. Bell, S. D. Clemans, and R. Oesterlin, *J. Med. Chem.* **13**, 389 (1970).

acid **147** by trifluoroacetic acid without destruction of the three-membered ring.[69]

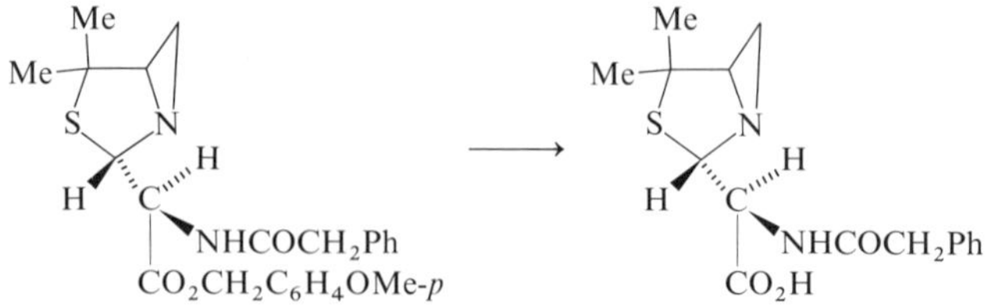

(146) → (147)

IV. Systems with Three Heteroatoms

A. 1,3,5-TRIAZABICYCLO[3.1.0]HEXANES

1. *Preparations of 1,3,5-Triazabicyclo[3.1.0]hexanes*

An aldehyde with chloramine (from *tert*-butyl hypochlorite and ammonia in methanol) gives a 1,3,5-triazabicyclo[3.1.0]hexane **148**[70-72] (27–80%). It has been pointed out by Nielsen[73] that the substituents at C2 and C4 of compounds formed in this reaction are normally trans to one another. However, on oxidation of the triazine **149** with *tert*-butyl hypochlorite, only the cis isomer of **148** was formed, provided that the alkyl group was small (i.e., R = Me, Et, n-C$_3$H$_7$). When the alkyl group was larger (i.e., R = n-C$_4$H$_9$, i-C$_4$H$_9$, n-C$_5$H$_{11}$, n-C$_6$H$_{13}$), only the trans isomers were formed. Since epimerization of cis to trans, when the alkyl group was small, occurred readily in methanol, it was assumed that epimerization occurred during the course of the reaction for the larger alkyl groups.[73] This reaction proceeded in only low yield when the substituent was 1-phenethyl and failed completely when it was benzhydryl.[74]

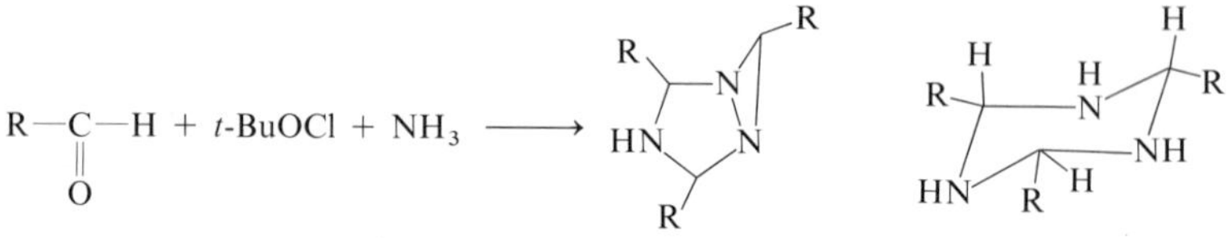

R—C(=O)—H + *t*-BuOCl + NH$_3$ ⟶ (148) (149)

R = Me, Et, Pr, CMe$_3$, or
C$_6$H$_4$X; X = H, Me, or MeO

[70] E. Schmitz, *Chem. Ber.* **95**, 688 (1962).
[71] E. Schmitz, D. Habisch, and C. Gründemann, *Chem. Ber.* **100**, 142 (1967).
[72] R. A. G. Smith and J. R. Knowles, *J. C. S. Perkin II*, 686 (1975).
[73] A. T. Nielsen, R. L. Atkins, D. W. Moore, D. Mallory, and J. M. LaBerge, *Tetrahedron Lett.*, 1167 (1973).
[74] A. T. Nielsen, R. L. Atkins, J. Di Pol, and D. W. Moore, *J. Org. Chem.* **39**, 1349 (1974).

Sec. IV.A]   1-AZABICYCLO[3.1.0]HEXANES AND ANALOGS

Izydore and McLean[75] have prepared the adduct **151** from ethyl diazoacetate and 4-phenyl-1,2,4-triazoline-3,5-dione (**150**).

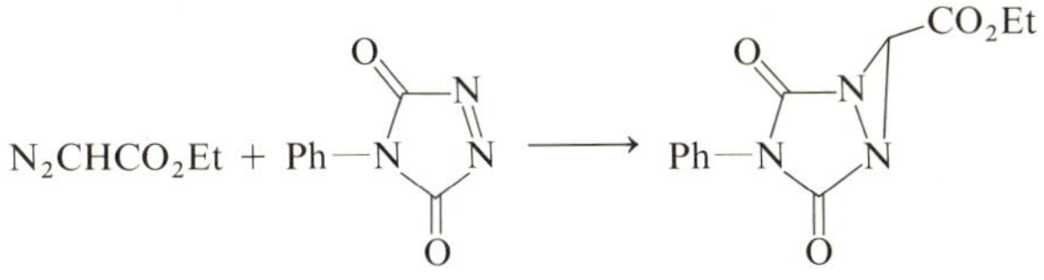

(150)   (151)

2. Reactions of 1,3,5-Triazabicyclo[3.1.0]hexanes

Oxidation of 1,3,5-triazabicyclo[3.1.0]hexanes **148** with potassium dichromate breaks the five-membered ring and forms a 3-substituted 3H-diazirine **152**.[76-78] Oxidation with *tert*-butyl hypochlorite has been reported to effect a similar transformation, but only when the substituent was phenyl.[72]

The triazabicyclohexanedione **151** reacts with acetic acid and acetic anhydride in methylene chloride provided that a small amount of acetate ion is added to give the diacetoxytriazole **153** in good yield.[75] When the dione **151** was heated to 150°C at low pressure, cleavage of the nitrogen–nitrogen bond gave **154** as the major product, with the triazolotriazole **155** also being formed.

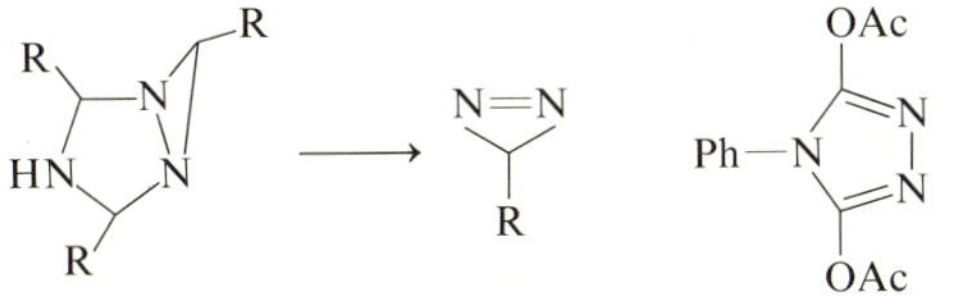

(148)   (152)   (153)

R = Et, Pr, or CMe$_3$

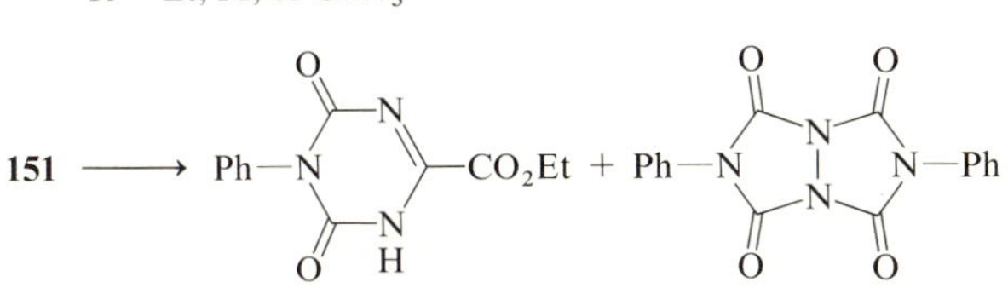

(154)   (155)

---

[75] R. A. Izydore and S. McLean, *J. Am. Chem. Soc.* **97**, 5611 (1975).
[76] E. Schmitz and R. Ohme, *Tetrahedron Lett.*, 612 (1961).
[77] E. Schmitz and R. Ohme, *Chem. Ber.* **95**, 795 (1962).
[78] H. M. Frey and I. D. R. Stevens, *J. Chem. Soc.*, 3101 (1965).

B. Oxadiazabicyclo[3.1.0]hexanes

Treatment of the amino oxime **156** with phosgene and triethylamine in tetrahydrofuran at −20°C gives 2-oxo-4,4,5-trimethyl-6-oxa-1,3-diazabicyclo[3.1.0]hexane (**158**) and the oxime **159**, presumed to form via the intermediate *N*-oxide **157**.[79]

(**156**) (**157**)

(**158**) (**159**)

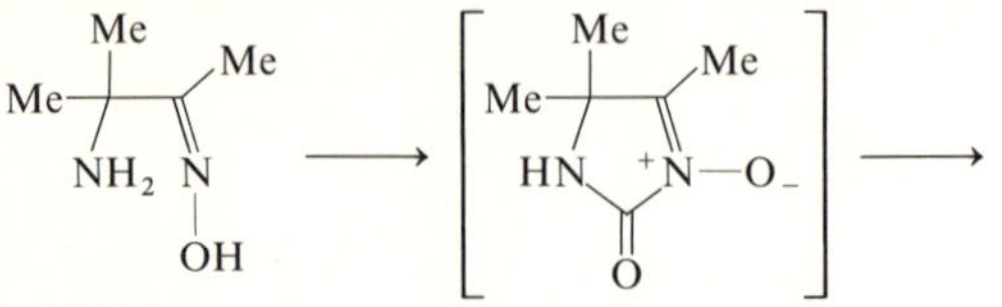

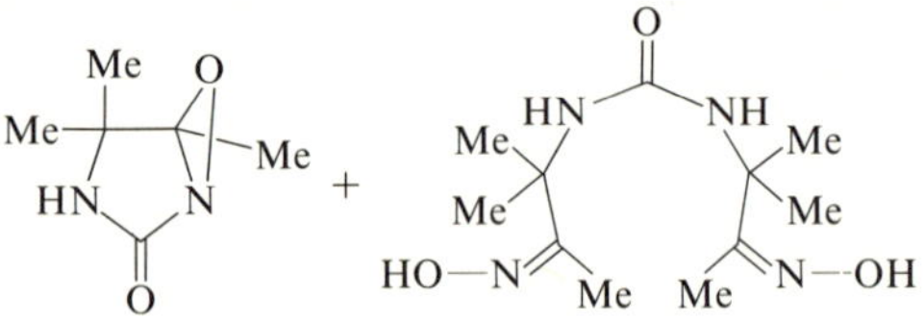

Irradiation of the nitrone nitroxide **160** in pentane gives the two isomeric radicals **162** and **163**, the formation of which was postulated to occur through the intermediate nitroxide **161**.[80]

(**160**) (**161**)

(**162**) (**163**)

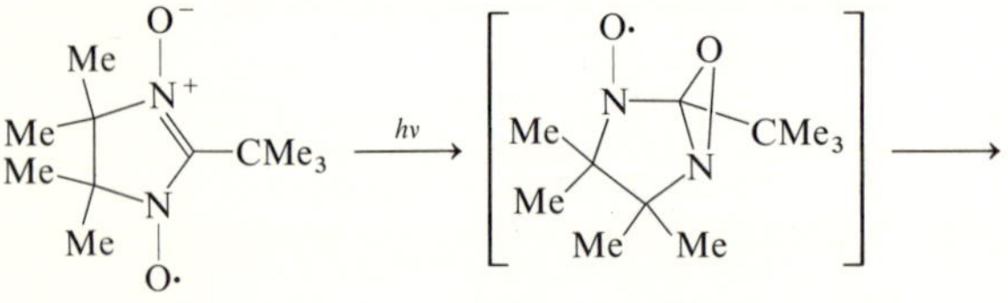

The oxadiazabicyclohexane **165** has been postulated[81] as an unstable

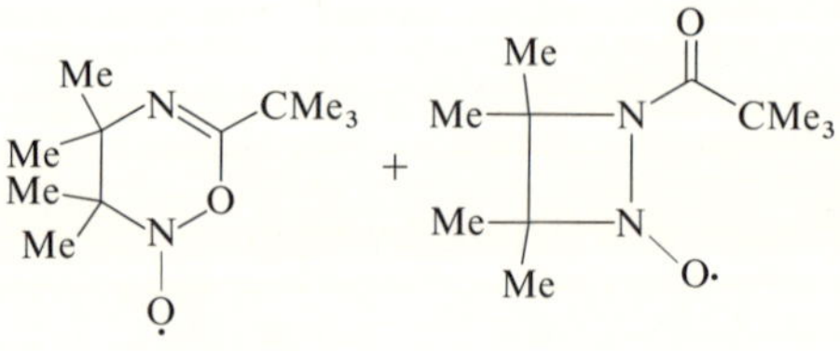

[79] H. Gnichtel, R. Walentowski, and K. E. Schuster, *Chem. Ber.* **105**, 1701 (1972).
[80] E. F. Ullman, L. Call, and S. S. Tseng, *J. Am. Chem. Soc.* **95**, 1677 (1973).
[81] J. B. Bapat, D. St.C. Black, and R. W. Clark, *Aust. J. Chem.* **25**, 1321 (1972).

intermediate in the peracid oxidation of 1,3,5-triphenyl-2-pyrazoline (**164**) to 1,3,5-triphenylpyrazole (**166**).

(**164**) (**165**) (**166**)

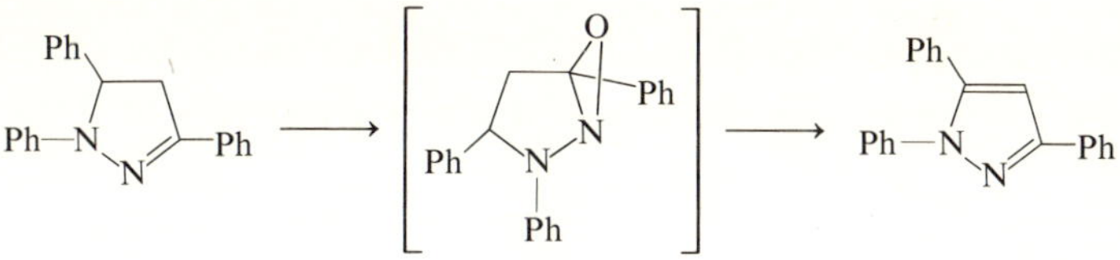

C. 4,6-Dioxa-1-azabicyclo[3.1.0]hexanes

Oxidation of 2-methyl-1,3-oxazoline (**167**) with *m*-chloroperbenzoic acid gives 5-methyl-4,6-dioxa-1-azabicyclo[3.1.0]hexane (**168**), which decomposes in solution to give the imine **169**, and the oxime **170**, in low yield.[60]

(**167**) (**168**) (**169**) (**170**)

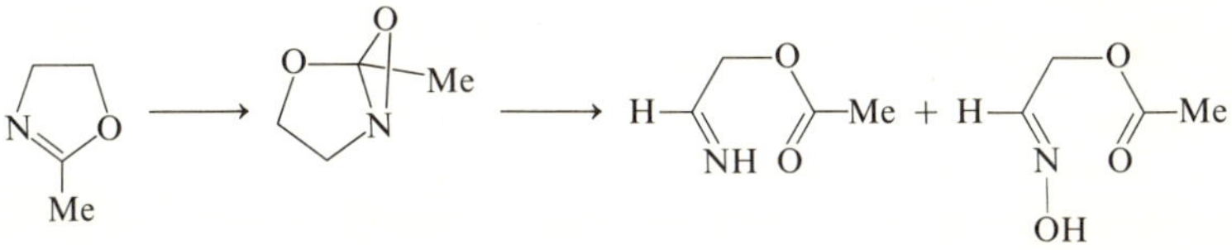

Similarly, oxidation of **171** gives the dioxazabicyclohexane **172** which was only stable below −20°C. Exposure of the compound to silica gel resulted in ring opening of the oxaziridine, to give the nitrone **173**.[82]

(**171**) (**172**) (**173**)

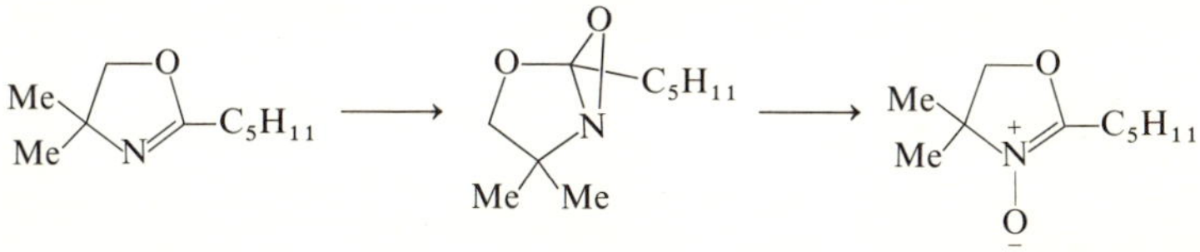

D. 2-Thia-3-oxa-1-azabicyclo[3.1.0]hexanes

Reaction of the hydroxyaziridines **174** with thionyl chloride and sodium hydride yields the 2-oxo-2-thia-3-oxa-1-azabicyclohexanes **175**.[83,84] The

[82] J. F. W. Keana and T. D. Lee, *J. Am. Chem. Soc.* **97**, 1273 (1975).
[83] Y. Diab, J. C. Duplan, and A. Laurent, *Tetrahedron Lett.*, 1093 (1976).
[84] P. Baret, M. Bourgeois, C. Gey, and J. L. Pierre, *Tetrahedron* **35**, 189 (1979).

TABLE II
^{1}H NMR Chemical Shifts of Selected Azabicyclo[3.1.0]hexanes^a

Structure	Ref.	Structure	Ref.
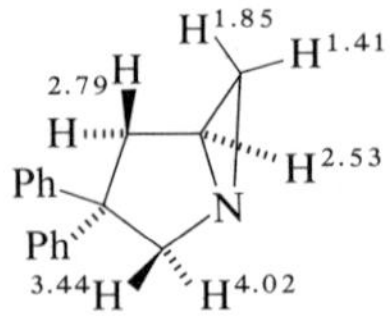 $H^{1.85}$, $H^{1.41}$, $_{2.79}H$, $H^{\cdots}$, Ph, Ph, $H^{2.53}$, $_{3.44}H$, $H^{4.02}$	(ref. 12)	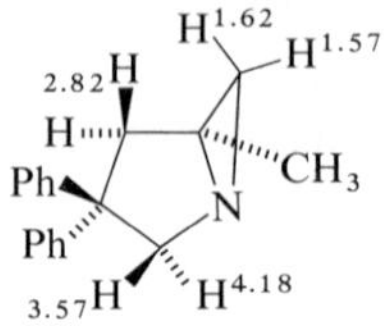 $H^{1.62}$, $H^{1.57}$, $_{2.82}H$, $H^{\cdots}$, Ph, CH_3, Ph, $_{3.57}H$, $H^{4.18}$	(ref. 12)
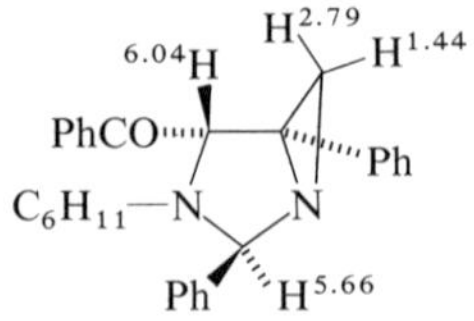 $_{6.04}H$, $H^{2.79}$, $H^{1.44}$, PhCO, Ph, $C_6H_{11}-N$, Ph, $H^{5.66}$	(ref. 30)	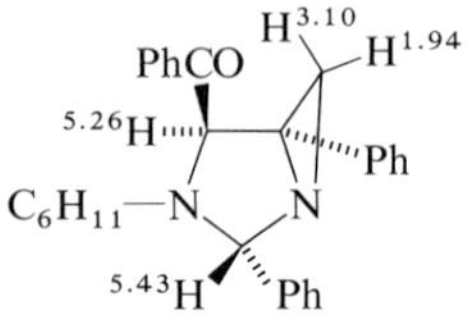 $H^{3.10}$, $H^{1.94}$, PhCO, $_{5.26}H$, Ph, $C_6H_{11}-N$, $_{5.43}H$, Ph	(ref. 30)
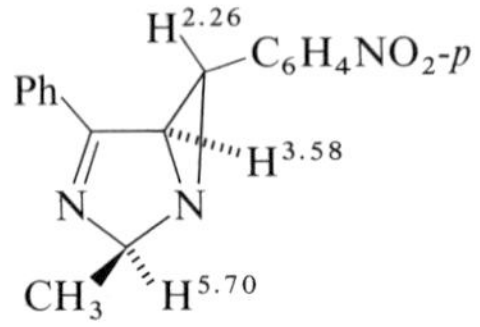 $H^{2.26}$, $C_6H_4NO_2$-p, Ph, $H^{3.58}$, CH_3, $H^{5.70}$	(ref. 31)	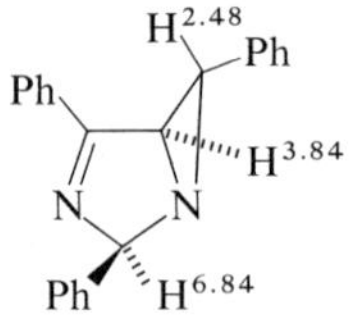 $H^{2.48}$, Ph, Ph, $H^{3.84}$, Ph, $H^{6.84}$	(ref. 31)
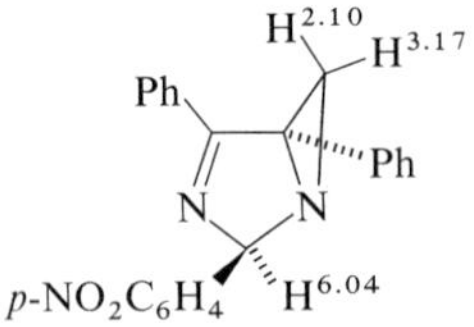 $H^{2.10}$, $H^{3.17}$, Ph, Ph, p-$NO_2C_6H_4$, $H^{6.04}$	(ref. 35)	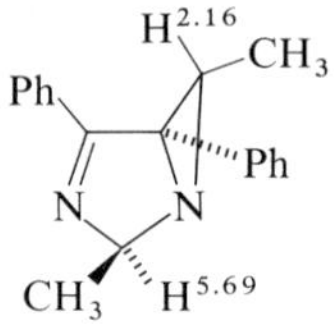 $H^{2.16}$, CH_3, Ph, Ph, CH_3, $H^{5.69}$	(ref. 35)
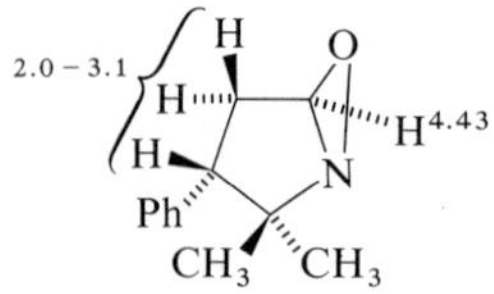 2.0–3.1, H, H, H, Ph, $H^{4.43}$, CH_3, CH_3	(ref. 55)	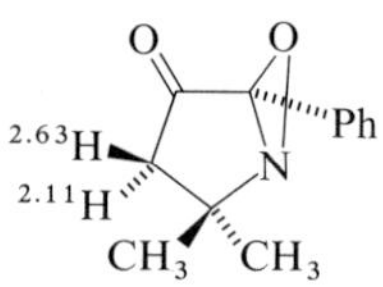 O, O, $_{2.63}H$, Ph, $_{2.11}H$, CH_3, CH_3	(ref. 64)
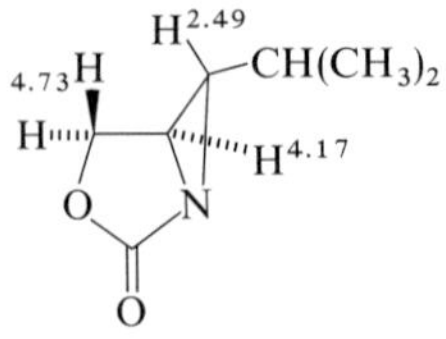 $H^{2.49}$, $_{4.73}H$, $CH(CH_3)_2$, $H^{\cdots}$, $H^{4.17}$, O, O	(ref. 49)	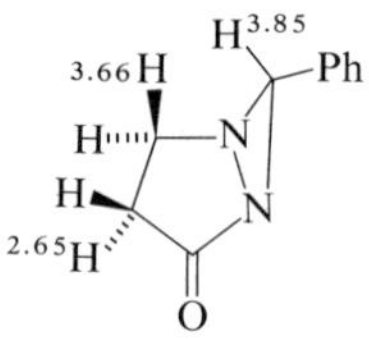 $H^{3.85}$, 3.66 H, Ph, H, $_{2.65}H$, O	(ref. 47)
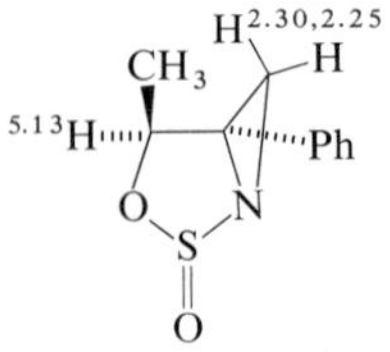 $H^{2.30, 2.25}$, CH_3, H, $_{5.13}H$, Ph, O-S-N, O	(ref. 83)	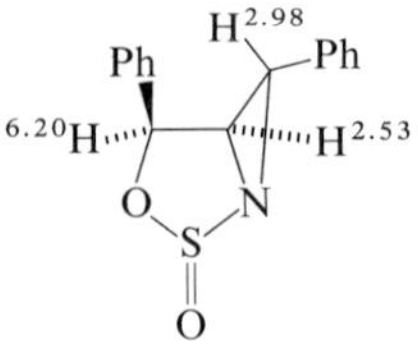 $H^{2.98}$, Ph, Ph, $_{6.20}H$, $H^{2.53}$, O-S-N, O	(ref. 84)

^{a} δ values in ppm downfield from tetramethylsilane (references in parentheses).

stereochemistry of these products was established by ^{1}H NMR spectroscopy with use of the nuclear Overhauser effects. The cyclization reaction thus provided confirmation of the stereochemistry of the precursor hydroxyaziridines.

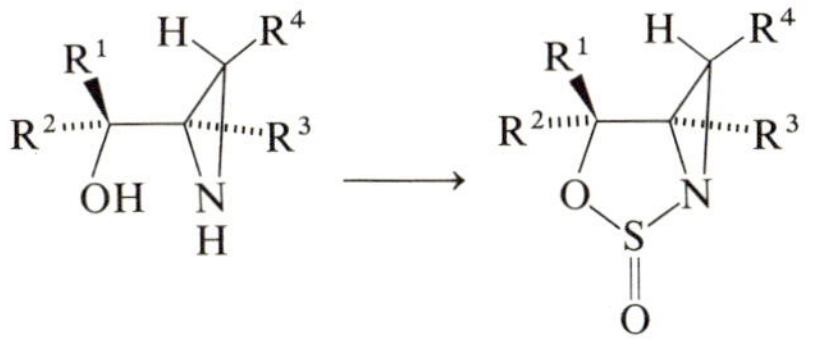

(174) (175)

R^1 = Me or Ph; R^2 = H or Ph; R^3 = H, Et, or Ph; R^4 = H or Ph

V. NMR Spectroscopic Data

^{1}H NMR data are available for a wide range of compounds of the classes considered previously. A selection of the more useful chemical shift values (δ, ppm with respect to tetramethylsilane $\delta = 0.00$ ppm) are shown in Table II.

Heteroaromatic Radicals, Part II: Radicals with Group VI and Groups V and VI Ring Heteroatoms

PETER HANSON

Department of Chemistry, University of York, Heslington, York, England

I. Introduction	32
II. Radicals from Oxygen Heterocycles	33
A. Radicals Containing a Five-Membered Heterocycle	33
1. Cation-Radicals Containing a Furan Ring and Isoelectronic Species . . .	33
2. Anion-Radicals Containing a Furan Ring and Isoelectronic Species . . .	35
3. Radicals from Oxygenated Furans	41
4. Radicals Containing a Dioxole Ring	45
B. Radicals Containing a Six-Membered Heterocycle	46
1. Pyranyl Radicals	46
2. Benzopyranyl Radicals	49
3. Polycyclic Pyranyl Radicals	50
4. Bipyryliumyl Radicals and Related Species	53
5. Radicals from Oxygenated Pyrans	54
6. Radicals Containing a Dioxin Ring	64
C. Radicals Containing Seven- and Eight-Membered Heterocycles	67
1. Radicals Containing an Oxepin Ring	67
2. Radicals Containing an Oxocin Ring	67
III. Radicals from Sulfur Heterocycles	68
A. Radicals Containing a Four-Membered Heterocycle	68
1,2-Dithiete Cation-Radicals	68
B. Radicals Containing a Five-Membered Heterocycle	69
1. Cation-Radicals Containing a Thiophene Ring	69
2. Neutral Radicals Containing a Thiophene Ring	71
3. Anion-Radicals Containing a Thiophene Ring	75
4. Oxygenated Thiophene Radicals	80
5. Sulfur *d*-Orbitals and Thiophene-Derived Radicals	84
6. Radicals Containing a 1,2-Dithiole Ring	86
7. Radicals Containing a 1,3-Dithiole Ring	90
C. Radicals Containing a Six-Membered Heterocycle	94
1. Thiopyranyl Radicals and Related Species	94
2. Bithiopyryliumyl Radicals	96
3. Radicals from Oxygenated Thiopyrans and Isoelectronic Species	96
4. Radicals Containing a Dithiin Ring	99
5. A Radical Containing a Tetrathiane Ring	106

    D. Radicals Containing Seven-Membered and Larger Heterocycles. 106
        1. Radicals from Dibenzo[b,f]thiepin 106
        2. 1,6-Dithiecin Cation-Radical 107
IV. Radicals from Selenium Heterocycles. 107
    A. Radicals Containing a Five-Membered Heterocycle. 107
        1. Radicals Containing a Selenophene Ring 107
        2. Radicals Containing a Diselenole Ring 108
    B. Radicals Containing a Six-Membered Heterocycle 109
        1. Radicals Containing a Selenopyran Ring 109
        2. Radicals Containing a 1,4-Diselenin Ring 110
V. Radicals from Heterocycles with Mixed Group VI Heteroatoms. 110
    A. Radicals Containing a Five-Membered Heterocycle. 110
        Radicals Containing a Thiaselenole Ring 110
    B. Radicals Containing a Six-Membered Heterocycle 110
        1. Phenoxathiin Cation-Radicals 110
        2. Phenoxaselenin Cation-Radicals 111
        3. Phenoxatellurin Radicals. 111
VI. Radicals from Heterocycles with Group V and Group VI Heteroatoms 112
    A. Radicals Containing a Five-Membered Heterocycle. 112
        1. Radicals Containing Oxazole and Thiazole Rings 112
        2. Violene Radicals from Oxazoles, Thiazoles, and Selenazoles 116
        3. Radicals from Oxadiazoles, Thiadiazoles, and Selenadiazoles 118
        4. 2,5-Diphenyl[1,2,4]dithiazolo[1,5-b][1,2,4]dithiazole Anion-Radical . . 124
    B. Radicals Containing a Six-Membered Heterocycle 124
        1. Radicals from Benzoxazines and Benzothiazines 124
        2. Dibenzo[c,e][1,2]thiazine 1,1-Dioxide Radical 125
        3. Radicals from Phenoxazines, Phenothiazines, and Phenoselenazines . . . 126
        4. Thia Analogs of Flavin and Pterin Radicals 147
        5. Radicals from Oxadiazines and Thiadiazines 148
VII. Conclusion 148

I. Introduction

This chapter concludes the review of heteroaromatic free radicals begun in Volume 25.[1] Here are discussed radicals containing heteroatoms from group VI and from groups V and VI together. The reader is referred to the introduction to Part I for a definition of terms, aims, and conventions. The only respect in which Part II differs is in the literature surveyed: here coverage is extended to the end of 1978. For convenience, each part has its own bibliography.

A considerable number of citations from the Russian literature are made. *Chemical Abstracts* and the various translations available are not consistent in their transliterations from the Cyrillic alphabet. In this review the spelling

---

[1] P. Hanson, *Adv. Heterocycl. Chem.* **25**, 205 (1979).

of surnames adopted is generally that of *Chemical Abstracts* and, it is hoped, is consistent within the chapter. Similarly, a consistent transliteration of forename initials is attempted. Original articles in Russian are cited and their *Chemical Abstracts* references given whether or not the article exists in translation.

II. Radicals from Oxygen Heterocycles

A. Radicals Containing a Five-Membered Heterocycle

1. *Cation-Radicals Containing a Furan Ring and Isoelectronic Species*

The "π-excessive" character of furan facilitates single-electron oxidation; however, the cation-radicals produced from simple furans are not, in general, persistent. The mediation of 2,5-dimethylfuran cation-radical (**1**) has been proposed in anodic oxidations yielding methoxylated, cyanomethoxylated, and acyloxylated products.[2-4] Although not specifically implicated, cation-radicals may be involved in very similar oxidative reactions.[5-7]

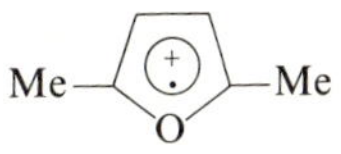

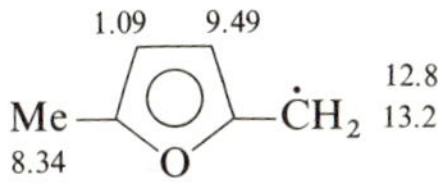

Deprotonation of **1** to give the neutral radical **2** under the conditions of anodic cyanomethoxylation has been proposed to account for the occurrence of minor products[3]; similar deprotonation has been suggested in a reaction where 2,5-dimethylfuran acts as a highly efficient scavenger for benzoyloxy radicals yielding 2-benzoyloxymethyl-5-methylfuran.[8,9] The ESR spectra of **2** and its 2-methyl analog have been recorded for the radicals generated

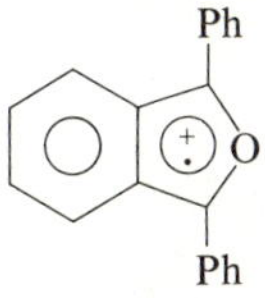

---

[2] S. D. Ross, M. Finkelstein, and J. J. Uebel, *J. Org. Chem.* **34**, 1018 (1969).
[3] K. Yoshida and T. Fueno, *J. Org. Chem.* **36**, 1523 (1971).
[4] N. L. Weinberg and H. R. Weinberg, *Chem. Rev.* **68**, 449 (1968).
[5] A. J. Baggaley and R. Brettle, *J. Chem. Soc. C*, 969 (1968).
[6] J. Froborg, G. Magnusson, and S. Thoren, *J. Org. Chem.* **40**, 122 (1975).
[7] I. Stibor, J. Srogl, and M. Janda, *J. C. S., Chem. Commun.*, 397 (1975).
[8] J. I. G. Cadogan, J. R. Mitchell, and J. T. Sharp, *J. Chem. Soc. D*, 1433 (1971).
[9] D. C. Nonhebel and J. C. Walton, "Free Radical Chemistry," p. 446. Cambridge Univ. Press, London and New York, 1974.

by γ-radiolysis of the heterocycles in an adamantane matrix[10]; **2** has also been formed in fluid solution by radiolysis of the parent heterocycle in strongly alkaline conditions.[11] Its ESR spectrum has been measured and assigned (splittings given in **2** are in gauss).

Cation-radicals from 2-arylfurans have been suggested to be involved in the oxidation of these compounds by ruthenium tetroxide.[12] The suggestion is based on the observation of the capture by the heterocycle of the chlorine from chlorinated solvents and of a broad ESR signal which persists for several days.

Anodic oxidation of 2,3,4,5-tetraphenylfuran in nitromethane leads to the formation of a stabilized cation-radical for which a low-resolution ESR spectrum has been recorded.[13] The splitting pattern indicates six major coupling protons which the authors suggest are the ortho and para protons of the 2- and 5-phenyl groups (cf. tetraphenylpyrrolyl, Part I: Section III,D,3). Depending upon conditions, the tetraphenylfuran cation-radical may either dimerize or disproportionate. The former process gives rise to a dimer possibly linked via the para positions of two of the phenyl rings, an inference made from the appearance in the product of an infrared absorption at 843 cm^{-1} which is taken to indicate para-substituted phenyl rings. The disproportionation gives rise to a dication which may be isolated as the perchlorate but which is also prone both to ring fragmentation and to attack by nucleophilic solvents.

The phenomenon of electrochemiluminescence (ECL) has been observed for tetraphenylfuran as well as for its nitrogen and sulfur analogs.[14] In the simplest form of this process, cation- and anion-radicals of the chosen substrate are generated together in solution by pulsing the working electrode between two appropriate potentials, one of which effects oxidation and the other reduction. If the energy released in the homogeneous electron-transfer reaction—whereby the ions of opposite charge annihilate each other—exceeds the energy required for promotion of the parent substrate to its first excited singlet state, the latter may then be populated with consequent fluorescence as the Boltzmann population is reestablished. Other processes may also occur. For example, the energy gained in radical-ion annihilation may be sufficient to populate only the first triplet level of the substrate, whence phosphorescence may occur or only weak fluorescence following triplet–triplet annihilation. The interacting ions may form excimers with the result that the fluorescence observed differs in wavelength from that of

[10] D. L. Winters and A. C. Ling, *Can. J. Chem.* **54**, 1971 (1976).
[11] R. H. Schuler, G. P. Laroff, and R. W. Fessenden, *J. Phys. Chem.* **77**, 456 (1973).
[12] D. C. Ayres and R. Gopalan, *J. C. S., Chem. Commun.*, 890 (1976).
[13] M. Libert and C. Caullet, *Bull. Soc. Chim. Fr.*, 805 (1974).
[14] M. Libert and A. J. Bard, *J. Electrochem. Soc.* **123**, 814 (1976).

the simple substrate. Frequently, it is arranged that the ions which interact are not derived from the same substrate but that the cation-radical of one material reacts with the anion-radical of another. On occasion, this occurs accidentally if impurity is present or if one of the ions reacts with solvent or some other component of the system before annihilation. The ECL phenomenon is not, of course, restricted to heterocycles. Reviews in the field have been given by Zweig,[15] Hercules,[16] and Bard et al.[17]; the reader is referred to the extensive work of Bard and co-workers[18-25] both for definitive treatment of the subject and for ingress to relevant literature away from the topic of present concern.

Isobenzofurans were studied for their ECL properties in the early development of the technique.[26-31] More recently, 1,3-diphenylisobenzofuran cation-radical (3) has been shown to react with superoxide anion $O_2^{-\cdot}$ to give initially singlet oxygen and 1,3-diphenylisobenzofuran which then undergo cycloaddition and yield ultimately 1,2-dibenzoylbenzene.[32]

2. *Anion-Radicals Containing a Furan Ring and Isoelectronic Species*

Even when isolated in an argon matrix, the anion-radical from furan undergoes ring scission at a C—O bond.[33] This is explicable in terms of extended Hückel molecular orbital (HMO) calculations[34] which show that

---

[15] A. Zweig, *Adv. Photochem.* **6**, 425 (1968).
[16] D. M. Hercules, *Acc. Chem. Res.* **2**, 301 (1969).
[17] A. J. Bard, C. P. Keszthelyi, H. Tachikawa, and N. E. Tokel, *in* "Chemiluminescence and Bioluminescence" (M. J. Cormier, D. M. Hercules, and J. Lee, eds.), p. 193. Plenum, New York, 1973.
[18] L. R. Faulkner and A. J. Bard, *J. Am. Chem. Soc.* **90**, 6284 (1968).
[19] C. P. Keszthelyi, H. Tachikawa, and A. J. Bard, *J. Am. Chem. Soc.* **94**, 1522 (1972).
[20] C. P. Keszthelyi and A. J. Bard, *Chem. Phys. Lett.* **24**, 300 (1974).
[21] H. Tachikawa and A. J. Bard, *Chem. Phys. Lett.* **26**, 246 (1974).
[22] H. Tachikawa and A. J. Bard, *Chem. Phys. Lett.* **26**, 568 (1974).
[23] D. Laser and A. J. Bard, *J. Electrochem. Soc.* **122**, 632 (1975).
[24] S. M. Park and A. J. Bard, *J. Am. Chem. Soc.* **97**, 2978 (1975).
[25] C. P. Keszthelyi, N. E. Tokel-Takvoryan, and A. J. Bard, *Anal. Chem.* **47**, 249 (1975).
[26] A. Zweig, G. Metzler, A. Maurer, and B. G. Roberts, *J. Am. Chem. Soc.* **88**, 2864 (1966).
[27] A. Zweig, G. Metzler, A. Maurer, and B. G. Roberts, *J. Am. Chem. Soc.* **89**, 4091 (1967).
[28] A. Zweig, A. K. Hoffmann, D. L. Maricle, and A. H. Maurer, *Chem. Commun.*, 106 (1967).
[29] A. Zweig, A. K. Hoffmann, D. L. Maricle, and A. H. Maurer, *J. Am. Chem. Soc.* **90**, 261 (1968).
[30] A. Zweig and D. L. Maricle, *J. Phys. Chem.* **72**, 377 (1968).
[31] E. A. Chandross and R. E. Visco, *J. Phys. Chem.* **72**, 378 (1968).
[32] E. A. Mayeda and A. J. Bard, *J. Am. Chem. Soc.* **95**, 6223 (1973).
[33] P. H. Kasai and D. MacLeod, *J. Am. Chem. Soc.* **95**, 4801 (1973).
[34] R. Hoffmann, *J. Chem. Phys.* **39**, 1397 (1963).

the LUMO of furan, in which the unpaired electron would reside in a hypothetical cyclic anion-radical, has strong antibonding character between the C and O atoms.[33] Accordingly, it is found that radiolysis of furans gives rise to a wide range of both nonaromatic cyclic adduct radicals and ring-opened radicals, but only in the presence of mesomerically electron-withdrawing substituents are aromatic anion-radicals observed.[11,35-37] Notwithstanding its probable nonexistence, calculations have been performed of the spin distribution in the furan anion-radical.[38] HMO spin-density calculations have also been made for a range of benzo[b]furan anion-radicals and the variation in spin density with substitution related to the reactivity of the benzofurans in photoreduction by aliphatic amines.[39]

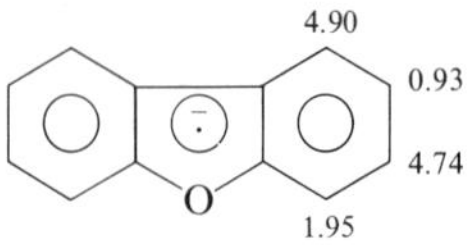

(4) 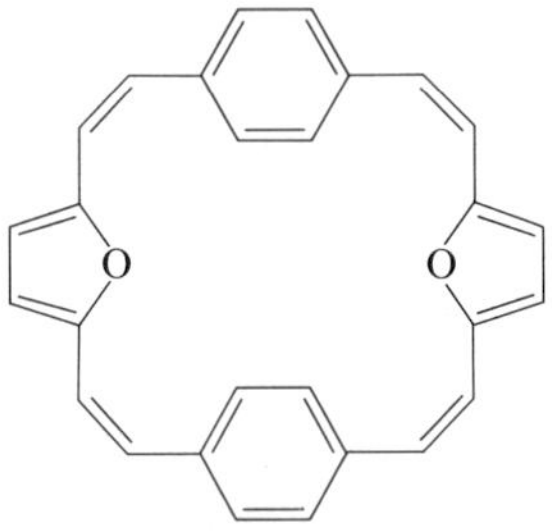   (5)

A persistent radical-anion (4) may be prepared from dibenzofuran, the reduction of which by alkali or alkaline earth metal in liquid ammonia was reported in 1964.[40] Gerdil and Lucken in 1965 also reported a preparation of 4 by reduction of dibenzofuran with potassium in DME; its ESR spectrum and those of its sulfur and selenium analogs were solved.[41] These authors also carried out a polarographic and spectroscopic study.[42] Evans and co-workers[43] reported ESR results for dibenzofuran closely similar to those of Gerdil and Lucken. The nature of the unimolecular decay of the anion-radical was studied and inferred to be C—O bond scission. The dependence of the rate of this reaction upon the counterion was investi-

[35] J. Lilie, Z. Naturforsch., Teil B **26**, 197 (1971).
[36] K. M. Bansal, A. Heinglein, and R. M. Sellers, Ber. Bunsenges. Phys. Chem. **78**, 569 (1974).
[37] T. Shiga and A. Isomoto, J. Phys. Chem. **73**, 1139 (1969).
[38] D. A. Morton-Blake, Proc. R. Ir. Acad., Sect. B **72**, 403 (1972).
[39] C. Parkanyi, A. Lablache-Combier, I. Marko, and H. Ofenberg, J. Org. Chem. **41**, 151 (1976).
[40] A. Maximadshy and F. Dörr, Z. Naturforsch., Teil B **19**, 359 (1964).
[41] R. Gerdil and E. A. C. Lucken, J. Am. Chem. Soc. **87**, 213 (1965).
[42] R. Gerdil and E. A. C. Lucken, J. Am. Chem. Soc. **88**, 733 (1966).
[43] A. G. Evans, P. B. Roberts, and B. T. Tabner, J. Chem. Soc. B, 269 (1966).

gated.[43,44] More recently, Canadian workers have obtained high-resolution ESR spectra for **4** and its sulfur analog and have resolved differences of splitting which were not apparent in the earlier work, confirming the result by specific deuteration.[45] Their splittings are shown (in gauss) in structure **4**. There have been calculations of the spin distribution in **4** by different MO methods.[41,46] Electronic absorption spectra of **4** generated by γ-irradiation of the heterocycle in organic glasses have been measured.[47] Bard and co-workers[48,49] have determined the rates of homogeneous electron transfer between **4** and its parent heterocycle and have correlated these with electroreduction rates.

Of other substances containing the furan ring in an otherwise hydrocarbon molecular environment, **5** forms an anion-radical on reduction with potassium. This radical has been the subject of a solution ESR and ENDOR study by Gerson and co-workers. Spin distribution, ion-pairing effects, and barriers to rotation of the phenylene groups were investigated.[50] As implied earlier in the discussion of ECL, isobenzofurans and tetraphenylfuran also form anion-radicals.[14,26-31]

Anion-radicals from 2-nitrofuran derivatives have been the subject of intensive study, particularly in the Soviet Union, on account of the wide-ranging biological activity of 2-nitrofurans and their mode of action involving perturbation of natural redox mechanisms.[51]

(6)   (7)

The ESR properties of the anion-radical of 2-nitrofuran (**6**) were recorded several years ago by Gavars et al.[52] This group has subsequently published a series of papers concerned with the π-electronic structure of nitrofuran

---

[44] A. G. Evans, *Proc. R. Soc. London, Ser. A* **302**, 331 (1968).
[45] F. C. Adam and C. R. Kepford, *Can. J. Chem.* **49**, 3529 (1971).
[46] N. K. Ray and P. T. Narasimhan, *Theor. Chim. Acta* **11**, 156 (1968).
[47] T. Matsuyama and H. Yamaoka, *Annu. Rep. Res. React. Inst., Kyoto Univ.* **8**, 75 (1975) [*CA* **84**, 128254 (1976)].
[48] B. A. Kowert, L. Marcoux, and A. J. Bard, *J. Am. Chem. Soc.* **94**, 5538 (1972).
[49] H. Kojima and A. J. Bard, *J. Am. Chem. Soc.* **97**, 6317 (1975).
[50] F. Gerson, W. Huber, and O. Wennerström, *Helv. Chim. Acta* **61**, 2763 (1978).
[51] Ya. P. Stradins, S. A. Hillers, R. A. Gavars, G. O. Reihmanis, and L. Kh. Baumane, *Experientia, Suppl.* **18**, 607 (1971).
[52] R. A. Gavars, Ya. P. Stradins, and S. A. Hillers, *Dokl. Akad. Nauk SSSR* **157**, 1424 (1964) [*CA* **62**, 3667 (1965)].

TABLE I
VARIATION WITH SOLVENT OF THE HYPERFINE SPLITTINGS (IN GAUSS) OF 2-NITROFURAN ANION-RADICAL (6)

Solvent	a(N)	a(H-3)	a(H-4)	a(H-5)	Reference
H_2O	13.2	6.1	0.9	4.7	56
MeCN	11.27	5.65	1.00	4.12	58
DMF	9.9	5.6	1.0	4.0	56

radicals.[53-57] The work has involved calculation of the spin distribution in 5-substituted 2-nitrofuran anion-radicals[54,55] as well as an investigation of the manner in which the spin distribution varies markedly with solvent, co-solutes, etc. and with extension of conjugation by vinylene substitution.[53,56,57] Italian workers have also solved the ESR spectra of **6** and its 5-methyl derivative for solutions generated electrochemically in acetonitrile and have performed INDO calculations of the spin distribution.[58] Table I indicates the variation in the spin distribution of **6** which occurs with change in solvation of the anion-radical.

The nitrogen splitting for **6** in acetonitrile may be compared with those for the nitro groups in the anion-radicals of nitrobenzene and the three nitropyridines in the same solvent (Part I: Section III,A,1,b). The relative orders of magnitude are 2-nitrofuran > nitrobenzene > 3-nitropyridine > 2-nitropyridine > 4-nitropyridine. The observation of the largest splitting for **6** is consistent with the mechanism of regulation of spin distribution discussed in Part I. The splitting from the nitrogen of the nitro group in these radicals depends upon the ability of the aromatic group to stabilize dipolar charge: if the aromatic ring is an acceptor of negative charge (e.g., pyridine), the substituent nitrogen splitting is relatively low as spin population is displaced onto oxygen; if the aromatic ring is a donor of negative charge (e.g., furan), the substituent nitrogen splitting is relatively high as the contribution from $-\overset{+\bullet}{N}(O)-\overset{-}{O}$ is relatively increased by virtue of the dipolar charge it bears.

[53] R. A. Gavars, V. K. Grins, G. O. Reihmanis, and Ya. P. Stradins, *Teor. Eksp. Khim.* **6**, 685 (1970) [*CA* **74**, 93818 (1971)].
[54] R. A. Gavars, V. A. Zilitis, Ya. P. Stradins, and S. A. Hillers, *Khim. Geterotsikl. Soedin.*, 294 (1970) [*CA* **73**, 38729 (1970)].
[55] R. A. Gavars, V. A. Zilitis, Ya. P. Stradins, and S. A. Hillers, *Khim. Geterotsikl. Soedin.*, 3 (1971) [*CA* **75**, 10082 (1971)].
[56] R. A. Gavars, L. Kh. Baumane, Ya. P. Stradins, and S. A. Hillers, *Khim. Geterotsikl. Soedin.*, 435 (1972) [*CA* **77**, 60977 (1972)].
[57] R. A. Gavars, L. Kh. Baumane, Ya. P. Stradins, and S. A. Hillers, *Khim. Geterotsikl. Soedin.*, 324 (1974) [*CA* **81**, 31555 (1974)].
[58] C. M. Camaggi, L. Lunazzi, and G. Placucci, *J. Org. Chem.* **39**, 2425 (1974).

In addition to the ESR work noted previously, Soviet groups have investigated the polarographic properties of nitrofurans, elucidating the mechanisms by which anion-radicals are formed, and their reactions, (e.g., acid–base equilibria and those which destroy the radicals), and the relationships of these properties to medium effects, ESR phenomena, and substituent character.[53,59-65] Kemula and Zawadowska[66] have reported a cyclic chronovoltammetric investigation of the reduction mechanism of 2-nitrofuran and various derivatives.

There has been particular interest in 2-nitrofurans bearing a 5-carbonyl substituent, especially the aldehyde and its derivatives. Laviron and coworkers[67,68] noted hydration and hemiacetal formation when 5-nitrofuran-2-carboxaldehyde is dissolved in water or alcohol. Equilibrium constants for these processes were determined polarographically. Hydration of the aldehyde was also observed in polarographic studies by Reihmanis and Stradins.[69] Stradins and co-workers[70,71] have investigated the polarography of oximes and hydrazones of 5-nitrofuran-2-carboxaldehyde and have demonstrated the intermediacy of radical-anions. Polarographic and pulse-radiolysis studies of nitroheterocyclic radiosensitizers, including 5-nitrofuran-2-carboxaldehyde and derivatives, have been reported by collaborating North American groups.[72,73] In this work, the anion-radical properties were related to the biological activity of the compounds; thus, either half-

[59] Ya. P. Stradins, G. O. Reihmanis, and R. A. Gavars, *Elektrokhimiya* **1**, 955 (1965)[*CA* **64**, 3055 (1966)].
[60] Ya. P. Stradins, R. A. Gavars, G. O. Reihmanis, and S. A. Hillers, *Abh. Dtsch. Akad. Wiss. Berlin, Kl. Med.*, 601 (1966)[*CA* **66**, 111011 (1967)].
[61] Ya. P. Stradins and G. O. Reihmanis, *Elektrokhimiya* **3**, 178 (1967) [*CA* **66**, 101084 (1967)].
[62] Ya. P. Stradins, R. A. Gavars, V. K. Grins, and S. A. Hillers, *Teor. Eksp. Khim.* **4**, 774 (1968) [*CA* **70**, 46707 (1969)].
[63] Ya. P. Stradins, I. Kravis, G. O. Reihmanis, and S. A. Hillers, *Khim. Geterotsikl. Soedin.*, 1309 (1972)[*CA* **78**, 37187 (1973)].
[64] V. N. Novikov, *Khim. Geterotsikl. Soedin.*, 1601 (1976)[*CA* **86**, 170284 (1977)].
[65] I. M. Sosonkin, G. N. Strogov, V. N. Novikov, and T. K. Ponomareva, *Khim. Geterotsikl. Soedin.*, 23 (1977)[*CA* **86**, 154848 (1977)].
[66] W. Kemula and J. Zawadowska, *Bull. Acad. Pol. Sci., Ser. Sci. Chim.* **16**, 419 (1968)[*CA* **69**, 112867 (1968)].
[67] E. Laviron, H. Troncin, and J. Tirouflet, *Bull. Soc. Chim. Fr.*, 524 (1962).
[68] E. Laviron and J. C. Lucy, *Bull. Soc. Chim. Fr.*, 2202 (1966).
[69] G. O. Reihmanis and Ya. P. Stradins, *Latv. PSR Zinat. Akad. Vestis, Kim. Ser.*, 23 (1967) [*CA* **67**, 60305 (1967)].
[70] Ya. P. Stradins and G. O. Reihmanis, *Latv. PSR Zinat. Akad. Vestis, Kim. Ser.*, 377 (1969) [*CA* **71**, 112048 (1969)].
[71] L. M. Baider, L. Kh. Baumane, R. A. Gavars, V. T. Glezer, and Ya. P. Stradins, *Nov. Polyarogr., Tezisy Dokl. Vses. Soveshch. Polyarogr.*, 6th, 1975, 87 (1975) [*CA* **86**, 62665 (1977)].
[72] C. L. Greenstock, G. W. Ruddock, and P. Neta, *Radiat. Res.* **66**, 472 (1976).
[73] G. W. Ruddock and C. L. Greenstock, *Biochim. Biophys. Acta* **496**, 197 (1977).

wave reduction potentials or hyperfine splitting values may be used as indices of biological activity.

Details of the ESR spectrum of 7 have been published in Russian with an indication that the radical, its thiophene counterpart, and the corresponding acetyl anion-radical exist as mixtures of rotamers.[74] Hyperfine splittings were reported to be solvent-dependent. The latter has been amply confirmed by others.[53] More recent work describes 7, generated by radiolytic means in aqueous medium. The hyperfine splittings reported (and given in structure 7 in gauss) are in good agreement with the earlier Soviet data for the same solvent.[53,75] Greenstock et al.[75] also give hyperfine splittings for the anion-radical of 5-nitrofuran-2-carboxylic acid, its conjugate acid, and conjugate base. The discrepancy between their splittings for the anion-radical of 5-nitrofuran-2-carboxaldoxime and those of the Soviet group probably derives from the difference of solvents, which is critical in the nitrofuran anion-radical family.[76]

Several furan aldehydes and ketones without a heterocyclic nitro group have been investigated polarographically and the intervention of ketyl radicals or their conjugate acids confirmed.[77-81] Their involvement in amalgam reductions of furan carbonyl derivatives was also shown.[82]

Radical species from furil and furoin, when treated with base in DMSO, were observed many years ago.[83] Subsequently, an ESR spectrum for the anion (8) was recorded, but at the time the assignment of hyperfine splittings was in doubt.[84] Very recently an assignment was made for 8 during a study of its ion-pairing properties.[85] The conjugate acid of 8, the neutral radical

---

[74] A. Sh. Mukhtarov, V. I. Savin, A. V. Il'yazov, and I. D. Morosova, *Mater. Nauchn. Konf., Inst. Org. Fiz. Khim., Akad. Nauk SSSR, 1969*, 85 (1970)[*CA* **76**, 58474 (1972)].
[75] C. L. Greenstock, I. Dunlop, and P. Neta, *J. Phys. Chem.* **77**, 1187 (1973).
[76] Ya. P. Stradins, L. M. Baider, L. Kh. Baumane, R. A. Gavars, and V. T. Glezer, *Elektrokhimiya* **13**, 759 (1977)[*CA* **87**, 124480 (1977)].
[77] Ya. P. Stradins, G. O. Reihmanis, and R. Frimm, *Khim. Geterotsikl. Soedin.*, 582 (1969) [*CA* **72**, 66222 (1970)].
[78] Ya. P. Stradins, I. Tutane, and O. Zarina, *Latv. PSR Zinat. Akad. Vestis, Kim. Ser.*, 431 (1970)[*CA* **73**, 136752 (1970)].
[79] I. G. Markova, M. K. Polievktov, A. F. Oleinik, and K. Yu. Novitskii, *Nov. Polyarogr., Tezisy Dokl. Vses. Soveshch. Polyarogr., 6th, 1975*, 90 (1975)[*CA* **86**, 48548 (1977)].
[80] M. K. Polievktov and I. G. Markova, *Khim. Geterotsikl. Soedin.*, 1607 (1976) [*CA* **86**, 80848 (1977)].
[81] M. K. Polievktov, I. G. Markova, and A. F. Oleinik, *Khim. Geterotsikl. Soedin.*, 165 (1977) [*CA* **87**, 22078 (1977)].
[82] G. N. Soltovets, V. A. Smirnov, and V. G. Kul'nevich, *Elektrokhimiya* **4**, 688 (1968)[*CA* **69**, 82895 (1968)].
[83] G. A. Russell, E. G. Janzen, and E. T. Strom, *J. Am. Chem. Soc.* **84**, 4155 (1962).
[84] E. T. Strom, G. A. Russell, and J. H. Schoeb, *J. Am. Chem. Soc.* **88**, 2004 (1966).
[85] K. S. Chen and J. K. S. Wan, *J. Am. Chem. Soc.* **100**, 6051 (1978).

9, has been observed during the photoreduction of furil and subjected to ESR and CIDEP study.[86]

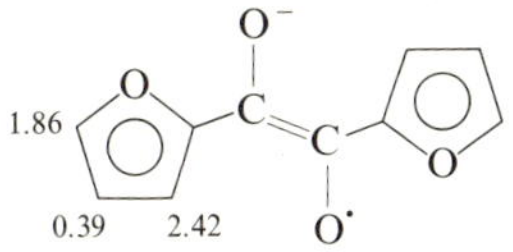

(8) 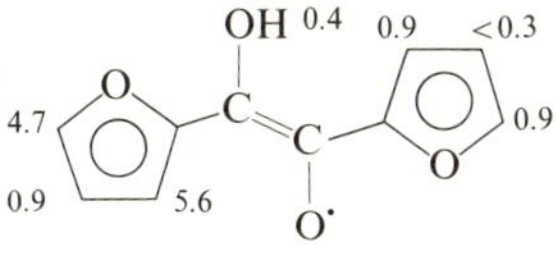 (9)

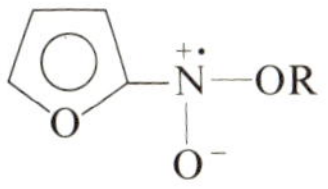

(10)

Alkoxynitroxyl radicals, isoelectronic with ketyls, are produced when aromatic nitro compounds trap other radicals [e.g., the nitroxyl **10** (R = tetrahydrofuran-2-yl)] when 2-nitrofuran is irradiated in THF[86a]; a similar radical, **10** (R = Et$_3$Si), is obtained when the solvent is triethylsilane.[58] The latter radical exhibits separate rotamers in proportions of 7:3 at low temperatures, and at −60°C two well-resolved spectra were measured. They were not, however, assigned.[58]

3. Radicals from Oxygenated Furans

a. *Dihydrofuranonyl Radicals (Furanoxyls).* A radical formulated as **11a**, but for which the aromatic furanoxyl contribution **11b** may be written, has been observed as a radiolysis product of furan under strongly alkaline conditions.[11] The mechanism of its formation is not clear, however. A similar carboxylated radical (**12**) is observed when either 5-nitro- or 5-bromofuran-2-carboxylic acid adds OH• under radiolytic conditions. *Ipso*-attack at the 5-position occurs with subsequent denitration or debromination,[75,87,88] reminiscent of the denitration of furans by methyl radicals.[89] The high ring–proton splittings indicated (in gauss) for **11** and **12** imply that the aromatic furanoxyl structure makes relatively little contribution and that the radicals are best regarded as cyclic allyl radicals.

There has been an ESR study of the persistent annelated radical (**13**), obtained by thermolysis of 2,2′-dioxo-3,3′-diphenyl-2,2′,3,3′-tetrahydro-

---

[86] A. J. Elliot and J. K. S. Wan, *Can. J. Chem.* **56**, 2499 (1978).
[86a] R. B. Sleight and L. H. Sutcliffe, *Trans. Faraday Soc.* **67**, 2195 (1971).
[87] P. Neta and C. L. Greenstock, *J. C. S., Chem. Commun.*, 309 (1973).
[88] H. Zemel and P. Neta, *Radiat. Res.* **55**, 393 (1973).
[89] U. Rudqvist and K. Torssel, *Acta Chem. Scand.* **25**, 2183 (1971).

bibenzo[*b*]furan-3-yl, and its derivatives.[90] The magnitudes of the splittings observed for the phenyl group, taken with various MO calculations, indicate the twist angle in the C3—Ph bond to be less than 25°.

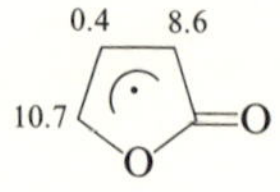

(11a) 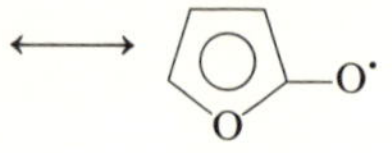 ↔ (11b) 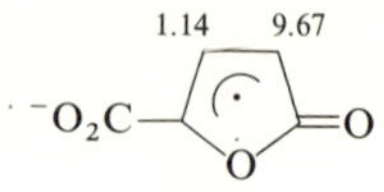 (12)

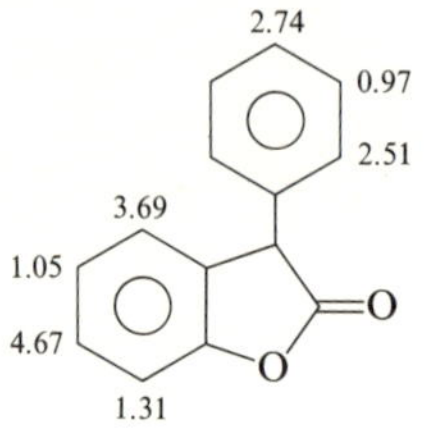

(13) 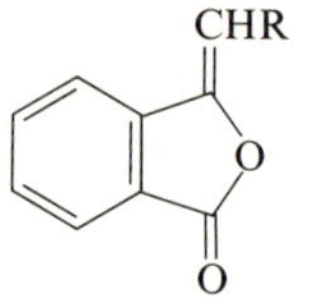 (14) 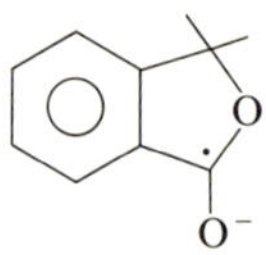 (15)

Interestingly, electrochemical reductions of alkylidenephthalides (14), which might conceivably give radicals isomeric in the heterocyclic moiety with 13, resulted either in the addition of two electrons or in ring scission, although phthalide radicals such as 15, which is not heteroaromatic, may be prepared.[91,92]

b. *Furan Semidiones.* Maleic anhydride (2,5-dihydrofuran-2,5-dione) is reduced by uptake of one electron to the anion-radical 16 (splitting in gauss)[93]; dimethylmaleic anhydride is similarly reduced to the corresponding 3,4-dimethylfuran-2,5-semidione which exhibits methyl splittings of 6.1 G.[94] The ready reduction of maleic anhydride had been noted by Peover[95] and by Takahashi and Elving.[96] The latter compared the polarographic reduction of maleic anhydride and maleic esters in pyridine. They noted that the reduction wave for the anhydride occurred at significantly more positive potentials than that for the diester, a fact they correlated with the conjugated cyclic (aromatic) nature of the radical produced from the anhydride.[96]

Furan-2,5-semidione has been incorporated as a spin-label into a wide variety of structures, both conformationally labile and rigid (e.g., 17–19)

[90] P. Karafiloglou, J.-P. Catteau, A. Lablache-Combier, and H. Ofenberg, *J. C. S. Perkin II*, 1545 (1977).
[91] A. Beno, P. Hrnciar, and M. Lacova, *Collect. Czech. Chem. Commun.* **37**, 3295 (1972).
[92] S. F. Nelsen, *J. Org. Chem.* **38**, 2693 (1973).
[93] G. A. Russell and R. L. Blankespoor, *Tetrahedron Lett.*, 4573 (1971).
[94] S. F. Nelsen and E. D. Seppanen, *J. Am. Chem. Soc.* **92**, 6212 (1970).
[95] M. E. Peover, *Trans. Faraday Soc.* **58**, 2370 (1962).
[96] R. Takahashi and P. Elving, *Electrochim. Acta* **12**, 213 (1967).

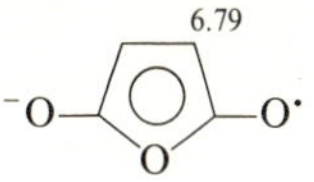

Sec. II.A] HETEROAROMATIC RADICALS, PART II 43

(16) 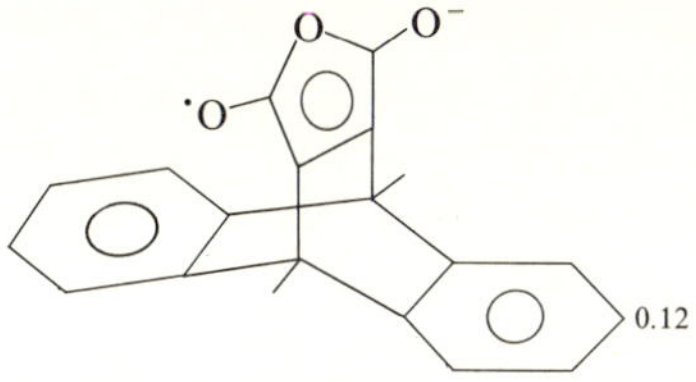 (17)

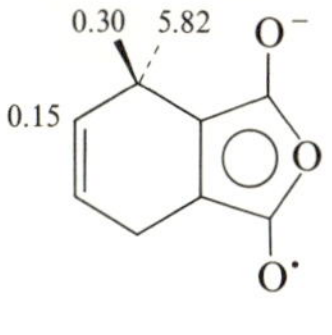

(18) 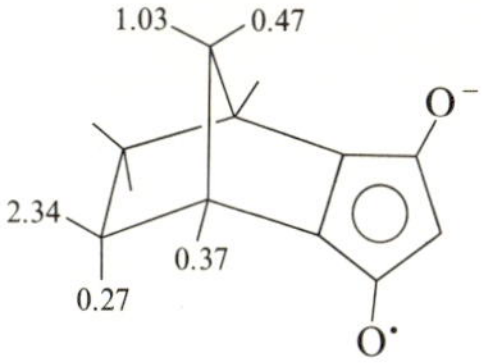 (19)

and has been of value in comprehending the mechanisms of spin transmission in such systems.[94,97–100] A recent low-temperature radiolytic study of 2-methyltetrahydrofuran solutions of maleic anhydride has shown the anion-radical to be formed and that it dimerizes giving products detectable by their intense IR absorptions.[101]

Phthalic anhydride forms a persistent anion-radical on reduction, exemplified by the isobenzofuran-1,3-semidione (20). The ESR spectrum of 20 was observed by several groups at about the same time and analyzed with comparable results.[102–104] The splittings indicated for 20 are those of Sioda and Koski.[102] Lasia has examined by polographic means the consequences of ion-pairing of 20 for its dimerization[105]; electronic absorption spectra of 20 and its derivatives and their MO interpretation have been studied by Shida et al.[106]

Pyromellitic dianhydride forms a radical (21) which has been the subject of several investigations.[102,107,108] On account of the low number and

---

[97] K. E. Anderson, D. Kosman, C. J. Mayers, B. P. Ruekberg, and L. M. Stock, *J. Am. Chem. Soc.* **90**, 7168 (1968).
[98] S. F. Nelsen and E. D. Seppanen, *J. Am. Chem. Soc.* **89**, 5740 (1967).
[99] S. F. Nelsen, E. F. Travecedo, and E. D. Seppanen, *J. Am. Chem. Soc.* **93**, 2913 (1971).
[100] G. A. Russell, G. W. Holland, K.-Y. Chang, R. G. Keske, J. Mattox, C. S. C. Chung, K. Stanley, K. Schmitt, R. Blankespoor, and Y. Kosugi, *J. Am. Chem. Soc.* **96**, 7237 (1974).
[101] S. Arai, A. Kiva, and M. Imamura, *J. Phys. Chem.* **81**, 110 (1977).
[102] R. E. Sioda and W. S. Koski, *J. Am. Chem. Soc.* **89**, 475 (1967).
[103] S. F. Nelsen, *J. Am. Chem. Soc.* **89**, 5256 (1967).
[104] A. V. Il'yasov, Y. M. Kargin, Ya. A. Levin, I. D. Morosova, and N. N. Sotnikova, *Izv. Akad. Nauk SSSR, Ser. Khim.*, 1030 (1968) [*CA* **69**, 48125 (1968)].
[105] A. Lasia, *J. Electroanal. Chem. Interfacial Electrochem.* **42**, 253 (1973).
[106] T. Shida, S. Iwata, and M. Imamura, *J. Phys. Chem.* **78**, 741 (1974).
[107] M. Hirayama, *Bull. Chem. Soc. Jpn.* **40**, 1557, (1967).
[108] D. C. McCain, *J. Magn. Reson.* **7**, 170 (1972).

magnetic equivalence of the protons in the radical, the proton hyperfine splitting of the ESR spectrum is very simple. Consequently, the splittings of other low-abundance magnetic nuclei may be observed. The splittings and signs in **21** are those of McCain.[108] The electronic absorption spectra of this radical have been studied,[106] as also has its formation in various photoexcitation processes, generally of charge-transfer complexes in which pyromellitic anhydride serves as the electron acceptor.[109–112]

$a(H)$ 0.68
$a(C-1)$ −5.55
$a(C-2)$ 3.61
$a(C-3)$ 2.50
$a(O-4)$ −3.2

(20)   (21)

(22)   (23)   (24)

There have been numerous studies of the radicals resulting from oxidation of ascorbic acid and related species.[113–127] Combining *in situ* radiolysis and ESR, Laroff *et al.*[124] have shown definitively that the principal radical has the highly conjugated but nonaromatic structure **22**, thus resolving

[109] R. L. Ward, *J. Chem. Phys.* **39**, 852 (1963).
[110] R. Potashnik, C. R. Goldschmidt, and M. Ottolenghi, *J. Phys. Chem.* **73**, 3170 (1969).
[111] Y. Achiba and K. Kimura, *Chem. Phys. Lett.* **19**, 45 (1973).
[112] M. Irie, S. Irie, Y. Yamamoto, and K. Hayashi, *J. Phys. Chem.* **79**, 699 (1975).
[113] I. Yamazaki, H. S. Mason, and L. H. Piette, *J. Biol. Chem.* **235**, 2444 (1960).
[114] I. Yamazaki and L. H. Piette, *Biochim. Biophys. Acta* **50**, 62 (1961).
[115] C. Lagercrantz, *Acta Chem. Scand.* **18**, 562 (1964).
[116] G. Foerster, W. Weis, and H. Staudinger, *Justus Liebigs Ann. Chem.* **690**, 166 (1965).
[117] G. A. Russell, E. T. Strom, E. R. Talaty, K.-Y. Chang, R. D. Stephens, and M. C. Young, *Rec. Chem. Prog.* **27**, 3 (1966).
[118] B. H. J. Bielski and A. O. Allen, *J. Am. Chem. Soc.* **92**, 3793 (1970).
[119] B. H. J. Bielski, D. A. Comstock, and R. A. Bowen, *J. Am. Chem. Soc.* **93**, 5624 (1971).
[120] Y. Kirino and T. Kwan, *Chem. Pharm. Bull.* **19**, 718 (1971).
[121] Y. Kirino and T. Kwan, *Chem. Pharm. Bull.* **19**, 831 (1971).
[122] Y. Kirino and T. Kwan, *Chem. Pharm. Bull.* **20**, 2651 (1972).
[123] H. H. Ruf and W. Weis, *Biochim. Biophys. Acta* **261**, 339 (1972).
[124] G. P. Laroff, R. W. Fessenden, and R. H. Schuler, *J. Am. Chem. Soc.* **94**, 9062 (1972).
[125] R. D. McAlpin, M. Cocivera, and H. Chen, *Can. J. Chem.* **51**, 1682 (1973).
[126] M. A. Schuler, K. Bhatia, and R. H. Schuler, *J. Phys. Chem.* **78**, 1063 (1974).
[127] Y. Kirino, *Chem. Lett.*, 153 (1974).

confusion over the state of protonation of the radical (it is anionic in the pH range 0 to 13) and confirming that the radical does not have the isomeric aromatic semidione structure **23**. A heteroaromatic furan-2,3-semidione does occur, however, in the coumaran-2,3-semidione **24**, whose ESR spectrum has been measured.[128]

c. *Semiquinones.* A number of semiquinones in the dibenzofuran system have been prepared by Hewgill and co-workers.[129,130] ESR spectra have been measured but not unambiguously assigned. The semiquinone **25** has been found to result from oxidative condensation of 3-isopropylcatechol and orcinol in alkali.[131] The ESR spectrum was assigned following deuteration experiments. It was shown that probably the C—C bond was formed before the C—O bond during the condensation process.

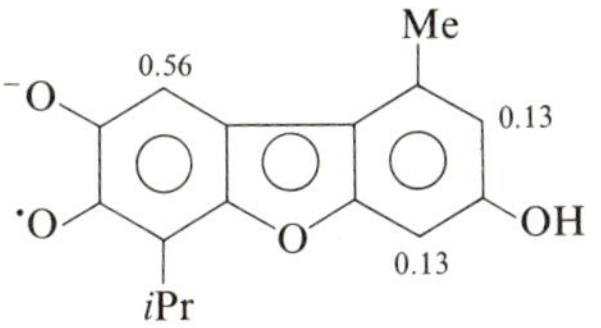

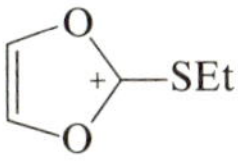

4. *Radicals Containing a Dioxole Ring*

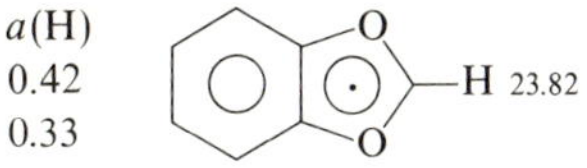

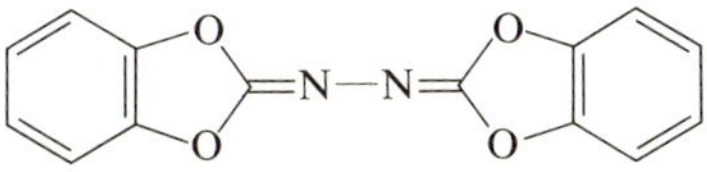

The electrochemical reduction of the 2-thioethoxydioxolylium ion (**26**) has been described.[132] Formation of the corresponding radical, however, is not unambiguously proved. The initial cation **26** is deliquescent and prone to reaction with adventitious water in the solvent; it is not apparent whether **26** is destroyed before electron transfer or whether the 2-thioethoxydioxolyl radical is first formed and then lost in a destructive reaction. A high reactivity for dioxolyl radicals would be consistent with the observation for the benzodioxolyl radical (**27**) of a large spin population at C-2, manifested as a large C-2 proton hyperfine splitting (given in structure **27** in gauss), with

---

[128] G. A. Russell, C. L. Myers, P. Bruni, F. A. Neugebauer, and R. Blankespoor, *J. Am. Chem. Soc.* **92**, 2762 (1970).
[129] F. R. Hewgill, T. J. Stone, and W. A. Waters, *J. Chem. Soc.*, 408 (1964).
[130] F. R. Hewgill and L. R. Mullings, *Aust. J. Chem.* **28**, 355 (1975).
[131] A. C. Waiss, J. A. Kuhnle, J. J. Windle, and A. K. Wiersema, *Tetrahedron Lett.*, 6251 (1966).
[132] R. D. Braun and D. C. Green, *J. Electroanal. Chem. Interfacial Electrochem.* **79**, 381 (1977).

very little spin delocalization into the benzo ring.[133] This paper clarifies an earlier anomalous result for **27**.[134]

The bis-dioxolone azine **28** failed to give reproducible and reversible electrochemical behavior upon anodic oxidation.[135]

B. Radicals Containing a Six-Membered Heterocycle

1. *Pyranyl Radicals*

In the early 1960s, it was found that pyrylium ions undergo single-electron reduction, either by chemical or electrochemical means, with the formation of pyranyl radicals which usually dimerize essentially entirely by coupling at the 4-position.[136–139] Reduction of 2,4,6-triphenylpyrylium salts, conveniently by zinc dust in inert solvents, led to the red radical **29** which persists for several days in a sealed tube.[140] The ESR spectrum of **29** was recorded and eventually analyzed by Degani *et al.*[141,142] Their hyperfine splittings (indicated in **29** in gauss) were obtained by experiments involving deuteration and spectrum simulation. To reconcile calculated splittings (McLachlan method) with the experimental values, it was found necessary to assume the 4-phenyl group and the 2(6)-phenyl groups to be twisted, respectively, 28 and 42° from the plane of the heterocycle.[142]

Other reductions of 2,4,6-triphenylpyrylium ion to give **29** have been examined. Tetramethyl-*p*-phenylene diamine (TMPD) transfers one electron and ESR spectroscopy shows **29** and TMPD$^{+\cdot}$ to result.[143] Chromium(II) ion was shown to reduce 2,4,6-triphenylpyrylium ion and other related cations.[144] A comparison of chemical reactivity with reduction potential suggested an outer-sphere activated complex. A scale of the relative stabilities of the various radicals was deduced. Since **29** results when 2,4,6-triphenyl-

---

[133] C. Gaze, B. C. Gilbert, and M. C. R. Symons, *J. C. S. Perkin II* 235 (1978).
[134] E. A. C. Lucken and B. Poncioni, *J. C. S. Perkin II*, 777 (1976).
[135] S. Hünig, G. Kiesslich, F. Linhart, and H. Schlaf, *Justus Liebigs Ann. Chem.* **752**, 196 (1971).
[136] K. Conrow and P. C. Radlick, *J. Org. Chem.* **26**, 2260 (1961).
[137] E. Gird and A. T. Balaban, *J. Electroanal. Chem.* **4**, 48 (1962).
[138] M. Feldman and S. Winstein, *Tetrahedron Lett.*, 853 (1962).
[139] A. T. Balaban, C. Bratu, and C. N. Rentea, *Tetrahedron* **20**, 265 (1964).
[140] V. A. Palchkov, Yu. A. Zhdanov, and G. N. Dorofeenko, *Zh. Org. Khim.* **1**, 1171 (1965) [*CA* **63**, 11276 (1965)].
[141] I. Degani and C. Vincenzi, *Boll. Sci. Fac. Chim. Ind. Bologna* **25**, 77 (1967).
[142] I. Degani, L. Lunazzi, and G. F. Pedulli, *Mol. Phys.* **14**, 217 (1968).
[143] L. A. Polyakova, K. A. Bilevich, N. N. Bubnov, G. N. Dorofeenko, and O. Yu. Okhlobystin, *Dokl. Akad. Nauk SSSR* **212**, 370 (1973) [*CA* **79**, 145660 (1973)].
[144] W. T. Bowie and M. R. Feldman, *J. Am. Chem. Soc.* **99**, 4721 (1977).

Sec. II.B] HETEROAROMATIC RADICALS, PART II 47

pyrylium hexachloroantimonate is dissolved in anhydrous pyridine, it has been suggested that a charge-transfer complex of pyridine with the cation decomposes, yielding **29** and pyridine cation-radical.[145]

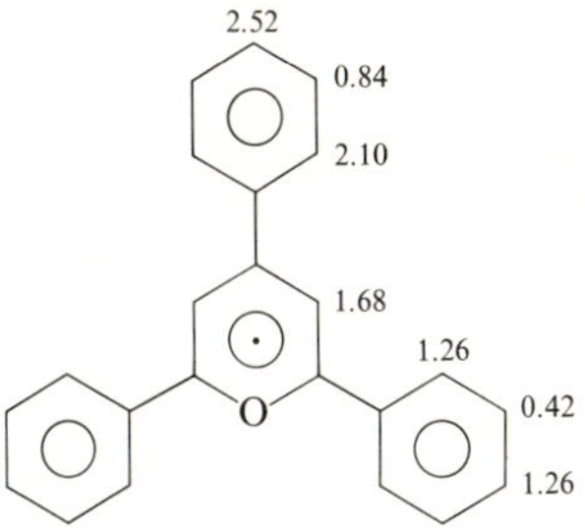

(29) 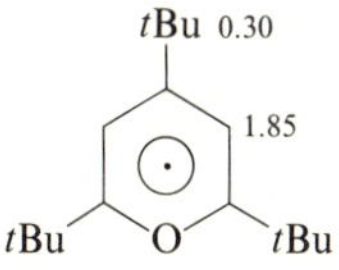   (30)

Radical **29** is formed from the precursor cation at different potentials in different solvents.[146] Donor solvents lower the reduction potential; the lowering has been suggested as a measure of the specific solvation of the triphenylpyrylium cation.

The ECL observed when **29** is generated in the presence of rubrene cation-radical arises via the triplet state of rubrene.[147] The triphenylpyranyl radical (**29**) has also been used in dynamic nuclear polarization experiments.[148,149]

2,4,6-Tri-*tert*-butylpyranyl (**30**) was one of a number of variously substituted pyranyls obtained electrochemically in DMF.[150] The bulky substituents confer a reversibility on radical formation which is absent from systems with smaller substituents such as methyl. The ESR spectrum of **30** has been analyzed by Hacquard and Rassat.[151] They resolved proton splittings for the 4-*tert*-butyl group as well as the pyranyl ring protons; their values are quoted (in gauss) in structure **30**. These authors also performed INDO, McLachlan, and HMO calculations for the radical and ascribed a ^{13}C splitting of 10.9 G to C-4.[151] Krumbholz and Steuber[152]

---

[145] M. Farcasiu and D. Farcasiu, *Chem. Ber.* **102**, 2294 (1969).
[146] N. T. Berberova, G. N. Dorofeenko, and O. Yu. Okhlobystin, *Khim. Geterotsikl. Soedin.*, 1574 (1976) [*CA* **86**, 88996 (1977)].
[147] F. Pragst, *Electrochim. Acta* **21**, 497 (1976).
[148] E. H. Poindexter, J. A. Potenza, D. D. Thomson, N. V. Nghia, and R. H. Webb, *Mol. Phys.* **14**, 385 (1968).
[149] R. L. Glazer and E. H. Poindexter, *J. Chem. Phys.* **55**, 4548 (1971).
[150] N. T. Berberova, G. N. Dorofeenko, and O. Yu. Okhlobystin, *Khim. Geterotsikl. Soedin.*, 318 (1977) [*CA* **87**, 67593 (1977)].
[151] C. Hacquard and A. Rassat, *Mol. Phys.* **30**, 1935 (1975).
[152] E. Krumbholz and F. W. Streuber, *Angew. Chem., Int. Ed. Engl.* **14**, 553 (1975).

have also analyzed the spectrum of **30** and assign very similar proton splittings but differ in the ^{13}C splitting ascribed to C-4 (21.40 G) by a factor of two.

(31)   (32)   (33)

The latter authors have, in addition, analyzed the spectra of the radicals **31–33** in search of evidence of C—N hyperconjugation.[152] In **31** and **32**, the planes of the two heterocycles are expected to be essentially perpendicular for steric reasons; thus, there should be no π-delocalization between the two moieties, unlike the case of **33**. The fact that **31** exhibits a larger nitrogen splitting than **33**, but a smaller N-proton splitting, is taken to imply effective spin transmission by hyperconjugation to the nitrogen atom in **31** and by homohyperconjugation to its attached proton, while the normal π-spin polarization mechanism obtains for **33** where conjugation exists. The hyperconjugative transmission mechanism in **31** is effective because the conformation causes the pyrrole C—N bond to eclipse the $2p(\pi)$ orbital on the pyranyl C-4. The lack of nitrogen splitting in **32** is taken as supporting evidence as are the relative magnitudes of ^{13}C splittings.[152]

Soviet workers have examined the polarographic properties of unsymmetrically substituted pyrylium ions (e.g., **34** and **35**).[153–157] As expected, pyranyl radicals intervene and dimerize. Adsorption phenomena at the dropping mercury electrode are important. The consequences for the electrode and homogeneous processes of variation in substitution, solvent, acidity, etc. were explored.

[153] M. M. Evstifeev, G. Kh. Aminova, G. N. Dorofeenko, and E. P. Olekhnovich, *Zh. Obshch. Khim.* **44**, 657 (1974) [*CA* **81**, 9099 (1974)].
[154] M. M. Evstifeev, G. Kh. Aminova, G. N. Dorofeenko, and E. P. Olekhnovich, *Zh. Obshch. Khim.* **44**, 2267 (1974) [*CA* **82**, 30704 (1975)].
[155] M. M. Evstifeev, G. Kh. Aminova, G. N. Dorofeenko, and E. P. Olekhnovich, *Nov. Polyarogr., Tezisy Dokl. Vses. Soveshch. Polyarogr., 6th, 1975*, 155 (1975) [*CA* **86**, 62671 (1977)].
[156] M. M. Evstifeev, G. N. Dorofeenko, E. P. Olekhnovich, and G. Kh. Aminova, *Zh. Obshch. Khim.* **46**, 1334 (1976) [*CA* **85**, 93513 (1976)].
[157] M. M. Evstifeev, G. Kh. Aminova, G. N. Dorofeenko, and E. P. Olekhnovich, *Zh. Obshch. Khim.* **46**, 2693 (1976) [*CA* **86**, 62698 (1977)].

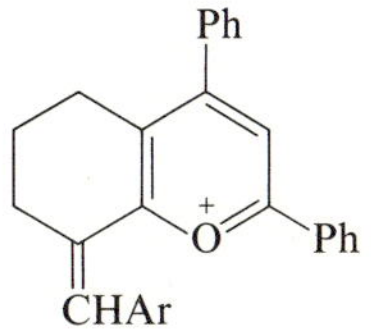

Sec. II.B] HETEROAROMATIC RADICALS, PART II 49

(34) Ph, CHAr substituent

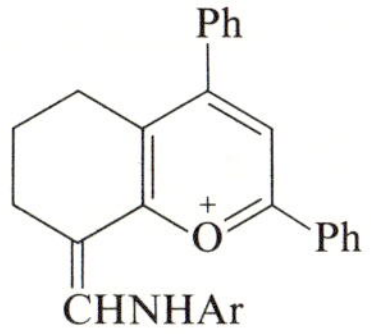

(35) Ph, CHNHAr substituent

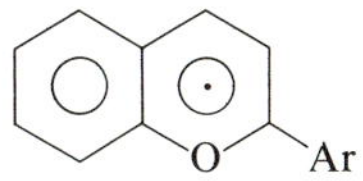

(36)

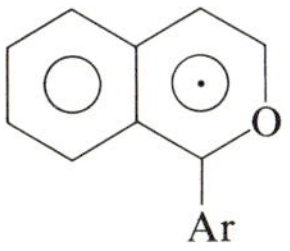

(37)

2. Benzopyranyl Radicals

The involvement of radicals **36** and **37** in reductions of 2-arylbenzopyrylium (flavylium) and 1-arylisobenzopyrylium ions is established.[150,153–165] In general, the radicals dimerize rapidly unless inhibited by pyranyl ring substitution; 2-(4-nitrophenyl)benzopyranyl is reported not to dimerize, however.[144] The kinetics of the formation of a series of **36** by reduction of the precursor flavylium ions by Cr(II) have been investigated; a Hammett plot of the second-order rate constants versus σ^+ has $\rho = 1.03$.[144] Most work has been concerned with the polarography of the parent cations; there appear to be no ESR results for benzopyranyl radicals. The mediation of benzopyranyl radicals in the oxidation of 2-phenyl-2H-benzopyrans by potassium permanganate has been suggested.[166] (See Section II,B,5,b for radicals from anthocyanidins.)

---

[158] J.-M. Meunier, M. Person, and P. Fournari, *Bull. Soc. Chim. Fr.*, 2872 (1967).
[159] M. M. Evstifeev, L. L. Pyshcheva, T. A. Tyagunova, and O. M. Orlova, *Nov. Polyarogr., Tezisy Dokl. Vses. Soveshch. Polyarogr., 6th, 1975*, 167 (1975) [*CA* **86**, 161369 (1977)].
[160] M. M. Evstifeev, O. M. Orlova, L. L. Pyshcheva, T. A. Tyagunova, T. N. Rastsvetaeva, and Z. Yu. Derispako, *Fiz.-Khim. Metody Anal. Kontrolya Proizvod., Mezhvuz. Sb.* **2**, 48 (1976) [*CA* **87**, 161232 (1977)].
[161] M. M. Evstifeev, L. L. Pyshcheva, A. I. Pyshchev, and D. N. Dorofeenko, *Fiz.-Khim. Metody Anal. Kontrolya Proizvod., Mezhvuz. Sb.* **2**, 133 (1976) [*CA* **87**, 126680 (1977)].
[162] M. M. Evstifeev, L. L. Pyshcheva, A. I. Pyshchev, and D. N. Dorofeenko, *Zh. Obshch. Khim.* **46**, 1340 (1976) [*CA* **85**, 93514 (1976)].
[163] M. Vajda, *Collect. Czech. Chem. Commun.* **25**, 1952 (1960).
[164] M. Vajda and A. Gelleri, *Ann. Univ. Sci. Budap. Rolando Eotvos Nominatae, Sect. Chim.* **3**, 107 (1961) [*CA* **57**, 8347 (1962)].
[165] M. Vajda and F. Ruff, *Abh. Dtsch. Akad. Wiss. Berlin, Kl. Chem., Geol. Biol.*, **112** (1964) [*CA* **62**, 5175 (1965)].
[166] Y. Ashihara, Y. Nagata, and K. Kurosawa, *Bull. Chem. Soc. Jpn.* **50**, 3298 (1977).

3. Polycyclic Pyranyl Radicals

a. *Xanthenyl (Dibenzopyranyl) Radicals.* Xanthenyl radical (**38**) itself is obtained in equilibrium concentrations when 9,9′-bixanthenyl is thermolyzed in suitable inert solvents. Sevilla and Vincow observed the ESR spectrum of **38** (their splittings indicated in gauss) by thermolysis of the dimer in *n*-tridecane at 180–260°C.[167] The assignment of the splittings was made following specific deuteration experiments, and in the light of Hückel and McLachlan calculations. Others have also performed calculations on xanthenyl radical.[168–170]

The ESR spectra of a range of 9-alkylxanthenyls (e.g., **39**) have also been analyzed in detail by Sevilla and Vincow.[171] The proton couplings of the heterocycle were found to be very little influenced by the nature of the alkyl substitution whereas the alkyl β-proton splittings are very variable, showing, for example, values of 12.177, 6.23, and 0.87 G, respectively, for the Me, Et and *i*Pr groups; the β-splittings of alkyl groups other than Me were also found to be temperature-dependent. This behavior was interpreted in terms of the rotation of the alkyl groups being highly restricted (see Part I: Section II,B,4). Each has a preferred equilibrium conformation and executes a torsional oscillation about this. A harmonic potential was used in modeling this motion, and force constants consistent with the experimental data and expectation were evaluated. Interestingly, long-range couplings were detected from γ-alkyl and benzyl aryl protons.[171] Similar long-range aryl splittings, given in **40** in gauss, have been observed for thiophenoxyxanthenyl.[172]

There have been several ESR studies of 9-phenylxanthenyl (**41**).[173–177] The splittings originally assigned by Sevilla and Vincow[173] have been confirmed. Those shown in **41** (in gauss) derive from solution ENDOR experiments.[177] In order adequately to reproduce the phenyl ring splittings in MO calculations, twist must be introduced into the C9—Ph bond. Different authors have deduced different values: Sevilla and Vincow suggested a value near 60°[173]; Maruyama *et al.*, whose splittings differ somewhat from those

---

[167] M. D. Sevilla and G. Vincow, *J. Phys. Chem.* **72**, 3635 (1968).
[168] N. K. Ray, *Chem. Phys. Lett.* **2**, 634 (1968).
[169] G. Orlandi, G. Poggi, F. Barigelletti, and H. Breccia, *J. Phys. Chem.* **77**, 1102 (1973).
[170] P. Devolder, *Theor. Chim. Acta* **39**, 277 (1975).
[171] M. D. Sevilla and G. Vincow, *J. Phys. Chem.* **72**, 3647 (1968).
[172] P. Devolder, *Can. J. Chem.* **54**, 1744 (1976).
[173] M. D. Sevilla and G. Vincow, *J. Phys. Chem.* **72**, 3641 (1968).
[174] K. Maruyama, M. Yoshida, and K. Murakami, *Bull. Chem. Soc. Jpn.* **43**, 152 (1970).
[175] L. Lunazzi, A. Mangini, G. Placucci, and C. Vincenzi, *J. C. S. Perkin I*, 2418 (1972).
[176] M. Hori, T. Kataoka, Y. Asachi, and E. Mizuta, *Chem. Pharm. Bull.* **21**, 1318 (1973).
[177] Y. Yamada, S. Toyoda, and K. Ouchi, *J. Phys. Chem.* **78**, 2512 (1974).

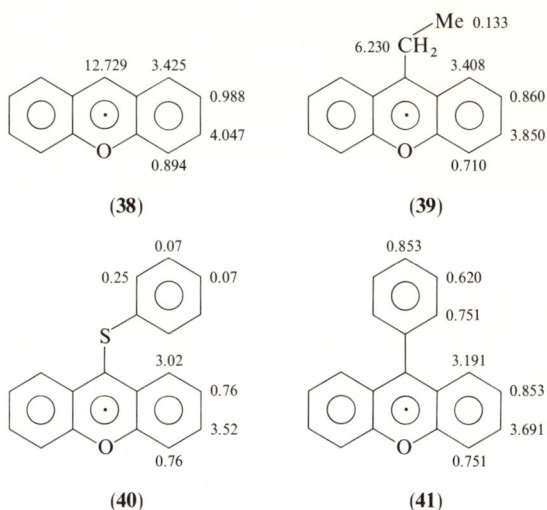

of other authors, preferred 63°[174]; Lunazzi et al. suggested 70–80°,[175] while Yamada et al.[177] required only 54°. Moreover, the last workers detected significant differences in the heterocyclic proton splittings between 41 on the one hand and the o-tolyl analog on the other which they ascribed to less effective delocalization through the C9—aryl bond for the more hindered radical. The increase in twist angle is suggested to be only four more degrees, giving the o-tolylxanthenyl a twist angle of 58°, however. These results may be compared with twist angles of ~65° found for the geometrically similar 9-phenylacridine anion-radical[178] and for 10-phenylphenoxazine and 10-phenylphenothiazine cation-radicals (see Section VI,B,3,b,iii and Part I: Section II,B,4).

Electronic absorption spectra have been recorded for 38 during flash photolysis of 9,9'-bixanthenyl. The spectra are distinguishable from those of the radical arising in photoexcitation of xanthen-9-one.[179]

Xanthenyl radicals have been implicated in the oxidation of xanthene in various conditions. Contrasting mechanisms have been advanced for the acid-catalyzed autoxidation of xanthene.[180,181] The result of oxidation of xanthene by superoxide ion in DMSO is consistent with the intermediacy of 38; it may also have a role in oxidation by ceric ammonium nitrate.[182,183]

[178] A. Lomax, L. S. Marcoux, and A. J. Bard, J. Phys. Chem. 76, 3958 (1972).
[179] J.-P. Marteel, P. Decock, P. Goudmand, and P. Devolder, Bull. Soc. Chim. Fr., 1767 (1975).
[180] A. P. ter Borg, H. R. Gersmann, and A. F. Bickel, Recl. Trav. Chim. Pays-Bas 85, 899 (1966).
[181] N. C. Deno, E. L. Booker, K. E. Kramer, and G. Saines, J. Am. Chem. Soc. 91, 5237 (1969).
[182] Y. Moro-Oka, P. J. Chung, H. Arakawa, and T. Ikawa, Chem. Lett., 1293 (1976).
[183] B. Rindone and C. Scolastico, J. C. S. Perkin I, 1398 (1975).

When xanthene is oxidized by certain 1,4-naphthoquinones, radical products are obtained which contain a xanthenyl moiety (e.g., **42**) for which splittings are quoted in gauss. The reactions have been investigated by the CIDNP technique.[184] Radical **41** has been involved as a substrate in Feldman and co-workers' series of papers on stabilities of trivalent carbon species.[144,185,186] (See Section II,B,5,e for radicals from xanthene dyes.)

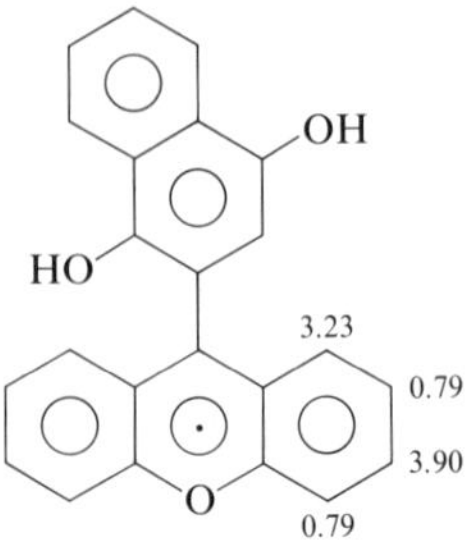

(**42**) 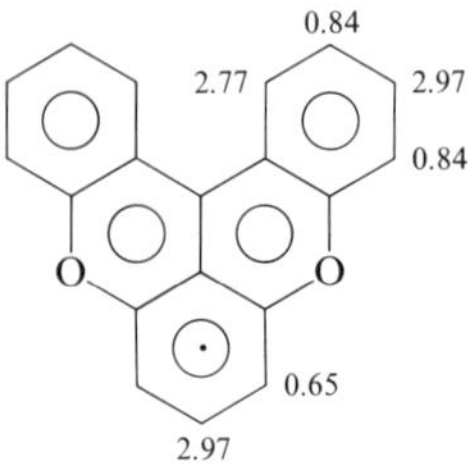 (**43**)

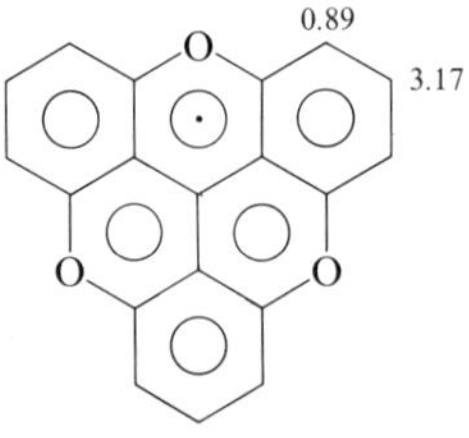

(**44**)

b. *Other Polycyclic Pyranyl Radicals.* The two highly symmetrical radicals [1]benzopyrano[2,3,4-*kl*]xanthen-13b-yl (**43**) and 4,8,12-trioxa-4*H*,8*H*-dibenzo[*cd,mn*]pyren-12c-yl (sesquixanthydryl) (**44**) have been synthesized and their ESR spectra assigned (splittings indicated in gauss).[174,187–190] The attraction of these species has been as "planar triarylmethyls." The electrochemistry of **44** has attracted various groups with an interest in its stabilization and the relationship of electrochemical and quantum-mechanical indices.[144,186,191,192] Hoffman and co-workers

---

[184] Z. Maruyama and S. Arakawa, *Bull. Chem. Soc. Jpn.* **47**, 1960 (1974).
[185] M. Feldman and W. C. Flythe, *J. Am. Chem. Soc.* **91**, 4577 (1969).
[186] M. Feldman and W. C. Flythe, *J. Org. Chem.* **43**, 2596 (1978).
[187] O. Neunhoeffer and H. Haase, *Chem. Ber.* **91**, 1801 (1958).
[188] J. C. Martin and R. G. Smith, *J. Am. Chem. Soc.* **86**, 2252 (1964).
[189] M. J. Sabacky, C. S. Johnson, R. G. Smith, H. S. Gutowsky, and J. C. Martin, *J. Am. Chem. Soc.* **89**, 2054 (1967).
[190] E. Müller, A. Moosmayer, A. Rieker, and K. Scheffler, *Tetrahedron Lett.*, 3877 (1967).
[191] I. Nemcova and I. Nemec, *Chem. Zvesti* **26**, 115 (1972) [*CA* **77**, 69348 (1972)].
[192] I. Nemcova and I. Nemec, *J. Electroanal. Chem. Interfacial Electrochem.* **30**, 506 (1971).

have examined theoretically the stabilization of triarylmethyl radicals and ions and the consequence for their stabilization of rotation from coplanarity of the phenyl rings. This paper is of relevance, therefore, for radicals dealt with in this and the previous section.[193]

4. Bipyryliumyl Radicals and Related Species

As indicated in Section II,B,1, pyranyl radicals dimerize by coupling at C-4 even when this position is substituted by small groups such as Me. In the absence of a 4-substituent, coupling is efficient to give a 4,4′-bipyran (**45**) which may be oxidized to a 4,4′-bipyranylidene (**46**) by transfer of hydride to a pyrylium ion (Scheme 1, R = tBu,Ph). This overall process may be effected from the initial pyrylium ion by either chemical or electrochemical reduction.[143,150,194,195]

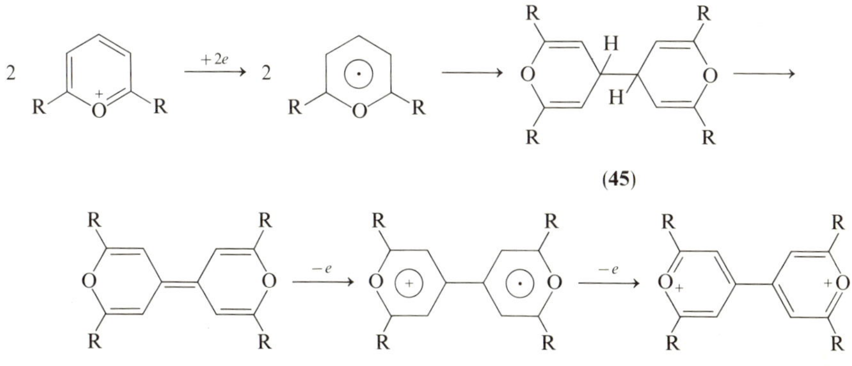

SCHEME 1

Bipyranylidenes may be oxidized in two single-electron steps via bipyryliumyl radicals (**47**) to bipyrylium ions (**48**).[196] Conversely, bipyrylium ions prepared by alternative routes may be reduced to bipyryliumyl radicals.[196,197] The radical **47** (R = tBu) has been characterized by ESR[143] and the electrochemical characteristics of **47** (R = Ph, tBu, and others) have been determined.[150,196] Decarboxylation of 4-carboxy-2,6-diphenylpyrylium ion in the presence of Vaska's compound [(Ph₃P)₂IrClCO] led to **47** (R = Ph) as the principal product, but whether the reaction involved

---

[193] R. Hoffmann, R. Bissell, and D. G. Farnum, *J. Phys. Chem.* **73**, 1789 (1969).
[194] F. Pragst and U. Seydewitz, *J. Prakt. Chem.* **319**, 952 (1977) [*CA* **88**, 104389 (1978)].
[195] C. Fabre, R. Fugnitto, and H. Strzelecka, *C. R. Acad. Sci., Ser. C* **282**, 175 (1976).
[196] S. Hünig, B. J. Garner, G. Ruider, and W. Schenk, *Justus Liebigs Ann. Chem.*, 1036 (1973) [*CA* **79**, 77634 (1973)].
[197] G. A. Reynolds and J. A. Van Allan, *J. Heterocycl. Chem.* **6**, 623 (1969).

prior formation of **46** (R = Ph) via dimerization of an intermediate carbene is uncertain.[198]

Phenylogs of bipyryliumyl ion (**49** and **50**) and a diazavinylog (**51**) have been described.[150,199] The biflavyliumyl radical **52** has been formed from the product of a coupling of 4-methylflavylium ion brought about by treatment with pyridine.[200]

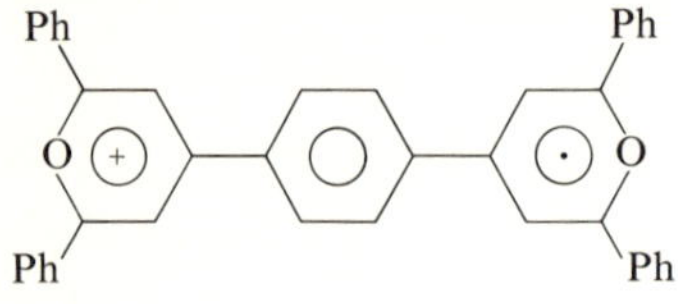

(49)

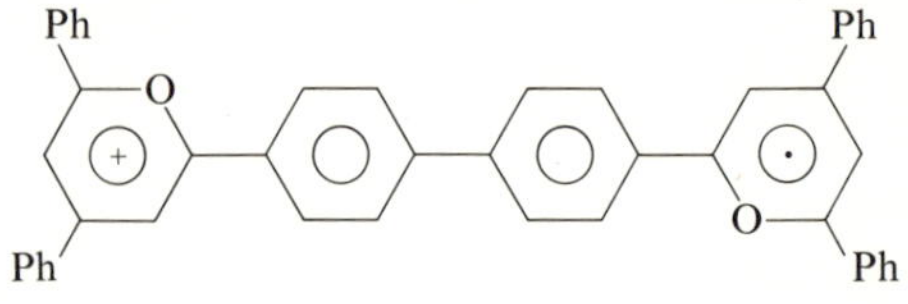

(50)

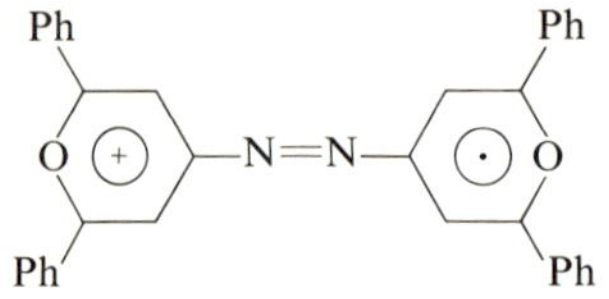

(51)

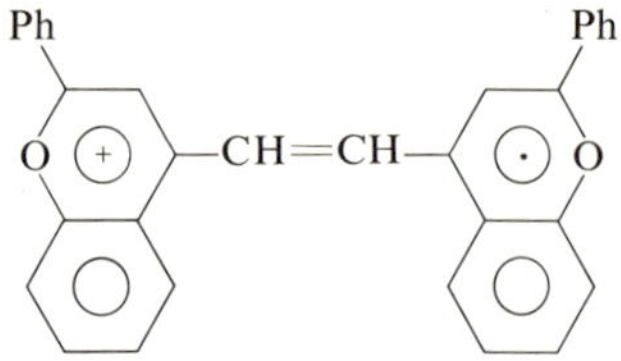

(52)

Other radicals containing two pyran rings are the polycyclic cation-radical **53**[201] and the neutral radical **54**, one from a series of polymethine dyes with terminal pyran rings.[202] ESR spectra have been recorded for all these radicals (quoted hyperfine splittings are in gauss).

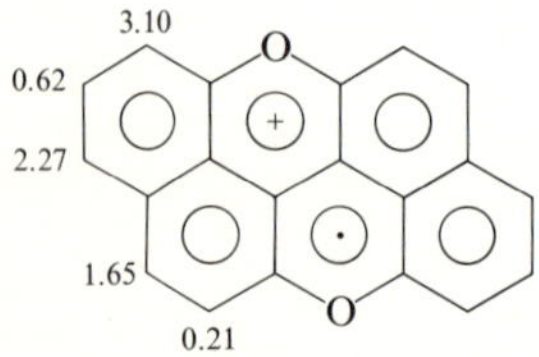

(53)

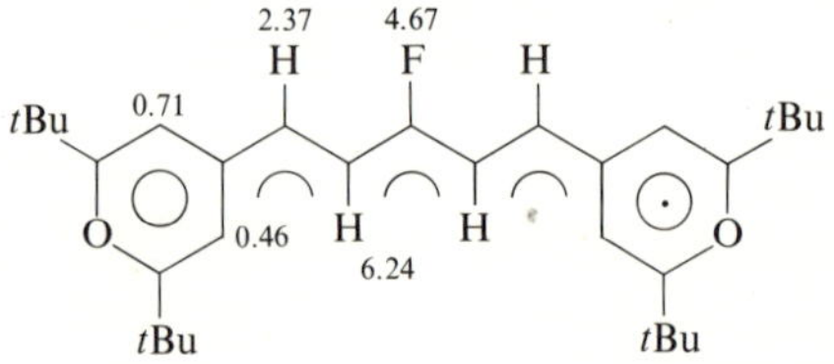

(54)

5. Radicals from Oxygenated Pyrans

a. *Radicals Derived from Pyranones.* Although 2,6-dimethylpyran-4-one failed to give a ketyl on electrochemical reduction in various organic

[198] V. V. Bessonov, O. Yu. Okhlobystin, T. P. Panova, and L. Yu. Ukhin, *Teor. Eksp. Khim.* **12**, 829 (1976) [*CA* **86**, 154984 (1977)].
[199] S. Hünig and G. Ruider, *Justus Liebigs Ann. Chem.*, 1415 (1974).
[200] J. A. Van Allan and G. A. Reynolds, *Tetrahedron Lett.*, 2047 (1969).
[201] K. Ikegami, Y. Matsunaga, K. Osafune, and E. Osawa, *Bull. Chem. Soc. Jpn.* **48**, 341 (1975).
[202] H. Oehling and F. Baer, *Org. Magn. Reson.* **8**, 623 (1976).

solvents,[203] the radical **55** was obtained by reduction by solvated electrons in liquid ammonia as proved by the observation of its ESR spectrum (splitting in **55** given in gauss).[204] The trianionic ketyl (**56**) from chelidonic acid and its first conjugate acid (**57**) were observed by Neta and Fessenden under radiolytic conditions.[205] They noted that the signal intensity for **56** was some 40-fold that of **57** under comparable conditions, indicating a large difference in the rates of the reactions destroying the two radicals (splittings in **56** and **57** are given in gauss). Oscillopolarographic reduction of 3-hydroxypyran-4-one would only occur in acidic conditions, indicating, perhaps, that protonation of the keto function is necessary to permit attachment of an electron in this case.[206] Whereas Parkanyi and Zahradnik observed only multiple electron reductions on polarography of various pyran-4- and -2-one derivatives,[207] Le Guillanton has recorded electroreductive dimerization of 4-methoxy-6-methylpyran-2-one and its 6-phenyl analog, probably indicative of the intermediacy of radicals.[208,209] Dixon and co-workers[210] observed that 3-hydroxy-6-methylpyran-2-one traps OH· to give ultimately 2,3,4-trihydroxy-6-methylpyranyl (**58**) which is the conjugate acid of a pyranone ketyl. The latter workers also found that certain hydroxypyranones are oxidized by Ce(IV) to give radicals (e.g. **59**) which are comparable with aryloxyls or semiquinones.[210]

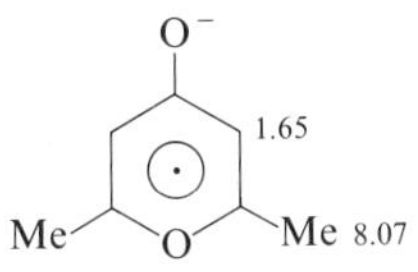

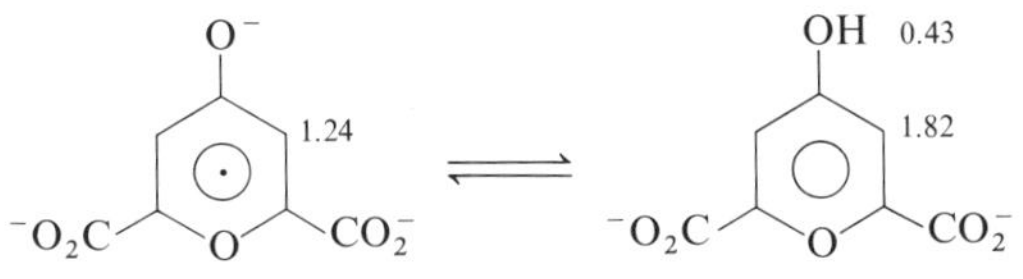

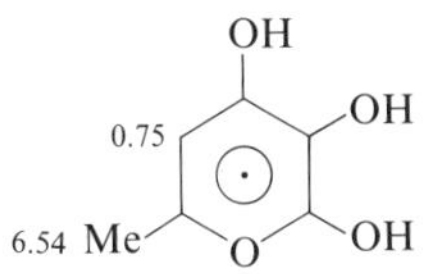

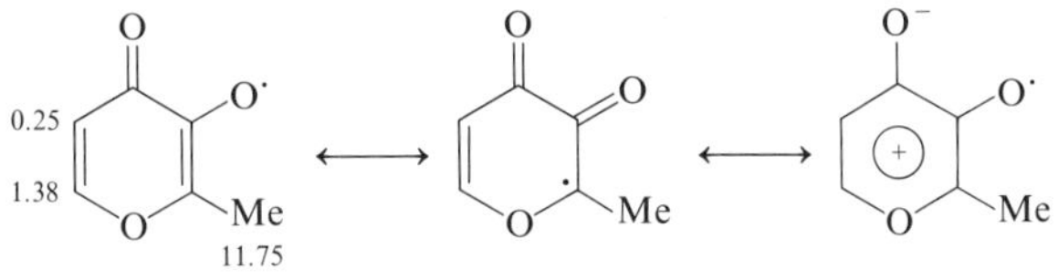

[203] V. E. Sahini, L. Ciurea, and E. Volanschi, *Rev. Roum. Chim.* **12**, 355 (1967).
[204] J. H. Elson, T. J. Kemp, D. Greatorex, and H. D. B. Jenkins, *J. C. S., Faraday 2* **69**, 665 (1973).
[205] P. Neta and R. W. Fessenden, *J. Phys. Chem.* **76**, 1957 (1972).
[206] V. P. Gladyshev, *Chem. Zvesti* **17**, 581 (1963) [*CA* **60**, 5077 (1964)].
[207] C. Parkanyi and R. Zahradnik, *Collect. Czech. Chem. Commun.* **27**, 1355 (1962).
[208] G. Le Guillanton, *C. R. Acad. Sci., Ser. C* **276**, 1131 (1973).
[209] G. Le Guillanton, *Bull. Soc. Chim. Fr.*, 627 (1974).
[210] W. T. Dixon, M. Moghimi, and D. Murphy, *J. C. S. Perkin II*, 101 (1975).

b. *Radicals Derived from Benzopyranones (Chromone and Coumarin)*. There is polarographic evidence that both the chromone (benzopyran-4-one) and coumarin (benzopyran-2-one) structures undergo single-electron reduction to give ketyl radicals. Many years ago, it was shown that various naturally occurring compounds are reduced to give short-lived radicals (e.g., quercetin gave **60**), and their half-wave potentials were related to variations in the pattern of hydroxyl substitution[211,212]; there have also been recent polarographic studies of flavones.[213,214] Tirouflet and Corvaisier[215] studied the polarography of chromones with 2-heteroaryl substituents. A recent Soviet study has reported the identification, by a combination of chromatography and mass spectrometry, of small amounts of radical-derived products from the electroreduction of chromone, coumarin, and other materials.[216] The participation of a radical, probably **61**, has been inferred from the observation of diethyl 4,4'-bis(3,4-dihydrocoumarin-3-carboxylate) among the products of the reaction of ethyl coumarin-3-carboxylate with a Grignard reagent.[217]

(60)   (61)

The polarography of anthocyanidins was studied several years ago by Zuman.[218,219] A more recent investigation of pelargonidin (**62**), cyanidin, and various model flavylium ions has been published by Harper and

---

[211] D. W. Engelkemeir, T. A. Geissman, W. R. Crowell, and S. L. Friess, *J. Am. Chem. Soc.* **69**, 155 (1947).
[212] T. A. Geissman and S. L. Friess, *J. Am. Chem. Soc.* **71**, 3893 (1949).
[213] H. Loth, G. Grelbig, G. Ketzler, H. Koehne, B. Kruse, and B. Lang, *Arch. Pharm. Ber. Dtsch. Pharm. Ges.* **305**, 317 (1972) [*CA* **77**, 61087 (1972)].
[214] E. F. Kalistratova, N. M. Dariglazov, and N. A. Tyukavkina, *Nov. Polyarogr., Tezisy Dokl. Vses. Soveshch. Polyarogr., 6th, 1975*, 169 (1975) [*CA* **86**, 23509 (1977)].
[215] J. Tirouflet and A. Corvaisier, *Bull. Soc. Chim. Fr.*, 535 (1962).
[216] Yu. E. Orlov, A. I. Udalov, V. Z. Lokshtanov, and T. E. Gulimova, *Nov. Polyarogr., Tezisy Dokl. Vses. Soveshch. Polyarogr., 6th, 1975*, 249 (1975) [*CA* **86**, 62679 (1977)].
[217] B. Gustafsson, *Finn. Chem. Lett.*, 49 (1975) [*CA* **83**, 178723 (1975)].
[218] P. Zuman, *Chem. Listy* **46**, 328 (1952) [*CA* **47**, 52 (1953)].
[219] P. Zuman, *Collect. Czech. Chem. Commun.* **18**, 36 (1953).

Sec. II.B] HETEROAROMATIC RADICALS, PART II 57

Chandler.[220-224] In acid conditions, **62** is reduced to the radical **63**. However, at higher pH **62** undergoes deprotonation to the tautomeric **64** which may be regarded as a phenylogous coumarin derivative. This, too, undergoes one-electron reduction to give a ketyl radical which is a conjugate base of **63**. At high pH certain anthocyanidins undergo heteroring scission to give chalcones which are also reducible. A complex array of radicals may therefore arise from anthocyanidins, not all of which are heteroaromatic.

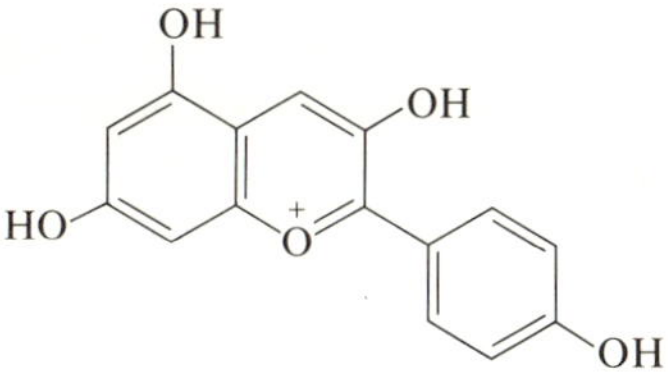

(62) 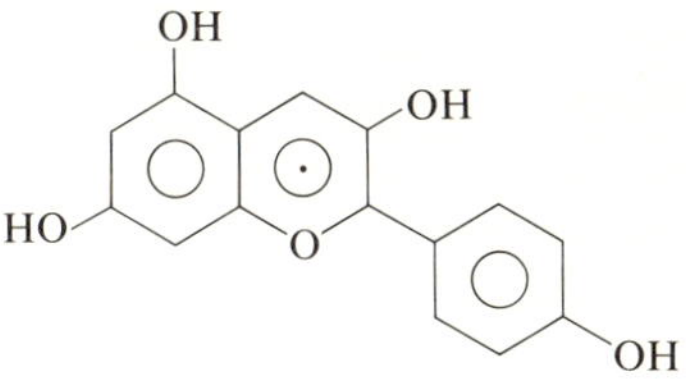   (63)

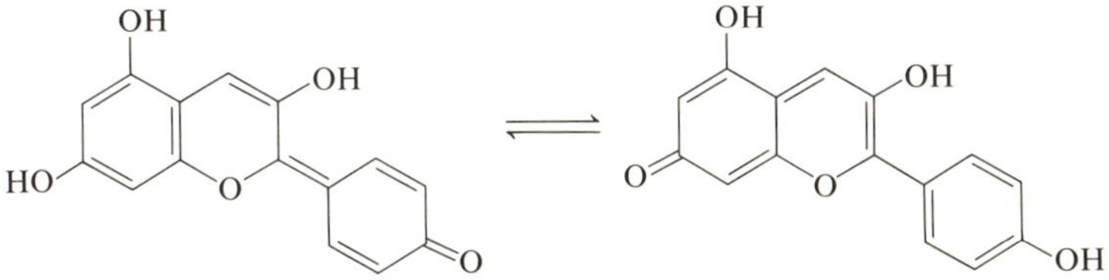

(64)

Oxidation of hydroxybenzopyrones leads to the formation of radicals which have been studied by ESR.[210,225,226] Thus, Ashworth[225,226] found that autoxidation of esculetin (6,7-dihydroxycoumarin) in alkaline conditions led to the initial formation of **65**, but that in due course a secondary nonheteroaromatic radical appeared following hydrolytic ring scission. Hyperfine splittings are indicated in gauss; those indicated parenthetically are for acid conditions (*vide infra*); 4-methyldaphnetin was autoxidized to the semiquinone **67** which was similarly hydrolyzed to **68**. 7-Hydroxycoumarin gave no heterocyclic radical, but in the presence of hydrogen peroxide hydroxylation of the cinnamic acid resulting from hydrolysis of the heterocycle occurred, giving rise to nonaromatic semiquinones. Several

---

[220] K. A. Harper and B. V. Chandler, *Aust. J. Chem.* **20**, 731 (1967).
[221] K. A. Harper and B. V. Chandler, *Aust. J. Chem.* **20**, 745 (1967).
[222] K. A. Harper, *Aust. J. Chem.* **20**, 2691 (1967).
[223] K. A. Harper, *Aust. J. Chem.* **21**, 221 (1968).
[224] K. A. Harper, *J. Appl. Chem. Biotechnol.* **23**, 261 (1973).
[225] P. Ashworth, *Tetrahedron Lett.*, 1435 (1975).
[226] P. Ashworth, *J. Org. Chem.* **41**, 2920 (1976).

other compounds were examined in these papers.[225,226] Dixon and coworkers[210] have oxidized several hydroxybenzopyranones in acidic conditions with Ce(IV). Esculetin and 4-methyldaphnetin were among their substrates too. Differences between their measured hyperfine splittings (indicated parenthetically in **65** and **66** in gauss) and those of Ashworth no doubt stem from the different states of protonation of the radicals in the two studies. Simple rules were given for assigning the splittings in a wide range of these unsymmetrical heterocyclic radicals. The observation by

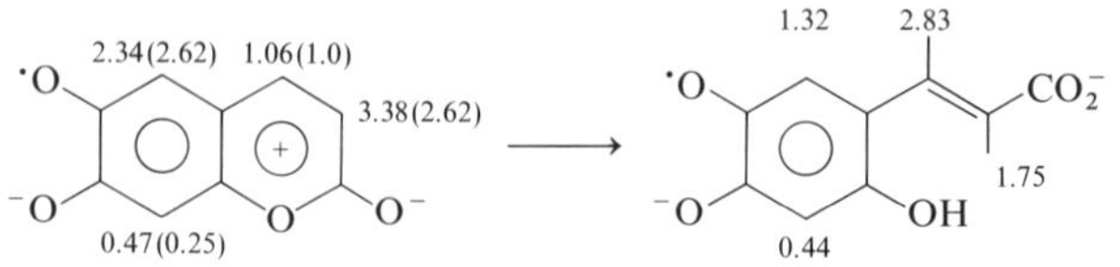

(**65**)   (**66**)

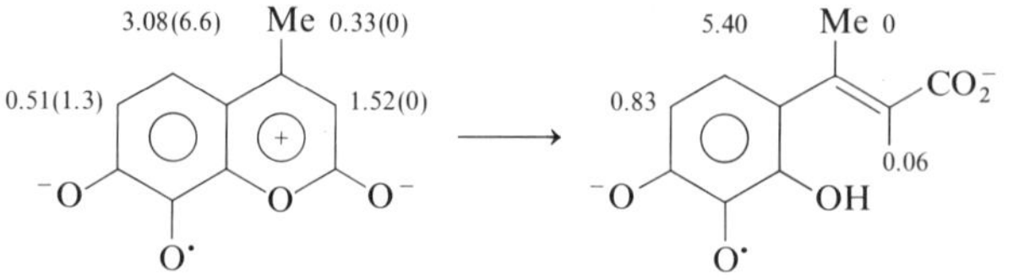

(**67**)   (**68**)

Sharma and Seshadri[227] of the oxidative dimerization of both esculetin and isoscopoletin (7-methylesculetin) upon treatment with $[Fe(DMF)_3Cl_2]\cdot$ $[FeCl_4]$ strongly suggests the participation of **65** and its 7-methylated analog, or their protonated forms. The coupling positions are consistent with the positions of high spin population indicated in **65**.

c. *1H,3H-Naphtho[1,8-cd]pyran-1,3-semidiones*. The dicarboxylic anhydride function can exist in a six-membered heteroaromatic ring when attached to the peri positions of naphthalene and similar structures. Such anhydrides may be reduced to semidione radicals. Thus 1*H*,3*H*-naphtho[1,8-*cd*]pyran-1,3-dione (naphthalic anhydride) has been reduced to radical **69**. The ESR spectrum of this radical has been analyzed by three groups.[102,107,228] The hyperfine splittings recorded by Nelsen[228] (given in gauss in **69**) agree with those of Hirayama[107]; the values given by Sioda and Koski[102] differ significantly. The difference is apparently due to an error of field calibration. Nelsen's assignment of hyperfine splitting was borne out by studies of alkylated radicals. He also characterized the radical

---

[227] D. K. Sharma and T. R. Seshadri, *Indian J. Chem.* **15B**, 939 (1977).
[228] S. F. Nelsen, *J. Am. Chem. Soc.* **89**, 5925 (1967).

70. The electronic absorption spectrum of **69** has been obtained by Shida et al.[106] and accounted for in MO terms. ESR spectra have also been measured and assigned for nitrated derivatives of **69**.[229]

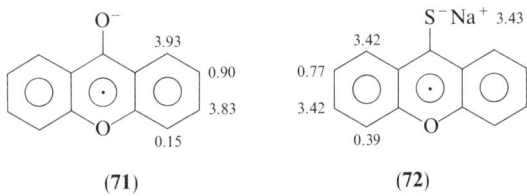

(69)   (70)

d. *Xanthen-9-one Ketyl and Isoelectronic Radicals.* Addition of an electron to xanthen-9-one may be brought about readily in aprotic solvents with the formation of the ketyl **71**. This heteroaromatic radical was observed by ESR early and has been used by several groups interested in ion-pairing phenomena.[82,203,230–239] The hyperfine splittings given in structure **71** (in gauss) are those of Tabner and Zdysiewicz[238] for the free ketyl, generated electrolytically in DMF. The proton splittings do not vary greatly upon formation of ion pairs; rather, it is the splitting of the counterion which is most susceptible to variation with conditions. Xanthene-9-thione ketyl (**72**) shows very large counterion splittings[233,239]; those given in **72** (in gauss) are from Aarons and Adam[239] and correspond to the ion-pair in DME at 22°C.

(71)   (72)

[229] C. H. J. Wells and J. A. Wilson, *J. Chem. Soc. B*, 1588 (1971).
[230] G. A. Russell, E. G. Janzen, H.-D. Becker, and F. J. Smentowski, *J. Am. Chem. Soc.* **84**, 2652 (1962).
[231] N. Hirota and S. I. Weissman, *J. Am. Chem. Soc.* **86**, 2537 (1964).
[232] N. Hirota and S. I. Weissman, *J. Am. Chem. Soc.* **86**, 2538 (1964).
[233] E. G. Janzen and C. M. DuBose, *J. Phys. Chem.* **70**, 3372 (1966).
[234] V. E. Sahini and L. Ciurea, *Rev. Roum. Chim.* **14**, 689 (1969).
[235] K. Maruyama, M. Yoshida, I. Tanimoto, and J. Osugi, *Rev. Phys. Chem. Jpn.* **39**, 117 (1969).
[236] K. Maruyama, M. Yoshida, and J. Osugi, *Rev. Phys. Chem. Jpn.* **39**, 123 (1969).
[237] K. S. Chen, S. W. Mao, K. Nakamura, and N. Hirota, *J. Am. Chem. Soc.* **93**, 6004 (1971).
[238] B. J. Tabner and J. R. Zdysiewicz, *J. Chem. Soc. B*, 1659 (1971).
[239] L. J. Aarons and F. C. Adam, *Can. J. Chem.* **50**, 1390 (1972).

(73)   (74)

Radical **71** is a strong base; consequently, in protic solvents it is protonated to the conjugate acid **73**. This radical may also be prepared photochemically: when xanthen-9-one n,π^* triplet is generated by irradiation in the presence of hydrogen-donor solvents, it abstracts hydrogen from the solvent, thus forming **73**. The acid–base equilibrium **71** ⇌ **73** has been investigated by Kalinowski et al.[240] They also studied the dimerization and dismutation reactions of **71** and **73** in comparison with other ketyls.[241,242] The ESR spectrum of **73** in various solvents has been reported by Wilson[243] whose hyperfine splittings, obtained in ethanol, are indicated (in gauss) in **73**. The OH proton splitting is variable with solvent. There has been a recent laser photochemical study of the triplet state of xanthen-9-one and the formation from it of **73**.[244] The electronic absorption spectra of both species were recorded and life-times of the triplet in various solvents determined. The spectroscopic result may be compared with those of Marteel et al.[179] noted earlier. There have been various other reports of the photoreduction of xanthen-9-one, all implicating **73**.[245–247] The interesting radical **74** (splittings indicated in gauss) has been synthesized by German workers; it is not decomposed thermally until well above 100°C although its counterpart with Sn in place of Si is less stable. Radical **74** is prepared by heating or irradiating xanthen-9-one with $(Me_3Si)_2Hg$.[248]

Ketyl radicals are involved in various chemical reductions of xanthen-9-one: reduction to xanthene by lithium in liquid ammonia,[249] reduction via 9,9′-bixanthenyl-9,9′-diol to xanthen-9-ol by zinc in alkaline ethanol.[250] Attempts to attach an electron to xanthene itself to form an anion-radical in DMF led instead to the formation of **71**[238]; xanthene *is* reducible at lower temperatures, but the radical which results is not, of

[240] M. K. Kalinowski, Z. R. Grabowski, and B. Pakula, *Trans. Faraday Soc.* **62**, 918 (1966).
[241] M. K. Kalinowski and Z. R. Grabowski, *Trans. Faraday Soc.* **62**, 926 (1966).
[242] M. K. Kalinowski, *Rocz. Chem.* **43**, 4139 (1969).
[243] P. Wilson, *J. Chem. Soc. B*, 1581 (1968).
[244] A. Garner and F. Wilkinson, *J. C. S., Faraday 2* **72**, 1010 (1976).
[245] R. S. Davidson and P. F. Lambeth, *Chem. Commun.*, 1265 (1967).
[246] D. R. G. Brimage, R. S. Davidson, and P. F. Lambeth, *J. Chem. Soc. C*, 1241 (1971).
[247] S. G. Cohen, A. Parola, and G. H. Parsons, *Chem. Rev.* **73**, 141 (1973).
[248] W. P. Neumann, B. Schroeder, and M. Ziebarth, *Justus Liebigs Ann. Chem.*, 2279 (1975).
[249] S. S. Hall, S. D. Lipsky, F. J. McEnroe, and A. P. Bartels, *J. Org. Chem.* **36**, 2588 (1971).
[250] G. E. Risinger and C. W. Eddy, *Chem. Ind. (London)*, 570 (1963).

course, heteroaromatic, conjugation being lost at C-9.[251] A scale of relative stabilities of ketyls, including **71**, has been constructed following an infrared study of the equilibration of ketones with the ketyls of other ketones.[252]

e. *Radicals from Xanthene Dyes.* The chemistry of free radicals from xanthene dyes goes back many years. There have been extensive photochemical studies both by continuous irradiation[253-265] and flash photolysis techniques.[266-279] These confirm that, on irradiation of the dyes, the photoexcited triplet states can react with other components of the system or with added reductants or oxidizing agents to form radicals. Radiolytic studies have generated both reduced and oxidized radicals by alternative means.[280-284] This work has yielded much information on the spectroscopic and kinetic properties of the radicals and their modes of reaction. The

---

[251] P. Lambelet and E. A. C. Lucken, *J. C. S. Perkin II*, 1652 (1975).
[252] I. N. Yukhnovski, I. B. Rashkov, and I. M. Paniotov, *Monatsh. Chem.* **101**, 1712 (1970) [*CA* **74**, 69725 (1971)].
[253] M. Imamura and M. Koizumi, *Bull. Chem. Soc. Jpn.* **28**, 117 (1955).
[254] M. Imamura and M. Koizumi, *Bull. Chem. Soc. Jpn.* **29**, 899 (1956).
[255] M. Imamura and M. Koizumi, *Bull. Chem. Soc. Jpn.* **29**, 913 (1956).
[256] G. Oster and A. H. Adelman, *J. Am. Chem. Soc.* **78**, 913 (1956).
[257] A. H. Adelman and G. Oster, *J. Am. Chem. Soc.* **78**, 3977 (1956).
[258] M. Imamura, *Bull. Chem. Soc. Jpn.* **30**, 249 (1957).
[259] M. Imamura, *Bull. Chem. Soc. Jpn.* **31**, 62 (1958).
[260] M. Imamura, *Bull. Chem. Soc. Jpn.* **31**, 962 (1958).
[261] K. Ushida, S. Kabo, and M. Koizumi, *Bull. Chem. Soc. Jpn.* **35**, 16 (1962).
[262] K. Uchida and M. Koizumi, *Bull. Chem. Soc. Jpn.* **35**, 1871 (1962).
[263] T. Ohno, Y. Usui, and M. Koizumi, *Bull. Chem. Soc. Jpn.* **38**, 1022 (1965).
[264] Y. Momose, K. Uchida, and M. Koizumi, *Bull. Chem. Soc. Jpn.* **38**, 1601 (1965).
[265] R. F. Bartholomew and R. S. Davidson, *J. Chem. Soc. C*, 2347 (1971).
[266] L. Lindqvist, *Ark. Kemi* **16**, 79 (1960).
[267] L. I. Grossweiner and E. F. Zwicker, *J. Chem. Phys.* **31**, 1141 (1959).
[268] S. Kato, T. Watanabe, S. Nakagi, and M. Koizumi, *Bull. Chem. Soc. Jpn.* **33**, 262 (1960).
[269] L. I. Grossweiner and E. F. Zwicker, *J. Chem. Phys.* **34**, 1411 (1961).
[270] E. F. Zwicker and L. I. Grossweiner, *J. Phys. Chem.* **67**, 549 (1963).
[271] L. I. Grossweiner and E. F. Zwicker, *J. Chem. Phys.* **39**, 2774 (1963).
[272] V. Kasche and L. Lindqvist, *J. Phys. Chem.* **68**, 817 (1964).
[273] L. Lindqvist, *Acta Chem. Scand.* **20**, 2067 (1966).
[274] T. Ohno, S. Kato, and M. Koizumi, *Bull. Chem. Soc. Jpn.* **39**, 232 (1966).
[275] B. Stevens, R. R. Sharpe, and W. S. W. Bingham, *Photochem. Photobiol.* **6**, 83 (1967).
[276] V. Kasche, *Photochem. Photobiol.* **6**, 643 (1967).
[277] U. Krüger and R. Memming, *Ber. Bunsenges. Phys. Chem.* **78**, 670 (1974).
[278] U. Krüger and R. Memming, *Ber. Bunsenges. Phys. Chem.* **78**, 679 (1974).
[279] U. Krüger and R. Memming, *Ber. Bunsenges. Phys. Chem.* **78**, 685 (1974).
[280] J. Chrysochoos, J. Ovadia, and L. I. Grossweiner, *J. Phys. Chem.* **71**, 1629 (1967).
[281] P. Cordier and L. I. Grossweiner, *J. Phys. Chem.* **72**, 2018 (1968).
[282] A. P. Rodde and L. I. Grossweiner, *J. Phys. Chem.* **72**, 3337 (1968).
[283] K. Kimura, T. Miwa, and M. Imamura, *Bull. Chem. Soc. Jpn.* **43**, 1329 (1970).
[284] K. Kimura, T. Miwa, and M. Imamura, *Bull. Chem. Soc. Jpn.* **43**, 1337 (1970).

radicals formed upon oxidation of xanthene dyes are less well characterized. They have been formed by the reaction of the triplet dye with Fe(III), with molecular oxygen, or with unexcited dye molecules and by two reactions involving an OH˙ radical: direct abstraction of hydrogen or elimination of water from an adduct of the dye with OH˙.[272-274,281,282] It has been suggested that the oxidized radical is a phenoxyl; thus, the radical from fluorescein would be **75**.[281] The absorbance at visible wavelengths by this radical is consistent with the highly delocalized structure. It is not a persistent species, however, and decays by a number of reactions, mainly reduction, but also by reactions to unknown products.

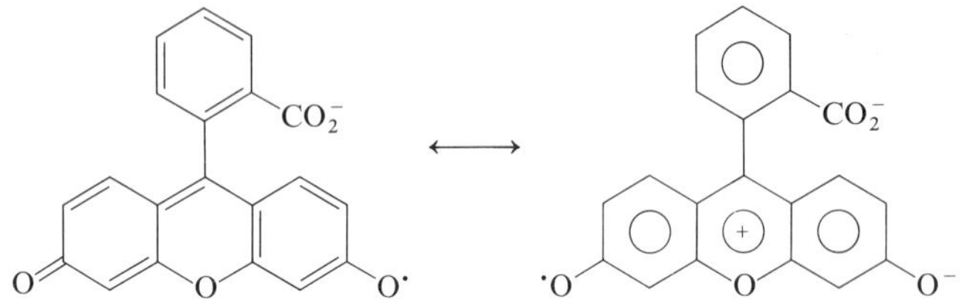

(75a)   (75b)

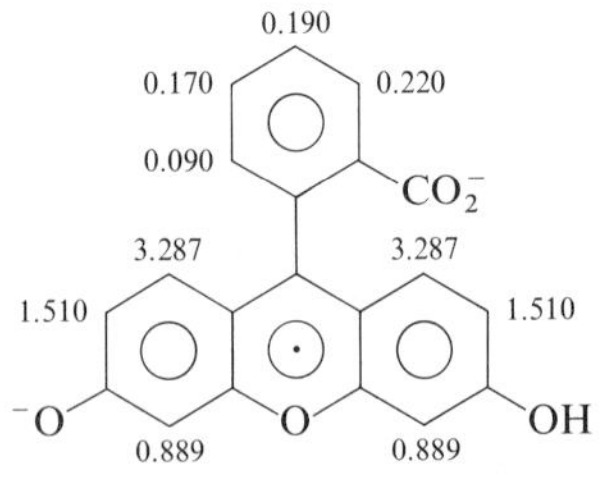

(76) 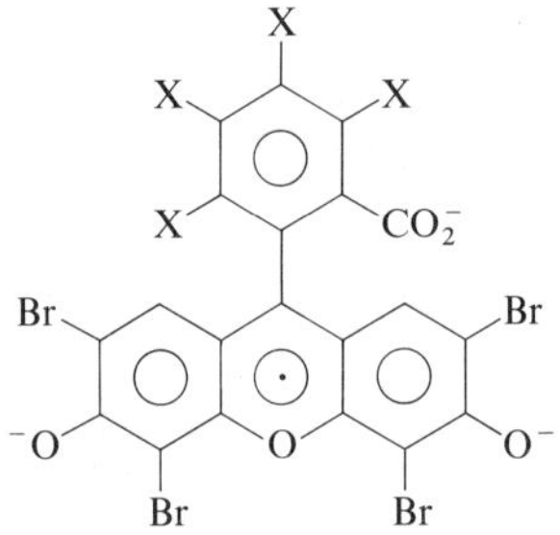   (77)

By contrast, the radicals formed upon single-electron reductions of xanthene dyes are persistent, at least in alkaline solution, and ESR spectra have been recorded. The early reports of persistent paramagnetism by Bubnov et al.[285] and by Lagercrantz and Yhland[286] have been followed by work reporting ESR data for fluorescein and various of its derivatives.[277-279,287-290] The hyperfine splittings indicated in **76** (in gauss) for the fluorescein radical-dianion are those of Niizuma et al.[290] It had been

[285] N. N. Bubnov, L. A. Kibalko, N. F. Tsepalov, and V. Ya. Shlyapintokh, *Opt. Spektrosk.* **7**, 117 (1959) [*CA* **54**, 23815 (1960)].
[286] C. Lagercrantz and M. Yhland, *Acta Chem. Scand.* **16**, 508 (1962).
[287] M. Okuda, Y. Momose, S. Niizuma, and M. Koizumi, *Bull. Chem. Soc. Jpn.* **40**, 1332 (1967).
[288] I. H. Leaver, *Aust. J. Chem.* **24**, 753 (1971).
[289] K. Kimura and M. Imamura, *Bull. Chem. Soc. Jpn*, **47**, 1358 (1974).
[290] S. Niizuma, Y. Sato, S. Konishi, and H. Kokubun, *Bull. Chem. Soc. Jpn.* **47**, 2121 (1974).

shown earlier that the hyperfine splittings vary with the state of protonation of the radicals[288]; the equivalence of heterocyclic ring splittings in **76**, which is not symmetrical, implies that they are averaged values; they may be compared with those in **41**. Halogenated fluorescein dye radicals have been shown to have interesting properties. Thus, radical **77** (X = Cl), formed on alkaline reduction of phloxine, gives an ESR spectrum which shows unusual linewidth effects: the linewidth decreases with decrease in temperature and increase in viscosity of the solvent. The authors suggest the phenomenon may arise by a modulation of hyperfine splitting constants due to bond distortions.[289] The radical **77** (X = H), formed on alkaline reduction of eosin, loses bromine upon photoexcitation. The suggested mechanism involves the mediation of σ-radicals formed by loss of bromide ion from **77** (X = H). Eventually, all the bromine atoms are eliminated.[284] Consistent with the photochemical nature of this elimination is the observation that polarographic reduction of a dibromofluorescein gives radicals in alkaline conditions, but without scission of bonds to bromine.[291] Other polarographic studies on dyes have included xanthene dyes[186,187]; there has been a recent polarographic study of the disproportionation kinetics of fluorescein anion-radicals.[291a]

f. *Miscellaneous Radicals Containing an Oxidized Xanthene Heterocycle.* The radical-anion formed on electrolytic reduction of 9,9'-bixanthenylidene has been characterized spectroscopically and distinguished from the dianion formed on reduction by alkali metals.[292] As the dianion is twisted about the 9—9' bond, it may be regarded as a diradical.

A low-resolution ESR spectrum has been observed of the aryloxyl **78**, a product of the oxidative condensation of 2-naphthol.[293]

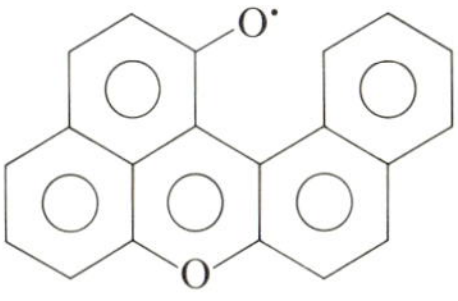

(78) 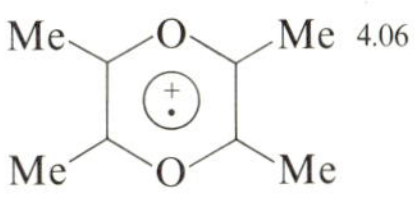   (79)

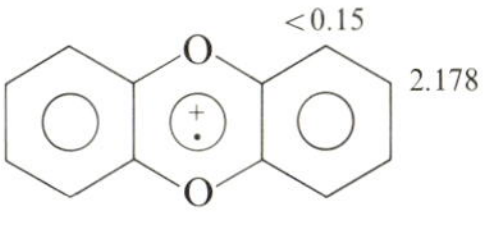

(80)

---

[291] N. R. Bannerjee and A. S. Negi, *Indian J. Chem.* **10**, 513 (1972).
[291a] S. K. Vig and G. P. Sato, *J. Electroanal. Chem. Interfacial Electrochem.* **91**, 71 (1978).
[292] M. Hishimo, M. Matsui, and M. Imamura, *Bull. Chem. Soc. Jpn.* **47**, 534 (1974).
[293] A. Rieker, N. Zeller, K. Schurr, and E. Muller, *Justus Liebigs Ann. Chem.* **697**, 1 (1966).

6. Radicals Containing a Dioxin Ring

It has been shown voltammetrically that oxidation of 1,4-dioxin and of its 2,3,5,6-tetraphenyl, benzo, and dibenzo derivatives takes place at a rotating platinum electrode in acetonitrile in single-electron steps with the intermediate formation of radical cations.[294] The simplest dioxin cation-radical for which an ESR spectrum has been obtained is **79**, for which hyperfine splittings are indicated in gauss.[295] The radical was prepared by treatment with sulfuric acid of the dimer which forms spontaneously on the storage of acetoin.

The cation-radical of dibenzo[1,4]dioxin (**80**) has been known for many years, as have several substituted derivatives; their ESR spectra were recorded in concentrated sulfuric acid.[296–301] The proton hyperfine splittings given, in gauss, in **80** are for a more recently obtained spectrum in trifluoromethanesulfonic acid.[302] The ESR spectrum of **80** is simple, consisting of five lines due to four equivalent interacting protons. The remaining four protons have splittings so close to zero that their hyperfine structure is masked by the linewidth. Because of the simplicity of the ESR spectrum of **80**, it has been possible to measure ^{13}C splittings on spectra obtained at high amplification.[300,302] Also, the spectrum has been found suitable for proving the validity for cation-radicals of formulas which relate line-broadenings to electron-exchange phenomena and which had previously been used only for measurements on anion-radicals.[303]

It is of interest to compare the distribution of spin population in **80** with that in anthracene anion-radical, with which it is isoelectronic, and also with those of similar heteroaromatic radicals. In Table II[304–306] are given the hyperfine splittings of the two types of proton in the radicals to be compared. It is evident that, as the electronegativity of the heteroatom increases, there is a redistribution of the spin population with that at C-2 increasing while that at C-1 decreases. Calculations have been performed which simulate this

---

[294] W. Schroth, R. Borsdorf, R. Herzschuh, and J. Seidler, *Z. Chem.* **10**, 147 (1970) [*CA* **73**, 10085 (1970)].
[295] G. A. Russell, R. Tanikaga, and E. R. Talaty, *J. Am. Chem. Soc.* **94**, 6125 (1972).
[296] M. Tomita, S. Ueda, Y. Nakai, Y. Deguchi, and H. Takaki, *Tetrahedron Lett.*, 1189 (1963).
[297] M. Tomita and S. Ueda, *Chem. Pharm. Bull.* **12**, 33 (1964).
[298] M. Tomita and S. Ueda, *Chem. Pharm. Bull.* **12**, 40 (1964).
[299] S. Ueda, *Chem. Pharm. Bull.* **12**, 212 (1964).
[300] T. N. Tozer and L. D. Tuck, *J. Chem. Phys.* **38**, 3035 (1963).
[301] B. Lamotte and G. Berthier, *J. Chim. Phys.* **63**, 369 (1966).
[302] G. C. Yang and A. E. Pohland, *J. Phys. Chem.* **76**, 1504 (1972).
[303] S. P. Sorensen and W. H. Bruning, *J. Am. Chem. Soc.* **94**, 6352 (1972).
[304] I. C. Lewis and L. S. Singer, *J. Chem. Phys.* **43**, 2712 (1965).
[305] A. Carrington and J. dos Santos-Veiga, *Mol. Phys.* **5**, 21 (1962).
[306] J. R. Bolton, A. Carrington, and J. dos Santos-Veiga, *Mol. Phys.* **5**, 465 (1962).

TABLE II
VARIATION OF PROTON HYPERFINE SPLITTINGS (IN GAUSS) WITH HETEROATOM IN RADICALS
ISOELECTRONIC WITH ANTHRACENE ANION-RADICAL

Radical	X	$a(H)_1$	$a(H)_2$	Reference
Anthracene anion	CH	2.74	1.51	304
Phenazine anion	N	1.93	1.61	305
Dihydrophenaziniumyl (cation)	NH	0.59	1.77	306
Dibenzo-1,4-dioxin cation (**80**)	O	~0	2.18	302

observable change and which also indicate the nature of spin-population changes at the heteroatom and the ring-junction carbons which are not necessarily manifest in ESR spectra.[301,304–310] Certain of these calculations have additionally taken account of the possibility of a fold in the radicals which bisects the central ring.[307–309]

There has been interest in radicals from halogenated dibenzo[1,4]dioxins on account of their extreme toxicity and concern over environmental pollution by them during the manufacture and use of organochlorine herbicides.[302,311,312] ESR and electronic absorption spectra have been recorded for use diagnostically; hyperfine splittings are observed from chlorines substituted on carbon of high spin population.[297–299,302,312,313] The ESR spectrum of the cation-radical of 2,7-difluorodibenzo[1,4]dioxin (**81**) has been recorded.[313] Hyperfine splittings (indicated in **81** in gauss) from fluorine and two types of proton were observed although the smallest proton splitting was not assigned; the relative magnitudes of $a(F)$ in **81** and $a(H)$ in **80** were discussed together with linewidth effects seen in the ESR spectrum of **81** in sulfuric acid.[313]

The same Romanian group has measured ESR spectra for the cation- and anion-radicals **82** and **83**, respectively, of 2-nitrodibenzo[1,4]dioxin.[314] As expected, the cation-radical is heterocycle-centered, showing a very small substituent-nitrogen hyperfine splitting, whereas the opposite is true for the

---

[307] J.-P. Malrieu, *J. Chim. Phys.* **62**, 485 (1965).
[308] T. S. Zhuraleva, *Zh. Strukt. Khim.* **7**, 516 (1966) [*CA* **65**, 17890 (1966)].
[309] R. J. Wratten and M. A. Ali, *Mol. Phys.* **13**, 233 (1967).
[310] V. Galasso, *Gazz. Chim. Ital.* **106**, 457 (1976) [*CA* **86**, 16122 (1977)].
[311] G. C. Yang and A. E. Pohland, *Adv. Chem. Ser.* **120**, 33 (1973).
[312] A. E. Pohland, G. C. Yang, and N. Brown, *Environ. Health Perspect.* **5**, 9 (1973).
[313] I. Baciu, M. Hillebrand, and V. E. Sahini, *J. C. S. Perkin II*, 986 (1974).
[314] I. Baciu, M. Hillebrand, V. E. Sahini, and E. Volanschi, *Rev. Roum. Chim.* **21**, 485 (1976).

anion-radical: spin and charge are mainly localized in the nitro group and the ring bearing it. (Splittings indicated in **82** and **83** are in gauss.) The cation-radical of 2,8-dinitrodibenzo[1,4]dioxin was also obtained and its ESR spectrum measured and assigned.[314]

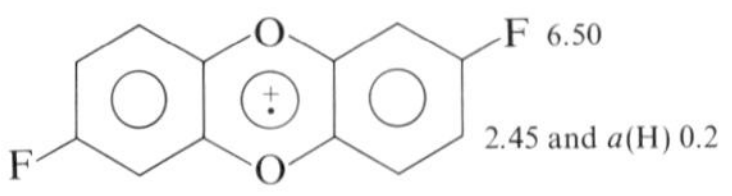

(81)

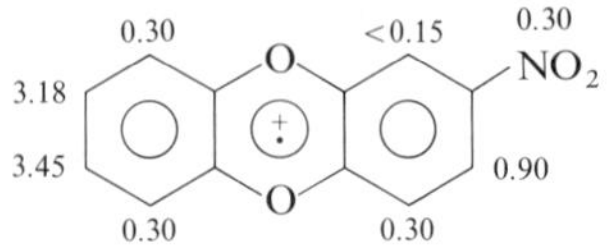

(82)

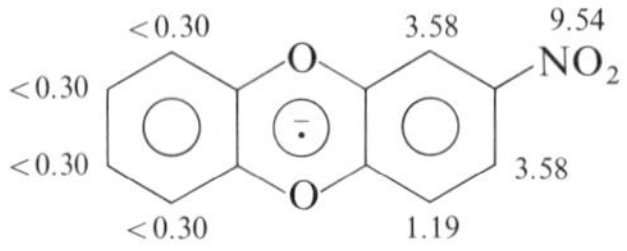

(83)

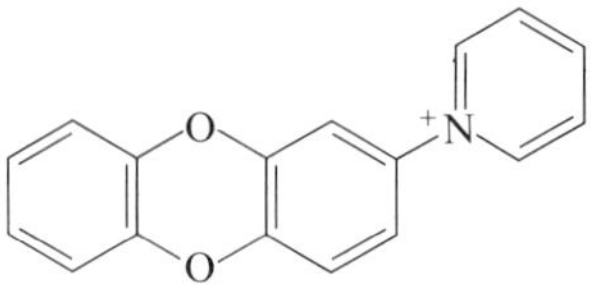

(84)

Radical **80** has been prepared as its perchlorate salt by anodic oxidation in ethyl acetate in the presence of lithium perchlorate.[315] The reactivity toward nucleophiles of material so prepared was investigated: nitrite and nitrate ions give 2-nitrodibenzo[1,4]dioxin although the mechanisms of the reactions are not clear. Pyridine gives *N*-(2-dibenzo[1,4]dioxinyl)pyridinium ion (**84**). Other nucleophiles acted as electron donors and largely reduced **80** back to the parent heterocycle; they included amines, cyanide ion and water.[315] In an earlier study, the reaction of **80** with water had been examined and the ultimate formation of catechol via dibenzo[1,4]dioxin-2,3-dione was inferred.[316,317] The cation-radical (**80**) has been found to accelerate the anisylation of thianthrene cation-radical (Section III,C,4,b); it has been found to participate in an electrochemiluminescence system with benzophenone involving phosphorescence of the latter in a fluid system,[318] and it has been used in a study of relative diffusion coefficients of aromatic cations which shows that it is justified to equate voltammetric potentials for these species with formal thermodynamic redox potentials.[319] The dibenzo[1,4]dioxin semiquinone **85** has been found to result from the alkaline autoxidation of catechol[129,320]; the same species may well be in-

[315] H. J. Shine and L. R. Shade, *J. Heterocycl. Chem.* **11**, 139 (1974).
[316] G. Cauquis and M. Maurey, *C. R. Acad. Sci., Ser. C* **266**, 1021 (1968).
[317] G. Cauquis and M. Maurey-Mey, *Bull. Soc. Chim. Fr.*, 3588 (1972).
[318] S. M. Park and A. J. Bard, *Chem. Phys. Lett.* **38**, 257 (1976).
[319] U. Svanholm and V. D. Parker, *J. C. S. Perkin II*, 755 (1975).
[320] N. F. Usacheva, Yu. G. Oranskii, R. S. Safiullin, E. F. Rul, and M. S. Khaikin, *Zh. Nauchn. Prikl. Fotogr. Kinematogr.* **13**, 459 (1968) [*CA* **70**, 56990 (1969)].

volved in the degradation of **80** to catechol upon reaction with water as noted earlier.[316,317] The radical-anion **86** has been produced by reduction of catechol oxalate in DMF (hyperfine splittings indicated in gauss).[321]

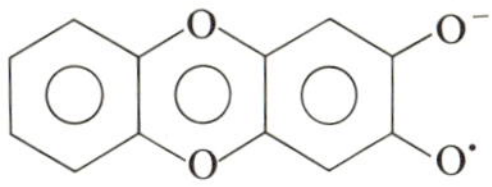

(85) 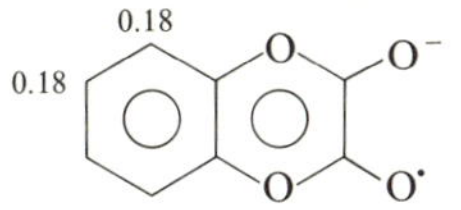   (86)

C. Radicals Containing Seven- and Eight-Membered Heterocycles

1. Radicals Containing an Oxepin Ring

The radical-anion of dibenz[*b,f*]oxepin (**87**) has been prepared and investigated by ESR spectroscopy. (Hyperfine splittings in **87** are in gauss.) The spin distribution was compared with that in the corresponding dibenzothiepin and its sulfoxidized derivatives[322] (see Section III,D,1). In view of this, an ESR investigation of simpler oxepins may be worthwhile although it has been shown that both 2,7-dimethyloxepin and 3-benzoxepin undergo ring cleavage on treatment with potassium metal in liquid ammonia–THF at low temperature on the small preparative scale.[323]

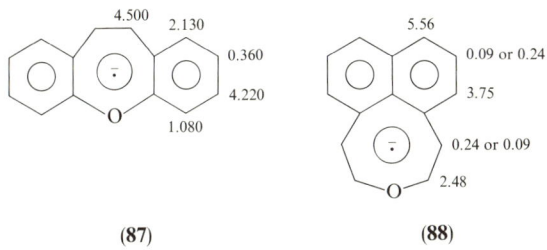

(87)   (88)

2. Radicals Containing an Oxocin Ring

The anion-radical of naphth[1,8-*de*]oxocin (**88**) has been obtained during a study of radicals from naphthalene with unsaturated peri substituents;

---

[321] G. A. Russell and S. A. Weiner, *J. Am. Chem. Soc.* **89**, 6623 (1967).
[322] R. Leardini and G. Placucci, *J. Heterocycl. Chem.* **13**, 277 (1976).
[323] L. A. Paquette and T. McCreadie, *J. Org. Chem.* **36**, 1402 (1971).

88 gave an exceptionally well resolved ESR spectrum when produced by reduction of the parent compound with sodium–potassium alloy in THF. Specific deuteration experiments confirmed the assignment of the major hyperfine splitting (indicated in **88** in gauss) in the heterocycle.[324]

III. Radicals from Sulfur Heterocycles

A. Radicals Containing a Four-Membered Heterocycle

1,2-Dithiete Cation-Radicals

Russell and co-workers have obtained cation-radicals in the 1,2-dithiete system. Such radicals are cyclic, conjugated within the heterocycle, and possess $(4n + 1)$ electrons, with $n = 1$; they are consequently heteroaromatic within the definition given in the introduction to Part I.[1] 3,4-Dimethyl-1,2-dithiete cation radical **89** ($R^1 = R^2 = Me$) was obtained by treatment of acetoin in sulfuric acid with sulfide ion. The radical is persistent at ambient temperature and unaffected by oxygen. This evident stability, and its formation from open-chain precursors to the exclusion of acyclic possibilities such as **90**, whose oxygen equivalent exists, implies that aromatic stabilization of **89** and similar radicals is a matter of fact and not merely definition.

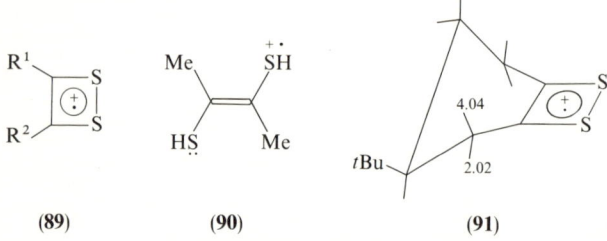

(89) (90) (91)

The 1,2-dithiete cation-radical ring was incorporated into various structures (e.g., by variation of R^1 and R^2 in **89** and by appending the 1,2-dithiete radical structure to various carbocyclic rings as in **91**). The conformational dependence of the proton hyperfine splittings was investigated (those given in **91** are in gauss) and the conformational requirements of the dithiete cation-radical moiety explored.[295]

---

[324] S. F. Nelsen and J. P. Gillespie, *J. Am. Chem. Soc.* **95**, 1874 (1973).

B. RADICALS CONTAINING A FIVE-MEMBERED HETEROCYCLE

Reviews have appeared in Italian and in English of radicals containing the thiophene ring system[325,326]; they deal with π- and σ-radicals, respectively.

1. Cation-Radicals Containing a Thiophene Ring

The electronic absorption spectrum of the cation-radical of thiophene itself has been observed following low-temperature γ-radiolysis of the heterocycle in a Freon matrix.[327] The radical has also been implicated in the oxidation of thiophene by dibenzoyl peroxide[328]; it is believed to be formed at the contact of certain transition metal layer-silicates with thiophene.[329] The anodic oxidation of 2,5-dimethylthiophene has been studied by Japanese workers who found strong evidence for the formation of the cation-radical as the primary oxidation product.[330] In the presence of strong nucleophiles such as cyanide ion, the cation-radical undergoes nucleophilic attack before further oxidation. In the presence of more basic species such as acetate ion, the cation-radical is deprotonated to give a thienylmethyl radical which undergoes further reaction. The results were compared with similar observations for the oxidation of 2,5-dimethylfuran.[3] Czech workers have also studied the anodic oxidation of substituted thiophenes.[221–335] This work has focused on the preparative value of anodic oxidations in acidified methanol. Cation-radical formation is implied for the primary step, but the value of the method lies in the fact that sulfur is ultimately eliminated from the substrate and functionalized γ-dicarbonyl compounds result.

---

[325] L. Lunazzi, A. Mangini, G. F. Pedulli, and M. Tiecco, *Gazz. Chim. Ital.* **101**, 10 (1971) [*CA* **75**, 28067 (1971)].
[326] M. Tiecco and A. Tundo, *Int. J. Sulfur Chem.* **3**, 295 (1973).
[327] A. Grimison and G. A. Simpson, *J. Phys. Chem.* **72**, 1776 (1968).
[328] C. E. Griffin and K. R. Martin, *Chem. Commun.*, 154 (1965).
[329] T. J. Pinnavia, P. L. Hall, S. S. Cady, and M. M. Mortland, *J. Phys. Chem.* **78**, 994 (1974).
[330] K. Yoshida, T. Saeki, and T. Fueno, *J. Org. Chem.* **36**, 3673 (1971).
[331] J. Srogl, M. Janda, and M. Valentova, *Collect. Czech. Chem. Commun.* **35**, 148 (1970).
[332] M. Janda, J. Srogl, A. Janousova, V. Kubelka, and M. Holik, *Collect. Czech. Chem. Commun.* **35**, 2635 (1970).
[333] M. Nemec, M. Janda, and J. Srogl, *Collect. Czech. Chem. Commun.* **38**, 3857 (1973).
[334] M. Janda, J. Srogl, M. Nemec, and A. Janousova, *Collect. Czech. Chem. Commun.* **38**, 1221 (1973).
[335] M. Janda, J. Srogl, M. Nemec, and I. Stibor, *Univ. Adama Mickiewicza Poznaniu, Wydz. Mat., Fiz. Chem.*, [*Pr.*], *Ser. Chem.* **18**, 157 (1975) [*CA* **85**, 32116 (1976)].

A persistent cation-radical **92** results from the oxidation of tetraphenylthiophene. An ESR spectrum and g-value were recorded for this and other 4-substituted phenyl derivatives, although no assignments were made.[336] The electrochemical characteristics of **92** have been investigated by Libert and Caullet,[337] and one of these authors has investigated the electrochemiluminescent properties of systems involving **92**.[14]

(92)   (93A)   (93B)

The simplest monocyclic thiophene cation-radicals which are persistent at ambient temperature and for which ESR spectra have been recorded and assigned are a series of 2,5-bis(alkylthio)thiophene cation-radicals.[338] These radicals exhibit conformational isomerism. Thus, 2,5-bis(methylthio)thiophene cation-radicals exists as a mixture of 75% of the symmetrical S-cis-cis isomer (**93A**) and 25% of the unsymmetrical S-cis-trans isomer (**93B**) with no detectable concentration of the symmetrical S-trans-trans isomer. The radicals were formed by treatment of the parent heterocycles in nitromethane with $AlCl_3$; the assignments were made on the basis of both S- and C-alkyl substitution experiments and supported by MO calculations (hyperfine splittings given for **93A** and **B** are in gauss).[338]

There have been a few oxidations of dibenzothiophene with indications that formation of a cation-radical is possible. Treatment with Lewis acids gave a product having an ill-resolved ESR signal[339]; anodic oxidation at a platinum electrode in acetonitrile suggests a quasi-reversible initial charge-transfer step followed by a rapid, irreversible chemical destruction of the ion produced[340]; sulfoxide is apparently the first isolable product produced.[341]

Cation-radicals have been observed for various thienothiophenes: thieno[3,2-*b*]thiophene forms a cation-radical **94** on treatment with $AlCl_3$ of a solution in nitromethane at $-20°C$ (hyperfine splittings are indicated for **94** in gauss). Similarly, dithieno[2,3*b*;2′,3′-*d*]thiophene forms a cation-

---

[336] U. Schmidt, K. Kabitzke, K. Markau, and A. Müller, *Justus Liebigs Ann. Chem.* **672**, 78 (1964).
[337] M. Libert and C. Caullet, *C. R. Acad. Sci., Ser. C* **278**, 439 (1974).
[338] C. M. Camaggi, L. Lunazzi, and G. Placucci, *J. C. S. Perkin II*, 1491 (1973).
[339] M. Kinoshita and H. Akamatu, *Bull. Chem. Soc. Jpn.* **35**, 1040 (1962).
[340] G. Bontempelli, F. Magno, G. A. Mazzocchin, and S. Zecchin, *J. Electroanal. Chem. Interfacial Electrochem.* **43**, 377 (1973).
[341] D. S. Houghton and A. A. Humffray, *Electrochim. Acta* **17**, 2145 (1972).

radical (**95**) whose ESR spectrum has also been measured but not unambiguously assigned.[342] Gleiter et al.[343] have carried out a detailed study of the heterocycle tetraphenylthieno[3,4-c]thiophene (**96**) in which the ESR spectrum of the cation-radical has been measured and assigned. This heterocycle is of interest on account of its "nonclassical" structures (**96a,b**) in which a valence of 4 is ascribed to the sulfur atoms[343] (see Section III,B,5).

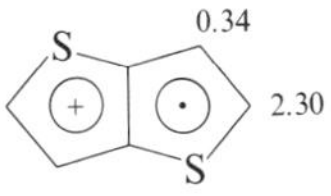

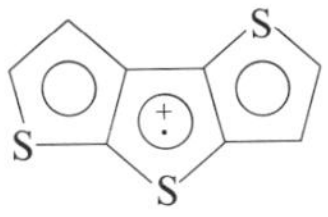

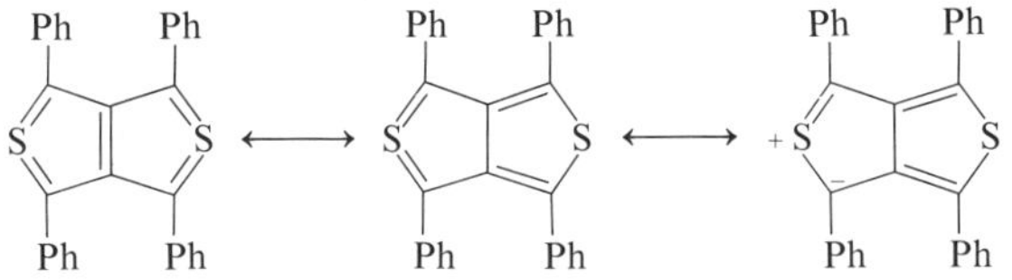

The heterocycle benzo[1,2-c:3,4-c′:5,6-c″]trithiophene (**97**), an example of a heteroradialene, has been reported to give a stable cation-radical on treatment of a solution in dichloromethane with $SbCl_5$.[344]

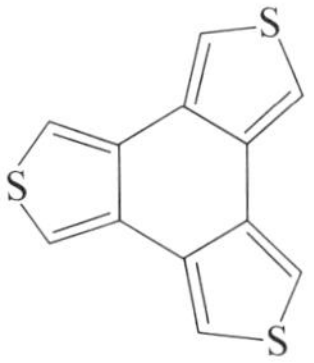

2. Neutral Radicals Containing a Thiophene Ring

The simple 2- and 3-thienylmethyl radicals **98** and **99** were observed by ESR spectroscopy by Hudson and co-workers[345,346] who generated them by steady-state photolysis of dibenzoyl peroxide in the appropriate methylthiophene. Their hyperfine splittings are indicated in gauss in the formulas;

---

[342] L. Lunazzi, G. Placucci, and M. Tiecco, *Tetrahedron Lett.*, 3847 (1972).
[343] R. Gleiter, R. Bartetzko, G. Brähler, and H. Bock, *J. Org. Chem.* **43**, 3893 (1978).
[344] H. Hart and M. Sasaoka, *J. Am. Chem. Soc.* **100**, 4326 (1978).
[345] A. Hudson, H. A. Hussain, and J. W. E. Lewis, *Mol. Phys.* **16**, 519 (1969).
[346] A. Hudson and J. W. E. Lewis, *Tetrahedron* **26**, 4413 (1970).

ring proton splittings were assigned with reasonable certainty on the basis of MO calculations. Interestingly, the methylene proton splittings were measurably different in each radical, as expected on symmetry grounds, but simple calculation could not assist in assigning them. There is a large difference in the magnitudes of the pairs of methylene splittings between **98** and **99** reflecting the greater spin population on the exocyclic function in **99**. Since the radicals are uncharged, this difference indicates a more effective delocalization of spin by the 2-thienyl group in comparison with the 3-isomer which, in turn, reflects the difference in magnitudes of the coefficients of the thiophene HOMO at the 2- and 3-positions.

These same radicals have been observed following γ-radiolysis of methylthiophenes in an adamantane matrix at room temperature.[10] In solid solution the differences in the methylene splittings were not resolved, but otherwise hyperfine splittings essentially similar to those found for fluid solution were observed.

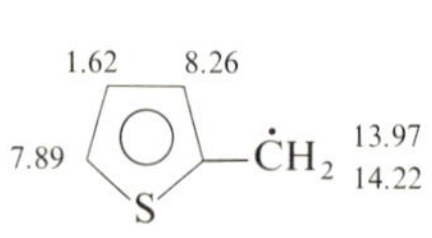

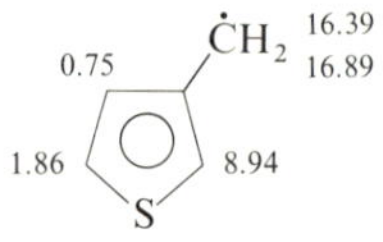

The relative homolytic reactivities of 2- and 3-methylthiophenes toward hydrogen abstraction from the side chain by *tert*-butoxyl radicals to form **98** and **99** have been investigated by the spin-trapping method.[347] The 2-isomer is 2.2-fold the more reactive. This finding is consistent with the ESR result that in **98** the spin population is more effectively delocalized, and hence the radical more effectively stabilized, than in **99** for, by Hammond's postulate, the transition state leading to **98** would be similarly better stabilized than that leading to **99**. A greater reactivity of 2-methylthiophene toward hydrogen abstraction by phenyl radicals has also been found[84]; the role suggested for **98** and **99** in the base-catalyzed autoxidation of methylthiophenes, where again the 2-isomer is the more reactive, is similarly plausible.[348]

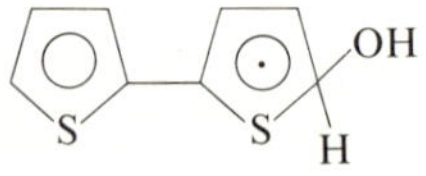

---

[347] L. Lunazzi, G. Placucci, M. Tiecco, and G. Martelli, *J. C. S. Perkin II*, 2215 (1972).
[348] T. J. Wallace and F. A. Baron, *J. Org. Chem.* **30**, 3520 (1965).

The 2-hydroxy-5-(2-thienyl)-2-hydrothienyl radical (**100**) has been observed during radiolysis of thiophene[349,350]; **100** (which may be regarded formally to be a derivative of **98**) arises by addition of hydroxyl radical to 2,2′-bithienyl, a product of the radiolysis.

Tri(2-thienyl)methyl (**101**), its isomers containing one, two, or three 3-thienyl groups, and various triarylmethyls containing mixed phenyl and thienyl groups have been studied in considerable detail.[325,351–354] Hyperfine splittings (indicated for **101** in gauss) have been assigned following deuteration experiments.[325] The radicals were prepared by reduction of the corresponding carbonium perchlorates, in turn prepared by dehydration of the appropriate carbinols.[351,352] The deformation of **101** that occurs when it is dissolved in the nematic mesophase of a liquid crystalline solvent has been the subject of a combined experimental and theoretical investigation.[353] It is necessary to invoke a deformation of the radical as the nature of the variation of the hyperfine splittings with temperature in the nematic phase cannot be explained by solvent ordering alone. On the assumption of a propeller shape for **101** in which the thiophene rings are twisted 25° from coplanarity in the unperturbed radical, the observed temperature variation of hyperfine splittings in the nematic solvent is accounted for by the redistribution of spin population which accompanies a flattening of the radical equivalent to a reduction by ~7° of the twist from coplanarity of the thiophene rings.

The isotropic ESR spectra of **101** at temperatures accessible to the radical show signs neither of the existence of separate rotamers nor of line-broadening effects suggestive of their rapid interconversion. An INDO MO study has shown that the radical probably exists in two preferred propeller-like conformations (that of C_3 symmetry is shown in **101**; that of C_1 symmetry is obtained by a rotation of any single thiophene ring in **101** through 180°).[354] The calculations indicate the C_3 conformer to be more stable by 0.75 kcal mol^{-1}, and the barrier to interconversion of the two forms, by a preferred two-ring-flip mechanism, to be about 6 kcal mol^{-1}. This low barrier to interconversion is consistent with the failure to observe effects of the isomerism in the ESR spectra at temperatures above those at which the radical dimerizes.

A number of thienyl nitroxyl radicals have been reported. Sleight and Sutcliffe irradiated 2-nitrothiophene in THF and obtained a radical **102**

---

[349] B. B. Saunders, P. C. Kaufman, and M. S. Matheson, *J. Phys. Chem.* **82**, 142 (1978).
[350] B. B. Saunders, *J. Phys. Chem.* **82**, 151 (1978).
[351] A. Mangini, G. F. Pedulli, and M. Tiecco, *Tetrahedron Lett.*, 4941 (1968).
[352] A. Mangini, G. F. Pedulli, and M. Tiecco, *J. Heterocycl. Chem.* **6**, 271 (1969).
[353] G. F. Pedulli, C. Zannoni, and A. Alberti, *J. Magn. Reson.* **10**, 372 (1973).
[354] F. Bernardi, M. Guerra, G. F. Pedulli, and K. Mislow, *Tetrahedron* **32**, 951 (1976).

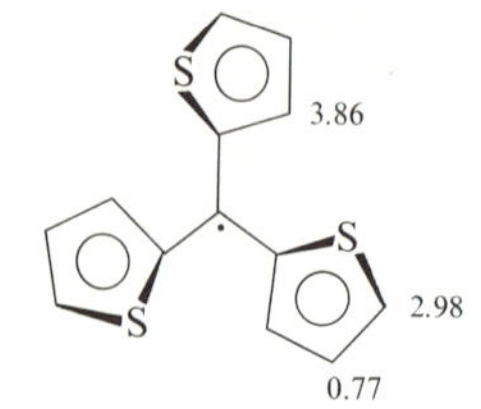

(101) 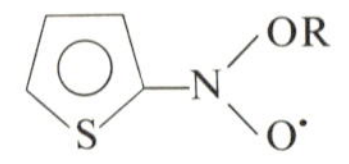 (102)

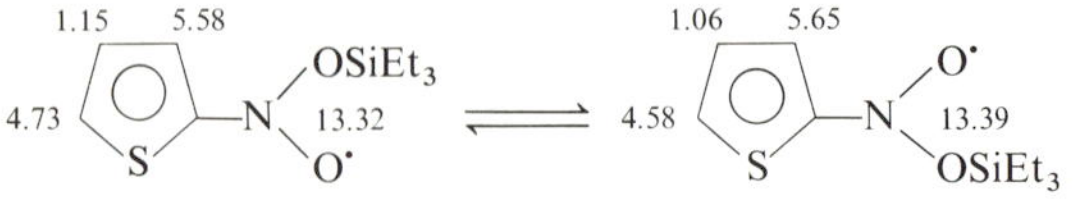

(103A) ⇌ (103B)

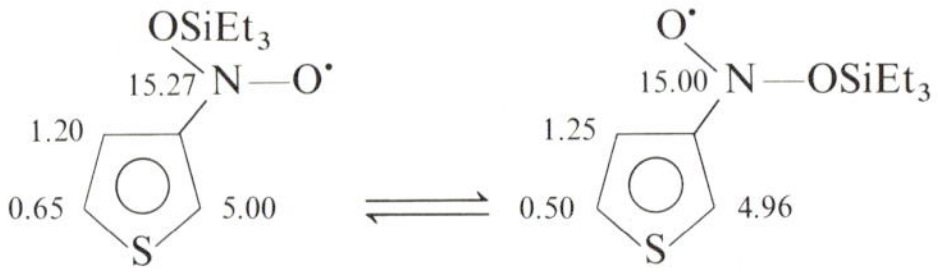

(104A) ⇌ (104B)

(R = tetrahydrofuran-2-yl) by trapping the tetrahydrofuran-2-yl radical (formed by reaction of the nitrothiophene triplet state with the solvent) with 2-nitrothiophene. An ESR spectrum was measured, including a proton hyperfine splitting from the tetrahydrofuran moiety R, but it was not assigned in detail.[86a] Camaggi et al.[355] have similarly trapped Et$_3$Si· with 2- and 3-nitrothiophenes and isomeric 2-nitrothienothiophenes. They found that for both 2- and 3-thienyl nitrosyls, **103** and **104**, spectra from two rotamers **A** and **B** could be measured at low temperatures. From an analysis of the spectral line-shapes at various temperatures, activation energies for the interconversions of the rotamers were determined. The barrier was higher for **103** than for **104**, indicating a higher C—N bond order in **103** which is consistent with the more effective delocalization occurring through the thiophene 2-position as noted earlier. The same group has investigated the influence of the substitutents on the spin distribution in 5-substituted derivatives of **103**, under temperature conditions where the spectra of the separate rotamers become averaged, and have compared the substituent effects with those in 2-nitrothiophene anion-radicals.[356]

---

[355] C. M. Camaggi, L. Lunazzi, G. F. Pedulli, G. Placucci, and M. Tiecco, *J. C. S. Perkin II*, 1226 (1974).
[356] C. M. Camaggi, R. Leardini, and G. Placucci, *J. C. S. Perkin II*, 1195 (1974).

The hydrazyl biradical **105** has been synthesized by Soviet workers and subjected to ESR study.[357,358]

(105)

3. Anion-Radicals Containing a Thiophene Ring

No persistent anion-radical of thiophene itself has been reported. On the contrary, Gerdil and Lucken reported failure to form the radical under conditions which permit the formation of dibenzothiophene anion-radical (*vide infra*)[359]; and recently it has been found that, in aqueous radiolytic conditions, the first product to be observed following attachment of an electron to thiophene is 2-hydrothienyl radical, which results from a rapid protonation of the anion-radical.[349] However, Soviet workers have described the possibility of the transfer of electrons to thiophene and methylthiophenes from hydrocarbon anions generated electrochemically in aprotic solvent.[360]

The anion-radical of benzo[*b*]thiophene has transient occurrence in the exciplex which results when the heterocycle is irradiated in amine solvents, and the reactivity of the exciplex is explicable in terms of the distribution of spin population calculated for the anion-radical.[361] The anion-radical **106** of benzo[*c*]thiophene has been formed by reduction of the parent heterocycle with potassium in DME.[343] Its ESR spectrum was measured and assigned. Hyperfine splittings in **106** are in gauss.

The extended conjugation present in dibenzothiophene permits the formation of a stabilized anion-radical **107** from this heterocycle. It was reported by several groups in the early 1960s.[40,42,45,359,362,363] The hyperfine splittings given (in gauss) in **107** are the most recent and are those of

---

[357] V. I. Koryakov, V. A. Gubanov, A. K. Chirkov, and R. O. Matevosyan, *Dokl. Akad. Nauk SSSR* **206**, 649 (1972) [*CA* **78**, 15115 (1973)].
[358] G. I. Yashchenko, V. I. Koryakov, and A. K. Chirkov, *Zh. Obshch. Khim.* **46**, 409 (1976) [*CA* **84**, 164553 (1976)].
[359] R. Gerdil and E. A. C. Lucken, *Proc. Chem. Soc., London*, 144 (1963).
[360] S. G. Mairanovskii, L. I. Kosychenko, and S. Z. Taits, *Elektrokhimiya* **13**, 1250 (1977) [*CA* **88**, 43133 (1978)].
[361] P. Grandclaudon, A. Lablache-Combier, and C. Parkanyi, *Tetrahedron* **29**, 651 (1973).
[362] E. Brünner and F. Dörr, *Ber. Bunsenges. Phys. Chem.* **68**, 468 (1964).
[363] D. H. Eargle and E. T. Kaiser, *Proc. Chem. Soc., London*, 22 (1964).

Adam and Kepford,[45] confirmed by deuteration experiments and obtained by potassium reduction in DME; the early literature contained a misassignment.[363] Bard and co-workers have correlated homogeneous electron-transfer rates between **107** and its parent heterocycle with heterogeneous rate constants for electroreduction of the heterocycle,[48,49] and Japanese workers have described optical absorption spectra for **107** and analogs. They conclude that the heteroatom has minimal perturbing effect on the π-electronic structure of the biphenyl skeleton consistent with the heteroatom lying at a node in the singly occupied MO.[47,364]

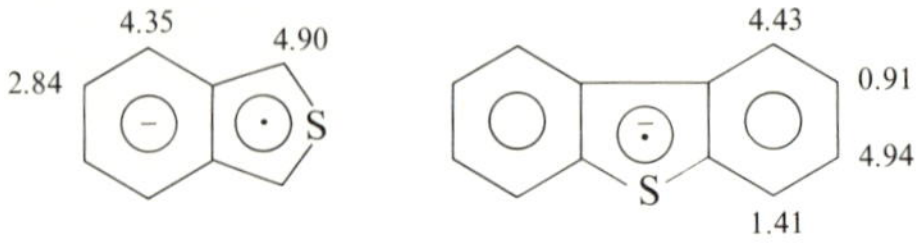

(106)   (107)

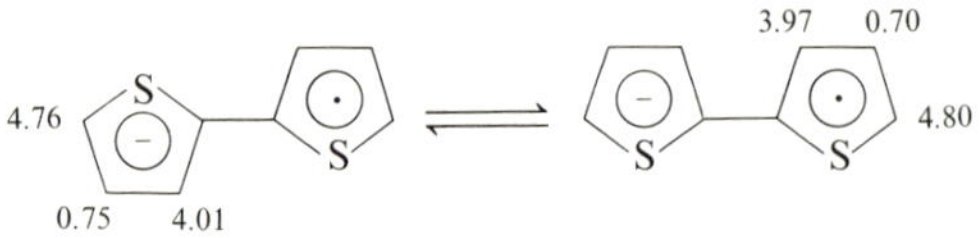

(108A)   (108B)

Anion-radicals have been produced by reduction of various arylthiophenes. Thus 2-phenyl- and 2,5-diphenylthiophene have yielded anion-radicals whose ESR spectra have been analyzed.[365,366] The observation of electrochemiluminescent phenomena for 2,3,4,5-tetraphenylthiophene confirms the occurrence of an anion-radical of this heterocycle, too.[14] The anion-radical **108** of 2,2'-bithienyl which may be formed by potassium reduction of the heterocycle at $-40°C$ exists in two conformations (**108A** and **108B**), separated by a high energy barrier (estimated at 26.8 kcal mol^{-1} by INDO MO calculation) which occurs because the LUMO of the bithienyl, occupied by the unpaired electron in **108**, has strong bonding character in the interannular region.[366] Interestingly, the anion radicals of 3,3'-bithienyl and 2,3'-bithienyl are reported to isomerize to **108** although the nature of the reactions is unknown. Anion-radicals of 5,5'-disubstituted derivatives of **108**, of 2-(4-pyridyl)thiophene, and of dithienylbenzenes were also described.[366]

---

[364] T. Matsuyama and H. Yamaoka, *Annu. Rep. Res. React. Inst., Kyoto Univ.* **9**, 37 (1976) [*CA* **86**, 120264 (1977)].

[365] P. Cavalieri d'Oro, A. Mangini, G. F. Pedulli, P. Spagnolo, and M. Tiecco, *Tetrahedron Lett.*, 4179 (1969).

[366] G. F. Pedulli, M. Tiecco, M. Guerra, G. Martelli, and P. Zanirato, *J. C. S. Perkin II*, 212 (1978).

Sec. III.B] HETEROAROMATIC RADICALS, PART II 77

The same Bologna group, who have contributed much to our knowledge of thiophene radicals, have also investigated the anion-radicals of isomeric *trans*-dithienylethylenes, 1-(2-thienyl)-2-phenylacetylene, and *trans*-1-(2-thienyl)-2-phenylethylene.[367] The ESR spectra indicate that rotation of the aryl groups in these molecules is slow on the time scale of the ESR experiment, and separate rotamers have been detected for the radical from 1,2-di(2-thienyl)ethylene (**109**) (hyperfine splittings given in gauss).[367]

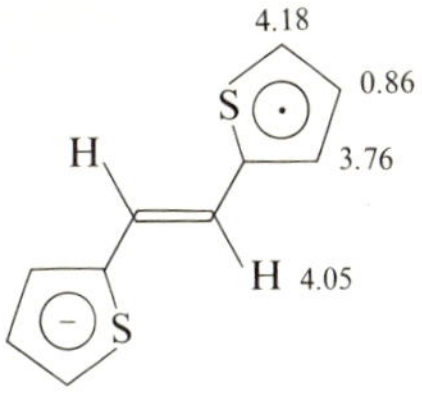

(**109**) 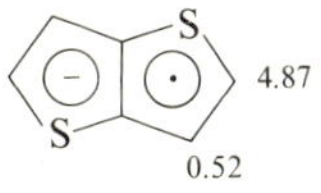 (**110**) 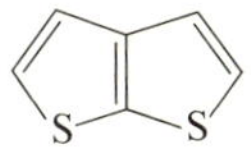 (**111**)

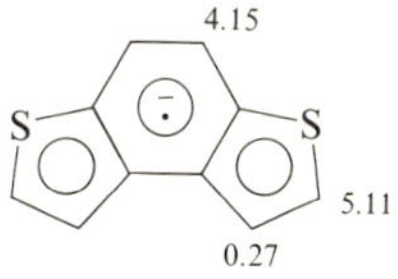

(**112**) 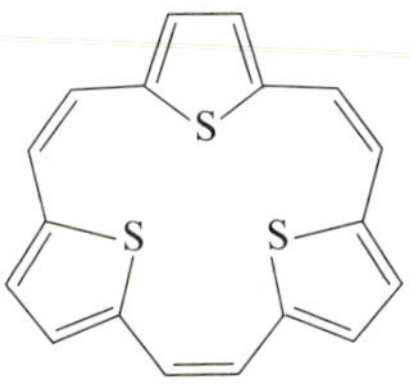 (**113**)

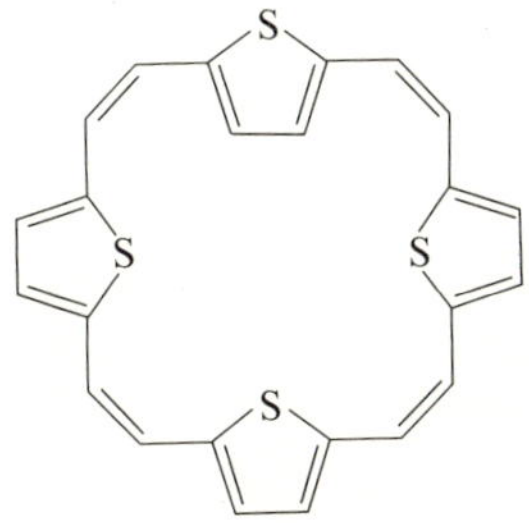

(**114**)

Other anion-radicals containing the thiophene heterocycle in an otherwise hydrocarbon molecular environment and whose ESR spectra have been analyzed are **110** (for which hyperfine splittings are given in gauss), the anionic counterpart of **94**, and the anion-radical of **96**[342,343]; thieno[2,3-*b*]thiophene (**111**) failed to give a persistent radical by alkali metal reduction,

[367] L. Lunazzi, A. Mangini, G. Placucci, P. Spagnolo, and M. Tiecco, *J. C. S. Perkin II*, 192 (1972).

even at $-100°C$.[368] Well-resolved spectra were obtained for three benzodithiophene anion-radicals (e.g., **112**).[368] These radicals, however, persisted only at very low temperatures ($-100°C$) in contrast to **108** and **109** which persist at higher temperatures but have a lesser extent of conjugation, and in marked contrast with **107** whose extent of conjugation is comparable. These examples nicely demonstrate the distinction between thermodynamic stabilization, which is reflected in the extent of conjugation, and kinetic stability—or lack thereof—which is principally a property of frontier orbitals, as discussed in Part I: Section II,B.

Gerson and co-workers[50,369] have prepared anion-radicals from the macrocyclic heterocycles **113** and **114** and the sulfur analog of **5**. Conformational and ion-pairing effects in the ESR spectra were investigated, and in the more recent work solution ENDOR spectroscopy was employed.

A considerable body of work has appeared concerning the anion-radicals of thiophenes bearing heteroatomic substituents. γ-Irradiation of crystalline 2-chlorothiophene at 77K yields an anion-radical in which the unpaired electron is confined mainly to an antibonding orbital localized on the C-1, C-2, and S atoms; the radical is a σ-radical which decays by C—S bond fission comparable with the ring fission of furan upon receipt of an electron, mentioned in Section II,A,2.[370] Several halogen-substituted thiophene aldehydes, ketones, and their derivatives have been reported to undergo C—halogen bond fission before the carbonyl function or its derivative is reduced on polarographic reduction.[371–373]

A long-standing analysis of the ESR spectrum of 2-nitrothiophene anion-radical **115** has been confirmed by more recent investigations[356,374,375]; the earlier paper also presented a treatment of the 3-isomer **116** and various methylated derivatives while one of the latter investigated the influence of 5-substituents on the spin distribution in **115**.[356] The hyperfine splittings in **115** and **116**, in gauss, are for solutions in acetonitrile. Splittings measured in other solvents are different and prove a solvent dependence of the spin distribution in **115** comparable with, but less than, that noted earlier for 2-nitrofuran anion-radical **6** (Section II,A,2 and Table I). Nitrothiophenes have been the subject of several polarographic studies made either for their

---

[368] L. Lunazzi, G. Placucci, M. Tiecco, and G. Martelli, *J. Chem. Soc. B*, 1820 (1971).
[369] F. Gerson and J. Heinzer, *Helv. Chim. Acta* **51**, 366 (1968).
[370] S. Nagai and T. Gillbro, *J. Phys. Chem.* **81**, 1793 (1977).
[371] M. Person, R. Guilard, and P. Fournari, *C. R. Acad. Sci., Ser. C* **264**, 1727 (1967).
[372] M. Person and R. Mora, *Bull. Soc. Chim. Fr.*, 521 (1973).
[373] M. Person and R. Mora, *Bull. Soc. Chim. Fr.*, 528 (1973).
[374] E. A. C. Lucken, *J. Chem. Soc. A*, 991 (1966).
[375] G. F. Pedulli, M. Tiecco, A. Alberti, and G. Martelli, *J. C. S. Perkin II*, 1816 (1973).

own sakes or to relate polarographic reduction potentials to ESR, quantum mechanical, or biological parameters.[68,72–74,374,376–378]

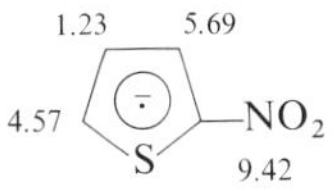

(115) 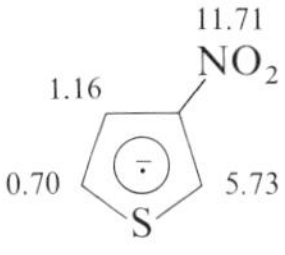 (116) 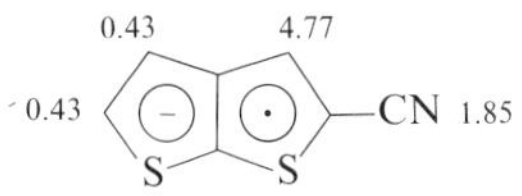 (117)

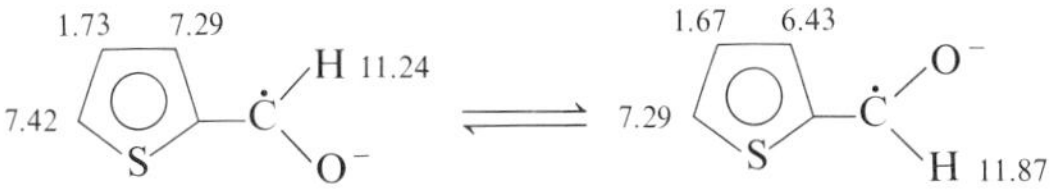

(118A) ⇌ (118B)

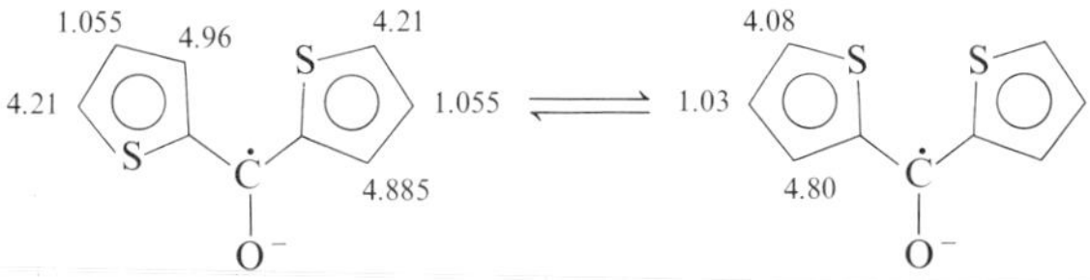

(119A) ⇌ (119B)

Anion-radicals have been obtained in solution from 2-cyanothiophene, 2-cyanothieno[3,2-b]thiophene, 2-cyanothieno[2,3-b]thiophene (117), and tetracyanothiophene[375,379]; 117 is noteworthy in that the hyperfine splittings (indicated in gauss) show the spin population to be largely confined to the substituted ring.[375] Thiophene-2-carboxylic acid has given an anion-radical in the solid phase upon γ-irradiation.[380]

Ketyl radicals containing the thiophene heterocycle have attracted wide interest. Hudson and Lewis[346] obtained the anion-radical of thiophene-2-carbaldehyde (118) by photolysis of the parent compound in alkaline methanol. The ESR spectrum showed the radical exists as an equilibrium mixture of rotamers.[346] Russian work has treated the polarographic properties of this radical and its derivatives.[74,381,382] Thiophene-3-carbaldehyde did not exhibit rotational isomerism.[346] Several workers have studied ketyls

---

[376] M. Person and J. Tirouflet, *C. R. Acad. Sci.* **258**, 4979 (1964).
[377] M. Person, *Bull. Soc. Chim. Fr.*, 1832 (1966).
[378] D. Meisel and P. Neta, *J. Am. Chem. Soc.* **97**, 5198 (1975).
[379] D. H. Eargle and M. de Conceicao Ramos de Carvalho, *J. Phys. Chem.* **77**, 1716 (1973).
[380] B. Eda, R. J. Cook, and D. H. Whiffen, *Trans. Faraday Soc.* **60**, 1497 (1964).
[381] L. N. Nekrasov and D. N. Soshchin, *Elektrokhimiya* **6**, 1219 (1970) [*CA* **74**, 8994 (1971)].
[382] L. N. Vykhodtseva and L. N. Nekrasov, *Elektrokhimiya* **13**, 1239 (1977) [*CA* **88**, 29461 (1978)].

from various thienyl ketones.[325,346,373,383–386] In general, the ketyls exist as equilibrium mixtures usually of two preferred conformational isomers. Several cases have been studied in detail[375,383,385]; 2,2'-dithienyl ketyl (**119**) is given as an example. The assignments of the hyperfine splittings to the 3- and 3'-positions in the unsymmetrical *cis-trans* conformer (**119A**) were made on the basis of expectation for the electrostatic influence of the ketyl oxygen atom on *g*-value and adjacent proton splittings.[385,387] The various thienyl ketyls show ion-pairing effects when formed by alkali metal reduction in ethereal solvents[375]; the effect on such ion-pairing, and on the energy barrier to interconversion of rotamers, of the presence macrocyclic polyethers has been investigated.[386]

Ketyls from dicarbonyl compounds in the thiophene series have also been examined. The Bologna group has studied the conformational equilibria in the anion-radical of thiophene-2,5-dicarbaldehyde.[388] They have also investigated the intramolecular cation exchange process within ion-pairs of this ketyl, and of analogous ketyls from the isomeric thienothiophenes, and the consequence for this process of variation in the size of the cation and its complexation by macrocyclic polyethers.[389,390] The anion-radicals of the isomeric 1,2-dithienylethane-1,2-diones (thenils) were recorded several years ago.[84]

4. Oxygenated Thiophene Radicals

a. *Sulfoxidized Thiophene Anion-Radicals.* Anion-radicals have been prepared by alkali metal reductions of both dibenzothiophene-5-oxide and the corresponding sulfone.[363] The sulfoxide does not appear to have been examined subsequently, but the partial assignment of its spectrum which was made[363] is probably erroneous, since the assignments made for the anion-radicals of both dibenzothiophene and its sulfone in the same work were also wrong. The hyperfine splittings of the dibenzothiophene sulfone anion radical **120** (indicated in gauss), are those of Gerdil and Lucken,[391] who confirmed their assignment by deuteration experiments. These authors

---

[383] P. Cavalieri d'Oro, G. F. Pedulli, P. Spagnolo, and M. Tiecco, *Boll. Sci. Fac. Chim. Ind. Bologna* **27**, 133 (1969) [*CA* **72**, 127117 (1970)].
[384] L. Kaper, J. U. Veenland, and T. J. de Boer, *Spectrochim. Acta, Part A* **24**, 1971 (1968).
[385] M. Guerra, G. F. Pedulli, M. Tiecco, and G. Martelli, *J. C. S. Perkin II*, 562 (1974).
[386] G. F. Pedulli and A. Alberti, *J. C. S. Perkin II*, 137 (1976).
[387] N. Steinberger and G. K. Fraenkel, *J. Chem. Phys.* **40**, 723 (1964).
[388] L. Lunazzi, G. F. Pedulli, M. Tiecco, C. Vincenzi, and C. A. Veracini, *J. C. S. Perkin II*, 751 (1972).
[389] M. Guerra, G. Pedulli, and M. Tiecco, *J. C. S. Perkin II*, 903 (1973).
[390] G. F. Pedulli, A. Alberti, and M. Guerra, *J. C. S. Perkin II*, 1327 (1977).

have also carried out polarographic studies on **120** and related sulfone radicals.[391,392] The electronic absorption spectra of **120** have been recorded at low temperature, following γ-irradiation of the heterocycle in a methyltetrahydrofuran glassy matrix.[393]

(120)   (121)

Anion-radicals from dithienothiophene dioxides have been studied (e.g., **121**) for which ESR and polarographic results have been presented.[394] Similar results were obtained for certain isomers of **121** but not for others. During electrolytic reduction of deuterated dithienothiophene dioxides, it was found that H–D exchange occurred in the anion-radicals.[395] The source of protons was either impurity from solvent decomposition (DMF) or an adventitious unknown present in acetonitrile. Janssen has also correlated electronic absorption spectra and polarographic reduction potentials with calculated molecular quantities for a wide range of thiophene S,S-dioxides.[396]

b. *Radicals from Thiophenes Oxidized at Carbon.* Hydroxythiophenes exist preferentially in a ketonized form; however, oxidation of such materials by ferricyanide leads to coupling products comparable with those from oxidative dimerization of phenols. Thus 5-*tert*-butyl-2,3-dihydrothiophen-2-one in oxidized to **122**.[397] The intermediacy of thienyloxyl radicals in such dimerizations is indicated by the observation of the O—C coupling product **123** from 2,5-*tert*-butyl-2,3-dihydrothiophen-3-one.[398] The corresponding dimethyl compound gives only C—C coupling through the 2-positions, a route presumably sterically impeded in the *tert*-butyl case.

Katritzky and co-workers[399,400] described a related benzothienyloxyl **124**, prepared by thermolysis of the corresponding 2,2′-dimer.

---

[391] R. Gerdil and E. A. C. Lucken, *Mol. Phys.* **9**, 529 (1965).
[392] R. Gerdil, *Helv. Chim. Acta* **56**, 196 (1973).
[393] O. Ito and M. Matsuda, *Chem. Lett.*, 909 (1974).
[394] P. B. Koster, M. J. Janssen, and E. A. C. Lucken, *J. C. S. Perkin II*, 803 (1974).
[395] P. B. Koster, M. J. Janssen, and E. A. C. Lucken, *Spectrosc. Lett.* **6**, 253 (1973).
[396] F. de Jong and M. Janssen, *J. C. S. Perkin II*, 572 (1972).
[397] A. B. Hörnfeldt, *Acta Chem. Scand.* **21**, 1952 (1967).
[398] A. B. Hörnfeldt and P. O. Sundberg, *Acta Chem. Scand.* **26**, 31 (1972).
[399] R. W. Baldock, P. Hudson, A. R. Katritzky, and F. Soti, *Heterocycles* **1**, 67 (1973).
[400] R. W. Baldock, P. Hudson, A. R. Katritzky, and F. Soti, *J. C. S. Perkin I*, 1422 (1974).

A recent German patent claims oxidative dimerization of substituted 3-hydroxybenzothiophenes for the synthesis of thioindigo dyestuffs.[401] Here also, a radical reaction is likely. Thioindigo itself exists in two geometrically isomeric forms. Each is capable of reduction to an anion-radical but the cis form rapidly isomerizes to the trans form **125**. The ESR spectrum of **125** has been observed on several occasions but not assigned.[402-404] The splittings indicated, in gauss, for **125** are from the recent measurement of Yeh and Bard[404] made following electroreduction of thioindigo in DMF.

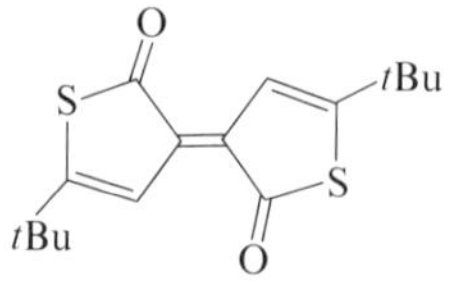

(122) 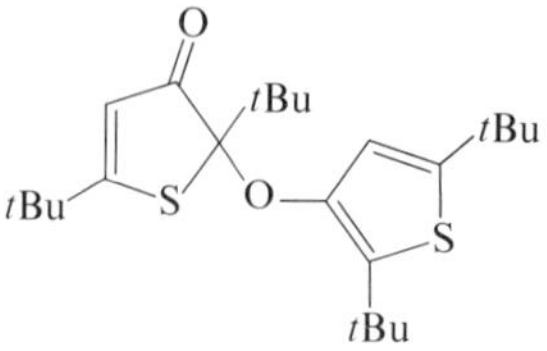   (123)

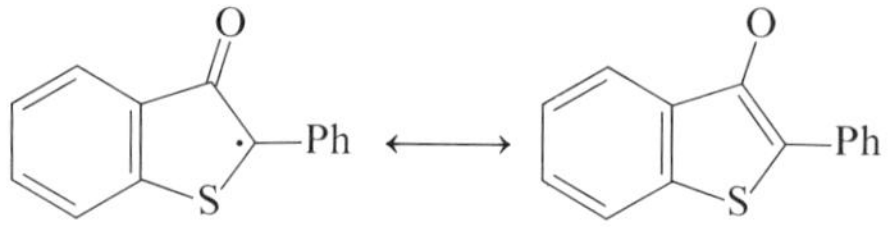

(124a) ⟷ (124b)

a(H)
1.375
1.163
0.450
0.330

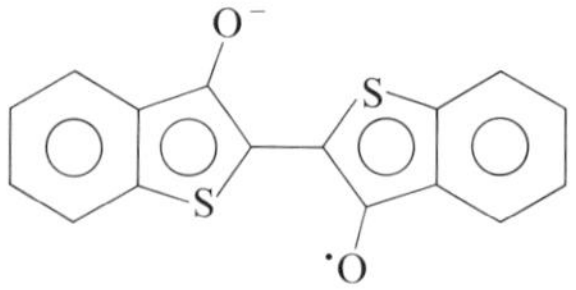

(125)

These latter workers have demonstrated, by electrochemical experiments, the different reactivities of **125** and its cis isomer.[404,405] Thus, whereas **125** is formed essentially reversibly and lost in a slow dimerization reaction and a slow coupling with unreduced thioindigo, its cis isomer is rapidly consumed in isomerization, radical–radical, and radical–parent coupling reactions.[404] Similar differential reactivities with carbon dioxide, acrylonitrile, and cinnamonitrile were also observed.[405] For 6,6'-diethoxythioindigo, it was found that the electroreduction of the cis and trans isomers occurred at different

---

[401] E. Spietschka and M. Urban, Ger. Offen. 2,504,935 (Cl. C09B7/10) (1976) [*CA* **85**, 144704 (1976)].
[402] M. Bruin, F. Bruin, and F. W. Heineken, *J. Chem. Phys.* **37**, 135 (1962).
[403] R. Chang, D. G. Kehres, and J. H. Markgraf, *J. Org. Chem.* **38**, 1608 (1973).
[404] L.-S. R. Yeh and A. J. Bard, *J. Electroanal. Chem. Interfacial Electrochem.* **70**, 157 (1976).
[405] L.-S. R. Yeh and A. J. Bard, *J. Electroanal. Chem. Interfacial Electrochem.* **81**, 319 (1977).

potentials, unlike thioindigo itself, and that the isomeric radicals showed different relative reactivities.[406]

Simultaneously with their work on sulfone anion-radicals, Lucken and co-workers[394,395] reported ESR and polarographic properties of six isomeric cyclopentadithiophenones (e.g., **126**) for which proton hyperfine splittings are indicated in gauss.

The anion-radical **127** formed upon reduction of benzo[*b*]thiophene-2,3-dione was first characterized by Russell and co-workers[128] and the result subsequently confirmed by Ciminale *et al*.[407] who found **127** to be the product of reaction of 5-halogeno benzo[*b*]thiophenediones with a range of nucleophiles. A mechanism was adduced whereby the 5-halogeno derivatives of **127** were formed upon transfer of an electron between the reactants. A subsequent loss of halide ion gave a σ-radical which abstracted hydrogen from solvent to give **127**. In the presence of benzenethiolate anion, the σ-radical was trapped to give the 5-phenylthio analog of **127**.

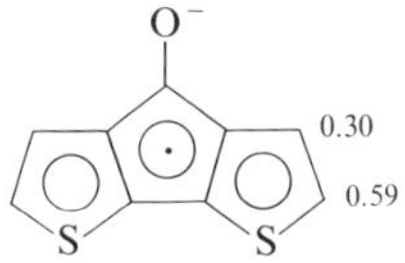

(**126**) 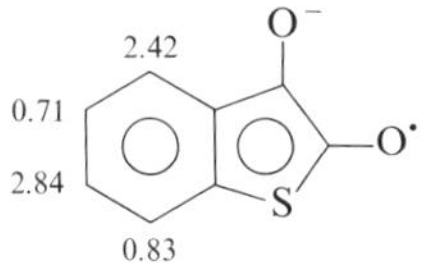   (**127**) 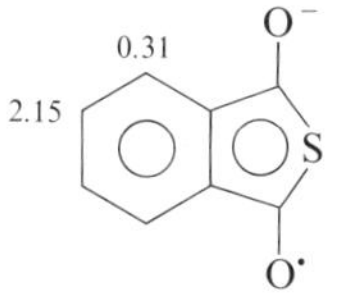   (**128**)

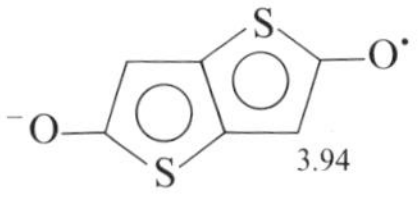

(**129**) 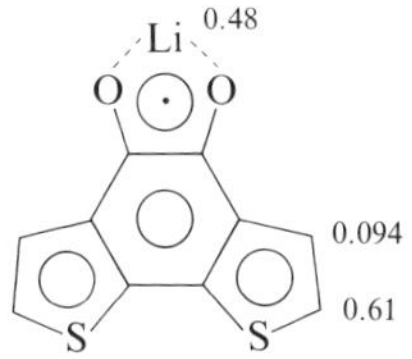   (**130**) 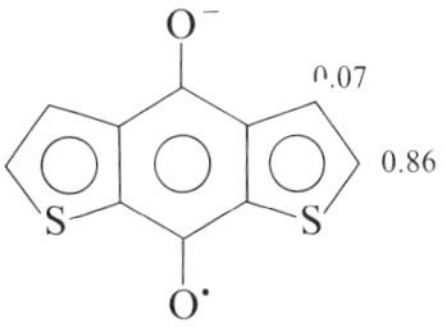   (**131**)

The anion-radical **128** of benzo[*c*]thiophene-1,3-dione was observed by ESR spectroscopy several years ago[103]; its electronic absorption spectrum has been recorded more recently.[106]

The thieno[2,3-*b*]thiophene-2,5-semidione (**129**) has been observed, together with several substituted derivatives, both symmetrical and unsymmetrical.[408] These radicals form ion-pairs with alkali-metal counterions in

---

[406] L.-S. R. Yeh and A. J. Bard, *J. Electroanal. Chem. Interfacial Electrochem.* **81**, 333 (1977).
[407] F. Ciminale, G. Bruno, L. Testaferri, M. Tiecco, and G. Martelli, *J. Org. Chem.* **43**, 4509 (1978).
[408] G. F. Pedulli, P. Zanirato, A. Alberti, M. Guerra, and M. Tiecco, *J. C. S. Perkin II*, 946 (1976).

ethereal solvents whose ESR spectra show a marked temperature dependence which arises from the exchange of the counterion between sites adjacent to the oxygen atoms. Nonequivalent ion-pairs exist when the substitution pattern is unsymmetrical. The consequences of the presence of macrocyclic polyethers in the system, which coordinate the counterion, for the exchange process have also been investigated for these ion-pairs.[390]

Pedulli and co-workers[409] have also examined the ion-pairing properties of several dithienobenzosemiquinones. When the radical is structurally analogous to phenanthrasemiquinone, (e.g., **130**), the counterion is chelated in a fixed position by the oxygen atoms. Thus, in **130** hyperfine splitting (indicated in gauss) is observed for lithium even in DMSO, a polar solvent, and the ESR spectra show only minor temperature dependence. By contrast, for the anthrasemiquinone analog **131** there are two sites of low energy, adjacent to the oxygen atoms, where the counterion may reside, and between which it may exchange in appropriate conditions. In DMSO the radical exists as the free ion (hyperfine splittings indicated in **131** in gauss); in THF an ion-pair may be formed with lithium which exhibits a counterion splitting. The oxygen sites in **131** are nonequivalent, and that adjacent to the sulfur atoms is preferred by the counterion.[409]

5. *Sulfur* d-*Orbitals and Thiophene-Derived Radicals*

There has been much work concerning the significance of the *d*-orbitals of sulfur for its chemistry, particularly its heterocyclic chemistry. This has been reviewed by Zahradnik[410] and Salmond.[411] Mitchell has discussed the role of *d*-orbitals in the chemistry of second-row elements generally,[412] and Coffen their role in the chemistry of bivalent sulfur compounds.[413] Clarke[414] has also discussed the structure and bonding of thiophene. The arguments for and against the necessity of involving sulfur *d*-orbitals in electronic descriptions of thiophene have been extended to the radicals derived from the heterocycle.[415] General consensus now has it that for the ground-state properties of thiophene the role of *d*-orbitals is small, but for excited states it may become more significant. Thus, anion-radicals from

---

[409] G. F. Pedulli, A. Alberti, L. Testaferri, and M. Tiecco, *J. C. S. Perkin II*, 1701 (1974).
[410] R. Zahradnik, *Adv. Heterocycl. Chem.* **5**, 1 (1965).
[411] W. G. Salmond, *Q. Rev., Chem. Soc.* **22**, 253 (1968).
[412] K. A. R. Mitchell, *Chem. Rev.* **69**, 157 (1969).
[413] D. L. Coffen, *Rec. Chem. Prog.* **30**, 275 (1969).
[414] D. T. Clark, *Int. J. Sulfur Chem.*, Part C **7**, 11 (1972).
[415] M. M. Urberg and E. T. Kaiser, *in* "Radical Ions" (E. T. Kaiser and L. Kevan, eds.), Ch. 8. Wiley (Interscience), New York, 1968.

thiophenes and their sulfones, where orbitals of higher energies are occupied, are likely to be the species where d-orbital participation becomes apparent. Several of the papers whose experimental content has been noted in previous sections also discuss the possibility of d-orbital involvement. In addition, there have been papers of entirely theoretical content which have considered the same problem. Three general types of conclusion have been drawn: that models including d-orbitals are (i) significantly worse than, (ii) significantly better than, or (iii) not significantly different from models which exclude them, in describing a particular property. MO methods of different types have been used, but, in general, it has been found that for anion-radicals containing the unoxidized thiophene ring, the inclusion of sulfur d-orbitals in the MO description worsens the description of the spin distribution, by comparison with the p-orbital model and experimental observation.[41,42,343,346,367,368] Noteworthy among these are the results of Gleiter and co-workers for **96**.[343] This work, which included photoelectron spectroscopic measurements besides the ESR data already noted, indicated that structures **96a–c** which imply d-orbital participation are, in fact, inaccurate; structures involving separated charges and no more than a shared octet of electrons at sulfur (e.g., **96c**) are preferable. The spin distribution in other anion-radicals has been adequately described by the p-model in both McLachlan and INDO MO calculations without the d-model being tested.[366,375,408] For nitrothiophene anion-radicals described by either the McLachlan or the INDO MO method, the inclusion of sulfur d-orbitals made no significant improvement to the description of the spin distribution.[356,374] The INDO MO method has also been used, without inclusion of d-orbitals, adequately to describe spin distribution and energy barriers to conformational change in thienylmethyl radicals.[352,353] Galasso,[416–418] who has examined theoretically the application of the INDO method to thiophene-derived radicals, showed that d-orbital inclusion was either unnecessary or detrimental to the adequate description of spin distribution. By contrast, Nanda and Narasimhan[419] reached the opposite conclusion in their unrestricted Hartree–Fock calculations.

Inclusion of d-orbital interactions made little improvement in the description of spin distribution in dithienothiophene sulfones[394]; Urberg and Tenpas showed that the sulfone group in dibenzothiophene sulfone anion-radical could be regarded as exercising merely an inductive effect and not a mesomeric effect,[420] although in earlier work Urberg and co-workers had

---

[416] V. Galasso and N. Trinajstic, *J. Chim. Phys.* **70**, 1489 (1973).
[417] V. Galasso, *Z. Naturforsch., Teil A* **28**, 1951 (1973).
[418] V. Galasso, *J. Chim. Phys.* **71**, 889 (1974).
[419] D. N. Nanda and P. T. Narasimhan, *Mol. Phys.* **24**, 1341 (1972).
[420] M. M. Urberg and C. Tenpas, *J. Am. Chem. Soc.* **90**, 5477 (1968).

used a *d*-model for the electronic influence of the sulfone group.[421] The results which are most convincing in suggesting a need to consider sulfur *d*-orbitals in describing thiophene sulfone radical properties are those concerning electronic absorption spectra.[393,396] Here good correlations have been found between observed transitions and calculated energies. As anticipated at the outset of this section, and by Koster et al.[394] previously, sulfur *d*-orbitals are most significant for properties involving high-energy MOs.

6. Radicals Containing a 1,2-Dithiole Ring

a. *1,2-Dithiolyl Radicals.* The simplest aromatic system containing the 1,2-dithiole ring is the 1,2-dithiolylium cation (**132**); 1,2-dithiolyl radicals are neutral radicals which result from one-electron reduction of this structure. MO calculations indicate that positions 3 and 5 of the radical are positions of high spin population,[422,423] and consistent with this 3,4-diaryl-1,2-dithiolylium ions dimerize rapidly upon electroreduction, without radicals being observable.[424] On the other hand, 3,5-disubstituted 1,2-dithiolylium ions are reduced to radicals (e.g., **133**) for which ESR measurements have been made.[425,426] Hyperfine splittings indicated in **133** are in gauss, and the assignment of the heterocyclic proton splitting was made following deuteration experiments.[426] On cooling, **133** and its congeners

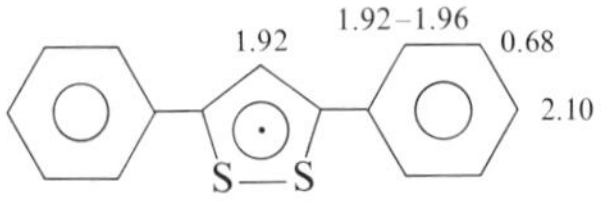

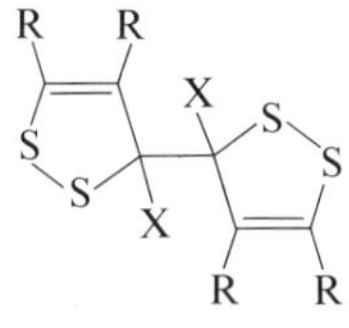

[421] E. T. Kaiser, M. M. Urberg, and D. H. Eargle, *J. Am. Chem. Soc.* **88**, 1037 (1966).
[422] R. Zahradnik, P. Carsky, S. Hünig, G. Kiesslich, and D. Scheutzow, *Int. J. Sulfur Chem., Part C* **6**, 109 (1971).
[423] C. Guimon, D. Gonbeau, G. Pfister-Guillouzo, K. Bechgaard, V. D. Parker, and C. T. Pedersen, *Tetrahedron* **29**, 3695 (1973).
[424] C. T. Pedersen and V. D. Parker, *Tetrahedron Lett.*, 767 (1972).
[425] C. T. Pedersen, K. Bechgaard, and V. D. Parker, *J. C. S., Chem. Commun.*, 430 (1972).
[426] K. Bechgaard, V. D. Parker, and C. T. Pedersen, *J. Am. Chem. Soc.* **95**, 4373 (1973).

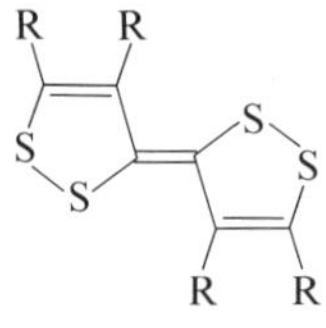

reversibly yield diamagnetic dimers, the position of the equilibrium depending critically upon the steric requirements of the substituents. Radicals such as **133** are further reducible, with ring opening to dithioketonate anions. In general, the dimers of dithiolyl radicals may be oxidized back to dithiolylium ions, but dimers of type **134** (X = H) may be dehydrogenated to 1,1′,2,2′-tetrathiafulvalenes (e.g., **135** or its cis isomer) with an ease which is related to the substitution pattern; **135** results directly from the reduction by zinc metal of 3-chloro-1,2-dithiolylium ions, presumably via **134** (X = Cl).[427]

Dithiolyl radicals are produced by photolysis of appropriate 1,2-dithiolylium ions in ethanolic solutions.[428,429] The reaction is a biphotonic process and involves ethanol as reductant.

b. *Annelated 1,2-Dithiole Radicals.* When the 1,2-dithiole ring is annelated as in naphtho[1,8-*cd*][1,2]dithiole, the resultant heterocycle bears no charge. Single-electron oxidation, therefore, gives a cation-radical (**136**). This radical has been studied by several groups, latterly by those with an interest in electrically conducting organic solids[430–433] (see Section III,B,7). All have quoted hyperfine splittings; those in **136** are due to Wudl and co-workers[433] who are the only group to measure ^{33}S hyperfine splittings for the radical. Cation-radicals from similar *peri*-disulfides have been observed both in solution and solid phase[433–435] (e.g., **137**).[433]

When the dithiole ring is annelated by phenalene, the laws of valence require the resultant aromatic heterocycle to be cationic; reduction thus gives the neutral phenaleno[1,9-*cd*][1,2]dithiolyl radical (**138**). This persistent radical has been examined in solution and its hyperfine splittings assigned on the basis of MO calculations.[436,437] The radical is remarkably stable against dimerization; it may be isolated as a paramagnetic solid whose mass

---

[427] H. Behringer and E. Meinetsberger, *Tetrahedron Lett.*, 3473 (1975).
[428] C. T. Pedersen and C. Lohse, *Tetrahedron Lett.*, 5213 (1972).
[429] C. T. Pedersen and C. Lohse, *Acta Chem. Scand.*, Ser. B **29**, 831 (1975).
[430] A. Zweig and A. K. Hoffmann, *J. Org. Chem.* **30**, 3997 (1965).
[431] B. I. Stepanov, W. Ya. Rodionov, A. Ya. Zheltov, and V. V. Orlov, *Tetrahedron Lett.*, 1079 (1971).
[432] G. F. Pedulli, P. Vivarelli, P. Dembech, A. Ricci, and G. Seconi, *Int. J. Sulfur Chem.* **3**, 255 (1973).
[433] F. B. Bramwell, R. C. Haddon, F. Wudl, M. L. Kaplan, and J. H. Marshall, *J. Am. Chem. Soc.* **100**, 4612 (1978).
[434] F. Wudl, D. E. Schafer, and B. Miller, *J. Am. Chem. Soc.* **98**, 252 (1976).
[435] V. Gailite and J. Freimanis, *Zh. Obshch. Khim.* **43**, 348 (1973) [*CA* **78**, 158611 (1973)].
[436] R. C. Haddon, F. Wudl, M. L. Kaplan, J. H. Marshall, and F. B. Bramwell, *J. C. S., Chem. Commun.*, 429 (1978).
[437] R. C. Haddon, F. Wudl, M. L. Kaplan, J. H. Marshall, R. E. Cais, and F. B. Bramwell, *J. Am. Chem. Soc.* **100**, 7629 (1978).

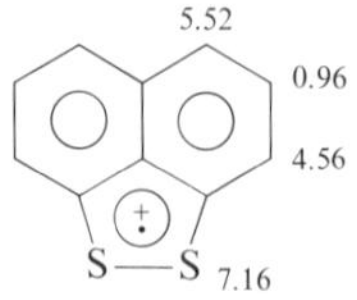

(136) 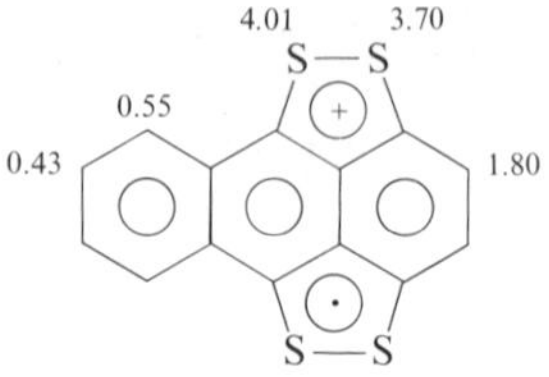   (137)

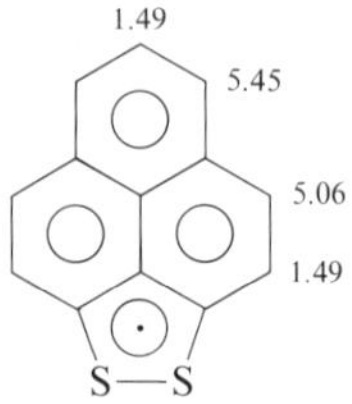

(138)

spectrum was recorded. In solution, **138** is reversibly oxidized to the parent dithiolylium ion and reversibly reduced to an anion of uncertain structure.

c. *Radicals from 1,2-Dithiole-3-thiones and Related Substances.* Although dimeric materials result from electroreduction of 5-methyl-1,2-dithiole-3-thione, Astruc et al.[438] reported failure to observe radicals by ESR. The reason probably lies in their temperature of operation, for others have observed the anion-radical (**139**) of 1,2-dithiole-3-thione itself, at low temperature (200K).[439] Hyperfine splittings are indicated in gauss. Electrochemical oxidation of 1,2-dithiole-3-thiones has also yielded dimeric products (e.g., **140**), indicative probably of radical intermediates, although none has been observed directly.[440] The benzo-1,2-dithiole azine **141** failed to give the behavior expected of a violene upon electrochemical oxidation.[135]

Ethylation of 1,2-dithiole-3-thione gives a 3-ethylthio-1,2-dithiolylium ion whose electroreductive dimerization has been investigated.[132] A series of aryl-substituted 1,2-dithiol-3-ylidene acetophenones underwent oxidative dimerization via cation-radicals (**142**) when electrolyzed; the dimeric products, which contain a new C—C bond, were elaborated chemically into a series of highly conjugated derivatives including bithiathiophthenes.[441]

---

[438] A. Astruc, M. Astruc, D. Gonbeau, and G. Pfister-Guillouzo, *Collect. Czech. Chem. Commun.* **39**, 861 (1974).
[439] H. Bock, G. Brähler, A. Tabatabai, A. Semkow, and R. Gleiter, *Angew. Chem. Int. Ed. Engl.* **16**, 724 (1977).
[440] C. T. Pedersen, and V. D. Parker, *Tetrahedron Lett.*, 771 (1972).
[441] C. T. Pedersen, V. D. Parker, and O. Hammerich, *Acta Chem. Scand., Ser. B* **30**, 478 (1976).

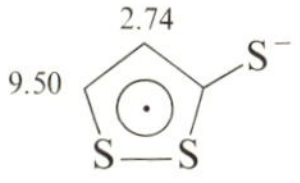

d. *Radicals from 6a-Thiathiophthenes (1,2-Dithiolo[1,5-b][1,2]dithioles)*. Although 2,5-diphenyl-6a-thiathiophthene (**143a**) may be represented also as a 1,2-dithiol-3-ylidene thioketone structure (**143b**), on anodic oxidation dimerization of the cation-radical formed results in the formation of the new S—S bond in **144**, and does not parallel the dimerization of **142**.[442] The dimer **144** is reversibly reduced back to the parent thiathiophthene, and the reaction appears general for phenyl- or anisyl-substituted compounds. Voltammetric peak potentials were determined for the redox processes described.

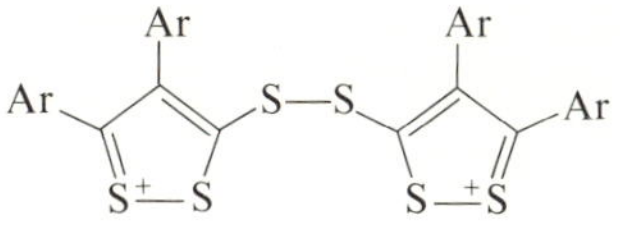

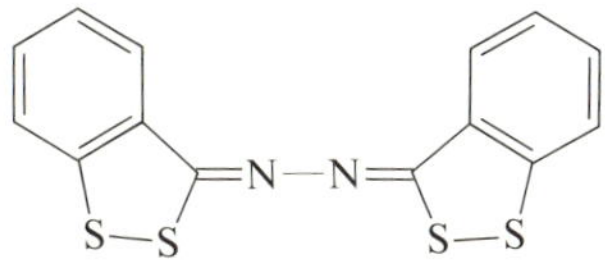

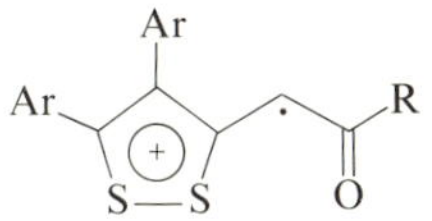

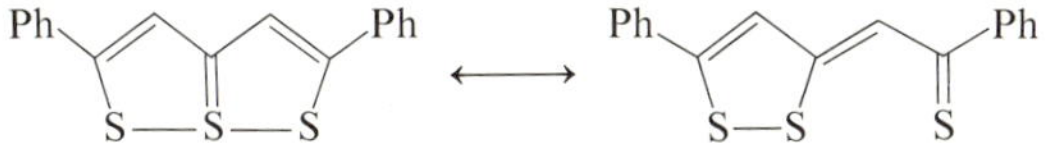

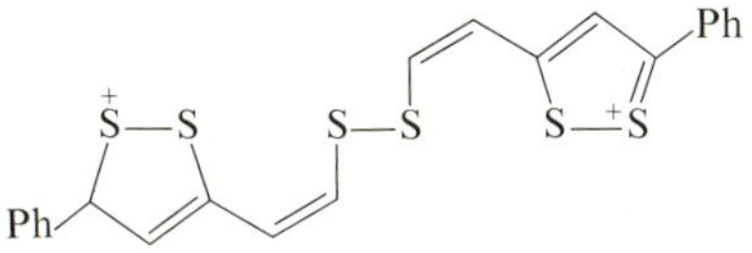

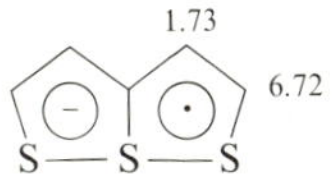

[442] C. T. Pedersen, O. Hammerich, and V. D. Parker, *J. Electroanal. Chem. Interfacial Electro-Chem.* **38**, 479 (1972).

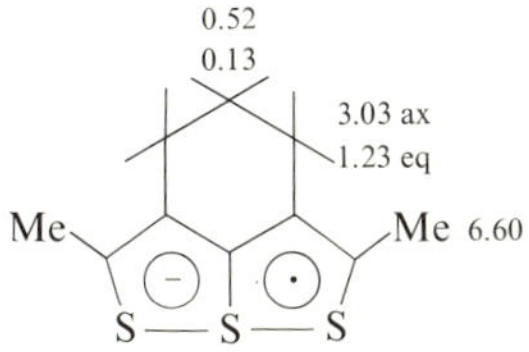

Formation of anion-radicals from simple 6a-thiathiophthenes also results in loss of ring integrity.[443,444] At low temperatures the primary anion-radicals may be observed but these are converted into the corresponding anion-radicals of 4H-thiapyran-4-thiones which exhibit closely similar hyperfine splittings but different g-values[443] (see Section III,C,3,b). The unsubstituted radical **145** subsequently undergoes further change into another unknown radical. As a result of these transformations, certain primary 6a-thiathiophthene anion-radicals were originally wrongly identified.[444] The hyperfine splittings indicated for **145**, in gauss, are the corrected values for the radical generated by electroreduction in DMF at $-60°C$.[443] Gerson and co-workers[443] suggest a plausible mechanism for the conversion of the primary radicals, a change which occurs with increasing difficulty as the bulk of the 2,5-substituents is increased and which is prevented completely by a bridging methylene chain as in **146**. The ESR spectra of **146** show marked variations with temperature; these arise from conformational inversion of the methylene chain.[445] The hyperfine splittings indicated in gauss for **146** are appropriate to the radical in DMF at $-60°C$. Under these conditions the methylene protons at the β- and γ-positions of the bridging chain show separate values; at room temperature these become averaged. When larger alkyl groups replace methyl in **146**, the ESR results indicate the ring inversion to be coupled with the rotation of the alkyl group.[445] The spectrum of the diethyl radical analogous to **146** has demonstrated the value of the computer program ESRCEX for the analysis of isotropic exchanged-broadened ESR line shapes.[446]

7. *Radicals Containing a 1,3-Dithiole Ring*

a. *1,3-Dithioyl Radicals and Related Species.* Calculation of the spin distribution in the 1,3-dithiolyl radical indicates a very high spin population at the 2-position.[422] Consistent with this, simple 1,3-dithiolyl radicals have not been observed directly although zinc metal reduction of 1,3-dithiolylium ions yields their dimerization product, 2,2'-bis(1,3-dithiole), in preparatively useful yield.[447]

The simplest 1,3-dithiolyl radicals to have been observed by ESR are, for example, **147–149**, obtained by trapping $CH_3{}^{\cdot}$, $CF_3{}^{\cdot}$, and $n\text{Bu}_3\text{Sn}^{\cdot}$ radicals

---

[443] F. Gerson, R. Gleiter, and H. Ohya-Nishiguchi, *Helv. Chim. Acta* **60**, 1220 (1977).
[444] F. Gerson, R. Gleiter, J. Heinzer, and H. Behringer, *Angew. Chem., Int. Ed. Engl.* **9**, 306 (1970).
[445] F. Gerson, J. Heinzer, and M. Stavaux, *Helv. Chim. Acta* **56**, 1845 (1973).
[446] J. Heinzer, *J. Magn. Reson.* **13**, 124 (1974).
[447] A. Kruger and F. Wudl, *J. Org. Chem.* **42**, 2778 (1977).

with 1,3-dithiole-2-thione.[448] (Hyperfine splittings are indicated in gauss.) The formation of 2-alkylthio-1,3-dithiolyl radicals during electrochemical reduction of 2-alkylthio-1,3-dithiolium ions has also been inferred.[132,449] These radicals are transient species which decay by second-order kinetics at rates approaching the diffusion-controlled limit; it was deduced from the magnitudes of the various hyperfine splittings that the exocyclic S—R bond eclipses the half-filled atomic orbital on C-2.[448]

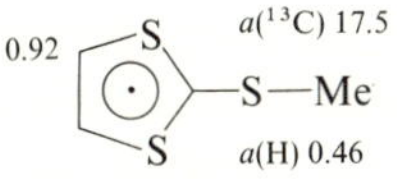

(147) 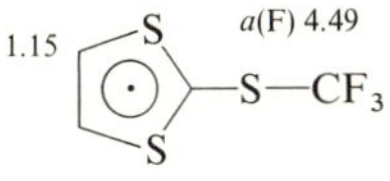 (148) 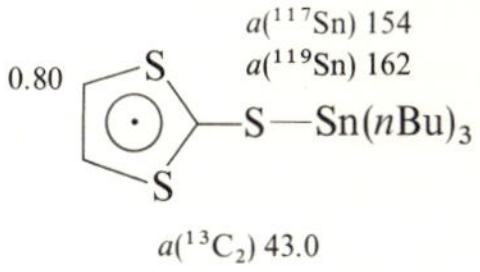 (149)

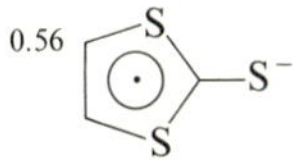

(150) 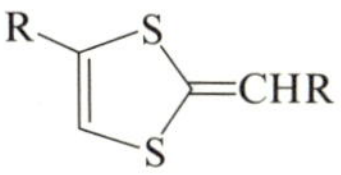 (151) 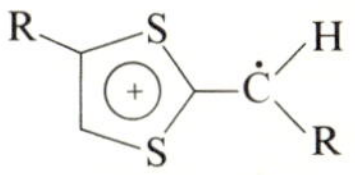 (152)

Reduction of 1,3-dithiole-2-thione yields an anion-radical (150) whose ESR spectrum at low temperature has been measured; the very low proton splitting (indicated in gauss) implies the spin population to be highly concentrated in the trithiocarbonate moiety.[439] Consistent with this, no proton hyperfine splittings were resolved in the ESR spectrum of the anion-radical of benzo-1,3-dithiole-2-thione.[449]

Electrochemical oxidation of 1,3-dithiol-2-one and -2-thione gave behavior consistent with the formation of dimers which were subsequently oxidized to radicals of uncertain structure.[449] Chemical oxidation of the isoelectronic dithiafulvenes 151 (R = phenyl, 4-chlorophenyl, 4-tolyl, or anisyl) led to dimeric exocyclic C—C-coupled products, indicating the intermediacy of the radical 152.[450]

b. *Tetrathiafulvalene Cation-Radicals* [*Bis(1,3-dithiolyliumyl) Radicals*] *and Related Species.* On account of the high electrical conductivity of salts of tetrathiafulvalene cation-radicals ("organic metals"), which is comparable with that of graphite, there has been intensive research into the solid-state properties of these materials. It is beyond the scope of this chapter to discuss this aspect of their character; the interested reader is

---

[448] D. Forrest and K. U. Ingold, *J. Am. Chem. Soc.* **100**, 3868 (1978).
[449] P. R. Moses, J. Q. Chambers, J. O. Sutherland, and D. R. Williams, *J. Electrochem. Soc.* **122**, 608 (1975).
[450] R. Mayer and H. Kroeber, *J. Prakt. Chem.* **316**, 907 (1974).

referred to specialist reviews.[451–454] Beer[455] has surveyed the area of dithiole chemistry generally, giving a comprehensive list of references to work on "organic metals," but excluding patent literature. The subject has also been reviewed in Japanese.[456] Several groups independently studied the redox properties of tetrathiafulvalene and its dibenzo derivative and discovered their persistent cation-radicals **153** and **154**.[422,457–460] The fundamental properties of the radicals in solution were defined: proton hyperfine splittings were measured (those indicated in gauss for **153** are from a recent determination in which the splitting due to ^{33}S was also measured[433]); redox potentials, comproportionation constants, and electronic absorption characteristics were also documented.[457–460] Subsequently, simply substituted analogs have been prepared (e.g., cation-radicals of variously alkylated tetrathiafulvalenes) for which hyperfine splittings as well as polarographic data are reported[461,462]; aryl,[463,464] vinyl,[465] tetracyano,[466] and tetramethylthio[467] derivatives of **153** have also been characterized, if not by ESR, then usually polarographically or by electronic absorption spectra. Cation-radicals have also been formed by one-electron oxidation of tetrathiafulvalene S-oxide.[468] Consequent upon the discovery of the electrical conducting properties of the chloride of **153**[469] and its salt with tetracyanoquinodimethane,[470] much

[451] A. F. Garito and A. J. Heeger, *Acc. Chem. Res.* **7**, 232 (1974).
[452] G. A. Thomas, D. E. Schafer, F. Wudl, P. M. Horn, D. Rimai, J. W. Cook, D. A. Glocker, M. J. Skove, C. W. Chu, R. P. Groff, J. L. Gillson, R. C. Wheland, L. R. Melby, M. B. Salamon, R. A. Craven, G. De Pasquali, A. N. Bloch, D. O. Cowan, V. V. Walatka, R. E. Pyle, R. Gemmer, T. O. Poehler, G. R. Johnson, M. G. Miles, J. D. Wilson, J. P. Ferraris, T. F. Finnegan, R. J. Warmack, V. F. Raaen, and D. Jerome, *Phys. Rev. B* **13**, 5105 (1976).
[453] M. J. Cohen, L. B. Coleman, A. F. Garito, and A. J. Heeger, *Phys. Rev. B* **13**, 5111 (1976).
[454] J. B. Torrance, *Acc. Chem. Res.* **12**, 79 (1979).
[455] R. J. S. Beer, *Org. Compd. Sulphur, Selenium, Tellurium* **4**, 308 (1977).
[456] F. Ishii and K. Akiba, *Kagaku No Ryoiki* **30**, 902 (1976).
[457] S. Hünig, H. Schlaf, G. Kiesslich, and D. Scheutzow, *Tetrahedron Lett.*, 2271 (1969).
[458] F. Wudl, G. M. Smith, and E. J. Hufnagel, *Chem. Commun.*, 1453 (1970).
[459] D. L. Coffen, J. Q. Chambers, D. R. Williams, P. E. Garrett, and N. D. Canfield, *J. Am. Chem. Soc.* **93**, 2258 (1971).
[460] S. Hünig, G. Kiesslich, H. Quast, and D. Scheutzow, *Justus Liebigs Ann. Chem.* 310 (1973).
[461] F. Wudl, A. A. Kruger, M. L. Kaplan, and R. S. Hutton, *J. Org. Chem.* **42**, 768 (1977).
[462] A. Mas, J.-M. Fabre, E. Torreilles, L. Giral, and G. Brun, *Tetrahedron Lett.*, 2579 (1977).
[463] G. Schukat, Le Van Hinh, and E. Fanghaenel, *Z. Chem.* **16**, 360 (1976) [*CA* **86**, 88560 (1977)].
[464] M. L. Kaplan, R. C. Haddon, F. Wudl, and E. D. Feit, *J. Org. Chem.* **43**, 4642 (1978).
[465] D. C. Green and R. W. Allen, *J. C. S., Chem. Commun.*, 832 (1978).
[466] M. G. Miles, J. D. Wilson, D. J. Dahm, and J. H. Wagenknecht, *J. C. S., Chem. Commun.*, 751 (1974).
[467] P. R. Moses and J. Q. Chambers, *J. Am. Chem. Soc.* **96**, 945 (1975).
[468] M. V. Lakshmikantham, A. F. Garito, and M. P. Cava, *J. Org. Chem.* **43**, 4394 (1978).
[469] F. Wudl, D. Wobschall, and E. J. Hufnagel, *J. Am. Chem. Soc.* **94**, 670 (1972).
[470] J. Ferraris, D. O. Cowan, V. Walatka, and J. H. Perlstein, *J. Am. Chem. Soc.* **95**, 948 (1973).

effort has been spent in the development of new synthetic routes to the tetrathiafulvalene heterocycle and its salts. Of relevance here are electrochemical methods for controlled synthesis of mixed-valence halide salts of **153** which contain both **153** and the parent tetrathiafulvalene.[471,472] A photooxidation of tetrathiafulvalene in carbon tetrachloride yields a similar nonintegrally stoichiometric complex.[473] Procedures for metathetical reactions of **153** have been described,[474] and also methods given for the synthesis of a wide range of salts, although not all exhibit high conductivity.[475–477] A comprehensive listing is not attempted.

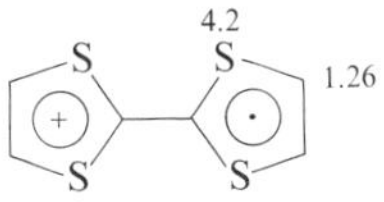

(153)

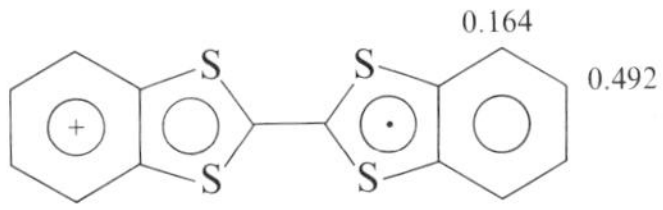

(154)

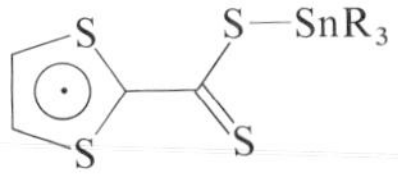

(155)

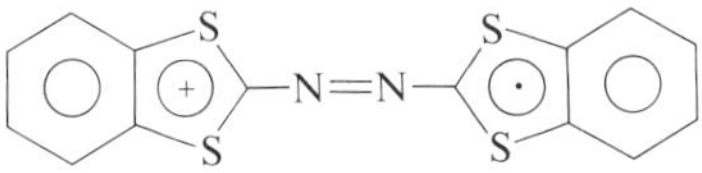

(156)

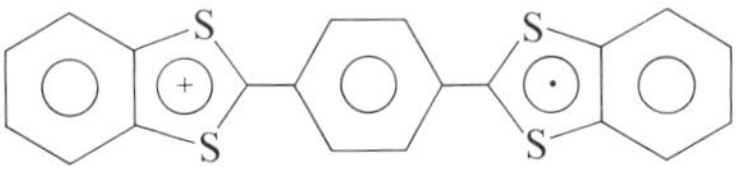

(157)

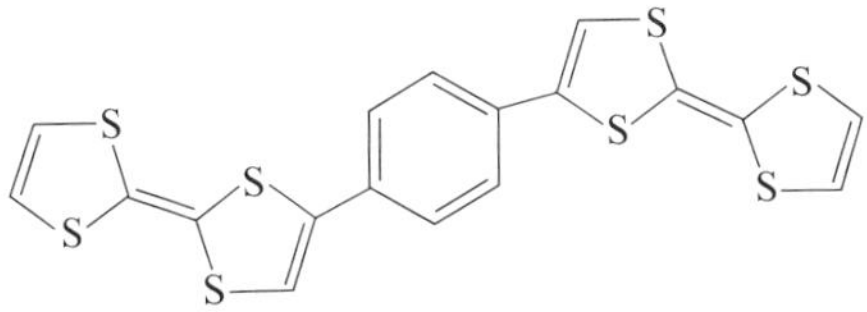

(158)

Besides oxidation to give **153**, tetrathiafulvalene has been shown to undergo attack by trialkyltin radicals in an S_H2 process which results in extrusion of a molecule of acetylene and formation of persistent radicals of type **155**.[448]

---

[471] F. B. Kaufman, E. M. Engler, and D. C. Green, *J. Am. Chem. Soc.* **98**, 1596 (1976).
[472] J. Q. Chambers, D. C. Green, F. B. Kaufman, E. M. Engler, B. A. Scott, and R. R. Schumaker, *Anal. Chem.* **49**, 802 (1977).
[473] B. A. Scott, F. B. Kaufman, and E. M. Engler, *J. Am. Chem. Soc.* **98**, 4342 (1976).
[474] F. Wudl, *J. Am. Chem. Soc.* **97**, 1962 (1975).
[475] R. C. Wheland and J. L. Gillson, *J. Am. Chem. Soc.* **98**, 3916 (1976).
[476] Y. Ueno and M. Okawara, *Chem. Lett.*, 1135 (1974).
[477] G. S. Bajwa, K. D. Berlin, and H. A. Pohl, *J. Org. Chem.* **41**, 145 (1976).

Variations on the theme of the tetrathiafulvalene structure have been investigated. Thus, the violene radical **156** has been characterized polarographically and by electronic absorption spectra[135]; **157**, the phenylog of **154**, has been reported[478,479]; the *p*-phenylenebis(tetrathiafulvalene) **158** gives an ESR spectrum upon oxidation which implies three equivalently coupling protons; i.e., it behaves as a monosubstituted **153**.[480] There have been various attempts to incorporate the tetrathiafulvalene structure into polymeric materials, but with limited success in producing conducting polymers.[464,465 481–484]

There have been several theoretical descriptions of **153** and analogs.[422,485,486]

C. Radicals Containing a Six-Membered Heterocycle

1. *Thiopyranyl Radicals and Related Species*

Consistent with prediction,[422] reduction of thiopyrylium ions with zinc metal in acetonitrile produces transient thiopyranyl radicals which dimerize by coupling at the 4-position.[487] The first thiopyranyl radical to have been characterized by ESR is 2,4,6-triphenylthiopyranyl (**159**) for which a geometry has been deduced which is closely similar to its oxygen analog (**29**)[141,488] (see Section II,B,1). The spin distribution in **159** is adequately described by a *p*-model, inclusion of *d*-electrons in the MO description resulting in no significant improvement. In order that calculations simulate the spin distribution adequately, it is necessary to allow twists of the 4-phenyl and 2,6-phenyl groups of 31 and 34°, respectively, from the plane of the heterocycle.[488]

Thioxanthenes cannot give radicals which are aromatic by the definition adopted here, by simple gain or loss of electrons, as conjugation within the heterocycle is incomplete; the cation-radical has been suggested to occur in

---

[478] Y. Ueno, A. Nakayama, and M. Okawara, *J. C. S., Chem. Commun.*, 74 (1978).
[479] M. Sato, M. V. Lakshmikantham, M. P. Cava, and A. F. Garito, *J. Org. Chem.* **43**, 2085 (1978).
[480] M. L. Kaplan, R. C. Haddon, and F. Wudl, *J. C. S., Chem. Commun.*, 388 (1977).
[481] G. Schukat, Le Van Hinh, E. Fanghaenel, and L. Libera, *J. Prakt. Chem.* **320**, 404 (1978).
[482] Y. Ueno, Y. Masuyama, and M. Okawara, *Chem. Lett.*, 603 (1975).
[483] C. U. Pittman, M. Narita, and Y. F. Liang, *Macromolecules* **9**, 360 (1976).
[484] W. R. Hertler, *J. Org. Chem.* **41**, 1412 (1976).
[485] P. Carsky, S. Hünig, D. Scheutzow, and R. Zahradnik, *Tetrahedron* **25**, 4781 (1969).
[486] M. A. Ratner, J. R. Sabin, and E. E. Ball, *Chem. Phys. Lett.* **28**, 393 (1974).
[487] Z. Yoshida, S. Yoneda, T. Sugimoto, and O. Kikukawa, *Tetrahedron Lett.*, 3999 (1971).
[488] I. Degani, L. Lunazzi, G. F. Pedulli, C. Vincenzi, and A. Mangini, *Mol. Phys.* **18**, 613 (1970).

the oxidation of thioxanthene to thioxanthylium ion, however,[489,490] and the cation-radical from 9,9-dimethylthioxanthene has been observed by ESR.[491] The aromatic thioxanthenyl (**160**) has been obtained by Devolder and co-workers, and its ESR and electronic absorption spectra were described.[170,172,179] Hyperfine splittings are indicated in gauss. The polarographic properties of the same radical have been studied by two groups[492,493]: it is formed upon electrochemical reduction of thioxanthylium ion and dimerization to 9,9′-bithioxanthene occurs; chemical reduction has the same result.[494]

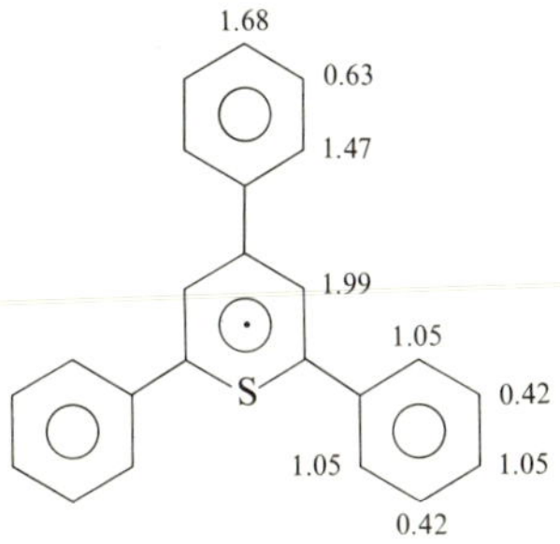

(**159**)

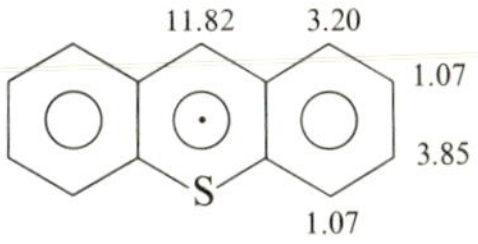

(**160**)

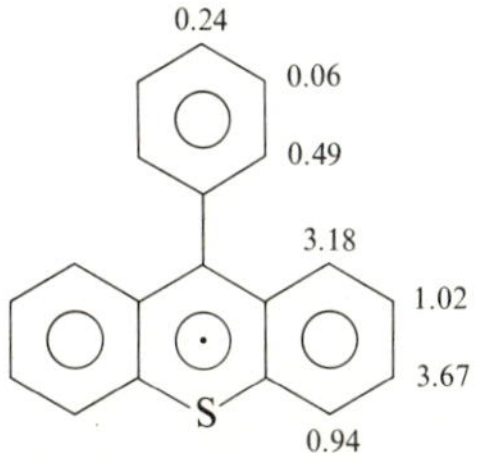

(**161**)

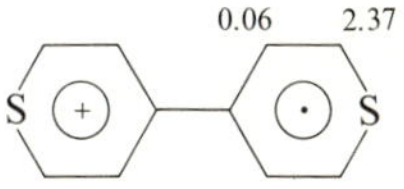

(**162**)

9-Phenylthioxanthenyl (**161**) has long been known[495] and has been the subject of several investigations.[169,175,496] There has been a measure of disagreement over the assignment of hyperfine splittings, but all workers

---

[489] H. J. Shine and L. Hughes, *J. Org. Chem.* **31**, 3142 (1966).
[490] H. J. Shine, L. Hughes, and D. R. Thompson, *Tetrahedron Lett.*, 2301 (1966).
[491] D. Deavenport, J. T. Edwards, A. L. Ternay, and E. T. Strom, *J. Org. Chem.* **40**, 103 (1975).
[492] P. T. Kissinger, P. T. Holt, and C. N. Reilley, *J. Electroanal. Chem. Interfacial Electrochem.* **33**, 1 (1971).
[493] V. A. Izmail'skii, G. E. Ivanov, and Yu. A. Davydovskaya, *Zh. Obshch. Khim.* **43**, 2503 (1973) [*CA* **80**, 70070 (1974)].
[494] C. C. Price, M. Siskin, and C. K. Miao, *J. Org. Chem.* **36**, 794 (1971).
[495] M. Gomberg and W. Minnis, *J. Am. Chem. Soc.* **43**, 1940 (1921).
[496] L. Lunazzi, A. Mangini, G. Placucci, and C. Vincenzi, *Mol. Phys.* **19**, 543 (1970).

are agreed that the preferred conformation of the radical has the phenyl ring twisted out of the plane of the heterocycle. The angle of twist is about 70°,[496] a value which may be compared with that for the oxygen analog **41** (see Section II,B,3,a), with the similarly shaped phenylacridine anion-radical,[178] and with arylphenoxazine and arylphenothiazine cation-radicals (see Section VI,B,3,iii). Hyperfine splittings for **161** are given in gauss.

9-Phenylthioxanthenyl dimerizes like **160**[494]; it has also been implicated in reactions purporting to yield S-substituted thiaanthracenes[497,498]; the observation of the radical is no doubt valid although the mechanistic inferences drawn have been refuted.[499-502] Incidentally, the work of Mislow and co-workers,[499-502] which clearly demonstrates the intrinsic instability and cyclic ylide character of S(IV)-thiabenzenes and analogs, possibly implies little chance of successfully preparing heteroaromatic radicals from this class of conjugated heterocycle.

2. Bithiopyryliumyl Radicals

Dimerization of thiopyranyl yields a product, 4,4'-bithiopyranyl, which may be dehydrogenated to the dication bithiopyrylium.[487] This dication is reducible, by uptake of one electron, to give 4,4'-bithiopyryliumyl (**162**). The reduction has been effected chemically for **162**[503] and electrochemically for **162** and its 2,2',6,6'-tetramethyl and -tetraphenyl derivatives.[196] Electronic absorption spectra were recorded in the latter work while ESR spectra were measured in the former. The hyperfine splittings in **162** are in gauss.

3. Radicals from Oxygenated Thiopyrans and Isoelectronic Species

a. *Sulfoxidized Thiopyran Radicals.* Single-electron transactions of sulfoxides or sulfones of heterocycles containing the thiopyran ring give

---

[497] M. Hori, T. Kataoka, H. Shimiu, and C.-F. Hsü, *Chem. Lett.*, 391 (1973).
[498] M. Hori, T. Kataoka, Y. Asahi, and E. Mizuta, *Chem. Pharm. Bull.* **21**, 1692 (1973).
[499] G. H. Senkler, J. Stackhouse, B. E. Maryanoff, and K. Mislow, *J. Am. Chem. Soc.* **96**, 5648 (1974).
[500] J. Stackhouse, B. E. Maryanoff, G. H. Senkler, and K. Mislow, *J. Am. Chem. Soc.* **96**, 5650 (1974).
[501] B. E. Maryanoff, G. H. Senkler, J. Stackhouse, and K. Mislow, *J. Am. Chem. Soc.* **96**, 5651 (1974).
[502] B. E. Maryanoff, J. Stackhouse, G. H. Senkler, and K. Mislow, *J. Am. Chem. Soc.* **97**, 2718 (1975).
[503] Z. Yoshida, T. Sugimoto, and S. Yoneda, *J. C. S., Chem. Commun.*, 60 (1972).

radicals which are not aromatic by present criteria: the presence of the methylene group breaks the conjugation within the heterocycle; however, such anion-radicals of thioxanthene sulfoxides and sulfones have been described.[504,505]

Maruyama et al.[174] have described the ESR characteristics of the aromatic 9-phenylthioxanthene sulfone radical (163) which was first made in 1921.[506] Their hyperfine splittings are indicated in guass.

b. *Thiopyran Ketyls.* Simple thiopyranone ketyls do not appear to have been investigated since an early report of failure to observe them polarographically[207]; their thione counterparts have been investigated, however. As already mentioned (see Section II,B,6,d), the anion-radicals of simple 6a-thiophthenes undergo a transformation with extrusion of sulfur, giving the ketyl radicals of the corresponding thiopyran-4-thiones.[443] Thus, the 6a-thiophthene anion-radical itself yields 164. The hyperfine splittings indicated in gauss in 164 are those of Gerson et al.[443] Gleiter and co-workers[439] have also reported hyperfine splittings for 164, but their values are close to those of an unknown, more persistent species also formed from 6a-thiophthene as described by Gerson.[443] Gleiter's group also reported the ESR characteristics of the isomeric ketyl (165) from thiopyran-2-thione.[439]

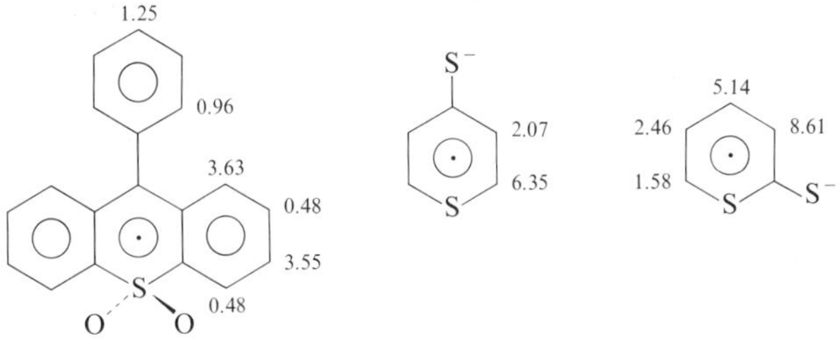

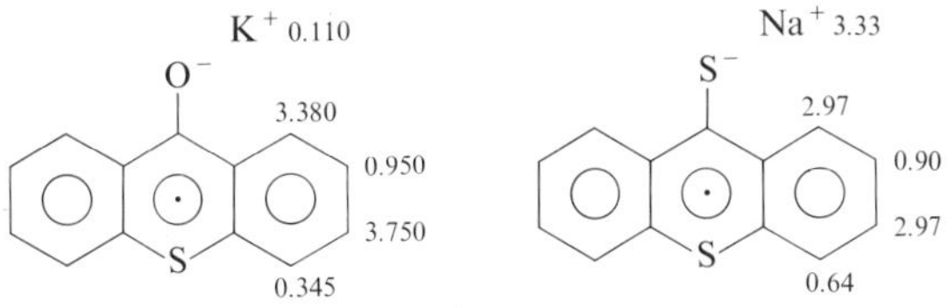

[504] P. Lambelet and E. A. C. Lucken, *J. C. S. Perkin II*, 164 (1976).
[505] P. Lambelet and E. A. C. Lucken, *J. C. S. Perkin II*, 617 (1978).
[506] M. Gomberg and E. C. Britton, *J. Am. Chem. Soc.* **43**, 1945 (1921).

The ketyl radicals of thioxanthen-9-one (**166**) and -9-thione (**167**) have been described by several groups, and various ion-pairing phenomena were reported.[233,235,236,238,239,507,508] The hyperfine splittings indicated in gauss, for the ion-pair of **166** with potassium in DME are those of Urberg and Kaiser,[508] and those for the ion-pair of (**167**) with sodium in the same solvent are due to Aarons and Adam,[239] who found a marked temperature dependence of the counterion hyperfine splittings; that indicated is for 23°C.

The conjugate acid of thioxanthen-9-one ketyl has been generated photolytically and characterized by both ESR and electronic absorption spectroscopy.[179,243]

c. *Sulfoxidized Thiopyran Ketyls.* The thioxanthen-9-one dioxide ketyl (**168**) was prepared in the 1960s by several groups.[415,507-510] Vincow[509] assigned the hyperfine splittings, indicated in **168** in gauss, for a solution of the radical in alcohol, on the basis of MO calculation and comparison with the spectrum of benzophenone ketyl. This assignment came to be doubted for two reasons; first, a spectrum obtained on impure material by others gave a different and spurious result,[507,508] and second the assignment was questioned by Gerdil and Lucken[391] who modeled the sulfone group differently in calculation and made comparison with the anion-radical of diphenyl sulfone. This led to further work involving the examination of specifically methylated derivatives.[508,511] The outcome vindicated the original assignment although some differences in the magnitudes of hyperfine splittings associated with solvent variation and ion-pairing in solvents less polar than alcohol came to light. This body of work as a whole contributes to an understanding of the nature of the electronic effect of the sulfone function. (See Section II,B,4,a and 5).

The ketyl **168** was included in a study which correlated the electron-density distribution in a set of radicals, as inferred from infrared measurements upon them and their diamagnetic parents, with ESR spin-distribution data.[512] The conclusion reached regarding **168** was that the sulfone group exerts very little mesomeric electron acceptance, a conclusion in harmony with Vincow's model of the radical which likened it to benzophenone ketyl.[509]

The thioxanthen-9-one sulfoxide ketyl (**169**) has been obtained as its ion-pair with potassium in DME.[513] Its hyperfine spectrum (splittings in

---

[507] E. T. Kaiser and D. H. Eargle, *J. Am. Chem. Soc.* **85**, 1821 (1963).
[508] M. M. Urberg and E. T. Kaiser, *J. Am. Chem. Soc.* **89**, 5179 (1967).
[509] G. Vincow, *J. Chem. Phys.* **37**, 2484 (1962).
[510] G. A. Razuvaev, V. S. Etlis, and A. N. Smirnov, *Zh. Obshch. Khim.* **33**, 3749 (1963) [*CA* **60**, 10511 (1964)].
[511] Sister J. P. Keller and R. G. Hayes, *J. Chem. Phys.* **46**, 816 (1967).
[512] D. H. Eargle, *J. Chem. Soc. B*, 1556 (1970).
[513] A. Trifunac and E. T. Kaiser, *J. Phys. Chem.* **74**, 2236 (1970).

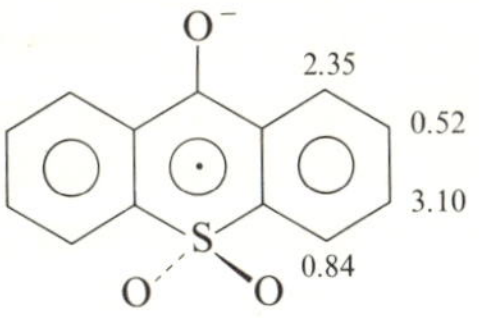

(168)
O⁻ 2.35
0.52
3.10
S
O O 0.84

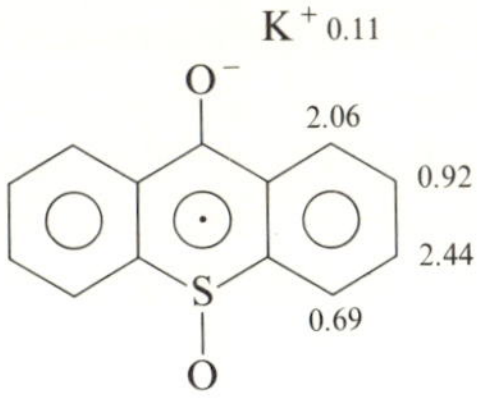

(169)
K⁺ 0.11
O⁻ 2.06
0.92
2.44
S
O 0.69

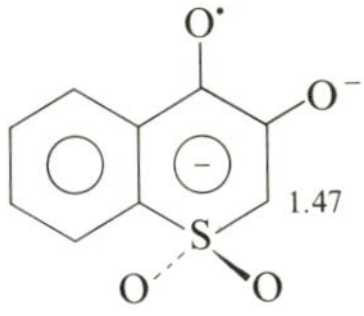

(170)
a(H)
3.96
2.47
1.00
0.49
O·
O⁻
1.47
S
O O

169 are indicated in gauss) was assigned following specific methylation experiments. It is noteworthy that the spin population in the flanking benzene rings, as implied by the indicated hyperfine splittings and by the experimental total spectrum width, is somewhat smaller in **169** than in **168**. It was suggested that the sulfoxide group may delocalize spin population rather more effectively than the sulfone group. Alternatively, this may reflect different relative weightings of the ketyl structures $\dot{C}\!-\!\bar{O} \leftrightarrow \bar{C}\!-\!\dot{O}$ in the two radicals caused by the different electrostatic effects at sulfur.

The dianion-radical **170** was prepared by Russell and co-workers,[514] along with several analogs. The proton hyperfine splitting for the 2-position was proved by specific deuteration; the other splittings were not assigned.[514]

4. *Radicals Containing a Dithiin Ring*

a. *Radicals from 1,4-Dithiin and Benzo-1,4-dithiin.* Although the heterocycle 1,4-dithiin is not generally considered to be aromatic, the cation-radical **171** formed upon its one-electron oxidation falls within the present definition of a heteroaromatic radical: conjugation is present in the heterocycle and **171** may alternatively arise by formal one-electron reduction of the aromatic dithiinium dication. The radical has been known for several years. Its ESR spectrum was first measured by Lucken[515] and subsequently

---

[514] G. A. Russell, R. L. Blankespoor, K. D. Trahanovsky, C. S. C. Chung, P. R. Whittle, J. Mattox, C. L. Myers, R. Penny, T. Ku, Y. Kosugi, and R. S. Givens, *J. Am. Chem. Soc.* **97**, 1906 (1975).
[515] E. A. C. Lucken, *Theor. Chim. Acta* **1**, 397 (1963).

by Sullivan[516] who also obtained ^{33}S hyperfine splittings for the isotope at natural abundance. Sullivan's hyperfine splittings are indicated for **171** and its benzo analog **172** in gauss. (The spin distributions of these two radicals were compared in general terms in Part I: Section II,B,2.) The electrochemical formation of **171**, **172**, and various of their substituted derivatives was described by Schroth *et al.*[294] The tetramethyl-1,4-dithiin cation-radical **173** has been found to occur when butane-2,3-dithione (or ene-thio analogs) is subjected to oxidative conditions.[295,517] The hyperfine splittings indicated for **173** are due to Russell and co-workers.[295]

There have been several theoretical descriptions of **171**, **172**, and related systems.[310,515,518] Both McLachlan and INDO MO calculations show modest improvement on inclusion of *d*-electrons in the basis set.

Dithiino[3,4-*b*]dithiin (**174**) has been synthesized and subjected to anodic oxidation.[519] Although the formation of a radical is implied, it occurs at much higher voltage than the oxidation of isomeric tetrathiafulvalene to **153**; electrically conducting charge-transfer salts are not formed by **174**.

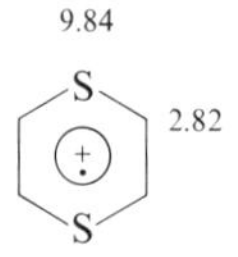

(**171**) 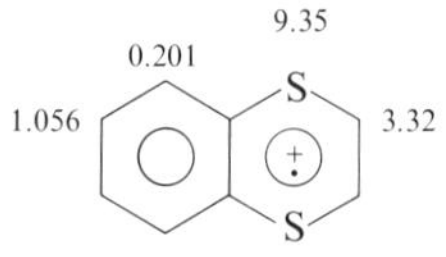 (**172**) 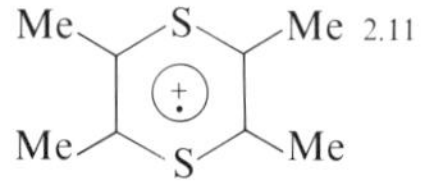 (**173**)

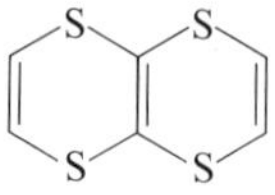

(**174**) 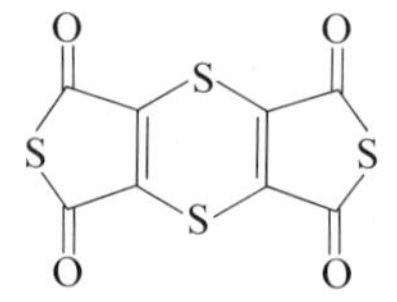 (**175**)

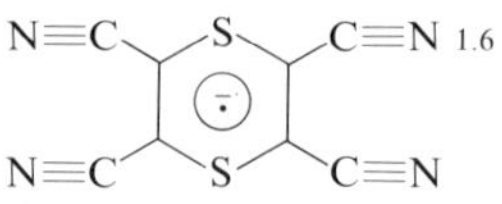

(**176**)

Gerson and co-workers[520] have formed both cation- and anion-radicals from the heterocycle dithieno[3,4-*b*:3′,4′-*e*][1,4]dithiin-1,3,5,7-tetraone

---

[516] P. D. Sullivan, *J. Am. Chem. Soc.* **90**, 3618 (1968).
[517] G. N. Schrauzer and H. N. Rabinowitz, *J. Am. Chem. Soc.* **92**, 5769 (1970).
[518] V. Galasso, *Mol. Phys.* **31**, 57 (1976).
[519] M. Mizuno, M. P. Cava, and A. F. Garito, *J. Org. Chem.* **41**, 1484 (1976).
[520] F. Gerson, C. Wydler, and F. Kluge, *J. Magn. Reson.* **26**, 271 (1977).

(175). These radicals are unusual in possessing no protons; the only hyperfine structure, therefore, in their ESR spectra is due to the low-abundance isotopes ^{13}C, ^{17}O, and ^{33}S. As expected intuitively, the spin population is largely localized in the dithiin moiety in the cation-radical and in the thiophenedione moieties in the anion-radical. The authors were able to make a detailed analysis of the ESR spectra including the signs of hyperfine splittings and to show the spin distributions to accord with the predictions of a simple MO model.[520]

The only other simple 1,4-dithiin anion-radical to have been documented is **176**.[379] This shows the ESR spectrum expected from four equivalent nitrogen nuclei. (The indicated hyperfine splitting is in gauss.) The authors draw attention to the low g-value of **176**, indicative of little spin population on S; they argue that the conjugation in the radical may well "short-circuit" the heteroatoms and arise from direct overlap of the π-systems on carbon, being possible on account of the nonplanarity (folding) of **176**.

b. *Radicals from Dibenzo[1,2]dithiin and Thianthrene (Dibenzo-[1,4]dithiin)*. Two groups have examined the ESR spectra of dibenzo-[1,2]dithiin cation-radical **177**.[431,432] The hyperfine splittings indicated (in gauss) for **177** are due to Pedulli *et al.*[432] who had the better spectral resolution and who quote a hyperfine splitting for ^{33}S; they also made their assignment by comparison of the spectra of **177** with those of methylated derivatives.[432] These latter workers found, however, that the McLachlan MO method, ignoring sulfur d-orbitals, gave a poor description of the spin distribution; Galasso has applied the INDO method to **177**.[418]

Thianthrene cation-radical **178** is one of the most studied of heteroaromatic radicals. Its individual chemistry has proved very rich; additionally, it is a persistent, readily recognized radical which has often been used as a well-behaved radical in systems where the prime interest has been the physical chemistry rather than the substrate itself. This chapter cannot deal fully with the chemistry of **178**; however, aspects have been reviewed on several and recent occasions.[521–523]

Lucken[524] first rationalized the nature of the paramagnetic species formed when thianthrene is dissolved in concentrated sulfuric acid, and isolated solid salts. He subsequently provided a solution and a theoretical account of the ESR spectrum.[515] About the same time Rundel and Scheffler

---

[521] H. J. Shine, *in* "Organosulfur Chemistry" (M. J. Janssen, ed.), Ch. 6, Wiley, New York, 1967.
[522] A. J. Bard, A. Ledwith, and H. J. Shine, *Adv. Phys. Org. Chem.* **13**, 155 (1975).
[523] H. J. Shine, *ACS Symp. Ser.* **69**, 359 (1978).
[524] E. A. C. Lucken, *J. Chem. Soc.*, 4963 (1962).

also prepared solid salts and obtained ESR spectra of a number of substituted derivatives of **178**,[525] as did Shine and co-workers.[526,527] Lamotte and Berthier, also reported the ESR spectrum of **178**[301] and, as did previous workers, noted the very small hyperfine splitting from four of the protons in the radical. Ultimately, Shine and Sullivan provided the high-resolution spectrum whose hyperfine splittings are noted, in gauss, in structure **178**. The splitting by naturally abundant ^{33}S was also measured, the first such measurement for an organic radical to be made.[516,528]

The marked difference in the hyperfine splittings of the 1- and 2-positions of **178** is notable and comparable with the similar difference in the dibenzodioxin cation-radical (**80**) (see Section II,B,6). The possibility that radicals of this type retain the fold present in the parent heterocycle has been considered theoretically[307]; other MO calculations have also been reported.[310,518]

Recently, the perchlorate of 2,3,7,8-tetramethoxythianthrenium dication has been prepared.[529] Interestingly, this material exists as a ground-state triplet[530]; as part of the study of this material its one-electron reduction product, the cation-radical **179**, has been well characterized by ESR (indicated hyperfine splittings are in gauss). Comparison of the splittings in **178** and **179** shows the substitution to bring about a redistribution of the spin

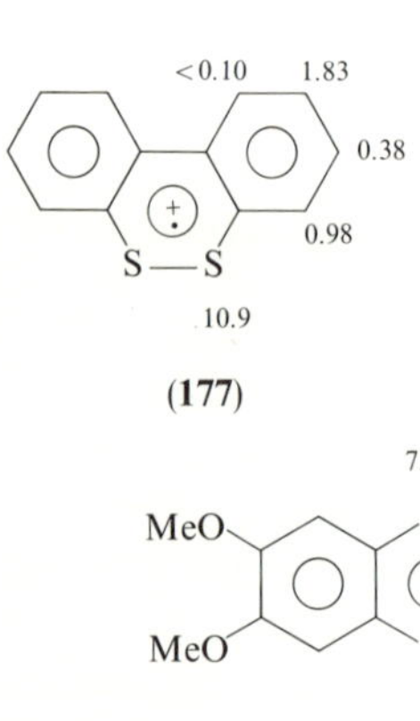

(**177**)

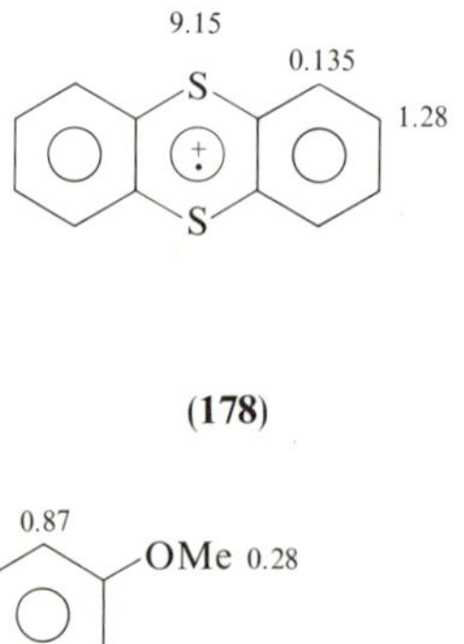

(**178**)

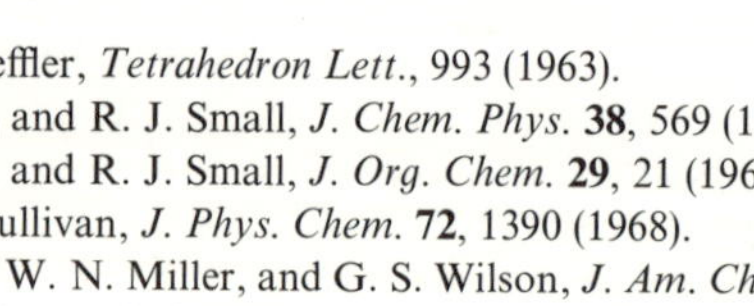

(**179**)

---

[525] W. Rundel and K. Scheffler, *Tetrahedron Lett.*, 993 (1963).
[526] H. J. Shine, C. F. Dais, and R. J. Small, *J. Chem. Phys.* **38**, 569 (1963).
[527] H. J. Shine, C. F. Dais, and R. J. Small, *J. Org. Chem.* **29**, 21 (1964).
[528] H. J. Shine and P. D. Sullivan, *J. Phys. Chem.* **72**, 1390 (1968).
[529] R. S. Glass, W. I. Britt, W. N. Miller, and G. S. Wilson, *J. Am. Chem. Soc.* **95**, 2375 (1973).
[530] I. B. Goldberg, H. R. Crowe, G. S. Wilson, and R. S. Glass, *J. Phys. Chem.* **80**, 988 (1976).

population, reducing it on sulfur and increasing it on the originally deprived 1-, 4-, 7-, and 9-positions.[530]

Various thermodynamic properties of **178** have been studied. Lucken[524] early recognized that the radical associates to a dimeric form in solid salts and certain solutions. This association has been the object of recent scrutiny.[531] Optical spectra of **178** free, in ion-paired form with perchlorate, and in a dimeric form associated with two perchlorate ions have been recognized. The exact nature of the forms present in a solution depends upon the solvent; equilibrium constants and activation parameters for the association were determined.

Redox processes involving **178** have also been studied.[294] Anodic oxidation of thianthrene has been effected in a wide variety of solvents. Use of trifluoroacetic acid gives stable solutions of **178**; and, if perchloric acid is included, the solid perchlorate salt may be isolated on evaporation of the solvent after electrolysis.[532] Dichloromethane at low temperatures has been used[533] and, at the opposite extreme, fused aluminum chloride–sodium chloride mixtures.[534] Propylene carbonate permits the ready formation of **178**,[535] whereas the inclusion of water in solvent mixtures gives an electrochemical means of sulfoxidizing thianthrene.[536] Reversible oxidation of **178** to thianthrenium dication may be brought about in customary solvents such as nitriles, nitro compounds, and dichloromethane if the solvent is treated with neutral alumina immediately before voltammetry[537]; addition of trifluoroacetic anhydride to trifluoroacetic acid equally ensures a water-free medium.[537] The availability of anhydrous solvent systems which permit the reversible oxidation and reduction of **178** has enabled the determination of the equilibrium constants for the disproportionation of the radical and for its equilibria with other aromatic materials.[319,538,539]

Whereas the reactivity of many aromatic radicals remains unexplored beyond the self-reactions, disproportionation and dimerization, and occasionally reaction with oxygen, the reactivity of **178** has been intensively

---

[531] M. de Sorgo, B. Wasserman, and M. Szwarc, *J. Phys. Chem.* **76**, 3468 (1972).
[532] O. Hammerich, N. S. Moe, and V. D. Parker, *J. C. S., Chem. Commun.*, 156 (1972).
[533] L. Byrd, L. C. Miller, and D. Pletcher, *Tetrahedron Lett.*, 2419 (1972).
[534] K. W. Fung, J. Q. Chambers, and G. Mamantov, *J. Electroanal. Chem. Interfacial Electrochem.* **47**, 87 (1973).
[535] C. Madec and J. Courtot-Coupez, *J. Electroanal. Chem. Interfacial Electrochem.* **84**, 169 (1977).
[536] H. E. Imberger and A. A. Humffray, *Electrochim. Acta* **18**, 373 (1973).
[537] O. Hammerich and V. D. Parker, *Electrochim. Acta* **18**, 537 (1973).
[538] O. Hammerich and V. D. Parker, *J. Electroanal. Chem. Interfacial Electrochem.* **36**, 13 (1972).
[539] U. Svanholm and V. D. Parker, *J. C. S. Perkin II*, 1594 (1973).

studied, largely by Shine and co-workers[540–557]; and Shine has regularly reviewed this work.[521–523] The principal reactivity of **178** is as an electrophile leading to substitution at S or at C-2. There has been controversy over the reaction mechanism in some cases. Initially, Shine and co-workers accounted for the second-order dependence of reaction rates upon **178** in terms of a disproportionation to give the thianthrenium dication as the actual kinetically active species, for example, in reaction with water[546] and with aromatic substrates.[547] This suggestion was faulted by electrochemists whose knowledge of the disproportionation equilibrium[538,558–560] informed them that, were this suggestion to be true, the dication would need to react at a rate greater than the diffusion-controlled limit in order to account for the observed rate. The primary reactivity of **178** was thus proved. Subsequently, there have appeared elegant studies which at once demonstrate both the value of electrochemical techniques in the investigation of ion-radicals and the subtleties of the reactivity of **178**.[561] As a result of this reactivity, new classes of compound have become accessible (e.g., **180–182**).[547,553,554,556]

Kim[562] has reported a reaction of **178** with α,α'-azobisbutyronitrile. Other reactions of **178** with radicals are the various reactions which occur in electrochemiluminescent systems.[19,20,25,318,563–565]

[540] H. J. Shine and L. Piette, *J. Am. Chem. Soc.* **84**, 4798 (1962).
[541] H. J. Shine and T. A. Robinson, *J. Org. Chem.* **28**, 2828 (1963).
[542] H. J. Shine and C. F. Dais, *J. Org. Chem.* **30**, 2145 (1965).
[543] H. J. Shine and D. R. Thompson, *Tetrahedron Lett.*, 1591 (1966).
[544] Y. Murata, L. Hughes, and H. J. Shine, *Inorg. Nucl. Chem. Lett.* **4**, 573 (1968).
[545] H. J. Shine and Y. Murata, *J. Am. Chem. Soc.* **91**, 1872 (1969).
[546] Y. Murata and H. J. Shine, *J. Org. Chem.* **34**, 3368 (1969).
[547] J. J. Silber and H. J. Shine, *J. Org. Chem.* **36**, 2923 (1971).
[548] H. J. Shine and J. J. Silber, *J. Am. Chem. Soc.* **94**, 1026 (1972).
[549] H. J. Shine, J. J. Silber, R. J. Bussey, and T. Okuyama, *J. Org. Chem.* **37**, 2691 (1972).
[550] H. J. Shine and K. Kim, *Tetrahedron Lett.*, 99 (1974).
[551] K. Kim, V. J. Hull, and H. J. Shine, *J. Org. Chem.* **39**, 2534 (1974).
[552] K. Kim and H. J. Shine, *J. Org. Chem.* **39**, 2537 (1974).
[553] K. Kim and H. J. Shine, *Tetrahedron Lett.*, 4413 (1974).
[554] K. Kim, S. R. Mani, and H. J. Shine, *J. Org. Chem.* **40**, 3857 (1975).
[555] B. K. Bandlish and H. J. Shine, *J. Org. Chem.* **42**, 561 (1977).
[556] B. K. Bandlish, S. R. Mani, and H. J. Shine, *J. Org. Chem.* **42**, 1538 (1977).
[557] B. K. Bandlish, W. R. Porter, and H. J. Shine, *J. Phys. Chem.* **82**, 1168 (1978).
[558] V. D. Parker and L. Eberson, *J. Am. Chem. Soc.* **92**, 7488 (1970).
[559] U. Svanholm, O. Hammerich, and V. D. Parker, *J. Am. Chem. Soc.* **97**, 101 (1975).
[560] U. Svanholm and V. D. Parker, *J. Am. Chem. Soc.* **98**, 997 (1976).
[561] J. F. Evans and H. N. Blount, *J. Org. Chem.* **42**, 976 (1977).
[562] K.-T. Kim, *Proc. Coll. Nat. Sci., Sect. 3 (Seoul Natl. Univ.)* **2**, 53 (1977) [*CA* **89**, 163512 (1978)].
[563] C. P. Keszthelyi and A. J. Bard, *J. Electrochem. Soc.* **120**, 241 (1973).
[564] C. P. Keszthelyi, *J. Am. Chem. Soc.* **96**, 1243 (1974).
[565] P. R. Michael and L. R. Faulkner, *J. Am. Chem. Soc.* **99**, 7754 (1977).

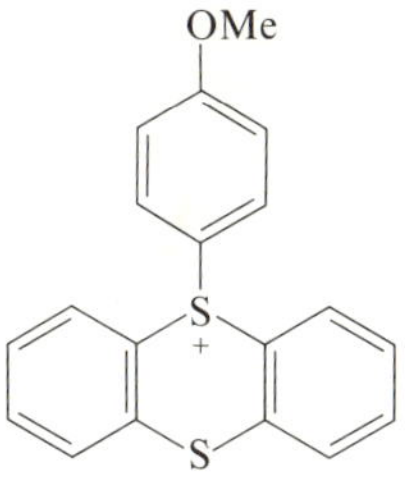

(180)

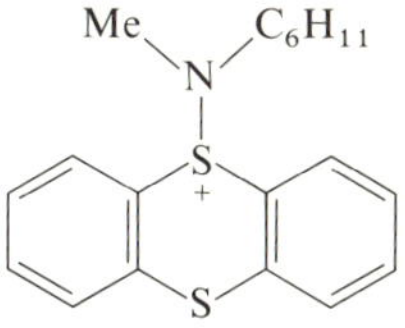

(181)

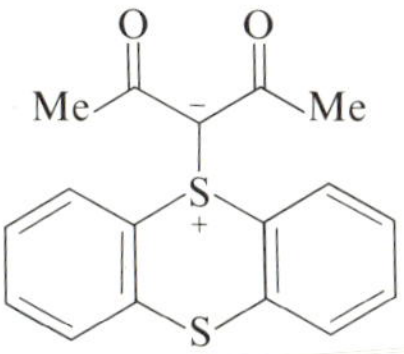

(182)

Several solid salts of **178** have been prepared,[339,524,525,544,555] some merely for convenience before solution studies while for others solid-state properties have been reported.[566–568]

c. *Anion-Radicals from Sulfoxidized Thianthrenes.* Whether the anion-radicals of sulfoxidized thianthenes are aromatic within the terms of our present definition depends on whether the sulfinyl and sulfonyl groups permit conjugation to occur in these systems. The reason for the study of these radicals has been partly to investigate just this point and the electronic nature of any conjugation provided (cf. Sections III,B,4,a and 5; III,C,3,a and c).[391,392,420,421] The anion-radical **183** of thianthrene tetroxide has been known for many years.[359,507] The proton hyperfine splittings shown in gauss in **183** are from a later ESR investigation in which ^{13}C hyperfine splittings were also measured.[569] Anion-radicals have been obtained from five other sulfoxidized thianthrenes, differing in their degree of oxidation.[570,571] Detailed assignments of the ESR spectra were not given, but from the number of lines obtained and from the widths of the spectra it is apparent that the electronic effect of the sulfinyl group depends upon its stereochemical orientation. Thus the anion-radical **184** of *cis*-thianthrene-5,10-dioxide has a spectrum which resembles that of **183**, whereas that of

[566] M. Kinoshita, *Bull. Chem. Soc. Jpn.* **35**, 1137 (1962).
[567] Y. Sato, M. Kinoshita, M. Sano, and H. Akamatu, *Bull. Chem. Soc. Jpn.* **40**, 2539 (1967).
[568] Y. Sato, M. Kinoshita, M. Sano, and H. Akamatu, *Bull. Chem. Soc. Jpn.* **42**, 548 (1969).
[569] D. H. Eargle, *J. Phys. Chem.* **73**, 1854 (1969).
[570] E. T. Kaiser and D. H. Eargle, *J. Chem. Phys.* **39**, 1353 (1963).
[571] E. T. Kaiser and D. H. Eargle, *J. Phys. Chem.* **69**, 2108 (1965).

trans-thianthrene-5,10-dioxide (**185**) has a considerably wider spectrum which shows more hyperfine structure. The interpretation assumes that the heterocycle retains the boat conformation of the parent thianthrene in all the radicals and suggests that pseudoequatorial S—O bonds effectively delocalize spin density while pseudoaxial S—O bonds do not.[570] On this basis the observations for the radical anions of thianthrene-5-oxide, -5,5-dioxide, and -5,5,10-trioxide were also explained.[571] A recent paper has described the use of organometallic reagents such as Grignard reagents and organolithiums in the presence of transition-metal ions for reductions to give anion-radicals; **183** and **184** are among radicals produced.[572]

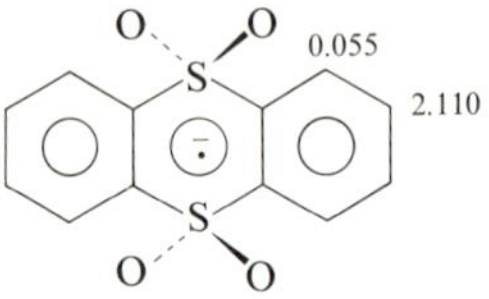

(**183**) 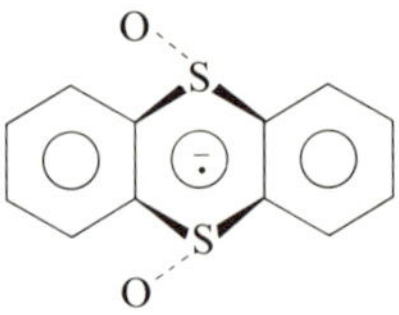   (**184**)

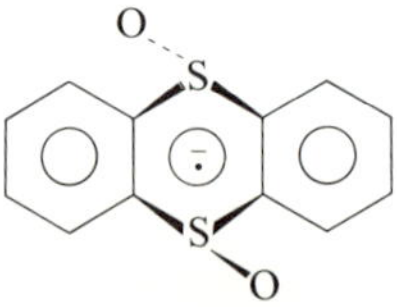

(**185**)

5. A Radical Containing a Tetrathiane Ring

An electrochemical oxidation of sodium 4-diethylaminodithiobenzoate in acetonitrile has been described which leads by dimerization ultimately to tetrathiane structures, one of which is the radical **186**.[573]

D. Radicals Containing Seven-Membered and Larger Heterocycles

1. Radicals from Dibenzo[b,f]thiepin

Two groups have examined ESR spectra of the anion-radical **187** of dibenzo[*b, f*]thiepin.[322,574] The hyperfine splittings indicated in gauss

[572] A. Stasko, L. Malik, A. Tkac, V. Adamcik, and M. Hronec, *Org. Magn. Reson.* **9**, 269 (1977).
[573] G. Cauquis and A. Deronzier, *J. C. S., Chem. Commun.*, 809 (1978).
[574] M. M. Urberg and E. T. Kaiser, *J. Am. Chem. Soc.* **89**, 5931 (1967).

for **187** are the more recent[322] and in broad agreement with the earlier values which were assigned following the examination of methylated analogs.[574] The sulfoxide and sulfone of dibenzo[b,f]thiepin have also yielded anion-radicals, **188** and **189**, respectively.[322,513] It was pointed out that the total spectrum width of **189** was broader than those of **188** and **187**,[322] which contrasts with findings in other families (cf. **166**, **168**, and **169**; Section III,C,3,c).

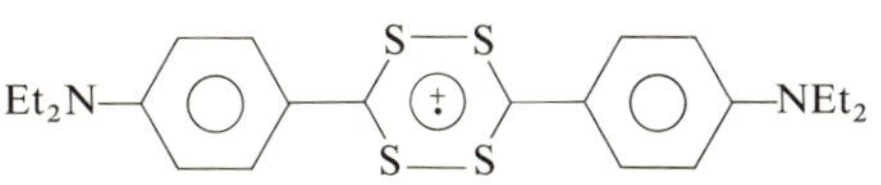

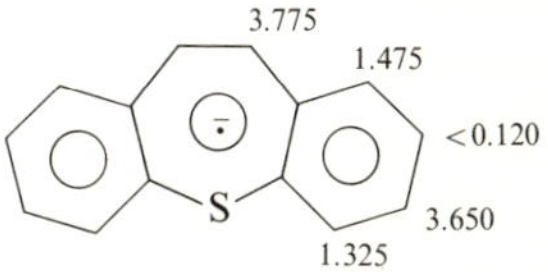

2. 1,6-Dithiecin Cation-Radical

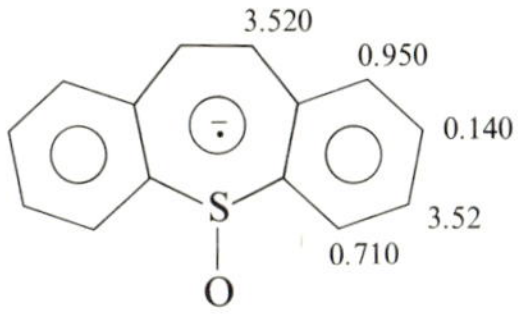

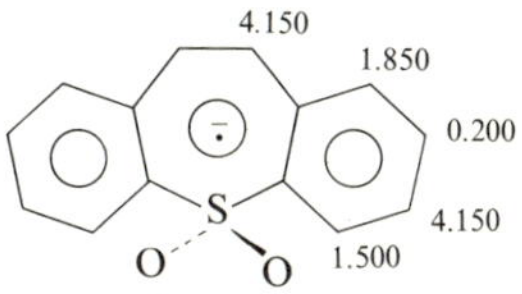

On the basis of MINDO/3 calculation, Dewar[575] expects that the 1,6-dithiecin cation-radical will have no special stability.

IV. Radicals from Selenium Heterocycles

A. Radicals Containing a Five-Membered Heterocycle

1. Radicals Containing a Selenophene Ring

Dibenzoselenophene anion-radical (**190**) is the only radical of this class whose ESR spectrum has been analyzed. Two groups found essentially the same proton hyperfine splittings.[41,576] Those indicated in gauss in **190** were given by Chiu and Gilbert who also measured the ^{77}Se splitting[576]; their spectrum was obtained by reduction of the heterocycle with potassium in DME. A low temperature ($\sim -50°$C) was necessary in order to obtain

---

[575] M. J. S. Dewar, *Pure Appl. Chem.* **44**, 767 (1975).
[576] M. F. Chiu and B. C. Gilbert, *J. C. S. Perkin II*, 258 (1973).

persistent radical concentrations. MO calculations have been performed for **190**.[576,577]

The occurrence of selenophene-2-carboxaldehyde ketyl, or its conjugate acid, during polarographic reduction of the parent aldehyde in aqueous ethanol at a mercury cathode is implied by the observation of two one-electron reduction waves and the formation of 1,2-di(selenol-2-yl)ethane-1,2-diol as principal product.[578]

Single-electron reduction steps have been observed for various heterocyclic quinones such as **191**.[579]

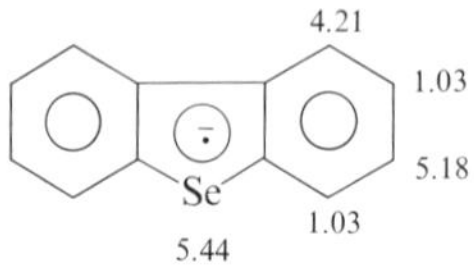

(190) 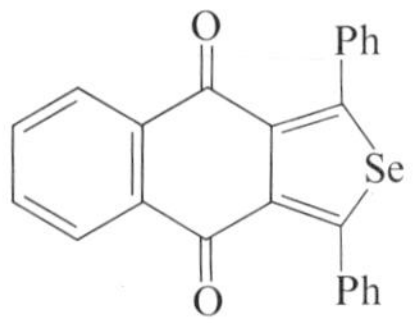   (191)

2. Radicals Containing a Diselenole Ring

Cyclic voltammetry on the cation **192** was not reversible[132]; attempted reoxidation of the reduction product of **192** produced the cation-radical of the tetraselenafulvalene **193**. Apparently, the initially formed radicals dimerize rapidly with elimination of the SeEt groups at ambient temperatures. The cation-radical of **193** and various substituted derivatives have been reported.[471-473,580-582] As with the sulfur analogs (see Section III,B,7,b), no comprehensive survey of the solid-state properties of these materials is attempted. In general, it is found that **193** and its derivatives, although rather more difficult to oxidize than tetrathiafulvalene counterparts, nevertheless do form electrically conducting salts and complexes.

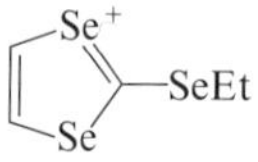

(192) 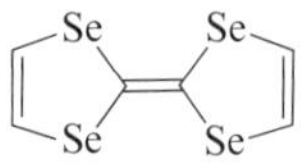   (193)

---

[577] V. Galasso and A. Bigotto, *Org. Magn. Reson.* **6**, 475 (1974).
[578] D. Guerout and C. Caullet, *C. R. Acad. Sci., Ser. C.* **281**, 643 (1975).
[579] E. Müller and W. Dilger, *Chem. Ber.* **106**, 1643 (1973).
[580] K. Bechgaard, D. O. Cowan, and A. N. Block, *J. C. S., Chem. Commun.*, 937 (1974).
[581] E. M. Engler, F. B. Kaufman, D. C. Green, C. E. Klots, and R. N. Compton, *J. Am. Chem. Soc.* **97**, 2921 (1975).
[582] E. M. Engler, D. C. Green, and J. Q. Chambers, *J. C. S., Chem. Commun.*, 148 (1976).

B. RADICALS CONTAINING A SIX-MEMBERED HETEROCYCLE

1. *Radicals Containing a Selenopyran Ring*

9-Phenylselenoxanthenyl (**194**) was studied along with its oxygen and sulfur counterparts[169,175]; as in the other cases, the phenyl ring was found to be twisted by about 70° from the plane of the heterocycle (cf. Sections, II,B,3,a; III,C,1 and 3) Japanese workers observed ESR spectra of **194** when selenoxanthylium salts were treated with organometallic reagents[497,583,584]; however, their conclusions drawn in this work have been refuted.[585] Hyperfine splittings for **194** are indicated in gauss.

Ion-pairs of selenoxanthone ketyl (**195**) have been generated with alkali-metal and alkaline-earth-metal counterions in THF.[235,236] The splittings indicated for **195** are for the ion-pair with lithium at room temperature. Comparing splittings in **195** with corresponding xanthone and thioxanthone ketyls, it was observed that the degree of covalence between ketyl oxygen and counterion increased along the series O < S < Se; the total width of the spectra decreased along the same series. These results were interpreted to mean a decreasing delocalization of the spin caused by an increasing deviation from coplanarity of the phenyl rings with increasing size of the heteroatom.[235] The structure of paramagnetic dimers formed by **195** and its analogs was also discussed.[236]

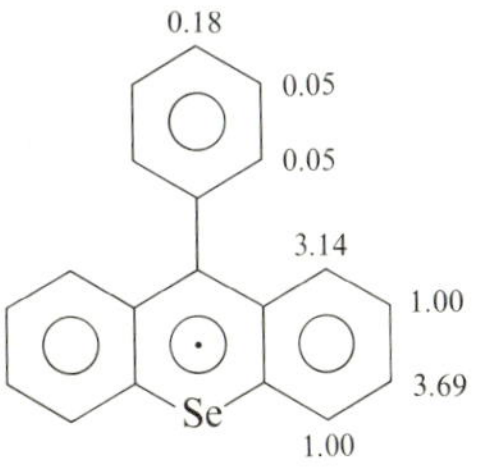

(**194**)

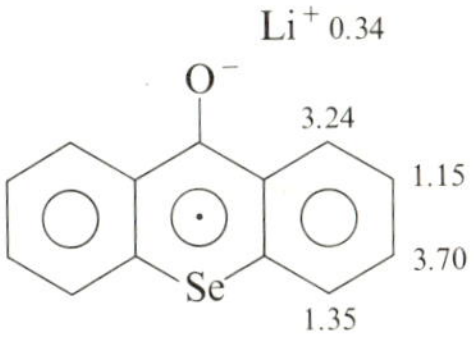

(**195**)

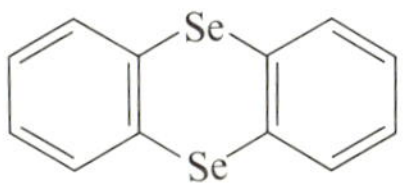

(**196**)

---

[583] M. Hori, T. Kataoka, and C.-F. Hsu, *Chem. Pharm. Bull.* **22**, 15 (1974).
[584] M. Hori, T. Kataoka, H. Shimizu, C.-F. Hsu, Y. Asahi, and E. Mizuta, *Chem. Pharm. Bull.* **22**, 32 (1974).
[585] J. Stackhouse, G. H. Senkler, B. E. Maryanoff, and K. Mislow, *J. Am. Chem. Soc.* **96**, 7835 (1974).

2. Radicals Containing a 1,4-Diselenin Ring

Cation-radicals of selenanthrene (**196**) are the only radicals in this category to have been observed. The ESR spectrum of the cation-radical of **196** itself, in concentrated sulfuric acid, shows no resolved proton hyperfine structure, which betokens a high spin population on Se.[301] Various MO calculations bear this out.[301,310,576]

The redox properties in solution of the cation-radical of **196** and its 2,3,7,8-tetramethoxy derivative have been described.[586,587] The methoxylated radical dimerizes in solution; its visible absorption spectrum, g-value, and ^{77}Se hyperfine splitting were reported.[587]

V. Radicals from Heterocycles with Mixed Group VI Heteroatoms

A. RADICALS CONTAINING A FIVE-MEMBERED HETEROCYCLE

Radicals Containing a Thiaselenole Ring

The only radicals to have been reported in this category are mentioned in work previously cited: the transient existence of **197** (X = S, Se) has been inferred from electrochemical measurements[132]; the cis and trans isomers of dithiadiselenafulvalene have been oxidized anodically to cation-radicals in acetonitrile and the electrochemical characteristics compared with those of tetrathia- and tetraselenafulvalene.[581]

B. RADICALS CONTAINING A SIX-MEMBERED HETEROCYCLE

1. Phenoxathiin Cation-Radicals

The cation-radical **198** of phenoxathiin and its derivatives have been studied in considerable detail. Various workers have measured ESR spectra over several years.[296,301,313,516,586,588–593] The hyperfine splittings indi-

---

[586] C. Barry, G. Cauquis, and M. Maurey, *Bull. Soc. Chim. Fr.*, 2510 (1966).
[587] A. W. Addison, T. H. Li, and L. Weiler, *Can. J. Chem.* **55**, 766 (1977).
[588] B. Lamotte, A. Rassat, and P. Servoz-Gavin, *C. R. Acad. Sci.* **255**, 1508 (1962).
[589] B. Lamotte, *Proc. Colloq. AMPERE* **12**, 282 (1963).
[590] U. Schmidt, K. Kabitzke, and K. Markau, *Chem. Ber.* **97**, 498 (1964).
[591] H. J. Shine and R. J. Small, *J. Org. Chem.* **30**, 2140 (1965).
[592] E. Volanschi and M. Hillebrand, *Rev. Roum. Chim.* **12**, 751 (1967).
[593] M. Hillebrand, O. Maior, V. E. Sahini, and E. Volanschi, *J. Chem. Soc. B*, 755 (1969).

cated in gauss in **198** are those of Sullivan[516] for solution in nitromethane. Papers already cited have considered **198** in MO calculations.[307,310,518]

Thermodynamic redox properties have also been given for **198** in previously cited papers,[294,319,539,586] and the rate of electron exchange between **198** and its parent heterocycle has been obtained by an electrochemical method[48]; the participation of **198** in electrochemiluminescent systems has been reported.[318]

The radical **198** parallels its thianthrene equivalent **178** in exhibiting an interesting chemical reactivity which has been explored by Shine's group.[522,523,554,556,557,591,594] In general, the radical is susceptible to attack at S by nucleophiles (cf. **180–182** for comparable products from **178**).

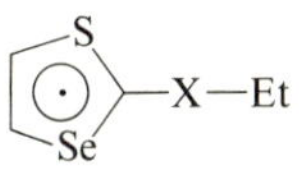

(197)

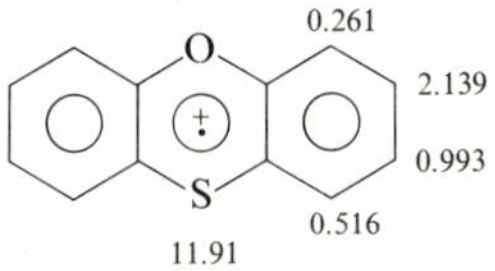

(198)

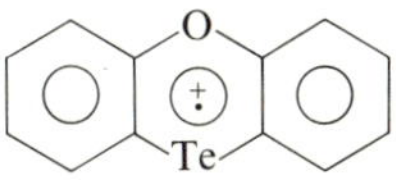

(199)

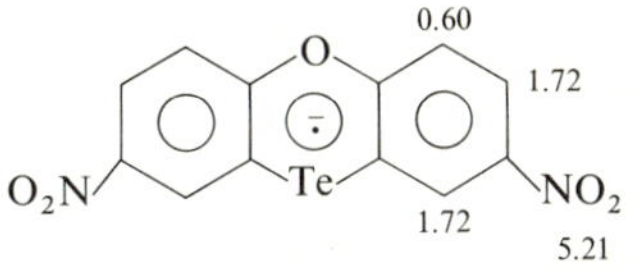

(200)

2. Phenoxaselenin Cation-Radicals

Phenoxaselenin is oxidized in acetonitrile at a platinum anode to give a cation-radical.[595] However, this radical gave a broad, featureless ESR spectrum from which only a ^{77}Se hyperfine splitting of 39.5 G could be measured.[301,576] The radical dimerizes to a diamagnetic product, a process which has been studied in acetonitrile[595] and in concentrated sulfuric acid.[596] The radical is also susceptible to nucleophilic attack at Se by water in these solvents.[595] Galasso's study of g-factors in group VI heterocyclic radicals included phenoxaselenin cation-radical.[310]

3. Phenoxatellurin Radicals

Bertier and Lamotte reported failure to observe the phenoxatellurin cation-radical **199** on dissolving the heterocycle in concentrated sulfuric

---

[594] S. R. Mani and H. J. Shine, *J. Org. Chem.* **40**, 2756 (1975).
[595] G. Cauquis and M. Maurey-Mey, *Bull. Soc. Chim. Fr.*, 291 (1973).
[596] M. Hillebrand, O. Maior, and V. E. Sahini, *Rev. Roum. Chim.* **16**, 1489 (1971).

acid.[301] Consistent with this, anodic oxidation of phenoxatellurin fails to give a radical observable by ESR; a diamagnetic dimer is rapidly formed which associates with a third unoxidized molecule; the perchlorate salt of the complex has been isolated.[597] The half-wave potential for the formation of **199** has been given[586] as has a calculation of its expected g-value.[310]

Anion-radicals have been obtained from 2-nitro- and 2,8-dinitrophenoxatellurin (e.g., **200**) for which hyperfine splittings are given in gauss.[598] As expected, the spin population is strongly associated with one nitro group and the carbocycle it substitutes. In the mono-nitro radical splittings are not observed for the unsubstituted ring. In **200** the ESR spectrum shows a solvent-dependent linewidth alternation explicable in terms of electron exchange between the two nitro sites.

VI. Radicals from Heterocycles with Group V and Group VI Heteroatoms

A. Radicals Containing a Five-Membered Heterocycle

1. *Radicals Containing Oxazole and Thiazole Rings*

Anion-radicals generated from oxazole and isoxazole in an argon matrix were shown, like their counterpart from furan, to have undergone ring scission.[33] Consistent with this, various attempts to reduce isoxazoles polarographically in both protic and aprotic media have also resulted in the loss of ring integrity, and indeed the reduction process has always involved two electrons.[599-601] However, electrochemical reduction of 2,5-diaryloxazoles has been shown to give anion-radicals as the initial product,[602-606] although the presence of proton sources may change

---

[597] G. Cauquis and M. Maurey-Mey, *Bull. Soc. Chim. Fr.*, 2870 (1973).
[598] A. Gioba, V. E. Sahini, and E. Volanschi, *J. C. S. Perkin II*, 529 (1977).
[599] H. Lund and A. D. Thomsen, *Acta Chem. Scand.* **23**, 3567 (1969).
[600] I. G. Markova, M. K. Polievktov, and S. D. Sokolov, *Zh. Obshch. Khim.* **46**, 398 (1976) [*CA* **84**, 120832 (1976)].
[601] R. N. Goyal and R. Jain, *J. Electroanal. Chem. Interfacial Electrochem.* **79**, 407 (1977).
[602] W. N. Greig and J. W. Rogers, *J. Electrochem. Soc.* **117**, 1141 (1970).
[603] S. L. Smith, L. D. Cook, and J. W. Rogers, *J. Electrochem. Soc.* **119**, 1332 (1972).
[604] N. P. Shimanskaya, L. A. Kotok, T. F. Alekhina, and V. D. Bezuglyi, *Zh. Obshch. Khim.* **43**, 1445 (1973) [*CA* **80**, 22104 (1974)].
[605] N. P. Shimanskaya and S. E. Kovalev, *Zh. Obshch. Khim.* **43**, 2355 (1973) [*CA* **80**, 59202 (1974)].
[606] N. P. Shimanskaya, L. A. Kotok, B. M. Krasovitskii, L. D. Shcherbak, and T. F. Alekhina, *Zh. Obshch. Kkim.* **46**, 2107 (1976) [*CA* **85**, 200076 (1976)].

this.[607–610] The anion-radicals from 2,5-diphenyloxazole and similar substances are sufficiently persistent for use in various electrochemiluminescent systems.[20,24,611–613]

Thiazole gives a transient radical in aqueous solution upon pulse radiolysis.[614] The radical formed under neutral conditions is the 3-hydrothiazolyl radical (**201**) for which electronic absorption spectra, acid–base characteristics, and second-order decay kinetics were measured. A derivative of this, the radical from thiamine (vitamin B_1) (**202**) was also characterized. A role is implied for related radicals (e.g., **203**) during the electrochemical dimerization reported for benzothiazolium salts.[615]

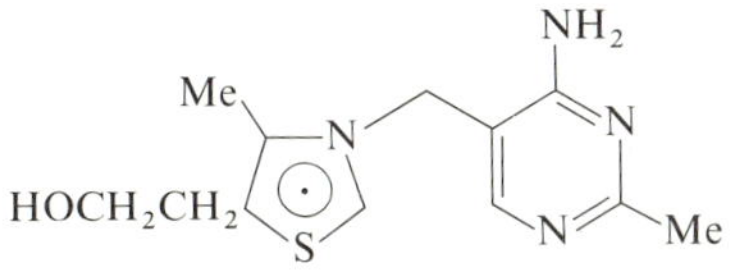

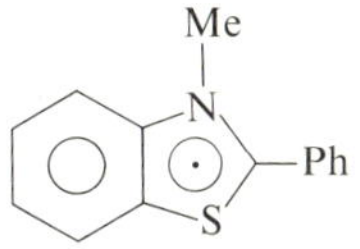

Various thiazole anion-radicals have been characterized by ESR. Tordo and co-workers have generated several nitrothiazole anion radicals (e.g., **204** and **205**) by reduction of the precursors with glucose in methanol containing

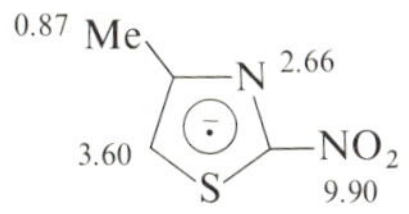

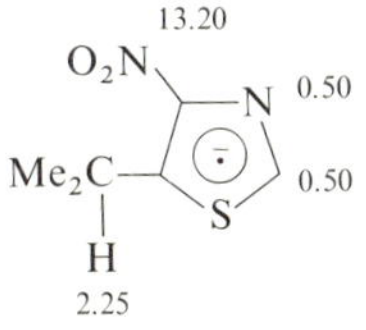

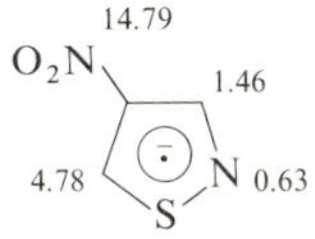

[607] V. D. Bezuglyi and N. P. Shimanskaya, *Zh. Obshch. Khim.* **31**, 3160 (1961) [*CA* **57**, 581 (1962)].
[608] V. D. Bezuglyi, N. P. Shimanskaya, and E. M. Peresleni, *Zh. Obshch. Khim.* **34**, 3540 (1964) [*CA* **62**, 8983 (1965)].
[609] C. Makkay, F. Makkay, and M. Ionescu, *Stud. Univ. Babes-Bolyai, Ser. Chem.* **15**, 71 (1970).
[610] F. Makkay, C. Makkey, and M. Ionescu, *Stud. Univ. Babes-Bolyai, Ser. Chem.* **15**, 119 (1970).
[611] F. Pragst, G. Fabian, R. Ziebig, D. Schmidt, and W. Jugelt, *Chem. Phys. Lett.* **36**, 630 (1975).
[612] L. M. Podgornaya and V. P. Leonov, *Zh. Prikl. Spektrosk.* **25**, 1006 (1976) [*CA* **86**, 98338 (1977)].
[613] L. M. Podgornaya, V. P. Leonov, V. I. Grigor'eva, L. P. Snagoshchenko, and R. N. Nurmukhametov, *Zh. Prikl. Spektrosk.* **26**, 285 (1977) [*CA* **86**, 163017 (1977)].
[614] P. N. Moorthy and E. Hayon, *J. Org. Chem.* **42**, 879 (1977).
[615] J. Nakaya, S. Kato, Y. Morimoto, and E. Imoto, *Bull. Univ. Osaka Prefect., Ser. A* **21**, 151 (1972).

strong base[616,617]; the authors made use of their nonlinear regression method in the analysis of the complex spectrum exhibited by **204** (hyperfine splittings are indicated in gauss).[616] They found that the magnitude of the hyperfine splitting from the substituent nitrogen depends critically on the ring position substituted. The order of magnitudes of $a(N)_{NO_2}$ was $4 > 5 > 2$, implying the reverse order for the effectiveness of the delocalization of the unpaired electron (cf. the different effectiveness of the thiophene ring positions in delocalizing spin; Section III,2 and 3). In addition, it was found that the larger alkyl groups adopt preferred conformations: the methine proton splitting in **205** is comparatively small, for example, since the isopropyl group is oriented in conformations in which the dihedral angles subtended between the bond to the methine proton and the $2p_z(\pi)$ orbital on C-5 are close to 90° (see Part I: Section II,B,4).

Pedulli and co-workers have also studied anion-radicals in the thiazole system, including those from 2-nitrothiazole, 2,2′-bithiazolyl, 2-arylthiazoles, and bis(2-thiazolyl) ketone. Here the anion-radicals were generated by alkali-metal reduction in ethereal solvents and found to exist in contact ion-pairs. Whenever the organic component contained two adjacent binding sites, chelation of the counterion occurred and was revealed by examining the variation of its hyperfine splitting with temperature and solvent. The interaction of the ion-pairs with macrocyclic polyethers was also studied by ESR for certain of the chelating radicals.[386]

Greenstock et al.[72] have measured ESR spectra for the 4-nitroisothiazole anion radical (**206**) and related species under conditions of pulse radiolysis in neutral aqueous solution. (Hyperfine splittings for **206** are given in gauss.) These authors also reported a polarographic investigation of the same set of materials and related ESR and polarographic characteristics to biological properties.[73] A transient existence has been suggested for 3-methylisothiazole anion-radical during the reaction of the exciplex formed on photolysis of the heterocycle in the presence of amines.[619]

Polarographic reduction of oxazole and thiazole aldehydes in alkaline conditions has been reported to give dimeric products, presumably by coupling of intermediate ketyls.[620]

---

[616] F. Humm, R. Romanetti, P. Tordo, L. Bouscasse, and R. Phan-Tan-Luu, *Org. Magn. Reson.* **5**, 365 (1973).

[617] P. Tordo, G. Pouzard, A. Babadjamian, H. J. M. Dou, and J. Metzger, *Nouv. J. Chim.* **1**, 493 (1977).

[618] G. F. Pedulli, P. Zanirato, A. Alberti, and M. Tiecco, *J. C. S. Perkin II*, 293 (1975).

[619] A. Lablache-Combier and A. Pollet, *Tetrahedron* **28**, 3141 (1972).

[620] I. Schwartz, R. D. Pop, H. Demian, A. Muresan, E. Chindris, and I. Simiti, *J. Electroanal. Chem. Interfacial Electrochem.* **59**, 209 (1975).

Various oxazole and thiazole nitroxyls have been prepared. Torssell photolyzed 2-iodothiazole and 2-iodobenzothiazole in the presence of tBuNO and trapped the appropriate 2-σ radicals as the nitroxyls **207** and **208**.[621] The hyperfine splittings indicated in gauss for **207** are due to Torssell. Those for **208** are due to Sutcliffe and co-workers[622] who generated the radical by oxidation of benzothiazole-2-hydrazine with HgO in benzene and achieved rather better resolution. The nitroxyl **209** was obtained by photolysis of 2-nitrothiazole in triethylsilane[618] (hyperfine splittings indicated in gauss); Sleight and Sutcliffe obtained a similar species, with an alkoxy rather than silyloxy substituent, by photolysis of the 2-nitrothiazole in tetrahydrofuran.[86a] The benzoxazole nitroxyl analogous to **208** was synthesized by Aurich and co-workers and nitrogen hyperfine splittings reported.[623,623a]

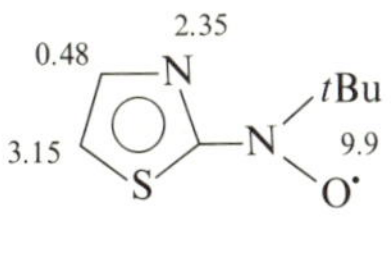

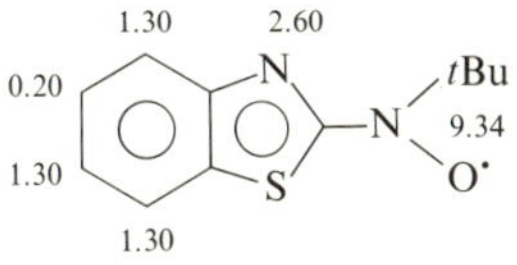

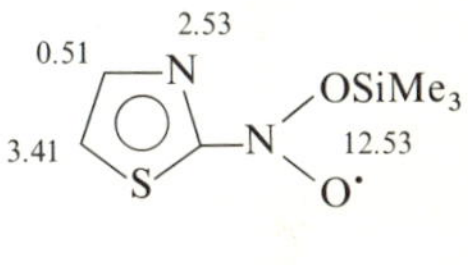

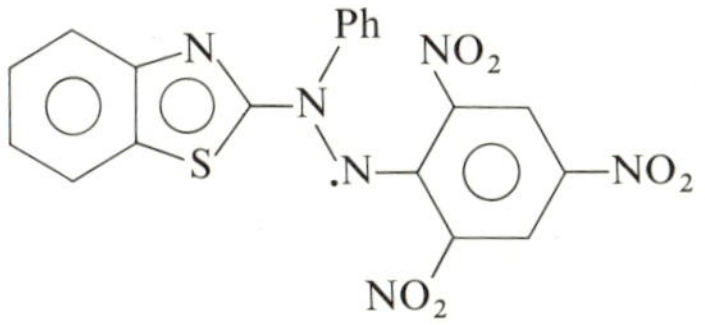

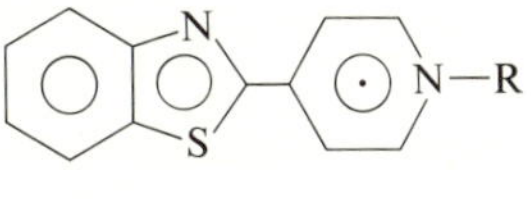

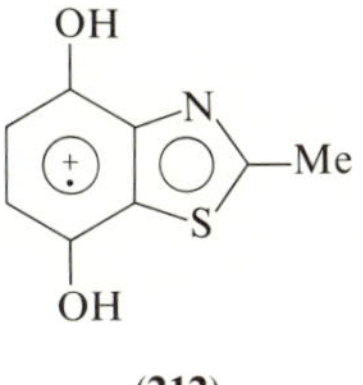

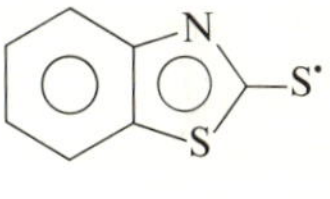

[621] K. Torssell, *Tetrahedron* **26**, 2759 (1970).
[622] T. W. Bentley, J. A. John, R. A. W. Johnstone, P. J. Russell, and L. H. Sutcliffe, *J. C. S. Perkin II*, 1039 (1973).
[623] H. G. Aurich, A. Lotz, and W. Weiss, *Chem. Ber.* **106**, 2845 (1973).
[623a] H. G. Aurich and K. Kabs, *Angew. Chem., Int. Ed. Engl.* **9**, 636 (1970).

Soviet workers have reported hydrazyl radicals such as **210**[624] and 1-alkyl-1-hydropyridinyl radicals **211**[625,626] containing benzothiazolyl or benzoxazolyl moieties, both types being persistent species.

The electron-transfer reactions which form **212** when the corresponding heterocycle is oxidized by various metal complexes in perchloric acid have been studied by a stopped-flow technique and the results interpreted in terms of Marcus theory.[627] The reaction forming the corresponding o-semiquinone was also studied.

Electrochemical oxidation of 2-mercaptobenzothiazole proceeds via the radical **213** which dimerizes to the corresponding disulfide; this electrochemical reaction is of synthetic value in preparing the disulfide.[628]

2. Violene Radicals from Oxazoles, Thiazoles, and Selenazoles

Deuchert and Hünig have recently reviewed violenes generally[629]; the review is pertinent to the limited range of violenes [e.g., **214–216** (X = O, S, R = alkyl)] considered in this section. Hünig and co-workers described biazoles which form persistent cation-radicals of type **214**[457,630,631]; syntheses of the parent heterocycles as well as their polarographic properties, the formation constants of the radicals, and their electronic absorption properties were given. Baldwin and co-workers[632] have reported an investigation of the mechanism whereby **217** rearranges thermally to **218**, concerted [1,3]-sigmatropic rearrangement being symmetry-forbidden. A radical mechanism was proved by the observation of cross-over products in a rearrangement of mixed deuterobenzyl and nondeuterated **217**. The intermediate radical is analogous to **214** (X = S) but lacking in one alkyl group and consequently, of course, the cationic charge.

---

[624] R. O. Matevosyan, N. I. Abramova, Yu, A. Abramov, V. N. Yakovleva, A. K. Chirkov, L. A. Perelyaeva, V. A. Gubanov, V. I. Koryakov, and O. B. Donskikh, *Khim. Geterotsikl. Soedin.* **7**, 462 (1971) [*CA* **76**, 24503 (1972)].

[625] V. Kadis, Ya. P. Stradins, E. Lavrinovich, and P. Zarins, *Khim. Geterotsikl. Soedin.*, 675 (1975) [*CA* **83**, 78165 (1975)].

[626] V. Kadis, E. Lavrinovich, P. Zarins, and Ya. P. Stradins, *Nov. Elektrokhim. Org. Soedin., Vses. Soveshch. Elektrokhim. Org. Soedin., Tezisy Dokl., 8th, 1973*, 112 (1973) [*CA* **82**, 66062 (1975)].

[627] E. Pelizzetti, E. Mentasti, and E. Barni, *J. C. S. Perkin II*, 623 (1978).

[628] H. Berge, H. Millat, and B. Straebing, *Z. Chem.* **15**, 37 (1975).

[629] K. Deuchart and S. Hünig, *Angew. Chem., Int. Ed. Engl.* **17**, 875 (1978).

[630] S. Hünig, D. Scheutzow, H. Schlaf, and H. Quast, *Justus Liebigs Ann. Chem.* **765**, 110 (1972).

[631] S. Hünig, D. Scheutzow, and H. Schlaf, *Justus Liebigs Ann. Chem.* **765**, 126 (1972).

[632] J. E. Baldwin, S. E. Branz, and J. A. Walker, *J. Org. Chem.* **42**, 4142 (1977).

Hünig's group studied violene radicals (e.g., **215**) derived from azines of thiazolin-2-ones and their benzo analogs[633–636]; azines of benzoxazolin-2-ones and benzoselenazolin-2-ones have also been examined.[135,637] Again, synthetic, polarographic, and spectroscopic data were given. Janata and Williams[638] have also investigated the electrochemistry of 2,2′-benzothiazolinone azines. They found that the apparent formation constants of the cation-radicals, used for characterizing their stability, depend on the acidity of the medium owing to protonation of both the neutral azine and the dication corresponding to it. The latter was found to decompose

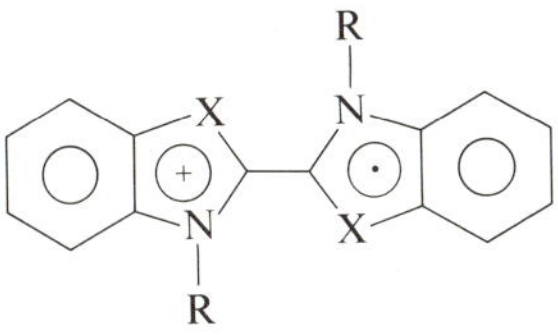

(214)

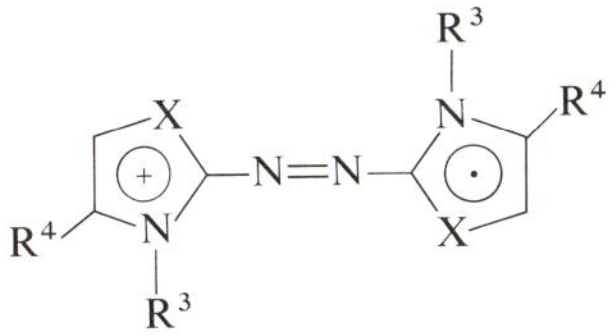

(215)

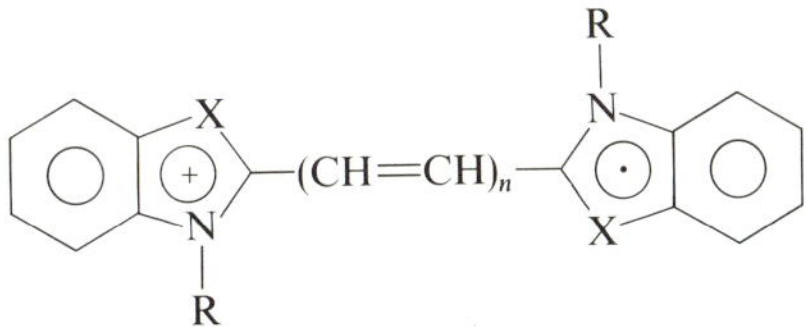

(216)

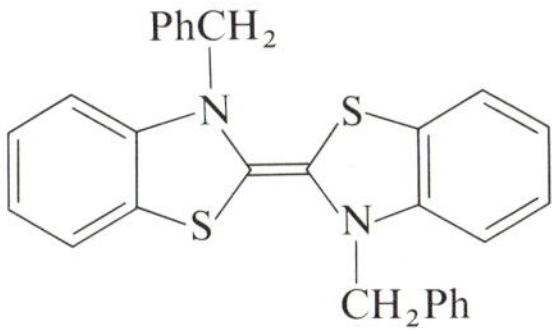

(217)

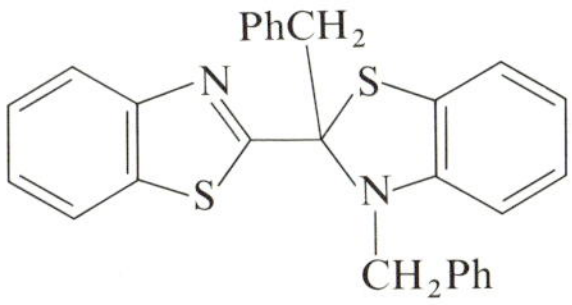

(218)

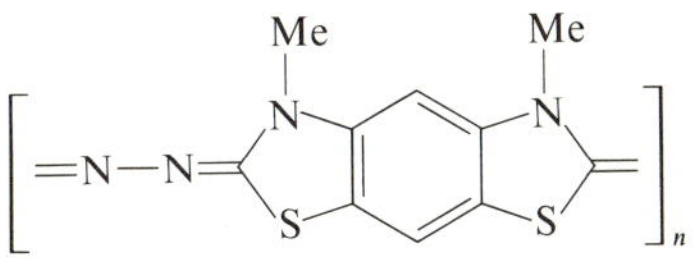

(219)

[633] S. Hünig, H. Balli, H. Conrad, and A. Schott, *Justus Liebigs Ann. Chem.* **676**, 36 (1964).
[634] S. Hünig, H. Balli, H. Conrad, and A. Schott, *Justus Liebigs Ann. Chem.* **676**, 52 (1964).
[635] S. Hünig and G. Sauer, *Justus Liebigs Ann. Chem.* **748**, 173 (1971).
[636] S. Hünig and G. Sauer, *Justus Liebigs Ann. Chem.* **746**, 189 (1971).
[637] S. Hünig, G. Kiesslich, F. Linhart, and H. Schlaf, *Justus Liebigs Ann. Chem.* **752**, 182 (1971).
[638] J. Janata and M. B. Williams, *J. Phys. Chem.* **76**, 1178 (1972).

in an autocatalytic reaction which is also acid-dependent. The work of Zahradnik and co-workers[422,485] relating violene radical formation constants and other properties to quantum-mechanical indices involved violenes of the present type.

The properties of benzothiazolinone azine redox systems have been studied with a view to their practical applications: Shelepin and co-workers studied a mixed system involving 3-ethylbenzothiazolin-2-one azine and methyl viologen as an electrochromic system activated by optically transparent electrodes[639]; Sharp has described perchlorate-sensitive electrodes also involving 3-ethylbenzothiazolin-2-one azine[640,641]; solid-state electrical properties of salts of the 3-methylbenzothiazolin-2-one azine radical-cation have been reported.[642]

The synthesis and properties of polymeric thiazolobenzothiazolinone azines (e.g., **219**) have been described[643,644]; oxidation leads to polymeric cation radicals.

Hünig and co-workers have described violene systems where the benzothiazole or benzoxazole parts are separated by a polyene chain[645,646]; the consequence for the formation constants of the radical-cation (e.g., **216**) of variation in n was studied. The consequences of aza substitution in the chain were also studied. In general, extension of the conjugation destabilizes the radical-cation relative to the upper and lower oxidation states.[485] (See also Part I: Section II,B,2.)

3. Radicals from Oxadiazoles, Thiadiazoles, and Selenadiazoles

a. *1,3,4-Oxadiazole Radicals.* Although polarographic reduction of 1,3,4-oxadiazoles in protic media results in the transfer of a total of six electrons in two waves corresponding to the attachment of first two and then four electrons,[608,647] one-electron reduction of 2,5-diaryl-1,3,4-oxadiazoles occurs in DMF.[648] The ESR spectrum of the anion-radical **220** of 2,5-

[639] I. V. Shelepin, O. A. Ushakov, N. I. Karpova, and V. A. Barachevskii, *Elektrokhimiya* **13**, 32 (1977) [*CA* **86**, 196919 (1977)].
[640] M. Sharp, *Anal. Chim. Acta* **62**, 385 (1972).
[641] M. Sharp, *Anal. Chim. Acta* **65**, 405 (1973).
[642] W. A. Barlow, G. R. Davies, E. P. Goodings, R. L. Hand, G. Owen, and M. Rhodes, *Mol. Cryst. Liq. Cryst.* **32**, 193 (1976).
[643] G. Manecke and J. Kautz, *Makromol. Chem.* **172**, 1 (1973).
[644] G. Manecke and J. Kautz, *Tetrahedron Lett.*, 629 (1972).
[645] S. Hünig, D. Scheutzow, H. Schlaf, and A. Schott, *Justus Liebigs Ann. Chem.*, 1423 (1974).
[646] S. Hünig, D. Scheutzow, H. Schlaf, and H. Pütter, *Justus Liebigs Ann. Chem.*, 1436 (1974).
[647] N. P. Shimanskaya and V. D. Bezuglyi, *Zh. Obshch. Khim.* **33**, 1726 (1963) [*CA* **59**, 9594 (1963)].
[648] G. L. Smith and J. W. Rogers, *J. Electrochem. Soc.* **118**, 1089 (1971).

Sec. VI.A] HETEROAROMATIC RADICALS, PART II 119

diphenyl-1,3,4-oxadiazole has been recorded. The assignment of hyperfine splittings, indicated in **220** in gauss, is based on MO calculation.[649] The authors offer no comment on the ortho proton splittings' being smaller than those for meta protons, which seems surprising, especially since it is inferred that the radical is close to planar. The consequences of variation of the aryl substituents on the polarographic properties of 2,5-diaryl-1,3,4-oxadiazoles and of the addition of proton donors to the solvent have been elucidated by both American and Soviet research groups.[603–605,648–651]

Several electrochemiluminescent systems have been described which involve **220**.[14,19,565,609,652]

Pirkle and Gravel[653,654] have described the synthesis of 3-substituted 1,3,4-oxazolidine-2,5-diones and the preparation from these of heterocyclic π-radicals (e.g., **221**) by oxidation of a solution in toluene with lead dioxide. The radicals are best considered as cyclic acylhydrazide radicals; they are relatively persistent, decomposing over 48 h in solution, but are not isolable.

(220)   (221)

(222)   (223)

b. *1,2,5-Oxadiazole (Furazan), -Thiadiazole, and -Selenadiazole Radicals and Their Annelated Derivatives.* A recent investigation of the chlorination of the alkyl groups in 3,4-dimethyl-1,2,5-oxadiazole has been reported.[655] The process is a chain reaction and involves **222** and its

---

[649] A. V. Il'yasov, Yu. M. Kargin, Ya. A. Levin, I. D. Morosova, A. A. Vafina, B. V. Mel'nikov, A. Sh. Mukhtarov, M. S. Skorobogatova, and E. I. Zoroatskaya, *Izv. Akad. Nauk SSSR, Ser. Khim.*, 2194 (1975) [*CA* **84**, 66726 (1976)].

[650] G. G. Kryukova, O. A. Yasimskii, V. A. Ustinov, G. S. Mironov, and Yu. E. Shapiro, *Nov. Polyarogr., Tezisy Dokl. Vses. Soveshch. Polyarogr.*, *6th, 1975*, 75 (1975) [*CA* **86**, 10020 (1977)].

[651] O. A. Yasimskii, V. A. Ustinov, V. V. Kopeikin, V. V. Plakhtinskii, E. R. Kofanov, G. G. Kryukova, and G. S. Mironov, *Zh. Obshch. Khim.* **47**, 211 (1977) [*CA* **86**, 155034 (1977)].

[652] M.-M. Chang, T. Saji, and A. J. Bard, *J. Am. Chem. Soc.* **99**, 5399 (1977).

[653] W. H. Pirkle and P. L. Gravel, *J. Org. Chem.* **41**, 3763 (1976).

[654] W. H. Pirkle and P. L. Gravel, *J. Org. Chem.* **42**, 1367 (1977).

[655] I. V. Vigalok, A. V. Ostrovskaya, G. G. Petrova, and Ya. A. Levin, *Zh. Org. Khim.* **14**, 1255 (1978) [*CA* **89**, 107402 (1978)].

partially chlorinated congeners as propagating radicals; relative reactivities of the different radicals were estimated. The same furazan undergoes polarographic reduction in DMF to form an anion-radical **223** as the first step in an overall six-electron reduction process.[656] Subsequent work has shown that 1,2,5-thiadiazoles and -selenadiazoles with various substituents in the 3- and 4-positions undergo comparable one-electron reductions in aprotic solvents.[657,658] The polarography of these heterocycles in aqueous solutions of differing pH has been investigated as has the consequence for the polarographic behavior of autoprotonation occurring in materials with acidic substituents.[659,660] No ESR results appear to have been reported for anion-radicals formed by these monocyclic diazoles.

Attempts to reduce furoxans (e.g., 3,4-dimethyl-1,2-5-oxadiazole 2-oxide) to anion-radicals led to paramagnetic products of indeterminate structure[661] or multielectron reduction products.[656]

Annelation of 1,2,5-X-diazole anion-radical structures stabilizes the radicals and gives species which are persistent in the absence of oxygen. The anion-radical **224** of benzo-2,1,3-oxadiazole was reported incidentally by Russell and co-workers in 1964[662]; a more detailed study was published the following year of **224** together with **225** and **226**.[663] The hyperfine splittings indicated for **224–226** in gauss are those of Atherton and co-workers[664] who improved the resolution of the spectra. Both groups agreed in the assignment of the larger proton splittings to the 4(7)-positions following MO calculations. Strom and Russell[663] chose a d-model in describing the bonding of S and Se although a p-model gave a similar order of hyperfine splittings. Their choice was made not so much for the d-model's success in describing the radicals, but rather because the optimum parameters evaluated from the study of the radicals then gave good account of the electron densities in the parent heterocycles. By contrast, Atherton and co-workers[664] selected the p-model after finding it adequate in describing the distribution

---

[656] E. S. Levin, Z. I. Fodiman, and Z. V. Todres, *Elektrokhimiya* **2**, 175 (1966) [*CA* **65**, 5022 (1966)].
[657] V. Sh. Tsveniashvili, V. M. Gaprindashvili, L. A. Tskalobadze, and V. A. Sergeev, *Zh. Obshch. Khim.* **43**, 2122 (1973) [*CA* **80**, 43444 (1974)].
[658] E. O. Sherman, S. M. Lambert and K. Pilgram, *J. Heterocycl. Chem.* **11**, 763 (1974).
[659] V. Sh. Tsveniashvili, V. N. Gaprindashvili, L. A. Tskalobadze, and V. A. Sergeev, *Zh. Obshch. Khim.* **42**, 2044 (1972) [*CA* **78**, 23219 (1973)].
[660] V. Sh. Tsveniashvili, V. M. Gaprindashvili, and L. A. Tskalobadze, *Elektrokhimiya* **11**, 523 (1975) [*CA* **83**, 67923 (1975)].
[661] S. P. Solodovnikov and Z. V. Todres, *Khim. Geterotsikl. Soedin.*, 811 (1967) [*CA* **68**, 114511 (1968)].
[662] G. A. Russell, E. G. Janzen, and E. T. Strom, *J. Am. Chem. Soc.* **86**, 1807 (1964).
[663] E. T. Strom and G. A. Russell, *J. Am. Chem. Soc.* **87**, 3326 (1965).
[664] N. M. Atherton, J. N. Ockwell, and R. Dietz, *J. Chem. Soc. A*, 771 (1967).

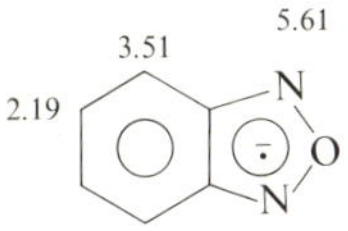

of the spin population in **224**–**226** and naphtho derivatives and also the polarographic properties of these heterocycles.

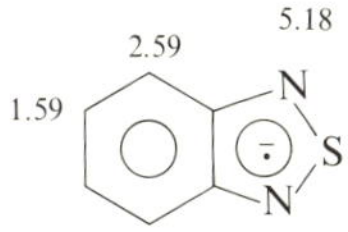

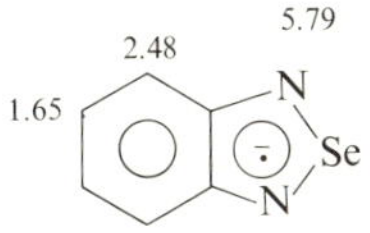

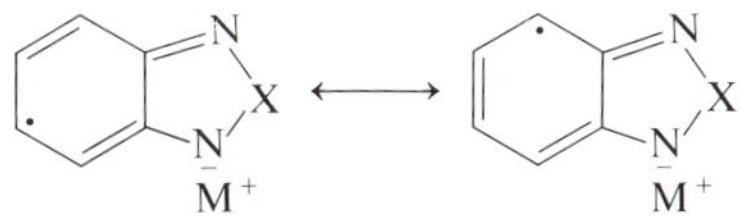

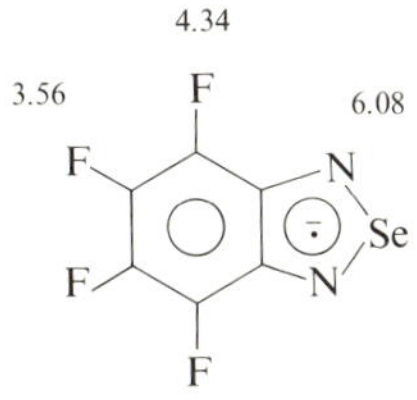

Solodovnikov and Todres[665] confirmed the assignment of proton hyperfine splittings made above for **225** and **226** by examining their chlorinated derivatives. Kamiyama and Akahori[666] studied the variation of the spin distribution in **225** and **226** effected by changes in solvent character. They generated the radicals by alkali-metal reductions and found that solvents which were less effective at solvating the counterions were associated with lower nitrogen hyperfine splittings, an observation interpreted in terms of increased weightings in such solvents for structures such as **227a** and **b**. The ^{77}Se hyperfine splitting in **226** has been found to be 4.9 G.[576]

Fajer et al.[667,668] have measured ESR parameters for the perfluorinated benzoselenadiazole anion-radical (**228**), together with electronic absorption spectra for this radical and for **225**, **226**, methylated derivatives, and isomeric naphthoselenadiazole anion-radicals. In this work it was shown that the radicals have no tendency to dimerize, that oxygen oxidizes the radicals back to their diamagnetic precursors with 80–90% efficiency, and that a p-model is, on balance, the best for describing the optical transitions of the radicals as well as their spin distributions.

A number of entirely theoretical papers have reported calculations on the anion-radicals discussed in this section[418,419,577,669,670]; the last includes calculations for the corresponding cation-radicals.

---

[665] S. P. Solodovnikov and Z. V. Todres, *Khim. Geterotsikl. Soedin.*, 360 (1968) [*CA* **69**, 72719 (1968)].
[666] M. Kamiya and Y. Akahori, *Bull. Chem. Soc. Jpn.* **43**, 268 (1970).
[667] J. Fajer, *J. Phys. Chem.* **69**, 1773 (1965).
[668] J. Fajer, B. H. J. Bielski, and R. H. Felton, *J. Phys. Chem.* **72**, 1281 (1968).
[669] N. K. Ray and P. T. Narasimhan, *Indian J. Chem.* **7**, 97 (1969).
[670] A. M. Gyul'maliev, I. V. Stankevich, and Z. V. Todres, *Khim. Geterotsikl. Soedin.*, 1055 (1975) [*CA* **84**, 30159 (1976)].

Besides the common alkali-metal reduction method for the production of anion-radicals, other electron-transfer reagents have been used for the production of radicals relevant in this section. Todres and co-workers have used cyclooctatetraene dianion[671,672] and have examined the redox equilibria of this reductant and various substituted substrates.[673] Radical **225** has also been produced by reduction of the precursor with organometallic reagents in the presence of transition-metal ions.[572]

Parallel with ESR studies, polarographic investigations of benzoxadiazoles and their sulfur and selenium isologs and derivatives have been carried out. Consistent with the ESR results, the common finding is that, in aprotic solvents, anion-radicals are formed in the first reversible reduction step.[656,658,661,664,674–679] Further reduction of the radicals leads to rupture of the heterocycle with ultimate formation of *o*-phenylenediamine via intermediate quinonoid products.[674,677] The group VI atom is expelled as hydroxide, sulfide, or selenide. It is through reaction with these inorganic anions that metal ions such as Cu^{2+} or Cd^{2+} influence the course of the polarography of 2,1,3-benzoxa-, -thia-, and -selenadiazoles.[680] Polarography of the heterocycles in aqueous media has been investigated as also have specific substituent effects such as autoprotolysis and halogen bond cleavage.[678,679,681–684] Single-electron reduction is not necessarily involved

---

[671] D. N. Kursanov and Z. V. Todres, *Dokl. Akad. Nauk SSSR* **172**, 1086 (1967) [*CA* **67**, 6148 (1967)].

[672] Z. V. Todres, V. Sh. Tsveniashvili, S. I. Zhdanov, and D. N. Kursanov, *Dokl. Akad. Nauk SSSR* **181**, 906 (1968) [*CA* **69**, 105759 (1968)].

[673] Z. V. Todres, Yu. I. Lyakhovetskii, and D. N. Kursanov, *Izv. Akad. Nauk SSSR, Ser. Khim.*, 1455 (1969) [*CA* **71**, 112131 (1969)].

[674] S. I. Zhdanov, V. Sh. Tsveniashvili, and Z. V. Todres, *J. Polarogr. Soc.* **13**, 100 (1967) [*CA* **68**, 118811 (1968)].

[675] V. Sh. Tsveniashvili, Z. V. Todres, and S. I. Zhdanov, *Zh. Obshch. Khim.* **38**, 1888 (1968) [*CA* **70**, 2985 (1969)].

[676] W. R. Fawcett, P. A. Forte, R. O. Loutfy, and J. M. Prokipcak, *Can. J. Chem.* **50**, 263 (1972).

[677] V. Sh. Tsveniashvili, Z. V. Todres, and S. I. Zhdanov, *Zh. Obshch. Khim.* **38**, 1894 (1968). [*CA* **70**, 3959 (1969)].

[678] V. Sh. Tsveniashvili, L. A. Tskalobadze, V. N. Gaprindashvili, and F. S. Mikhailitsyn, *Izv. Akad. Nauk Gruz. SSR, Ser. Khim.* **1**, 174 (1975) [*CA* **83**, 199441 (1975)].

[679] V. Sh. Tsveniashvili, *Izv. Akad. Nauk Gruz. SSR, Ser. Khim.* **2**, 86 (1976) [*CA* **85**, 132792 (1976)].

[680] V. Sh. Tsveniashvili, V. N. Gaprindashvili, and N. S. Khavtasi, *Zh. Obshch. Khim.* **42**, 2049 (1972) [*CA* **78**, 23232 (1973)].

[681] V. Sh. Tsveniashvili, S. I. Zhdanov, and Z. V. Todres, *Fresenius' Z. Chem.* **224**, 389 (1967) [*CA* **66**, 52057 (1967)].

[682] V. Sh. Tsveniashvili, S. I. Zhdanov, and Z. V. Todres, *Khim. Geterotsikl. Soedin.* **4**, 712 (1969) [*CA* **70**, 25181 (1969)].

[683] V. Sh. Tsveniashvili, *Zh. Obshch. Khim.* **43**, 1203 (1973) [*CA* **79**, 66263 (1973)].

[684] V. Sh. Tsveniashvili, L. A. Tskalobadze, and V. N. Gaprindashvili, *Zh. Obshch. Khim.* **45**, 1090 (1975) [*CA* **83**, 87283 (1975)].

in these processes. The electrochemical reduction of benzofuroxan (2,1,3-benzoxadiazole 2-oxide) in aqueous solutions has been described without implication of radical intermediates although other intermediates such as *o*-benzoquinone dioxime and diimine were recognized.[685]

Single-electron reduction of benzotrifurazan gave the anionic π-radical **229** whose ESR spectrum consisted of 13 equally spaced lines due to hyperfine splitting from six equivalent nitrogens; by contrast, the radical obtained from benzotrifuroxan exhibited a 5-line spectrum due to coupling of the unpaired electron with only two equivalent nitrogen nuclei.[686] The σ-radical structure **230** was proposed. (Hyperfine splittings in **229** and **230** are indicated in gauss.)

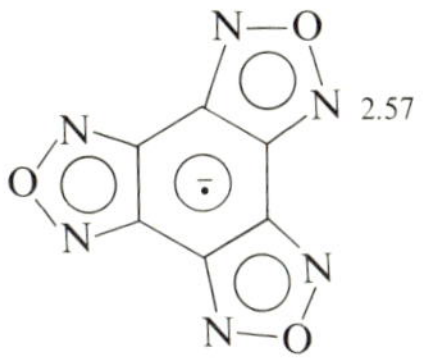

(229) 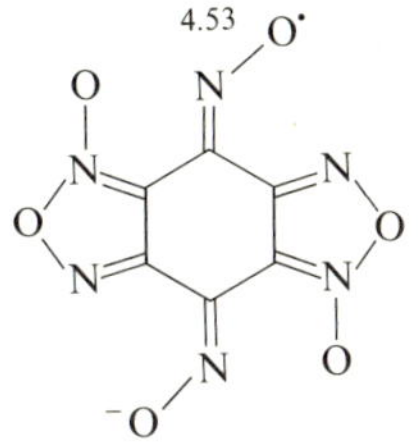   (230)

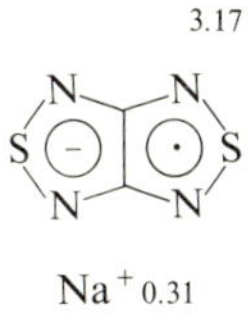

(231) Na$^+$ 0.31   (232) K$^+$ 0.15

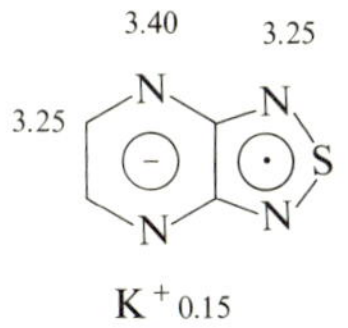

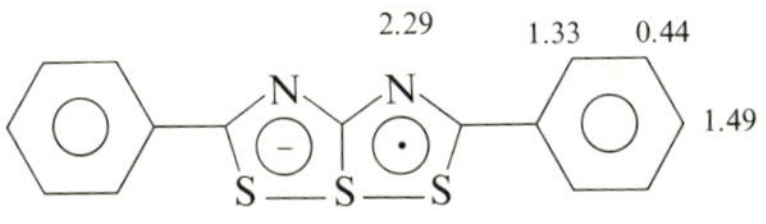

(233)

Kochi and co-workers[687] have prepared the anion-radical **231** of [1,2,5]thiadiazolo[3,4-*c*][1,2,5]thiadiazole. Also prepared were the anion-radicals of [1,2,5]thiadiazolo[3,4-*b*]pyrazine (**232**) and [1,2,5]thiadiazolo[3,4-*b*]quinoxaline. Hyperfine splittings are indicated for **231** and **232** in gauss. The radicals were prepared by alkali-metal reduction in THF; the observation of counterion splittings indicates their occurrence in ion-pairs. Radical **231** is of interest from the point of view of the nature of the

---

[685] C. D. Thompson and R. T. Foley, *J. Electrochem. Soc.* **119**, 117 (1972).
[686] A. S. Bailey, C. J. W. Gutch, J. M. Peach, and W. A. Waters, *J. Chem. Soc. B*, 681 (1969).
[687] C. L. Kwan, M. Carmack, and J. K. Kochi, *J. Phys. Chem.* **80**, 1786 (1976).

bonding of sulfur. Unless charge-separated structures are drawn for the parent heterocycle, S(IV) structures, implying d-orbital participation, must be drawn. The authors found that a d-model of the bonding gave good agreement with the experimental data for **231** and that the same set of MO parameters gave good account of the spin distribution in the whole family of radicals considered. The improvement over the p-model was only marginal, however (cf. the findings for the radicals from **96**; Section II,B,1 and 3).

4. *2,5-Diphenyl[1,2,4]dithiazolo[1,5-b][1,2,4]dithiazole Anion-Radical*

Gerson and co-workers[444] some years ago reported an analysis of the ESR spectrum of the anion-radical **233** of 2,5-diphenyl[1,2,4]dithiazolo-[1,5-*b*][1,2,4]dithiazole. Some of the work reported at the same time concerning anion-radicals of analogous 6a-thiathiophthenes was subsequently reinterpreted.[443] However, the inference from footnote 7 in Gerson *et al.*[443] is that the results for **233** are not superseded. (Hyperfine splittings indicated for **233** are in gauss.)

B. RADICALS CONTAINING A SIX-MEMBERED HETEROCYCLE

1. *Radicals from Benzoxazines and Benzothiazines*

Russell and co-workers[514] prepared the 1,2-benzothiazine 1,1-dioxide semidione (**234**) along with structurally analogous radicals. Hyperfine splittings are indicated for **234** in gauss. The small nitrogen and proton splittings are in accordance with expectation for a semidione radical where the spin population is strongly concentrated on the ketyl oxygen atoms.

Treatment of 3-oxo-3*H*-indol-2-yl nitroxyls (e.g., **235**) with OH⁻ results in the formation of corresponding 4-oxo-4*H*-3,1-benzoxazin-2-yl nitroxyls (e.g., **236**), unless the C-5 carbon atom is substituted by a group sufficiently large to inhibit sterically initial attack of the nucleophile upon the carbonyl group.[688] The mechanism of ring expansion was discussed by Aurich and Weiss,[688] and an independent synthesis of **236** and analogs was given together with ESR data.

In general, 2*H*- or 4*H*-benzo-1,4-thiazines are readily autoxidized by a mechanism which involves radicals such as **237** in the propagating step. Work published mainly in Italian, which describes the course and products

[688] H. G. Aurich and W. Weiss, *Chem. Ber.* **105**, 2389 (1972).

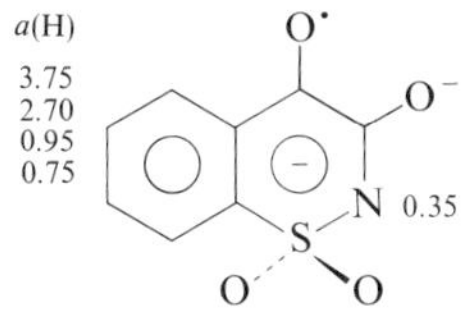

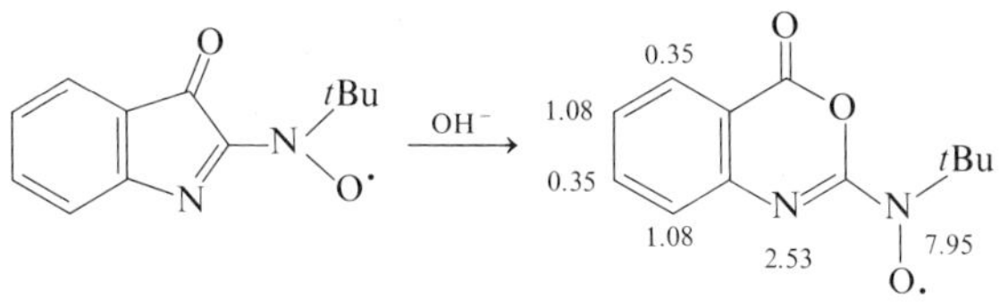

(234)   (235)   (236)

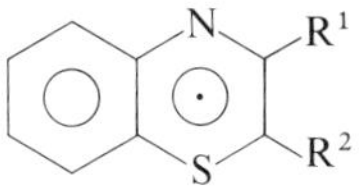

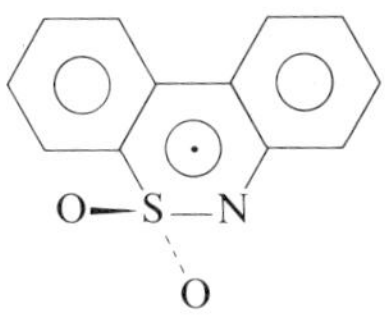

(237)   (238)

of the reaction, has been summarized in English.[689] Intermediate 2*H*-benzo-1,4-thiazine 2-hydroperoxides react to give, ultimately, 4*H*-benzo-1,4-thiazine sulfoxides and ring-contracted benzo-1,3-thiazolin-2-yl ketones. Other Italian workers have studied the oxidative dimerization of 3-aryl-2*H*-benzo-1,4-thiazines and -oxazines.[690,691] Again, radicals such as 237 (R^1 = Ar, R^2 = H) and the oxazine analogs are involved. The primary coupling products are, predominantly, meso diastereoisomers[690]; these may be dehydrogenated to isomeric $\Delta^{2,2'}$-bi-(2*H*-benzo-1,4-thiazines).[692,693] The dimerization of 237 (R^1 = CO_2H, R^2 = H) has been employed in studies of model reactions for the biosynthesis of pheomelanin pigments.[694,695] Thomson has reviewed the benzothiazine pigments of red hair and feathers[696]; although the chemistry described does not involve benzothiazine radicals, it relates to their dimerization products.

2. Dibenzo[c,e][1,2]thiazine 1,1-Dioxide Radical

Oxidation of biphenyl-2-sulfonamide with persulfate gives a sulfonamidyl radical which cyclizes intramolecularly to yield, after further oxidation,

---

[689] V. Carelli, F. Moracci, F. Liberatore, M. Cardellini, M. G. Lucarelli, P. Marchini, G. Liso, and A. Reho, *Int. J. Sulfur Chem.* **3**, 267 (1973).
[690] D. Sica, C. Santacroce, and G. Prota, *J. Heterocycl. Chem.* **7**, 1143 (1970).
[691] F. Chioccara, E. Ponsiglione, G. Prota, and R. H. Thomson, *Tetrahedron* **32**, 2033 (1976).
[692] F. Giordano, L. Mazzerella, G. Prota, C. Santacroce, and D. Sica, *J. Chem. Soc. C*, 2610 (1971).
[693] G. Prota, E. Ponsiglione, and R. Ruggiero, *Tetrahedron* **30**, 2781 (1974).
[694] S. Crescenzi, G. Misuraca, E. Novellino, and G. Prota, *Chim. Ind. (Milan)* **57**, 392 (1975) [*CA* **83**, 178961 (1975)].
[695] G. Prota and E. Ponsiglione, *Tetrahedron Lett.*, 1327 (1972).
[696] R. H. Thomson, *Angew. Chem., Int. Ed. Engl.* **13**, 305 (1974).

dibenzo[c,e][1,2]thiazine 5,5-dioxide.[697] The prevailing oxidizing conditions then result in the formation of the radical **238** whose presence is betrayed by the formation of a dimeric product linked by an N—C bond, the C being para to N, thus confirming the π nature of **238**.

3. Radicals from Phenoxazines, Phenothiazines, and Phenoselenazines

The radicals of this section have a 40-year history[698]; this review is concerned with the latter half of this. Despite this restriction, further limitation of coverage is necessary in the interests of brevity. Thus, no consideration is given to biological studies, even though radicals may be explicitly involved; the incidental occurrence of radicals from this group during use of the heterocycles or their derivatives as redox indicators, or during the analysis of phenothiazine drugs, is not covered systematically; nor is the treatment of the photochemistry of oxazine and thiazine dyes exhaustive. As in the remainder of the chapter, solid-state studies are given only passing mention.

a. *General Redox Relationships.* Various families of persistent radicals exist for each of the parent heterocycles. Before considering their chemistry, their redox relationships relative to their parents are set in context. In Scheme 2, passage horizontally from column to column represents

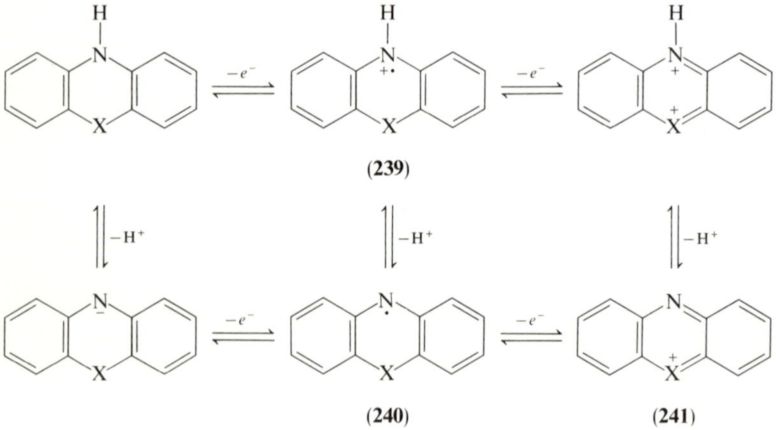

(239)

(240)   (241)

SCHEME 2

---

[697] P. S. Dewar, A. R. Forrester, and R. H. Thomson, *J. C. S. Perkin I*, 2862 (1972).
[698] See, for example, L. Michaelis, S. Granick, and M. P. Schubert, *J. Am. Chem. Soc.* **63**, 351 (1941).

single-electron redox steps, whereas passage vertically from row to row represents the protolytic equilibria of the various species. It is seen that the heterocycles may give rise each to a cation-radical (**239**) on single-electron oxidation and that these have, as isoelectronic conjugate bases, the neutral radicals **240** (see Scheme 2).

Formal hydration of the quinonoid structures (**241**) of Scheme 2, by addition of water at C-3 and proton transfer, gives the 3-hydroxy heterocycles which, like the parent heterocycles, may be oxidized in single-electron steps to quinonoid forms, each entity entering into acid–base equilibria through change in the number of its ionizable protons. As indicated in Scheme 3, cationic **242**, tautomeric neutral **243**, and anionic **244** radical forms may, in principle, occur according to the state of protonation, the key point being that they are isoelectronic and in the third level of oxidation relative to the parent molecules.

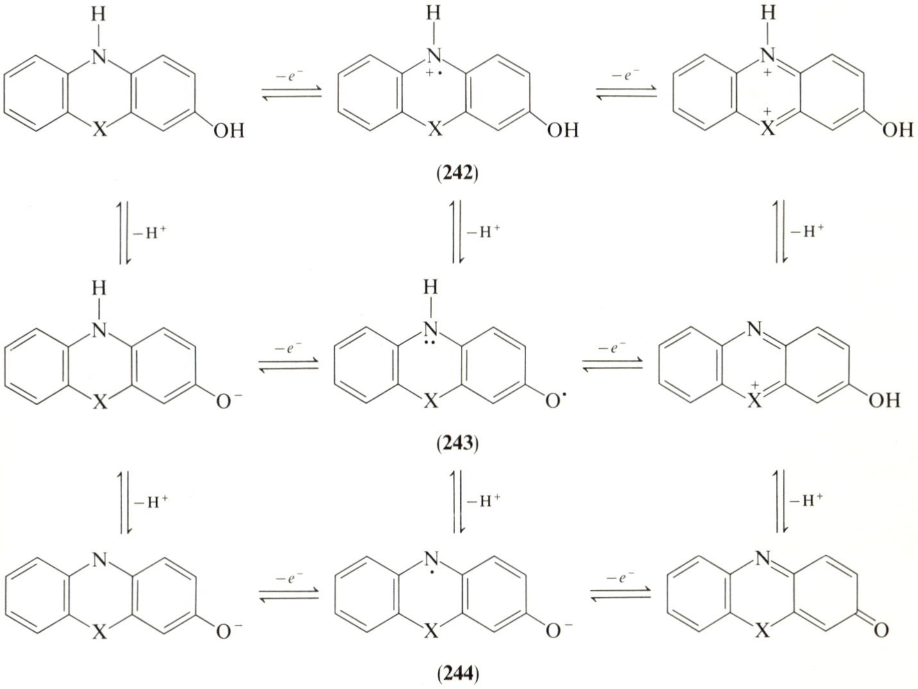

Scheme 3

Further formal hydration at C-7 of the quinonoid forms of Scheme 3 leads analogously via 3,7-dihydroxy heterocycles and corresponding ionizable radical intermediates to structures of the form **245** which are dyes [e.g., **245**: (X = O), resorufin; (X = S), thionol].

A formal hydration of **241** at N would yield *N*-hydroxy heterocycles. While these species are not well characterized, their radical oxidation products the nitroxyls (**246**) are well known (*vide infra*). It should be noted that these radicals, like **242**–**244**, are in the third level of oxidation relative to the parent heterocycles.

A formal hydration of **241** at X, only possible for X = S or Se, yields the corresponding sulfoxide (e.g., **247**) or selenoxide of the heterocycle. No radicals have been substantiated, in the writer's view, which arise by further oxidation of heterocycles with oxidized group VI heteroatoms.

(245)   (246)   (247)

b. *ESR and Other Spectroscopic Properties. i. Unsubstituted radicals.* Various groups recorded ESR spectra for the cation-radicals of phenoxazine (**248**)[699] and phenothiazine (**249**)[699–705] in the early 1960s. Commonly, the solutions were in concentrated sulfuric acid, which also served as oxidant; or an initial oxidation mixture in concentrated sulfuric acid was diluted with water or organic solvent.[699–703] Electrochemical oxidation of the phenothiazine in acetonitrile was also used.[704,705] In general, these early ESR spectra were not resolved much beyond the four-line structure indicative of the interaction of the unpaired electron with a ^{14}N and a ^{1}H nucleus having comparable magnitudes of hyperfine splitting. In the late 1960s, however, well resolved spectra were analyzed for both **248** and **249**[706–709]; the ESR spectrum of the phenoselenazine cation-radical

[699] L. D. Tuck and D. W. Schieser, *J. Phys. Chem.* **66**, 937 (1962).
[700] E. Crosignani, P. Franzosini, G. Siragusa, and L. Zanotti, *Arch. Sci.* **24**, 153 (1961) [*CA* **61**, 9077 (1964)].
[701] D. Gagnaire, H. Lemaire, A. Rassat, and P. Servoz-Gavin, *C. R. Acad. Sci.* **255**, 1441 (1962).
[702] C. Lagercrantz, *Acta Chem. Scand.* **15**, 1545 (1961).
[703] H. J. Shine and E. E. Mach, *J. Org. Chem.* **30**, 2130 (1965).
[704] J.-P. Billon, G. Cauquis, J. Combrisson, and A.-M. Li, *Bull. Soc. Chim. Fr.*, 2962 (1960)
[705] J.-P. Billon, G. Cauquis, and J. Combrisson, *J. Chim. Phys.* **61**, 374 (1964).
[706] B. C. Gilbert, P. Hanson, R. O. C. Norman, and B. T. Sutcliffe, *Chem. Commun.*, 161 (1966).
[707] J. M. Lhoste and F. Tonnard, *J. Chim. Phys.* **63**, 678 (1966).
[708] P. D. Sullivan and J. R. Bolton, *J. Magn. Reson.* **1**, 356 (1969).
[709] M. F. Chiu, B. C. Gilbert, and P. Hanson, *J. Chem. Soc. B*, 1700 (1970).

250 was also analyzed.[709] The hyperfine splittings indicated for **248** are from Sullivan and Bolton,[708] while those for **249** and **250** are due to Gilbert and co-workers who also measured the respective splittings of the low abundance group VI nuclei ^{33}S and ^{77}Se.[709] The splittings indicated are in gauss and pertain to the cation-radicals in solution in nitroalkanes. The assignments were made on the basis of MO calculations of which several have been performed for these radicals.[576,577,706–711] The ESR spectrum of 3-deuterophenothiazine cation-radical has been recorded by Cadogan and co-workers[712] during study of a nitrene-induced molecular rearrangement which leads to formation of the phenothiazine heterocycle from phenyl 2-nitrophenylsulfide.

(**248**)   (**249**)   (**250**)

Owing to their having the same magnetically interacting nuclei, phenothiazinyl (**251**) and phenothiazine 10-oxyl (**252**) were initially confused; the problem was compounded by both radicals being formed by oxidative processes from the parent heterocycle (though to different degrees, see previous section) and by their ESR spectra being subject to variation with solvent. As early as 1961, Baird and Thomas obtained a radical from a treatment of phenothiazine with decomposing azobisisobutyronitrile in benzene which they took to be **252**[713]; in fact it was **251**.[714] Conversely, Shine and co-workers[703,715] irradiated phenothiazine in ethanol and obtained a radical which was thought to be **251** but, in fact, was **252**.[714] Indeed, irradiations of ethanol solutions of phenothiazine have given both **251** and **252**[714,715]; different wavelengths of radiation were used in the two investigations and neither group of workers made clear the state of oxygenation of their solutions. Photochemical oxidations of phenothiazines will be discussed subsequently (Section VI,B,3,c,i). A claim to generate the anion-radical of phenothiazine by alkali-metal reduction[716] in fact gave **251**.

[710] S. Odiot and F. Tonnard, *J. Chim. Phys.* **61**, 382 (1964).
[711] J.-P. Malrieu and B. Pullman, *Theor. Chim. Acta* **2**, 293 (1964).
[712] J. I. G. Cadogan, S. Kulik, and C. Thomson, *J. Chem. Soc. D*, 436 (1970).
[713] J. C. Baird and J. R. Thomas, *J. Chem. Phys.* **35**, 1507 (1961).
[714] C. Jackson and N. K. D. Patel, *Tetrahedron Lett.*, 2255 (1967).
[715] H. J. Shine, C. Veneziani, and E. E. Mach, *J. Org. Chem.* **31**, 3395 (1966).
[716] M. Bruin, F. Bruin, and F. W. Heineken, *J. Org. Chem.* **29**, 507 (1964).

The ESR spectrum of **251** was first obtained unambiguously by the writer and co-workers following deprotonation of **249** in acetonitrile by addition of aqueous buffer[706]; the hyperfine splittings given in **251** are from a recent measurement on the radical formed by direct oxidation of phenothiazine with PbO_2 in benzene.[717] The assignment of splittings is made by comparison with **249** and following a measurement on 3-deuterated derivative which showed that the largest proton hyperfine splittings derive from protons on C-3 and C-7, thus disproving an earlier assignment, based on MO calculations, which attributed them to the protons at C-1 and C-9.[709] This correction is of relevance to the ENDOR measurements made by McDowell and co-workers on **251** and **253**[718]: their assignments were made on the basis of the now superseded MO calculations.[709] The electronic absorption spectrum of **251** was reported by Hanson and Norman.[719]

The analysis of the ESR spectrum of **252** was correctly given early by Buchachenko[720]; the hyperfine splittings given in **252** are from the work of Aurich and co-workers[721] who also determined the hyperfine splitting due to ^{17}O in a labeled sample. This determination allows experimental confirmation of calculations on the distribution of spin in the nitroxyl radical function and has allowed the formulation of McLachlan MO parameters of general applicability in this type of radical.[722]

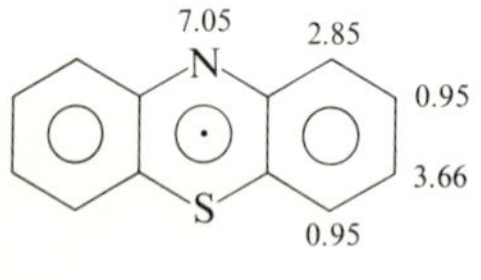

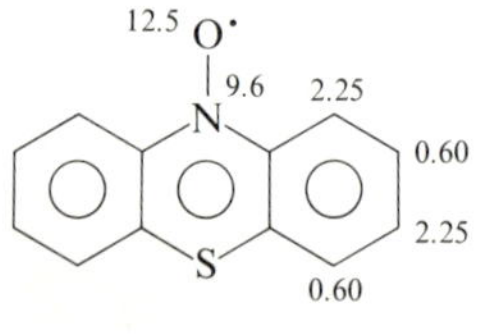

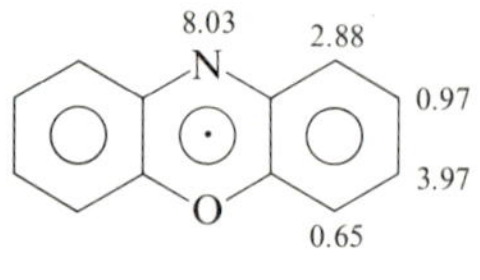

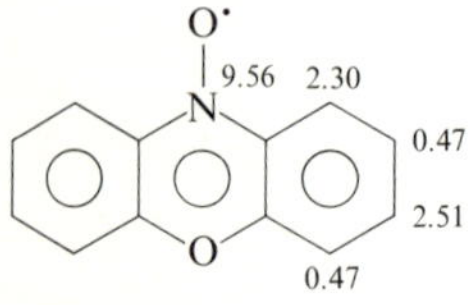

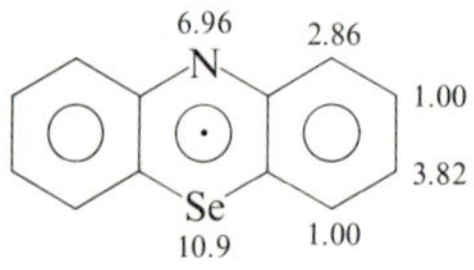

Unlike their phenothiazine counterparts, phenoxazinyl (**253**) and phenoxazine 10-oxyl (**254**) were not confused. They were properly distinguished,

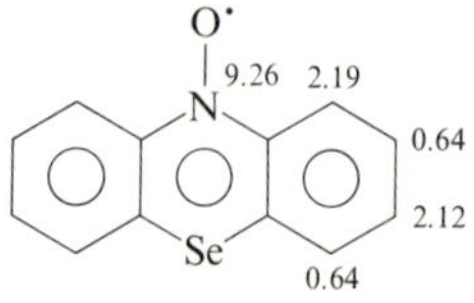

---

[717] D. Clarke, B. C. Gilbert, and P. Hanson, *J. C. S. Perkin II*, 517 (1977).
[718] D. E. Kennedy, N. S. Dalal, and C. A. McDowell, *Chem. Phys. Lett.* **29**, 521 (1974).
[719] P. Hanson and R. O. C. Norman, *J. C. S. Perkin II*, 264 (1973).
[720] A. L. Buchachenko, *Opt. Spektrosk.* **13**, 795 (1962).
[721] H. G. Aurich, K. Hahn, K. Stork, and W. Weiss, *Tetrahedron* **33**, 969 (1977).
[722] H. G. Aurich, E. Deuschle, and I. Lotz, *J. Chem. Res.(S)*, 248 (1977).

and their ESR spectra correctly assigned by Scheffler and Stegmann.[723] Their hyperfine splittings are indicated in gauss for **253** and **254**; Aurich and co-workers[721] have also recorded the ^{17}O hyperfine splitting for **254** as 13.1 G.

The ESR spectra of phenoselenazinyl (**255**) and phenoselenazine 10-oxyl (**256**) were measured by Gilbert and co-workers.[576,709] The hyperfine splittings indicated (in gauss) for **256** follow the original papers, but for **255** the assignments of splittings for the 1- and 3-protons have been interchanged (with the assent of Dr. B. C. Gilbert) in view of the fact that they were made in the light of MO calculations which, for **251** and **253**, were misleading. The calculations of Galasso and Bigotto[577] provide no additional insight; they too predict $a(H)_1 > a(H)_3$ and, in addition, give very unsatisfactory account of the nitrogen hyperfine splitting.

Within the families of cation-radicals and neutral radicals, it is apparent that the nitrogen splitting decreases in the order $O > S > Se$, which is consistent with an increased spin, and concomitant positive charge, on the group VI atom the less electronegative it becomes. The nitrogen splitting in the nitroxyls is not dependent on the group VI atom in such an obvious way, probably because the principal contributing structures in this family distribute the spin population mainly between N and the nitroxyl O. In all three families, changes in ring proton splittings are small and not well ordered; it is apparent, however, that the 3(7)- and 1(9)-proton splittings are the greatest in each family for the oxazine radicals, which may reflect a closer approach to planarity in the radicals of this heterocycle. Direct evidence is sparse, but crystal structures have been determined for cis-bis(trifluoromethylethylene-1,2-dithioloato)nickelate salts of **248** and **249**.[724] The former is planar while the latter is folded with a dihedral angle of ~170°.

ii. C-Substituted radicals. Most studies of substituted radicals have been in the phenothiazine radical families, owing, probably, both to the greater accessibility of substituted phenothiazines and the greater interest in this heterocycle for the therapeutic value of its derivatives.

The consequences of methylation and hydroxylation of phenothiazine for the ESR spectra of its cation-radicals were noted early by Billon et al.[725] Bodea and Silberg[726-728] have described a series of halogenated phenothiazinyls and the effect of progressive halogenation on their g-values.

---

[723] K. Scheffler and H. B. Stegmann, *Tetrahedron Lett.*, 3619 (1968).
[724] A. Singhabhandu, P. D. Robinson, J. H. Fang, and W. E. Geiger, *Inorg. Chem.* **14**, 318 (1975).
[725] J.-P. Billon, G. Cauquis, and J. Combrisson, *C. R. Acad. Sci.* **253**, 1593 (1961).
[726] C. Bodea and I. Silberg, *Nature (London)* **198**, 833 (1963).
[727] C. Bodea and I. Silberg, *Rev. Roum. Chim.* **9**, 505 (1964).
[728] C. Bodea and I. Silberg, *Rev. Roum. Chim.* **10**, 887 (1965).

Other Romanian workers have studied solid-state properties of octachlorophenothiazinyl (**257**).[729,730] The ENDOR spectrum of 2-chlorophenothiazine has been measured.[718]

Hanson and co-workers[717] have studied substituent effects on the distribution of spin in cation, neutral, and nitroxyl radicals of phenothiazine. The substituent effects were described quantitatively using Taft's σ_I and various σ_R values,[731] and are understandable in terms of a redistribution of spin population principally between N and S in the cation-radicals and phenothiazinyls, and between N and O in the nitroxyls.

Substituted cation and neutral radicals have been obtained by reduction of phenoxazine and phenothiazine dyes. ESR spectra were obtained, but not analyzed in detail, by Heineken et al.[716,723] In alkaline media the spectra showed an overall triplet structure consistent with the radicals being phenoxazinyls or phenothiazinyls; in acidic medium the extra proton splitting of the corresponding cation-radicals was evident. Recently, Park and Bard[733] have analyzed the ESR spectrum of the phenoxazinyl (**258**) produced on electrochemical reduction of the laser dye Oxazine-1. Hyperfine splittings are indicated in gauss and are for solution in acetonitrile.

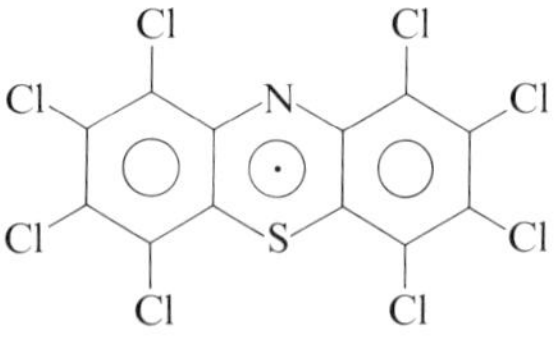

(**257**) 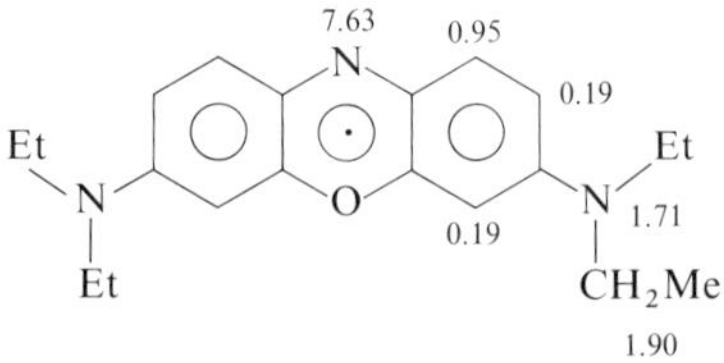   (**258**)

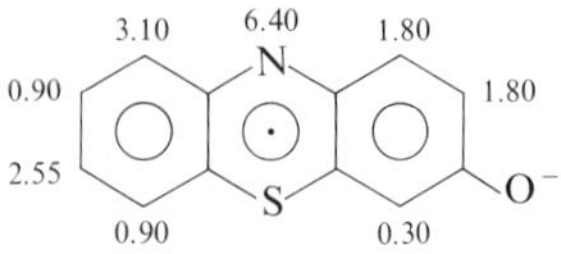

(**259**) 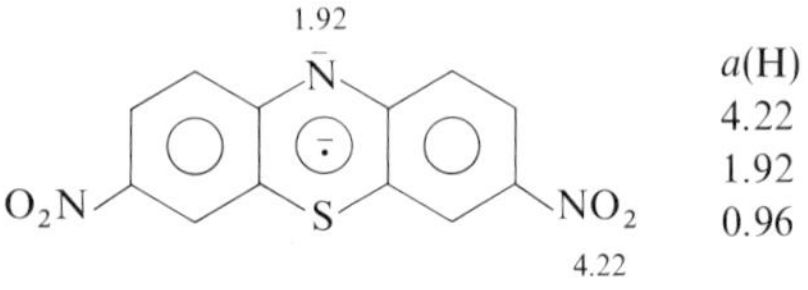   (**260**)

Soviet researchers have reported ESR and spectroscopic studies on the

---

[729] R. Baican, D. Demco, and A. Bodi, *Phys. Status Solidi* **32**, K89 (1969).
[730] D. Demco and R. Baican, *Stud. Univ. Babes-Bolyai, Ser. Math.-Phys.* **13**, 113 (1968) [*CA* **71**, 17307 (1969)].
[731] S. Ehrenson, R. T. C. Brownlee, and R. W. Taft, *Prog. Phys. Org. Chem.* **10**, 1 (1973).
[732] F. W. Heineken, M. Bruin, and F. Bruin, *J. Chem. Phys.* **37**, 1479 (1962).
[733] S. M. Park and A. J. Bard, *J. Electroanal. Chem. Interfacial Electrochem.* **77**, 137 (1977).

radicals produced by reduction of phenothiazine dyes including Methylene Blue and Thionine.[734–736]

Optical and ESR spectra have been reported for 1-substituted phenothiazinyls and MO calculations performed of the optical transitions.[737]

There have been reports of various phenothiazine and phenoxazine anion-radicals which are species which correspond to structure **244** and its substituted derivatives.[738–740] Thus, reduction of phenothiazin-3-one in DMF gives **259** for which hyperfine splittings are indicated in gauss.[738] The assignment of splittings was made following study of halogenated derivatives of **259** and McLachlan MO calculations. Similar anions from phenoxazin-3-ones, and their conjugate acids, have been described in the Russian literature.[739,740]

All the heterocyclic radicals described thus far in this section have been, irrespective of their charge, isoelectronic with equivalently substituted anthracene anion-radicals. The failure of an attempt to reduce phenothiazine itself to an anion-radical (which would be isoelectronic with anthracene trianion-radical) has been noted already[716]; however, anion-radicals have been successfully prepared by reduction of nitrophenothiazines and their sulfoxides.[741] In unsymmetrical species such as 3-nitrophenothiazine anion-radical, the unpaired electron is strongly concentrated in the substituent and only small splittings from the heterocyclic moiety are observed. As further nitro groups are symmetrically introduced, the spin is distributed more evenly over the heterocycle and the heterocyclic nitrogen splitting increases. No heterocyclic *N*-proton splitting is observed in these radicals. It is presumably ionized; the radicals are thus dianion-radicals (e.g., **260**) for which hyperfine splittings are indicated in gauss.

iii. N-Substituted radicals. *N*-Substitution in phenothiazine and its isologs permits only the formation of cation-radicals. That of 10-methylphenothiazine (**261**) was investigated early along with **249**.[700,705,707] The

---

[734] G. N. Fadeev, G. V. Fomin, and L. A. Nikolaev, *Zh. Fiz. Khim.* **44**, 2703 (1970) [*CA* **74**, 77383 (1971)].

[735] Kh. L. Arvan and G. A. Korsunovskii, *Zh. Prikl. Spektrosk.* **20**, 235 (1974) [*CA* **80**, 146960 (1974)].

[736] Kh. L. Arvan, A. Bobrovskii, and G. A. Korsunovskii, *Teor. Eksp. Khim.* **10**, 692 (1974) [*CA* **82**, 59879 (1975)].

[737] M. Kamiya, T. Mitsui, and Y. Akahori, *Bull. Chem. Soc. Jpn.* **46**, 1577 (1973).

[738] M. Hillebrand, M. Raileanu, and V. E. Sahini, *Rev. Roum. Chim.* **20**, 581 (1975).

[739] I. I. Bil'kis, E. G. Bouguslavskii, T. S. Viktorova, G. B. Afanas'eva, I. Ya. Postovskii, and S. M. Shein, *Tezisy Dokl. Vses. Sovshch. Kompleksam Perenosom Zaryada Ion-Radikal 'nym Solyam*, 3rd, 1975, 92 (1976) [*CA* **87**, 67477 (1977)].

[740] B. E. Geogrievich, *Mater. Vses. Nauchn. Stud. Konf., Khim., 13th*, 1975, 6 (1975) [*CA* **86**, 106502 (1977)].

[741] M. Hillebrand and M. Raileanu, *Rev. Roum. Chim.* **19**, 1227 (1974).

first analysis of the ESR spectrum of **261** is due to Sullivan and Bolton,[708] who generated the radical in nitromethane on treatment of the solution with AlCl$_3$. The hyperfine splittings indicated for **261** are from recent measurements made by Hanson and co-workers for solution in nitromethane[742], the assignment is made on the basis of Sullivan and Bolton's McLachlan MO calculations.[708]

A recent investigation has been made of the influence of the structure and conformation of 10-alkyl and similar side chains on the spin distribution in phenothiazine cation-radicals.[742] The nitrogen hyperfine splitting was found to vary in a manner which depended on the polarizability of the side chain, and the hyperfine splittings observed for the first protons in the side chain (β-protons) indicated that while the methyl group is free to rotate, larger alkyl groups have strongly preferred conformations. There is also evidence that the fold in the phenothiazine moiety responds to the steric requirements of the group attached to N. Phenothiazines with allylic side chains on nitrogen were found to undergo loss of the side chain on attempted oxidation, although it is not clear whether this is merely an artifact of the acidic oxidation conditions or an intrinsic property of allylic phenothiazine cation-radicals themselves. The observation provides an alternative explanation to *N*-protonation of the cation-radicals which had been suggested by others to account for the form of ESR spectra from allyl and similar phenothiazines.[743,744]

Substituent effects in the optical spectra of simple alkyl and arylphenothiazine radicals have been reported by Biehl *et al.*[745]

The important drugs derived from phenothiazines are substituted at N by a functionalized alkyl side chain. The cation-radicals from these materials have been the subject of various ESR investigations[742,746-752]; those of Fenner have been the most extensive.[749-752] He has partially analyzed the ESR spectra of several drugs with substituents at C-2. The hyperfine split-

---

[742] D. Clarke, B. C. Gilbert, P. Hanson, and C. M. Kirk, *J. C. S. Perkin II*, 1103 (1978).

[743] L. L. Kamenov, D. Simov, and V. B. Golubev, *Teor. Eksp. Khim.* **7**, 129 (1971) [*CA* **75**, 48008 (1971)].

[744] L. Kamenov and D. Simov, *God. Sofii. Univ., Khim. Fak.* **65**, 301 (1970-1971) [*CA* **80**, 150772 (1974)].

[745] E. R. Biehl, H.-S. Chiou, J. Keepers, S. Kennard, and P. C. Reeves, *J. Heterocycl. Chem.* **12**, 397 (1975).

[746] D. W. Schieser and L. D. Tuck, *J. Pharm. Sci.* **51**, 694 (1962).

[747] P. Machmer, *Z. Naturfosch., Teil B* **21**, 934 (1966).

[748] F. N. Pirnazorova, A. P. Poltorakov, and N. M. Emanuel, *Uzb. Khim. Zh.* **14**, 62 (1970) [*CA* **74**, 79525 (1971)].

[749] H. Fenner and H. Möckel, *Tetrahedron Lett.*, 2815 (1969).

[750] H. Fenner, *Arch. Pharm. (Weinheim, Ger.)* **303**, 919 (1970).

[751] H. Fenner, *Arch. Pharm. (Weinheim, Ger.)* **304**, 36 (1971).

[752] H. Fenner, *Arch. Pharm. (Weinheim, Ger.)* **304**, 47 (1971).

tings, indicated in gauss, for the cation-radical **262** of the drug promazine hydrochloride, whose symmetry permits a fairly complete analysis, are from the work of Hanson and co-workers.[742]

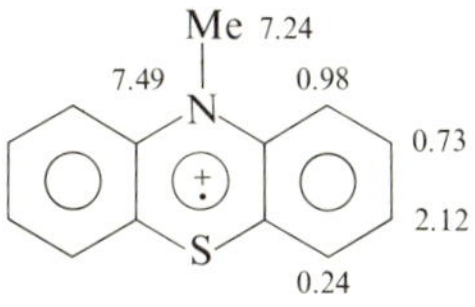

(261) phenothiazine cation-radical: Me 7.24; N 7.49; ring 0.98; 0.73; 2.12; S 0.24.

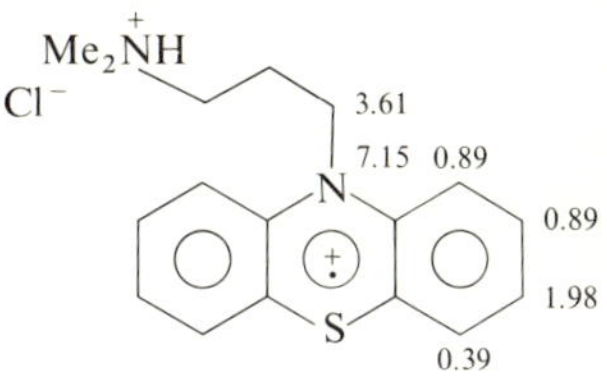

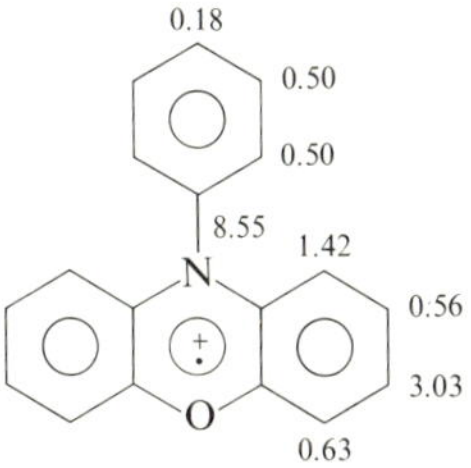

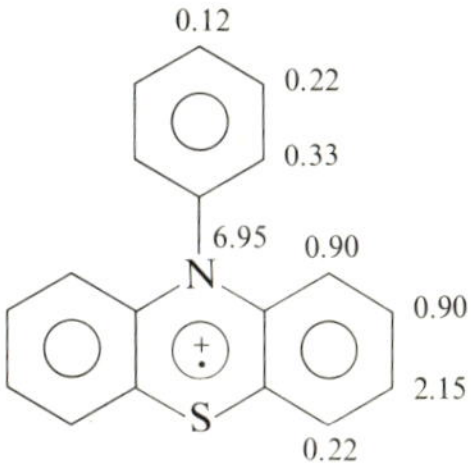

(262) promazine cation-radical: Me₂NH⁺ Cl⁻ chain 3.61; N 7.15, 0.89; ring 0.89, 1.98; S 0.39.

(263) 10-phenylphenoxazine cation-radical: phenyl 0.18, 0.50, 0.50; N 8.55, 1.42; ring 0.56, 3.03; O 0.63.

(264) 10-phenylphenothiazine cation-radical: phenyl 0.12, 0.22, 0.33; N 6.95, 0.90; ring 0.90, 2.15; S 0.22.

The same group has investigated 10-phenylphenoxazine (**263**) and -phenothiazine (**264**) cation-radicals and the influence of substituents in the phenyl group upon the spin distribution.[753,754] The interest here lay in the effect of the conformation on electronic effects. Earlier studies on similarly shaped radicals had indicated that a twist of about 65° must exist in the *N*-phenyl bond (cf. phenylxanthenyl and isologs; Sections II,B,3,a; III,C,1; IV,B,1). This value may be calculated trigonometrically for a model of such radicals which has planar aromatic components, each C—C and C—H bond equal to those in benzene, and in which the phenyl ortho and heterocyclic 1- and 9-protons are at van der Waals separation. A rather similar angle has to be allowed in MO calculations in order correctly to reproduce experimental hyperfine splittings.

For **263** and **264**, the assignment of hyperfine splittings to protons in the phenyl ring was validated by selective deuteration experiments. It is noteworthy that the para proton splitting is the smallest in each case, reflecting the diminished mesomeric interaction through the twisted bond. In McLachlan MO calculations on these radicals, a twist angle of ~70° had to be allowed in order to reproduce the correct magnitude of para proton splitting.[753] The influence of substituents in the meta or para positions of

---

[753] D. Clarke, B. C. Gilbert, and P. Hanson, *J. C. S. Perkin II*, 1078 (1975).
[754] D. Clarke, B. C. Gilbert, and P. Hanson, *J. C. S. Perkin II*, 114 (1976).

the phenyl rings upon the nitrogen hyperfine splittings was analyzed quantitatively using Taft's σ_I and $\sigma_R{}^+$ substituent constants. A twist angle of ~65° was again adduced from the differential dependence of the nitrogen hyperfine splittings upon the inductive and mesomeric effects of the substituents.[754]

The substituent NMe_2 has a unique effect when in the para position of the phenyl group of **264**. The character of the ESR spectrum changes from that of a 10-phenylphenothiazine cation-radical becoming like that of a Würster's salt. Evidently, this substituent causes a gross change in spin distribution, becoming itself a major site of unpaired spin population; geometry change is also probable.[754]

iv. Annelated radicals. Zander and co-workers have analyzed ESR spectra for annelated phenothiazinyls (e.g., **265**[755] and **266**[756]), and Brandt and Zander[757] have reviewed this and related work. Hyperfine splittings, indicated in gauss for **265** and **266**, were assigned following Hückel MO calculations.

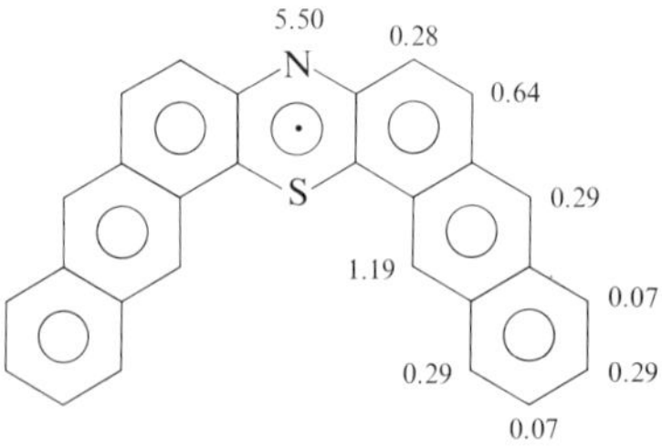

(265)

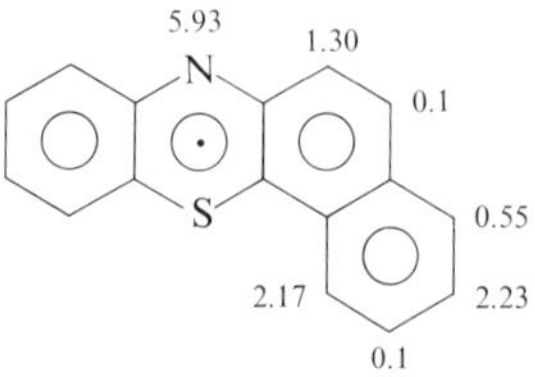

(266)

c. *Radical-Producing Processes.* The very ready formation of radicals by the heterocycles of present concern has led to a number of investigations of the processes involved, both physical and chemical. These are reviewed in this section.

i. Physical processes. Electrochemistry has played an important role in the study of the cation-radicals in particular, and French workers have been preeminent in the field. Billon[758-759] reported a detailed study of the anodic oxidation of phenothiazine and a number of its methylated and other derivatives in acetonitrile. This work laid the foundations of the

---

[755] J. Brandt, G. Fauth, W. H. Franke, and M. Zander, *Chem. Ber.* **104**, 519 (1971).
[756] J. Brandt and M. Zander, *Chem. Ber.* **105**, 3500 (1972).
[757] J. Brandt and M. Zander, *Chem.-Ztg.* **99**, 272 (1975).
[758] J.-P. Billon, *Bull. Soc. Chim. Fr.*, 1784 (1960).
[758a] J.-P. Billon, *Bull. Soc. Chim. Fr.*, 1923 (1961).
[759] J.-P. Billon, *Ann. Chim. (Paris)* **7**, 183 (1962).

unambiguous recognition of the various species which result from oxidation of phenothiazines. It was primarily concerned with oxidations in acidic conditions which led to **249** as initial product. Subsequent chemical work (see later) elucidated the conditions of formation of **251** from **249** and the reactions of **251** to give dimeric products, and these latter processes have also been the subject of electrochemical investigations.[760–762] Others reported consistent anodic behavior of phenothiazine.[763]

Phenothiazine drugs have been the subjects of various electrochemical investigations.[764–769] The motivation for these studies has been the search for means of analyzing drugs and their derivatives. Tozer et al.[770] examined the correlation between C-substitution, standard electrode potential, cation-radical formation constants, and the anthelmintic activity of the heterocycles.

Polarographic studies of dyes have been described: Pamfilov et al. reported the polarography of Methylene Blue and Thionine at a dropping mercury electrode,[771] and other Soviet workers have reported a polarographic study of 3H-phenoxazin-3-one and its benzo derivatives.[772]

Pragst and co-workers[773] have prepared a phenothiazine and a phenoxazine with 10-[3-(9-anthracenyl)propyl] side chains. The molecules form heterocycle-centered cation-radicals and anthracene-centered anion-radicals, but there is no evidence of interaction between the two aromatic moieties in either type of ion-radical.[773]

There has been intensive study of the photooxidation of, mainly, phenothiazine systems. Allusion has already been made to the formation of both **251** and **252** upon photolysis of phenothiazine in ethanol in the previous section.[714]

---

[760] G. Cauquis, A. Deronzier, and D. Serve, *J. Electroanal. Chem. Interfacial Electrochem.* **47**, 193 (1973).
[761] G. Cauquis, A. Deronzier, J.-L. Lepage, and D. Serve, *Bull. Soc. Chim. Fr.*, 295 (1977).
[762] G. Cauquis, A. Deronzier, J.-L. Lepage, and D. Serve, *Bull. Soc. Chim. Fr.*, 303 (1977).
[763] W. Kemula and M. K. Kalinowski, *Fresenius' Z. Anal. Chem.* **224**, 383 (1967) [*CA* **66**, 51666 (1967)].
[764] P. Kabasakalian and J. McGlotten, *Anal. Chem.* **31**, 431 (1959).
[765] F. H. Merkle and C. A. Discher, *J. Pharm. Sci.* **53**, 620 (1964).
[766] F. H. Merkle and C. A. Discher, *Anal. Chem.* **36**, 1639 (1964).
[767] G. J. Patriarche and J. J. Lingane, *Anal. Chim. Acta* **49**, 25 (1970).
[768] K. Sykut and J. Macias, *Chem. Anal.* (Warsaw) **22**, 501 (1977) [*CA* **88**, 5917 (1978)].
[769] M. Neptune and R. L. McGreery, *J. Med. Chem.* **21**, 362 (1978).
[770] T. N. Tozer, L. D. Tuck, and C. J. Cymerman Craig, *J. Med. Chem.* **12**, 294 (1969).
[771] A. V. Pamfilov, Ya. S. Mazurkevich, and E. P. Pakhomova, *Ukr. Khim. Zh.* **34**, 48 (1968) [*CA* **68**, 101228 (1968)].
[772] N. P. Shimanskaya, G. B. Afanas'eva, I. I. Pashkevich, L. A. Kotok, I. Ya. Postovskii, and V. D. Bezuglyi, *Zh. Obshch. Khim.* **41**, 1106 (1971) [*CA* **75**, 83505 (1971)].
[773] H. J. Hamann, F. Pragst, and W. Jugelt, *J. Prakt. Chem.* **318**, 369 (1976).

Koizumi and co-workers[774–776] have carried out mechanistic studies of the photochemical oxidation of phenothiazine, by oxygen, in ethanol. The mechanism deduced involves the formation of an exciplex between triplet phenothiazine and molecular oxygen which breaks down to give **251** and $HO_2^{\bullet}$. Exciplexes formed between triplet phenothiazine and other electron acceptors were found to undergo ionization to give **249** and a corresponding counterion in polar solvents. Kholmogorov reached a comparable conclusion,[777] whereas Burrows and co-workers prefer a mechanism where **249** is the initial photoproduct.[778] Koizumi and co-workers[774] had reported evidence of photoionization of phenothiazine in an alcoholic glass at 77K. Consistent with this, Alkaitis et al.[779] reported higher quantum yields of photoionization and lower quantum yields of triplet phenothiazine in methanolic micellar solutions than in simple methanolic solutions. Phenothiazine triplet was also found to be effectively oxidized to **249** by electron transfer to Cu^{2+} and Eu^{3+} ions.

Kholmogorov[780] reported that photooxidation of phenothiazine and diphenylamine could be sensitized by tetrapyrrole pigments. Phenothiazine was reported to give **251**, while diphenylamine gave its nitroxyl. This puzzling result is clarified by more recent work of Rosenthal and Poupko[781] on the dye-sensitized photooxidation of phenothiazine. Here it was observed that dye-sensitized reactions in methanol gave either **252** alone or a mixture of **252** and **251**. The oxygen of the nitroxyl was proved to be derived from molecular oxygen by isotopic labeling; no radicals were formed in the reaction sensitized by Methylene Blue if the solution was deoxygenated; direct irradiation of phenothiazine in alcoholic solution led to a mixture of **251** and **252**, although with continued irradiation the latter was selectively destroyed. It appears that **251** is not a secondary product from the destruction of **252** since other nitroxyls, photostable alone, were destroyed on irradiation in the presence of phenothiazine, but without formation of **251**. These facts were taken to imply the operation of an oxidation mechanism involving the attack of singlet oxygen upon phenothiazine, leading to the formation of an unstable 10-hydroperoxide which may decompose either to **251** or **252** depending upon whether N—O or O—O bond fission occurs.

---

[774] T. Iwaoka, H. Kokubun, and M. Koizumi, *Bull. Chem. Soc. Jpn.* **44**, 341 (1971).
[775] T. Iwaoka, H. Kokubun, and M. Koizumi, *Bull. Chem. Soc. Jpn.* **44**, 3466 (1971).
[776] T. Iwaoka, H. Kokubun, and M. Koizumi, *Bull. Chem. Soc. Jpn.* **45**, 73 (1972).
[777] V. E. Kholmogorov, *Khim. Vys. Energ.* **4**, 28 (1970) [*CA* **72**, 110439 (1970)].
[778] H. D. Burrows, J. Dias da Silva, and M. I. Ventura Batista, *Excited States Biol. Mol., Proc. Int. Conf., 1974*, 116 (1976).
[779] S. A. Alkaitis, G. Beck, and M. Grätzel, *J. Am. Chem. Soc.* **97**, 5723 (1975).
[780] V. E. Kholmogorov, *Biofizika* **16**, 378 (1971) [*CA* **75**, 13476 (1971)].
[781] I. Rosenthal and R. Poupko, *Tetrahedron* **31**, 2103 (1975).

Among this array of somewhat conflicting evidence on the photooxidation of phenothiazine lies a truth which accounts for the initial confusion of **251** and **252**. There are several reports that the stable ultimate product of the photooxidation of phenothiazine is the sulfoxide (**247**),[775,781,782] although its mechanism of formation is unknown.

A flash-photolytic study of phenoxazine has been reported.[783] Three transient species were observed: the first triplet state, **253**, and **248**. The results are consistent with the formation of **248** in a biphotonic process involving the triplet. In the presence of oxygen, neither the triplet nor **248** was observed; continuous-irradiation experiments indicated that probably **253** reacts with oxygen, via unstable hydroperoxide intermediates, to give, as stable photooxidation product, 3H-phenoxazin-3-one.

The photooxidation of 10-substituted phenothiazines, in particular the drug chlorpromazine, has been investigated, again with conflicting conclusions about mechanisms.[784–789] The sulfoxide is again a principal oxidation product,[782] although 10-methylphenothiazine has also been shown to undergo dealkylation and oxidation to 3H-phenothiazin-3-one.[790] The effect of substituents on the photostability of phenothiazine drugs has been investigated by Polish workers.[791,792]

Dyes from phenoxazine, phenothiazine, and phenoselenazine undergo photoreduction in various conditions yielding radical products. The subject has a long history and, as intimated at the outset, treatment here is cursory. Parker reported an early flash-photolytic study of the photobleaching of Methylene Blue,[793] which was followed by others in the 1960s.[794–796] This technique permitted the recognition of transient species in the system,

---

[782] T. M. Zaitseva, A. V. Kotov, and L. A. Gribov, *Izv. Timiryazevsk. S-kh. Akad.*, 195 (1975) [*CA* **81**, 177237 (1974)].
[783] D. Gegiou, R. J. Huber, and K. Weiss, *J. Am. Chem. Soc.* **92**, 5058 (1970).
[784] T. Iwaoka and M. Kondo, *Bull. Chem. Soc. Jpn.* **47**, 980 (1974).
[785] T. Iwaoka and M. Kondo, *Chem. Lett.*, 1105 (1976).
[786] T. Iwaoka and M. Kondo, *Bull. Chem. Soc. Jpn.* **50**, 1 (1977).
[787] T. Iwaoka and M. Kondo, *Chem. Lett.*, 731 (1978).
[788] A. K. Davies, S. Navaratnam, and G. O. Phillips, *J. C. S. Perkin II*, 25 (1976).
[789] S. Navaratnam, B. J. Parsons, G. O. Phillips, and K. Davies, *J. C. S., Faraday 1* **74**, 1811 (1978).
[790] C. D. M. Ten Berge and C. H. P. Bruins, *Pharm. Weekbl.* **104**, 1433 (1969) [*CA* **72**, 26990 (1970)].
[791] E. Pawelczyk and B. Marciniec, *Pol. J. Pharmacol. Pharm.* **29**, 143 (1977).
[792] E. Pawelczyk and B. Marciniec, *Pol. J. Pharmacol. Pharm.* **29**, 137 (1977).
[793] C. A. Parker, *J. Phys. Chem.* **63**, 26 (1959).
[794] S. Matsumoto, *Bull. Chem. Soc. Jpn.* **37**, 491 (1964).
[795] S. Matsumoto, *Bull. Chem. Soc. Jpn.* **37**, 499 (1964).
[796] S. Kato, M. Morita, and M. Koizumi, *Bull. Chem. Soc. Jpn.* **37**, 117 (1964).

whether photoexcited states or radicals. The photochemistry of thiazine and related dye molecules holds continuing interest for physical chemists.[797-800] One particular photoredox system, that involving Thionine (**245**; X = S) and ferrous ions, originally subjected to investigation by the flash-photolysis technique by Hatchard and Parker,[801] is of current interest for photogalvanic systems.[802,803]

Phenothiazine radicals have been generated in various radiolytic experiments. Pulse radiolysis of phenothiazine in aerated cycloalkane solutions gives rise to phenothiazinyl (**251**).[804] The mode of oxidation is believed to involve abstraction of H˙ from the heterocycle by cycloalkylperoxyl radicals. Under similar conditions, 10-methylphenothiazine gives **261**. Pulse radiolysis of phenothiazine in acetonitrile has also been documented.[805] The reduction of phenothiazine dyes under radiolytic conditions has been reported. Several years ago, Ayscough and Thomson[806] obtained paramagnetic products from X-irradiation of acid alcoholic solutions of Methylene Blue and other thiazine dyes. Low-resolution ESR spectra of the corresponding cation-radicals were recorded.

More recent work has investigated the bleaching of dyes, including Methylene Blue and Thionine, by electron-transfer from radicals such as ˙CH_2OH generated under radiolytic conditions.[807] There is a strong dependence of the bleaching reaction, both with respect to efficiency and rate, upon the redox potential of the dye.

ii. Chemical processes. A wide range of chemical oxidants has been used to prepare the radicals of phenothiazine and its isologs; attention here is focused on those which are synthetically convenient or which are of mechanistic interest.

Phenothiazine cation-radical (**249**) perchlorate salt is readily prepared by the method of Billon[759]: an equimolar mixture of phenothiazine and its sulfoxide (**247**) is treated with 70% $HClO_4$, whereupon reaction occurs

---

[797] E. Vogelmann, H. Schmidt, U. Steiner, and H. E. A. Kramer, *Z. Phys. Chem.* (*Frankfurt am Main*) **94**, 101 (1975).
[798] E. Vogelmann and H. E. A. Kramer, *Photochem. Photobiol.* **23**, 383 (1976).
[799] L. V. Romashov, Yu. I. Kiryukhin, and Kh. S. Bagdasar'yan *Dokl. Akad. Nauk SSSR* **230**, 1145 (1976) [*CA* **87**, 31838 (1977)].
[800] U. Steiner, G. Winter, and H. E. A. Kramer, *J. Phys. Chem.* **81**, 1104 (1977).
[801] C. J. Hatchard and C. A. Parker, *Trans. Faraday Soc.* **57**, 1093 (1961).
[802] M. Z. Hoffmann, *Curr. State Knowl. Photochem. Form. Fuel, Rep. Workshop, 1974*, 86 (1975) [*CA* **87**, 204264 (1977)].
[803] K. Shigehara and E. Tsuchida, *J. Phys. Chem.* **81**, 1883 (1977).
[804] H. D. Burrows, T. J. Kemp, and M. J. Welbourn, *J. C. S. Perkin II*, 969 (1973).
[805] J. L. Baptista and H. D. Burrows, *J. C. S., Faraday 1* **70**, 2066 (1974).
[806] P. B. Ayscough and C. Thomson, *J. Chem. Soc.*, 2055 (1962).
[807] P. S. Rao and E. Hayon, *J. Phys. Chem.* **77**, 2753 (1973).

according to Scheme 4. The salt is precipitated by addition of acetone. The tetrafluoroborate and halides may be prepared analogously; the method has been applied to benzophenothiazines.[808] Oxidation of the heterocycles with nitrosonium tetrafluoroborate[555] is also a convenient method, and applicable to phenoxazine for which Billon's method is not possible.

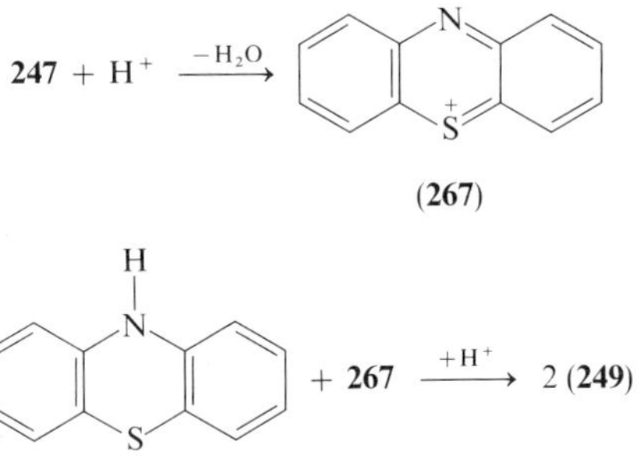

$$247 + H^+ \xrightarrow{-H_2O} (267)$$

$$\text{(phenothiazine)} + 267 \xrightarrow{+H^+} 2\,(249)$$

SCHEME 4

Salts have been prepared of **249** by oxidation of the heterocycle with $SbCl_5$,[567] UCl_6^-,[809] and $MoOCl_4$.[810] The perchlorate of 3,7-dimethoxyphenothiazine cation-radical was prepared by oxidation using iminoxyl radicals in organic solvents followed by treatment with perchloric acid.[811] Various salts of **261** and its benzo derivatives and of 10-phenylphenothiazine were prepared by oxidation with iodine and the appropriate silver salt,[812,813] and radical salts of phenothiazine drugs have been made by oxidation of the heterocycles with the acid whose counterion was used for the salt: perchlorate[814] and sulfate.[815]

The reaction of phenothiazine with halogens is a complex subject; certainly radicals are involved. Alcais and Rau[816] showed that the first step of interaction between phenothiazines without an N-substituent and bromine, in a variety of solvents, is irreversible electron transfer to give the

---

[808] Y. Matsunaga and Y. Suzuki, *Bull. Chem. Soc. Jpn.* **46**, 719 (1973).
[809] J. Selbin, D. G. Durrett, H. J. Sherrill, G. R. Newkome, and M. Collins, *J. Inorg, Nucl. Chem.* **35**, 3467 (1973).
[810] J. Selbin and H. J. Sherrill, *J. C. S., Chem. Commun.*, 120 (1973).
[811] A. A. Medzhidov, E. G. Rozantsev, and M. B. Neimann, *Dokl. Akad. Nauk SSR* **168**, 348 (1966) [*CA* **65**, 8707 (1966)].
[812] M. H. Litt and J. Redovic, *J. Phys. Chem.* **78**, 1750 (1974).
[813] B. K. Bandlish, A. G. Padilla, and H. J. Shine, *J. Org. Chem.* **40**, 2590 (1975).
[814] F. H. Merkle, C. A. Discher, and A. Felmeister, *J. Pharm. Sci.* **53**, 965 (1964).
[815] M. K. Kalinowski, K. Leibler, T. Piekut, and M. Jaworski, *Adv. Mol. Relaxation Processes* **5**, 179 (1973).
[816] P. Alcais and M.-C. Rau, *Bull. Soc. Chim. Fr.*, 3390 (1969).

bromide of the cation-radical. More recently, Chiou et al.[817] have examined the bromination of 10-alkylphenothiazines and infer that here the initial attack of the halogen is electrophilic substitution, although radical formation becomes the preferred course of reaction after two such substitutions. Various solids of different stoichiometries have been obtained from phenothiazine and its isologs with halogens and their solid-state properties examined.[818,823] The interest in these solids has been for their semiconductor properties. The reader is referred to the review by Bodea and Silberg[824] for an account of the early chemistry of the halogenation of phenothiazine.

Several years ago, Shine and co-workers reported on reactions whereby sulfuric acid treatment of phenothiazine sulfoxides results in the formation of cation-radicals which are in a lower level of oxidation than the starting materials[702,703,825] (i.e., **247** → **249**; see Section VI,B,3,a). Similar formation of cation-radicals has been reported on treatment of phenothiazine sulfoxides with alcoholic HCl.[826] Shine and Mach also reported the apparent reduction of 3H-phenothiazin-3-one to 3-hydroxyphenothiazine cation-radical upon dissolution in 59% H_2SO_4[703]; others have made a similar report.[827] The means whereby reduction occurs is not obvious; Shine and Mach[703] originally suggested loss of OH˙ from the protonated sulfoxide. Hanson and Norman[719] presented arguments why this is unlikely during a discussion of the reaction in which the phenothiazinium cation (**267**) undergoes apparent reduction on treatment of its solutions in acetonitrile with water. In the writer's view, the rationale is similar in all these cases: in strong acid solutions sulfoxides suffer dehydration to phenothiazinium ions (cf. Scheme 4) (N-substituted sulfoxides give dicationic phenothiazinium ions), and phenothiazin-3-ones undergo protonation to 3-hydroxyphenothiazinium ions[703,825] (cf. Scheme 3; X = S). Attack on carbon in such phenothiazinium ions by nucleophiles present—water or whatever—results in the formation

---

[817] H. S. Chiou, P. C. Reeves, and E. R. Biehl, *J. Heterocycl. Chem.* **13**, 77 (1976).
[818] Y. Sato, M. Kinoshita, M. Sano, and H. Akamatu, *Bull. Chem. Soc. Jpn.* **42**, 548 (1969).
[819] Y. Matsunaga and K. Shono, *Bull. Chem. Soc. Jpn.* **43**, 2007 (1970).
[820] Y. Iida, *Bull. Chem. Soc. Jpn.* **44**, 663 (1971).
[821] S. N. Bhat and H. Kuroda, *Bull. Chem. Soc. Jpn.* **46**, 3585 (1973).
[822] Y. Matsunaga, *Energy Charge Transfer Org. Semicond., Proc. U.S.-Jpn. Semin., 1973*, 189 (1974) [*CA* **83**, 177966 (1975)].
[823] S. Doi and Y. Matsunaga, *Bull. Chem. Soc. Jpn.* **48**, 3747 (1975).
[824] C. Bodea and I. Silberg, *Adv. Heterocycl. Chem.* **9**, 321 (1968).
[825] H. J. Shine, D. R. Thompson, and C. Veneziani, *J. Heterocycl. Chem.* **4**, 517 (1967).
[826] D. Simov, L. Kamenov, G. Georgiev, V. Koleva, and L. Mladenova, *God. Sofii Univ., Khim. Fak.* **67**, 252 (1976) [*CA* **86**, 154898 (1977)].
[827] M. Raileanu, I. Radulian, O. Duliu, and S. Radu, *Collect. Czech. Chem. Commun.* **37**, 478 (1972).

of a substituted phenothiazine which may serve as reductant. Mixed radical products may conceivably occur; or, depending upon relative redox potentials, the reducing secondary phenothiazine may be oxidized to diamagnetic products, in which case the only radical products to be observed by ESR will be the one in the oxidation level below that of the initial sulfoxide, (viz. Scheme 5, for the production of **249** from **247**). Shine and Mach[703] identified **268** in their reaction mixture, and indeed invoked nucleophilic attack of water on **267** to account for it, but did not recognize the reducing role of its precursor.

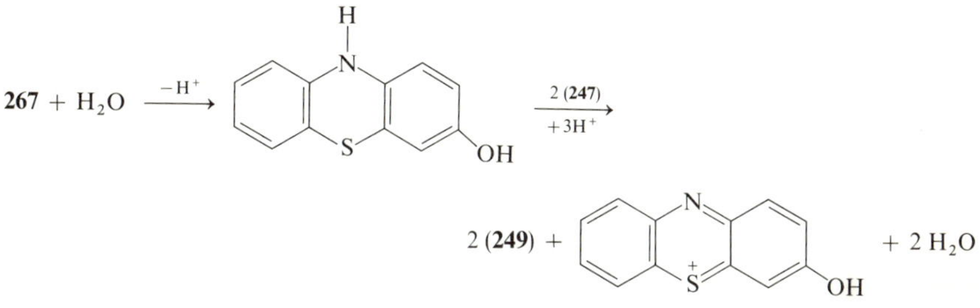

SCHEME 5

There have been kinetic investigations of reactions which produce **249**. Phenothiazines form charge-transfer complexes with electron acceptors, and with certain strongly oxidizing quinones complete electron transfer occurs forming a pair of ion-radicals.[828,829] The latter step is slow in the case of phenothiazine and chloranil, and is part of a suite of reactions leading to a bimolecular addition compound[830–832]; the kinetics of the equilibration of the charge-transfer complex with the pair of ion-radicals has been studied by Barigand et al.[833]

Diazonium ions also oxidize phenothiazines at a measurably slow rate to their cation-radicals, and this reaction has been the subject of kinetic measurements.[834] The mechanism deduced involves the rapid formation of a σ-complex, by attack of the diazonium ion at the heterocyclic nitrogen,

---

[828] R. Foster and P. Hanson, *Biochim. Biophys. Acta* **112**, 482 (1966).
[829] R. Foster and C. A. Fyfe, *Biochim. Biophys. Acta* **112**, 490 (1966).
[830] M. Barigand, J. Orszagh, and J. J. Tondeur, *Bull. Soc. Chim. Belg.* **79**, 177 (1970).
[831] J. J. Tondeur, M. Barigand, and J. Orszagh, *Bull. Soc. Chim. Belg.* **79**, 401 (1970).
[832] P. C. Dwivedi, K. G. Rao, S. N. Bhat, and C. N. R. Rao, *Spectrochim. Acta, Part A* **31**, 129 (1975).
[833] M. Barigand, J. Orszagh, and J. J. Tondeur, *Bull. Soc. Chim. Fr.*, 51 (1973).
[834] J. M. Bisson, P. Hanson, and D. Slocum, *J. C. S. Perkin II*, 1331 (1978).

which then undergoes homolysis of the new N—N bond in the rate-determining step, the transition state of which resembles the radical products of the reaction. Methyl groups at nitrogen or on carbon atoms adjacent to heteroatoms hinder the reaction, whereas elsewhere in the molecule they enhance it.

Since phenothiazinyls and their isologs are not, in general, isolable (exceptions are **257**[726-728] and certain annelated phenothiazinyls[757]), they are customarily prepared *in situ* before requirement. Methods of preparation have involved, in addition to the physical methods mentioned previously (Section VI,B,3,c,i), oxidation by PbO_2, AgO, HgO, and $KMnO_4$[714,757] and hydrogen abstraction by other radicals (e.g., diphenylpicrylhydrazyl or tri-*tert*-butylphenoxyl)[714]; oxidation of both phenothiazine and phenoxazine with a mixture of acetic anhydride and DMSO also yielded neutral radicals.[835,836] Phenoxazinyls have been formed during oxidative coupling of 2-aminophenols[837] and during oxidative cleavage of 2,3-dihydro-1,3,2λ^5-benzoxazastiboles.[838]

Besides the earlier mentioned photolytic routes to the nitroxyl radicals, these radicals are also generally accessible by reaction of the parent heterocycles with hydroperoxides.[709,720,839] Phenoxazine-10-oxyls have also been obtained by the trapping of phenoxyl radicals with nitrosobenzenes.[840-842]

d. *Radical Reactivity*. There have been several product studies on the nonphotolytic oxidative degradation of phenothiazines and isologs. Tsujino reported oxidations with acetic anhydride and DMSO,[835,836] with iodine and DMSO,[843-845] and with concentrated sulfuric acid.[846] It was shown by ESR spectroscopy that the first oxidation mixture contained the neutral radicals whereas the latter two contained the cation-radicals. Products isolated from these reaction mixtures included 3,10′- and 1,10′-biphenothiazinyls, indicative of dimerization of the radicals concerned. Similar

---

[835] Y. Tsujino, *Tetrahedron Lett.*, 4111 (1968).
[836] Y. Tsujino, *Nippon Kagaku Zasshi* **90**, 304 (1969) [*CA* **70**, 115093 (1969)].
[837] H. B. Stegmann. K. Scheffler, F. Stöcker, and H. Bürk, *Chem. Ber.* **101**, 262 (1968).
[838] G. Bauer, K. Scheffler, and H. B. Stegmann, *Chem. Ber.* **109**, 2231 (1976).
[839] S. M. Kavun and A. L. Buchachenko, *Zh. Fiz. Khim.* **42**, 818 (1968) [*CA* **69**, 86931 (1968)].
[840] S. Terabe, K. Kuruma, and R. Konaka, *Chem. Lett.*, 115 (1972).
[841] S. Terabe and R. Konaka, *J. C. S. Perkin II*, 2163 (1972).
[842] E. L. Zhuzhgov, N. M. Bazhin, M. P. Terpugova, and Yu. D. Tsvetsko, *Izv. Akad. Nauk SSSR, Ser. Khim.*, 2530 (1974) [*CA* **82**, 97779 (1975)].
[843] Y. Tsujino, *Tetrahedron Lett.*, 763 (1969).
[844] Y. Tsujino, *Nippon Kagaku Zasshi*, **90**, 490 (1969) [*CA* **71**, 49876 (1969)].
[845] Y. Tsujino, *Nippon Kagaku Zasshi* **90**, 809 (1969) [*CA* **71**, 101795 (1969)].
[846] T. Constantinescu and S. Enache, *Farmacia (Bucharest)* **19**, 731 (1971) [*CA* **76**, 140679 (1972)].

Sec. VI.B]   HETEROAROMATIC RADICALS, PART II   145

results have been observed by groups in Romania[846,847] and The Netherlands.[848-851] The nature of a dimeric product (**269**) from the oxidation of phenothiazine which gives a characteristic green color was suggested by Foster and Hanson,[828] and proved by unambiguous synthesis by Hanson and Norman.[719] The biphenothiazinyls are now well characterized materials, whose spectroscopic, electrochemical, and radical-forming properties have been investigated.[760,762,848-853]

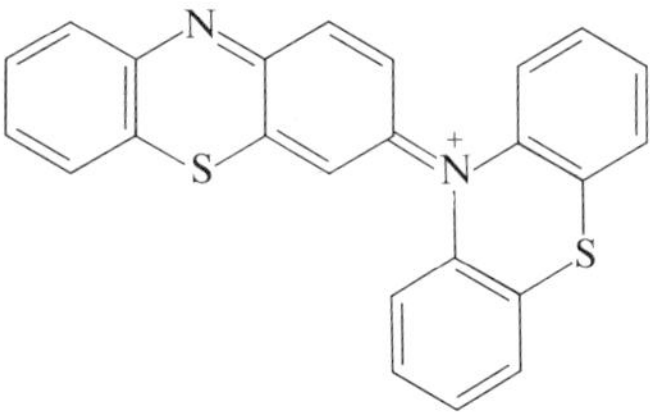

(**269**) 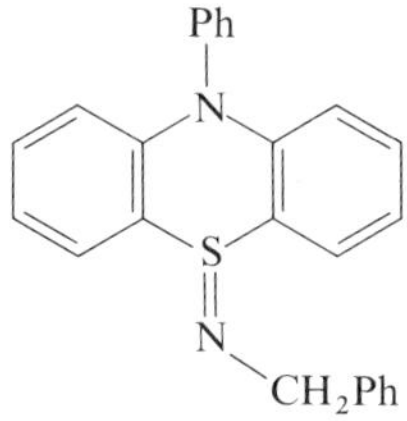   (**270**)

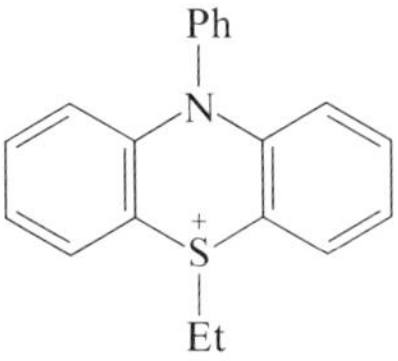

(**271**) 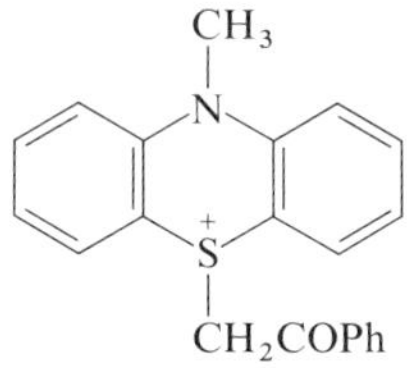   (**272**)

Shine and co-workers have carried out product studies of the reactions of N-substituted phenothiazine cation-radicals with nucleophilic reagents.[549,557,813,854,855] As with the cation-radicals of thianthrene (**178**) and phenoxathiin (**198**), addition of certain nucleophiles at S may occur to give, ultimately, such adducts as **270–272** (see Sections III,C,4,b; V,B,1). Attack by other nucleophiles may result in reduction by electron transfer or substitution at position 3. Shine[522,523] has reviewed much of this work. The mechanisms of such reactions have been controversial (see Section

[847] T. Constantinescu and S. Enache, *Farmacia (Bucharest)* **20**, 591 (1972) [*CA* **78**, 163996 (1973)].
[848] H. Roseboom and J. A. Fresen, *Pharm. Acta Helv.* **50**, 55 (1975).
[849] H. Roseboom and J. A. Fresen, *Pharm. Acta Helv.* **50**, 64 (1975).
[850] H. Roseboom and J. H. Perrin, *J. Pharm. Sci.* **66**, 1392 (1977).
[851] H. Roseboom and J. H. Perrin, *J. Pharm. Sci.* **66**, 1395 (1977).
[852] Y. Tsujino, *Tetrahedron Lett.*, 2545 (1968).
[853] Y. Tsujino, K. Nishikida, and T. Naito, *Nippon Kagaku Zasshi* **91**, 1080 (1970) [*CA* **74**, 125598 (1971)].
[854] A. G. Padilla, B. K. Bandlish, and H. J. Shine, *J. Org. Chem.* **42**, 1833 (1977).
[855] B. K. Bandlish, K. Kim, and H. J. Shine, *J. Heterocycl. Chem.* **14**, 209 (1977).

II,C,4,b); for many reactions which result in addition at sulfur the consensus now seems against prior disproportionation and in favor of a "half-regenerative" mechanism in the simplest form of which the radical adduct of the initial cation-radical and the nucleophile is oxidized by a second cation-radical, thus yielding a diamagnetic sulfur adduct and a phenothiazine molecule in equimolar amounts. The work of Blount and co-workers on the hydrolysis and pyridination of **178**[561] and on the pyridination of **264**[856] is an elegant demonstration of the use of electrochemical techniques in unraveling the subtleties of the reaction kinetics of cation-radicals. A kinetic study of the hydrolysis of chlorpromazine cation-radical has also indicated that here too simple disproportionation of the cation-radical does not occur.[857] These kinetic studies, taken together, imply that earlier work on the kinetic stability of phenothiazine cation-radicals in aqueous acid requires reappraisal.[858,859] A kinetic study of the reduction of chlorpromazine cation-radical by ascorbic acid has also led to the conclusion that a prior complex of the reactants participates in a rate-determining redox step rather than that the radical-cation disproportionates before reduction of the upper oxidation state.[860]

Disproportionation equilibria have been studied for various systems. Cauquis and co-workers[761,861] investigated by electrochemical means the matrix of equilibria corresponding to Scheme 2 for 3,7-dimethoxyphenothiazine and its derivatives, and applied the measurement of the response of the equilibria to different conditions of basicity to the definition of a scale of basicity in acetonitrile. The disproportionation kinetics of the iron–thionine system were measured several years ago[801]; solvent effects on the disproportionation rate constant have been examined,[862] and, lately, an indirect measurement of the synproportionation rate constant of thionine and leucothionine has been made.[863]

Other electron-transfer reaction rates which have been measured are those between phenothiazine cation-radical (**249**) and dimethyl- and tetramethyl-*p*-phenylenediamines.[864] Electron transfer takes place at the diffusion-controlled rate; equilibrium constants were determined which demonstrated the greater stability of the Würster's radicals over **249**. This

---

[856] J. F. Evans, J. R. Lenhard, and H. N. Blount, *J. Org. Chem.* **42**, 983 (1977).
[857] H. Y. Cheng, P. H. Sackett, and R. L. McCreery, *J. Am. Chem. Soc.* **100**, 962 (1978).
[858] T. N. Tozer and L. D. Tuck, *J. Pharm. Sci.* **54**, 1169 (1965).
[859] L. Levy, T. N. Tozer, L. D. Tuck, and D. B. Loveland, *J. Med. Chem.* **15**, 898 (1972).
[860] N. A. Klein and D. L. Toppen, *J. Am. Chem. Soc.* **100**, 4541 (1978).
[861] G. Cauquis, A. Deronzier, D. Serve, and E. Vieil, *J. Electroanal. Chem. Interfacial Electrochem.* **60**, 205 (1975).
[862] P. D. Wildes, N. N. Lichtin, and M. Z. Hoffmann, *J. Am. Chem. Soc.* **97**, 2288 (1975).
[863] P. D. Wildes and N. N. Lichtin, *J. Phys. Chem.* **82**, 981 (1978).
[864] A. Yamagishi, *Bull. Chem. Soc. Jpn.* **48**, 3475 (1975).

is consistent with the effect, noted earlier, of the para NMe_2 substituent in the 10-phenylphenothiazine cation-radical in bringing about a marked redistribution of spin population and concomitant geometry change[754] (see Section VI,B,3,b,iii.) The homogeneous electron-transfer reactions of **248**, **249**, and **261** with their respective unoxidized heterocycles were investigated by Bard and co-workers[48]; the influence on the rate of ion-pairing effects was studied for **249** by Sorensen and Bruning.[865] The luminescent electron transfer from fluoranthene anion to **261** has been included in several researches by Faulkner and co-workers.[866–870] The electrochemiluminescent behavior of a selection of laser dyes from the oxazine and thiazine families has been investigated.[733]

Phenoselenazine cation-radical **250** undergoes reversible dimerization in aqueous sulfuric acid. The equilibrium constant and thermodynamic parameters for the dimerization of **250** and its benzo[c] derivative have been determined by Matsunaga and Tanaka.[871]

4. Thia Analogs of Flavin and Pterin Radicals

Fenner and co-workers have described radicals in which an NH from the pyrazine ring of lumazine or isoalloxazine derived radicals is replaced by S (e.g., **273–275**)[872–874] (cf. Part I: Section III,C,5). Cation-radicals were generated by solution of the parent heterocycles in concentrated sulfuric acid or by oxidation by dibenzoyl peroxide in trifluoracetic acid-containing

(273)   (274)   (275)

---

[865] S. P. Sorensen and W. H. Bruning, *J. Am. Chem. Soc.* **95**, 2445 (1973).
[866] D. J. Freed and L. R. Faulkner, *J. Am. Chem. Soc.* **93**, 2097 (1971).
[867] D. J. Freed and L. R. Faulkner, *J. Am. Chem. Soc.* **93**, 3565 (1971).
[868] L. R. Faulkner, H. Tachikawa, and A. J. Bard, *J. Am. Chem. Soc.* **94**, 691 (1972).
[869] D. J. Freed and L. R. Faulkner, *J. Am. Chem. Soc.* **94**, 4790 (1972).
[870] R. Bezman and L. R. Faulkner, *J. Am. Chem. Soc.* **94**, 6331 (1972).
[871] Y. Matsunaga and T. Tanaka, *Bull. Chem. Soc. Jpn.* **48**, 1043 (1975).
[872] H. Fenner, *Tetrahedron Lett.*, 617 (1970).
[873] H. Fenner, H. Motschall, S. Ghisla, and P. Hemmerich, *Justus Liebigs Ann. Chem.*, 1793 (1974).
[874] H. Fenner, R. W. Grauert, and P. Hemmerich, *Justus Liebigs Ann. Chem.*, 193 (1978).

benzene. The neutral radical **275** was obtained by treatment of 1,2,3,4-tetrahydro-1,3-dimethyl-2,4-dioxo-10H-pyrimido[5,4-b][1,4]benzothiazine with biacetyl in DMSO. Comparison of these radicals with their lumazine and alloisoxazine counterparts indicates that the substitution of S for NH does not have great effect on the spin distribution.

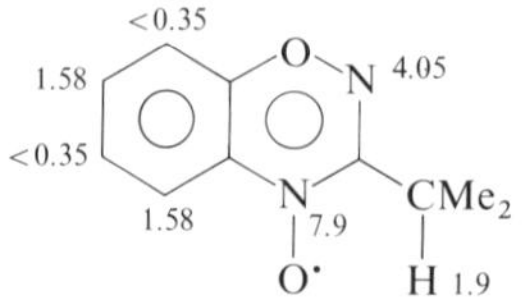

(276) 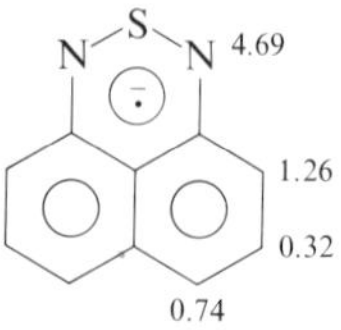   (277) 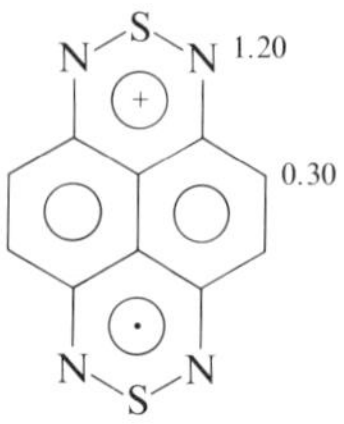   (278)

5. Radicals from Oxadiazines and Thiadiazines

Aurich and Stork[875] have described benzo-1,2,4-oxadiazine oxyls (e.g., **276**) which arise via cycloaddition of phenylhydroxylamine to nitrile oxides. (Hyperfine splittings are indicated for **276** in gauss.)

The anion-radical of naphtho[1,8-cd][1,2,6]thiadiazine (**277**) has been known for several years[664]; it has also been the subject of various MO calculations.[418,419] Recently both the cation-radical (**278**) and the anion-radical of naphtho[1,8-c,d;4,5-c'd']bis[1,2,6]thiadiazine have been prepared as part of an exploration of the aromatic character, or otherwise, of the heterocycle.[876] (Hyperfine splittings are indicated for **277** and **278** in gauss.)

VII. Conclusion

Between the two parts of this review over 1400 references have been cited which are relevant to heteroaromatic radicals. The majority have described work in which the characterization of the radicals has been the primary concern although it is evident that an increasing amount of work is now being devoted to their reactivity. Necessarily, the thrust of most of the papers cited is lost in a review of this kind and, no doubt, some workers have been unwittingly misrepresented, or their data inaccurately transcribed. For these shortcomings the author proffers apologies but hopes that, nevertheless,

---

[875] H. G. Aurich and K. Stork, *Chem. Ber.* **108**, 2764 (1975).
[876] R. C. Haddon, M. L. Kaplan, and J. H. Harshall, *J. Am. Chem. Soc.* **100**, 1235 (1978).

Sec. VII] HETEROAROMATIC RADICALS, PART II 149

the review as a whole may serve some useful purpose in illustrating the diversity of heteroaromatic radical structures and in providing for the first time a comprehensive compendium of papers in this field.

ACKNOWLEDGMENT

I am indebted to colleagues for helpful comments during the preparation of this chapter and especially to my wife for her assistance with the collation of references.

ADVANCES IN HETEROCYCLIC CHEMISTRY VOL. 27

The 1,2- and 1,3-Dithiolium Ions

NOËL LOZAC'H AND MADELEINE STAVAUX

*Institut des Sciences de la Matière et du Rayonnement,
Université de Caen, Caen, France*

I. Introduction	152
II. The 1,2-Dithiolium Ion	153
A. Methods of Preparation	153
1. With Formation of the 1,2-Dithiole Ring	153
2. From 1,2-Dithiole Derivatives	159
B. Chemical Properties	168
1. General Remarks	168
2. Hydrogen Exchange	169
3. Reduction	170
4. Attack by Oxygen Nucleophiles	171
5. Attack by Sulfur or Selenium Nucleophiles	173
6. Attack by Nitrogen Nucleophiles	174
7. Attack by Carbon Nucleophiles	183
8. Electrophilic Attack	190
C. Physical Properties	192
1. Quantum Mechanical Calculations	192
2. Molecular Structure	193
3. Electronic Absorption Spectra	194
4. NMR, ESR, and ESCA Data	194
5. Conductivity	195
6. Mass Spectra	198
III. The 1,3-Dithiolium Ion	199
A. Methods of Preparation	199
1. With Formation of the 1,3-Dithiole Ring	199
2. From 1,3-Dithiole Derivatives	206
B. Chemical Properties	212
1. General Remarks	212
2. Hydrogen Exchange	214
3. Reduction	214
4. Attack by Oxygen Nucleophiles	216
5. Attack by Sulfur Nucleophiles	219
6. Attack by Nitrogen Nucleophiles	220
7. Attack by Carbon Nucleophiles	221
8. Attack by Phosphorus Nucleophiles	228
9. Electrophilic Attack	229
10. Addition Reactions of 2,5-Disubstituted 1,3-Dithiolium-4-olates	230

C. Physical Properties 231
  1. Quantum Mechanical Calculations 231
  2. Molecular Structure 232
  3. IR and Electronic Absorption Data 233
  4. NMR Data 235
  5. Conductivity 237
  6. Miscellaneous Physical Properties 238

I. Introduction

The object of this chapter is to bring up to date the review by Prinzbach and Futterer[1] published earlier in this series. For this reason, except when necessary for a good understanding, only papers appearing later than this review will be included.

In the last 12 years, the chemistry of 1,2- and 1,3-dithiole compounds has developed considerably for several reasons. One reason is that bis(1,3-dithiol-2-ylidene), often called tetrathiafulvalene, and its derivatives are important components of organic metals. Another is that the discovery of 1,6,6aλ^4-trithiapentalenes has brought attention to their partial bonding with quasi-covalent bonds whose bonding character, in terms of "usual" covalent bonds, is between zero and unity.

In these fields, as elsewhere, dithiolium cations are useful synthetic intermediates. In the case of 1,6,6aλ^4-trithiapentalenes, some synthetic applications of 1,2-dithiolium cations were reviewed by one of us[2] in this series in 1971. For these particular topics we shall not refer to papers already considered in that review (the corresponding information will be found in Ref. 2: pp. 173–177, 181, 187–190, 193, 197, 201–202, 217, and 223).

Most known neutral dithiole derivatives may be written as polar compounds, and these dipolar formulas give a correct picture of some properties of these compounds. As typical examples we may cite the 1,2-dithiole-3-thiones and 1,6,6aλ^4-trithiapentalenes to which formulas **1** and **2**, respectively, may be given.

(1)   (2)

Accordingly, the scope of this review must be defined relatively arbitrarily when deciding what should be described as a 1,2-dithiolium cation. For instance, let us consider the protonation of compounds having the general

---

[1] H. Prinzbach and E. Futterer, *Adv. Heterocycl. Chem.* **7**, 39 (1966).
[2] N. Lozac'h, *Adv. Heterocycl. Chem.* **13**, 161 (1971).

formula 3 where X may be O, S, NR, or CR_2. When the medium is not very strongly acidic, the dithiolone (X = O) or dithiolethione (X = S) is predominant, while protonation of the compounds where X is NR or CRR' is generally rather easy. Accordingly, we shall consider these last compounds as "normally" being in the 1,2-dithiolium form, which is not the case for dithiolones and dithiolethiones.

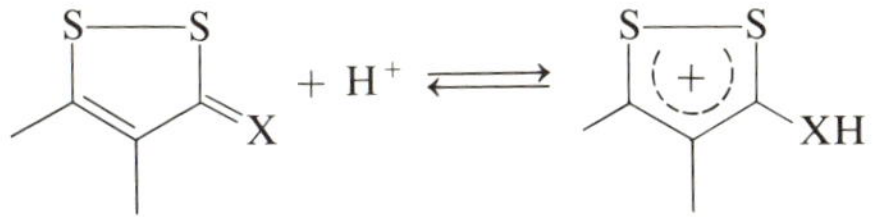

(3)

For compounds with partial bonding, such as 1,6,6aλ^4-trithiapentalenes, the "1,2-dithiolium" character is low, and these compounds will not be considered in this review. The reader may refer to another review[2] of this series.

II. The 1,2-Dithiolium Ion

A. METHODS OF PREPARATION

1. *With Formation of the 1,2-Dithiole Ring*

a. *From α-Unsaturated Ketones or Aldehydes* Reaction of phosphorus pentasulfide on α-ethylenic ketones, followed by iodine oxidation, leads to the formation of 1,2-dithiolium salts in poor yield, as shown in Eq. (1).[3]

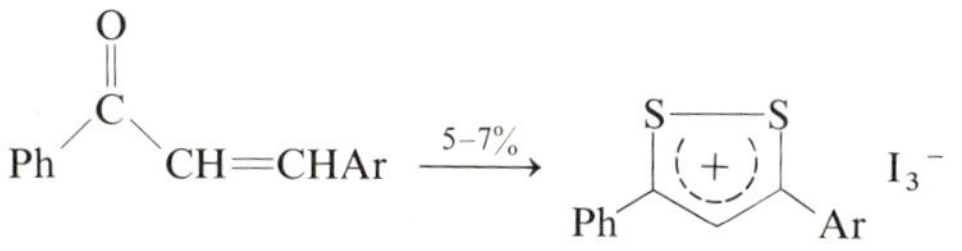

(1)

In the presence of hydrogen chloride, α-acetylenic ketones react with thioacetic acid to give a monothio-β-diketone which is converted to a 1,2-dithiolium ion by phosphorus pentasulfide or hydrogen sulfide.[4] These reactions, which involve some sort of oxidation, are discussed in the next paragraph.

b. *From β-Diketones and Their Functional Derivatives.* These methods, already cited in the previous review (Ref. 1: pp. 42–45) have been widely used, with many variations in experimental details.

---

[3] J.-P. Guemas and H. Quiniou, *Bull. Soc. Chim. Fr.*, 592 (1973).
[4] H. Behringer and A. Grimm, *Justus Liebigs Ann. Chem.* **682**, 188 (1965).

Hydrogen sulfide has been used in the presence of oxidative agents such as bromine,[5] iodine,[3,5,6] and ferric salts.[7–9] In these reactions, there is always some strong acid present, mostly hydrogen halide. Acid catalysis is probably essential, as it appears that high concentrations of hydrogen chloride improve the yield.

Hydrogen disulfide or polysulfide, in strongly acidic medium, also lead to 1,2-dithiolium cations. Some of these procedures have already been reviewed.[1] Other studies may be cited, which employ hydrogen disulfide with hydrogen chloride,[10–12] hydrogen disulfide with hydrogen bromide,[13,14] hydrogen disulfide with perchloric acid,[13–16] and hydrogen polysulfides with hydrogen chloride.[17]

The mechanisms of these reactions are not completely elucidated, but some reaction paths have been substantiated by isolation of intermediates. The simplest case is apparently the reaction of hydrogen disulfide in the presence of strong protonic acids (Scheme 1).

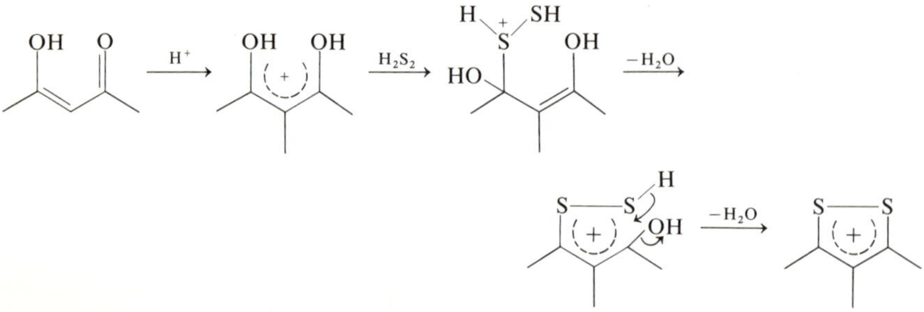

SCHEME 1

Owing to preparative problems arising from the instability of hydrogen disulfide and polysulfides, more readily available reagents seem to be

---

[5] A. R. Hendrickson and R. L. Martin, *J. Org. Chem.* **38**, 2548 (1973).
[6] J.-P. Guemas and H. Quiniou, *C. R. Hebd. Seances Acad. Sci., Ser. C* **268**, 1805 (1969).
[7] K. Knauer, P. Hemmerich, and J. D. W. van Voorst, *Angew. Chem., Int. Ed. Engl.* **6**, 262 (1967).
[8] G. A. Heath, R. L. Martin, and I. M. Stewart, *J. C. S., Chem. Commun.*, 54 (1969).
[9] G. A. Heath, R. L. Martin, and I. M. Stewart, *Aust. J. Chem.* **22**, 83 (1969).
[10] D. Leaver, W. A. H. Robertson, and D. M. McKinnon, *J. Chem. Soc.*, 5104 (1962).
[11] D. Barillier, P. Rioult, and J. Vialle, *Bull. Soc. Chim. Fr.*, 3031 (1973).
[12] D. Barillier, P. Rioult, and J. Vialle, *Bull. Soc. Chim. Fr.*, 444 (1976).
[13] J. C. Poite, S. Coen, and J. Roggero, *Bull. Soc. Chim. Fr.*, 4373 (1971).
[14] S. Coen, J. C. Poite, and J. P. Roggero, *Bull. Soc. Chim. Fr.*, 611 (1975).
[15] J. G. Dingwall, S. McKenzie, and D. H. Reid, *J. Chem. Soc.*, 2543 (1968).
[16] J. G. Dingwall, A. S. Ingram, D. H. Reid, and J. D. Symon, *J. C. S., Perkin 1*, 2351 (1973).
[17] M. Schmidt and H. Schulz, *Chem. Ber.* **101**, 277 (1968).

generally preferred, such as hydrogen sulfide in the presence of a strong protonic acid and an oxidizing agent. Various mechanisms may be considered, the main question being whether the S—S bond is made before or after the C—S bonds. In the first case, a mechanism of the hydrogen disulfide type may be assumed, but this pathway is certainly not unique. Other mechanisms have been suggested and some intermediates isolated, such as monothio-β-diketones and the corresponding disulfides.[3]

A possible mechanism is given in Scheme 2.

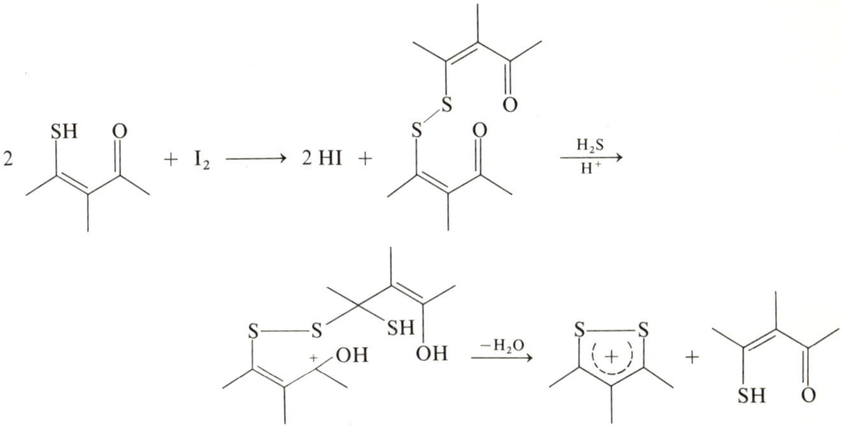

SCHEME 2

Another possible intermediate is the β-dithioketone or the corresponding disulfide shown in Scheme 3.

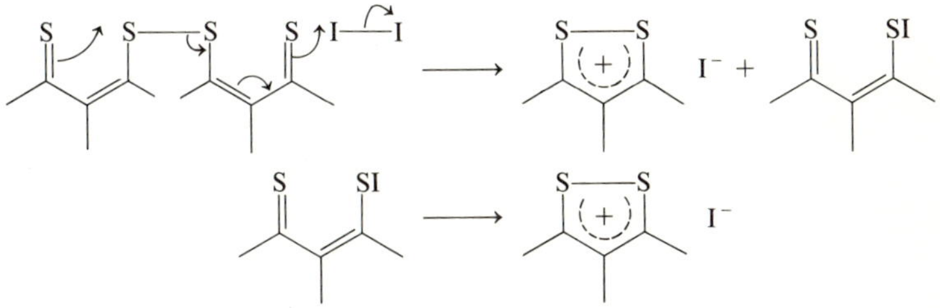

SCHEME 3

Sulfurization of 1,3-diphenylpropanetrione by hydrogen sulfide and hydrogen chloride affords a mixture of 3,5-diphenyl-1,2-dithiolium chloride and 4-hydroxy-3,5-diphenyl-1,2-dithiolium chloride.[18] Thus, the ring formation, realized as in the preceding reactions, is accompanied by some

---

[18] A. Chinone, K. Inouye, and M. Ohta, *Bull. Chem. Soc. Jpn.* **45**, 213 (1972).

reduction of the central carbonyl. In the presence of a base such as triethylamine, the 4-hydroxy-3,5-diphenyl-1,2-dithiolium cation (**4**) is deprotonated to give the corresponding mesoionic compound **5**.

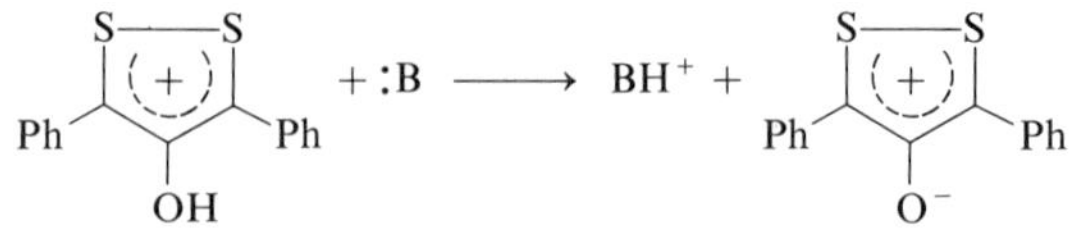

When hydrogen disulfide sulfurization leads to a 1,2-dithiolium cation having a labile hydrogen on a substituent, water is sufficient to deprotonate the 1,2-dithiolium ion (Eq. 2).[19] A similar deprotonation was observed with a similarly placed isopropyl substituent.[14]

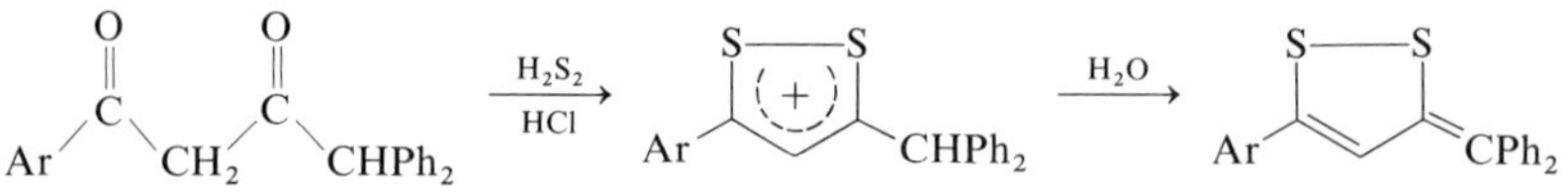

Diacetyl disulfide in acidic medium has also been used[20,21]; Scheme 4 shows a likely pathway.

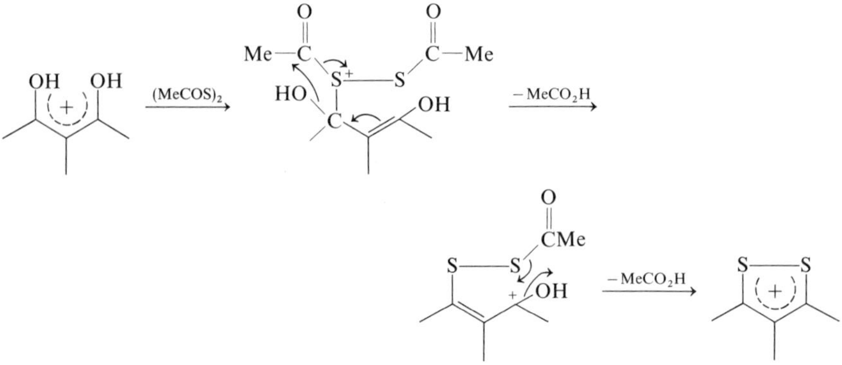

SCHEME 4

A widely used sulfurization agent is tetraphosphorus decasulfide, commonly named phosphorus pentasulfide. This reagent has been used with β-diketones, leading to 1,2-dithiolium salts.[11,22,23] Yields seem to be improved when a mild oxidizing agent such as iodine is added.[3] A side reaction, common when phosphorus pentasulfide is used, is the dealkyla-

[19] P. Rioult and J. Vialle, *Bull. Soc. Chim. Fr.*, 2883 (1967).
[20] H. Hartmann, K. Fabian, B. Bartho, and J. Faust, *J. Prakt. Chem.* **312**, 1197 (1970).
[21] K. Hartke and D. Krampitz, *Chem. Ber.* **107**, 739 (1974).
[22] G. Duguay and H. Quiniou, *Bull. Soc. Chim. Fr.*, 1918 (1970).
[23] G. Brisset, L. Morin, D. Paquer, and P. Rioult, *Recl. Trav. Chim. Pays-Bas.* **96**, 161 (1977).

tion of alkoxy or alkylthio groups. As shown in Scheme 5, this dealkylation does not take place when hydrogen disulfide is used.[24]

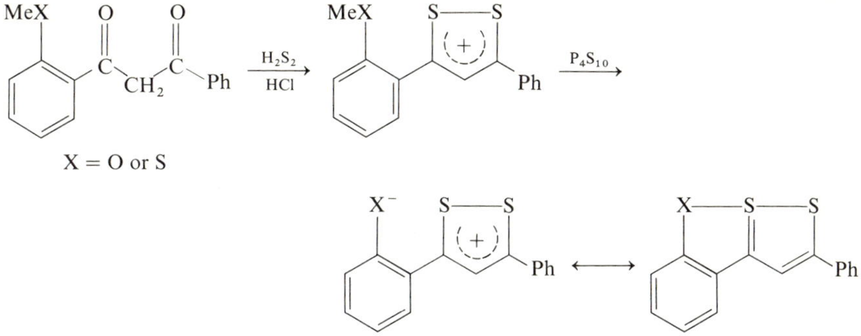

SCHEME 5

Various functional derivatives of β-diketones have been converted to 1,2-dithiolium cations by diacyl disulfides (Eq. 3).[20] Similarly, 9-ethoxyphenalen-1-one and phosphorus pentasulfide gave the phenaleno[1,9-cd]-1,2-dithiolium cation,[25] and thioamide vinylogs the corresponding 1,2-dithiolium cations (Eq. 4).[22]

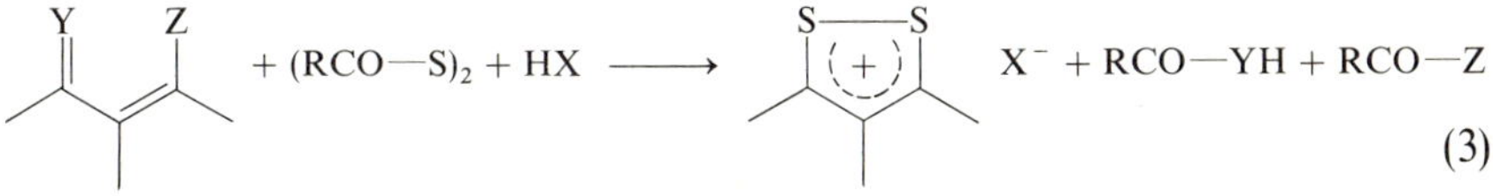

$$Y = O, S, (OEt)_2, \overset{+}{NMe_2}$$
$$Z = OH, Cl, OEt$$

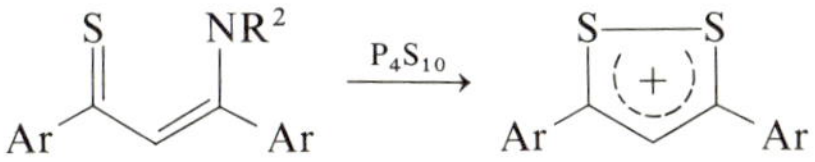

The product obtained by reacting potassium O-ethyl dithiocarbonate with 1,1,3,3-tetrabromo-1,3-diphenylpropan-2-one in dimethylformamide, initially described as a dithiotriketone, is more correctly represented by the mesoionic 1,2-dithiolium structure **5**.[26]

Monothiophosphates derived from the enol form of a β-dithioketone (**6**) give, when treated with perchloric acid, a 1,2-dithiolium cation and a dialkylphosphite (Eq. 5).[27]

---

[24] D. Barillier, C. Gy, P. Rioult, and J. Vialle, *Bull. Soc. Chim. Fr.*, 277 (1973).
[25] R. C. Haddon, F. Wudl, M. L. Kaplan, and J. H. Marshall, *J. C. S., Chem. Commun.*, 429 (1978); R. C. Haddon, F. Wudl, M. L. Kaplan, J. H. Marshall, R. E. Cais, and F. B. Bramwell, *J. Am. Chem. Soc.* **100**, 7629 (1978).
[26] A. Schönberg and E. Frese, *Chem. Ber.* **103**, 3885 (1970).
[27] B. A. M. Oude-Alink, U.S. Patent 3,957,925 (1976) [*CA* **85**, 177427 (1976)].

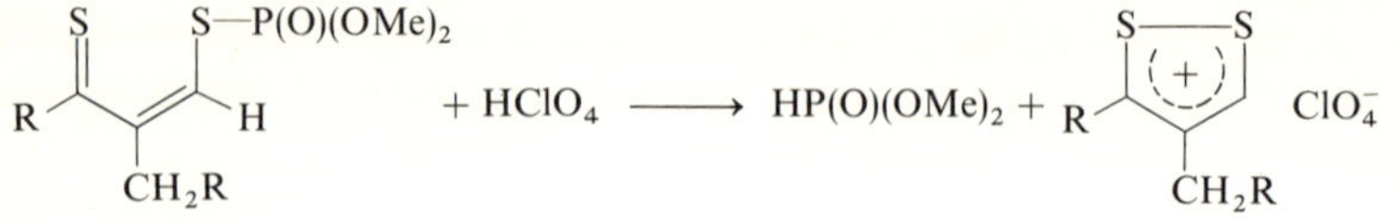

$$R\overset{S}{\underset{CH_2R}{\diagdown}}\!\!\!=\!\!\!\overset{S-P(O)(OMe)_2}{\underset{H}{\diagup}} + HClO_4 \longrightarrow HP(O)(OMe)_2 + R\overset{S\!-\!S}{\underset{CH_2R}{(+)}} ClO_4^- \quad (5)$$

(6) R = t-Bu

c. *From β-Oxocarboxylic Acids and Related Compounds.* When α-(bismethylthio)methylene ketones (7) are treated with phosphorus pentasulfide in boiling xylene, 1,2-dithiole-3-thiones are obtained in good yield. However, at lower temperatures various intermediates including 1,2-dithiolium ions are obtained.[28-31]

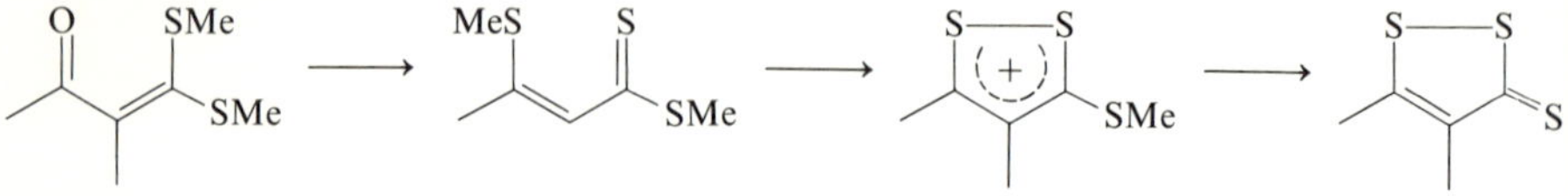

(7)

3-Amino-1,2-dithiolium salts are obtained from either β-oxocarboxamides[29,32] or β-oxocarbonitriles[33] by reaction with phosphorus pentasulfide, followed by addition of perchloric acid (Scheme 6). These cations, which may also be considered as iminium salts, are deprotonated to the corresponding 1,2-dithiol-3-imine by ammonia.[29]

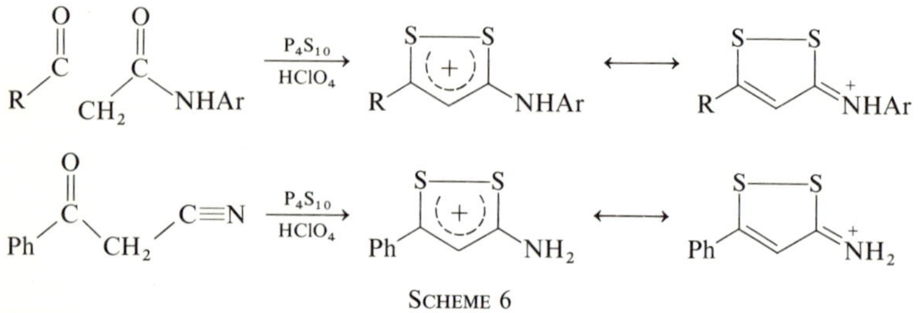

SCHEME 6

Formation of 3,5-diamino-1,2-dithiolium cations by iodine oxidation of dithiomalonamides, was cited in the earlier review.[1] More recently, the same oxidation has been performed by bromine[34] or hydrogen peroxide[35] and extended, with iodine, to diselenomalonamides.[36]

[28] G. Duguay, J. P. Biton, and H. Quiniou, *C. R. Hebd. Seances Acad. Sci., Ser. C* **266**, 1715 (1968).
[29] G. Duguay and H. Quiniou, *Bull. Soc. Chim. Fr.*, 637 (1972).
[30] J. Maignan and J. Vialle, *Bull. Soc. Chim. Fr.*, 2388 (1973).
[31] D. Barillier, P. Rioult, and J. Vialle, *Bull. Soc. Chim. Fr.*, 659 (1977).
[32] J. P. Biton, G. Duguay, and H. Quiniou, *C. R. Hebd. Seances Acad. Sci., Ser. C* **267**, 586 (1968).
[33] P. Condorelli, G. Pappalardo, and B. Tornetta, *Ann. Chim. (Rome)* **57**, 471 (1967).
[34] A. D. Grabenko, L. N. Kulaeva, and P. S. Pel'kis, *Khim. Geterotsikl. Soedin.*, 924 (1974).
[35] L. Menabue and G. C. Pellacani, *J. C. S., Dalton*, 455 (1976).
[36] K. A. Jensen and U. Henriksen, *Acta Chem. Scand.* **21**, 1991 (1967).

Sec. II.A]   THE 1,2- AND 1,3-DITHIOLIUM IONS   159

d. *Miscellaneous Methods.* By reacting carbon disulfide with sodium, in hexamethylphosphorotriamide, in the presence of methyl iodide, the tris(methylthio)-1,2-dithiolium iodide (**8**) is obtained.[37]

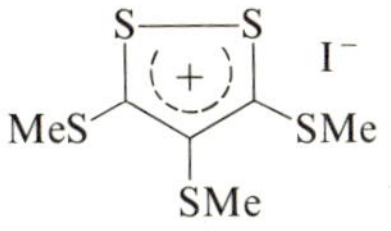

(**8**)

Reaction of some isothiazolium salts (**9**) with hydrogen sulfide leads to unstable compounds which have been described as bis(1,2-dithiol-3-yl) sulfides (**10**). These compounds are converted into 1,2-dithiolium salts (**11**) by strong acids such as hydrogen chloride or iodide, or perchloric acid.[38]

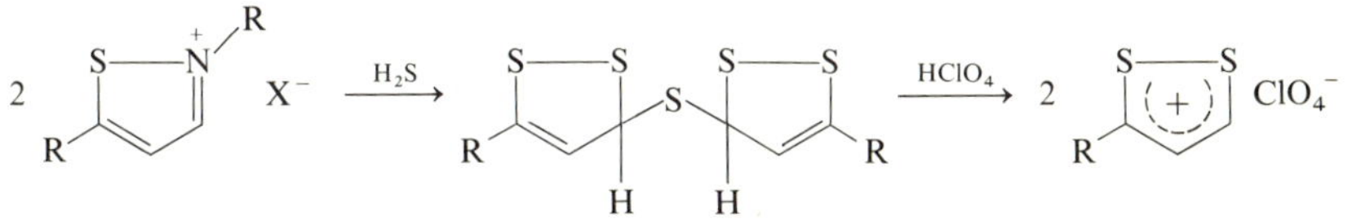

(**9**) R = H or Ph   (**10**)   (**11**)

2. *From 1,2-Dithiole Derivatives*

a. *Protonation.* Many 1,2-dithiole derivatives (**12**) with an exocyclic double bond may be considered as the conjugate bases of 1,2-dithiolium cations (**13**).

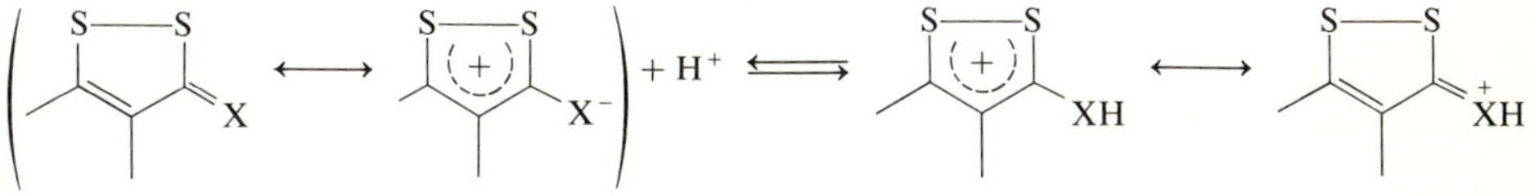

(**12**) X = O, S, Se, NR, CRR'   (**13**)

For the protonated species, various limiting forms may be considered, and their contributions to the resonance hybrid have been discussed in a previous review (Ref. 1: p. 54). The protonation equilibrium depends on the nature of X. When X is O, S, or Se, the protonated species exists in noticeable amounts only in strong acids such as concentrated sulfuric acid. On the other hand, when X is NR[39] or CRR', the protonation is easy and the formation of the 3-imino- or 3-alkylidene-1,2-dithiole needs the action of a base. The protonation of 3-alkylidene-1,2-dithioles is discussed in the earlier review (Ref. 1: p. 57).

[37] G. Kiel, U. Reuter, and G. Gattow, *Chem. Ber.* **107**, 2569 (1974).
[38] P. Sykes and H. Ullah, *J. C. S., Perkin I*, 2305 (1972).
[39] H. Behringer and D. Bender, *Chem. Ber.* **100**, 4027 (1967).

An example of the formation of 1,2-dithiolium salts by protonation of an alkylidene-1,2-dithiole is the decarboxylation of 1,2-dithiol-3-ylidene malonic acids (**14**) in the presence of perchloric acid.[40] In this particular case, a strongly acidic reaction medium is necessary for obtaining satisfactory yields because the corresponding 3-aryl-5-methylene-1,2-dithiole is unstable and undergoes an autocondensation reaction leading to an α-(thiopyran-2-ylidene)thioketone (cf. Ref. 1: p. 83).

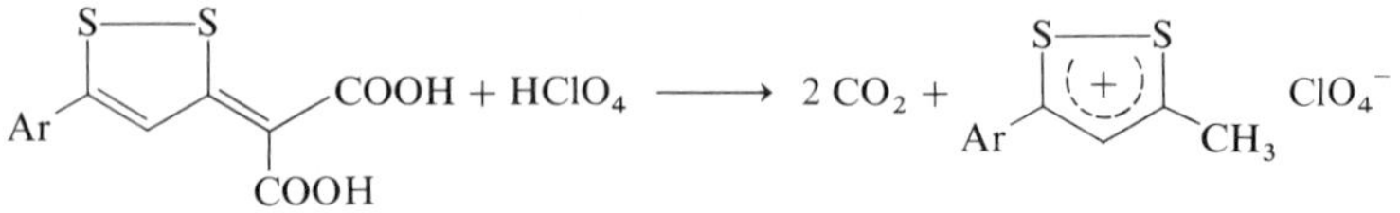

Ar	Yields (%)
Ph	71
o-MeOC$_6$H$_4$	94
p-MeOC$_6$H$_4$	60

(**14**)

b. *Oxidation by Electron Loss.* Oxidation of "tetrathiotetracene" (TTT) (**15**) to 1,2-dithiolium radical-cation or dication was discussed in an earlier review (Ref. 1: p. 48). The chemistry of this type of compound has recently been developed in view of their possible use as organic metals (see Section II,C,5).

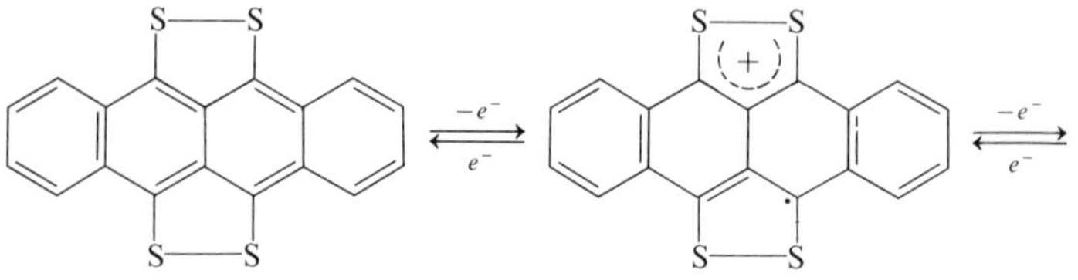

(**15**)   (Radical-cation)

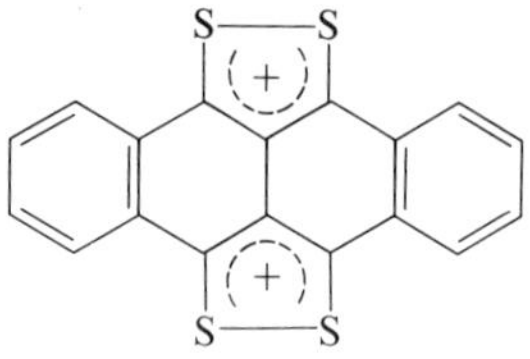

(Dication)

---

[40] M. Bard and G. Duguay, *C. R. Hebd. Seances Acad. Sci., Ser. C* **275**, 905 (1972).

In mineral acids, with or without added oxidizer, according to the oxidizing power of the acid, radical-cation and/or dication salts are formed.[41,42] The ion-radical halides and thiocyanates of nearly 1:1 stoicheiometry can be prepared by mixing solutions of TTT acetate and solutions containing an excess of the sodium or potassium salt of the anion.[43] Simply by mixing TTT with iodine in nitrobenzene, with different proportions of the reagents, two kinds of crystals are obtained with the composition (TTT)I and (TTT)$_2$I$_3$.[44]

TTT is a powerful electron donor which forms radical-cation salts with many strong electron acceptors such as 2,5-cyclohexadiene-1,4-diylidene dimalononitrile (commonly named tetracyanoquinodimethane or TCNQ),[45] o-chloranil, o-bromanil, o-iodanil, and tetracyanoethylene. These salts are prepared by mixing the components in an appropriate solvent and recovering the product as precipitate. With TCNQ the complex is a 1:1 association; however it has the stoicheiometry (TTT)$_3$A$_n$ where $n = 1$ when A is o-chloranil, o-bromanil, or o-iodanil and where $n = 2$ when A is tetracyanoethylene.[46]

Similarly, the "dithiotetracene" (DTT) and the "tetrathionaphthalene" (TTN) react with the π-electron acceptor TCNQ to give the 1:1 adducts 16,[47] and 17.[48]

(16) (17)

Other types of oxidation by electron loss are related to the formation of a disulfide linkage between two molecules of dithiole compounds. For instance, dications are formed by anodic oxidation of 1,2-dithiole-3-thiones,[49] and of 1,6,6aλ^4-trithiapentalenes,[50] which may be considered as vinylogs of 1,2-dithiole-3-thiones (Scheme 7).

[41] C. Marschalk, *Bull. Soc. Chim. Fr.*, 147 (1952).
[42] E. P. Goodings, D. A. Mitchard, and G. Owen, *J. C. S., Perkin 1*, 1310 (1972).
[43] E. A. Perez-Albuerne, H. Johnson, and D. J. Trevoy, *J. Chem. Phys.* **55**, 1547 (1971).
[44] L. I. Buravov, G. I. Zvereva, V. F. Kaminskii, L. P. Rosenberg, M. L. Khidekel, R. P. Shibaeva, I. F. Shchegolev, and E. B. Yagubskii, *J. C. S., Chem. Commun.*, 720 (1976).
[45] J. H. Perlstein, J. P. Ferraris, V. V. Walatka, D. O. Cowan, and G. A. Candela, *AIP Conf. Proc.* **10**, 1494 (1972).
[46] Y. Matsunaga, *J. Chem. Phys.* **42**, 2248 (1965).
[47] P. J. Nigrey and A. F. Garito, *J. Chem. Eng. Data* **23**, 182 (1978).
[48] F. Wudl, D. E. Schafer, and B. Miller, *J. Am. Chem. Soc.* **98**, 252 (1976).
[49] C. T. Pedersen and V. D. Parker, *Tetrahedron Lett.*, 771 (1972).
[50] C. T. Pedersen and V. D. Parker, *Tetrahedron Lett.*, 767 (1972).

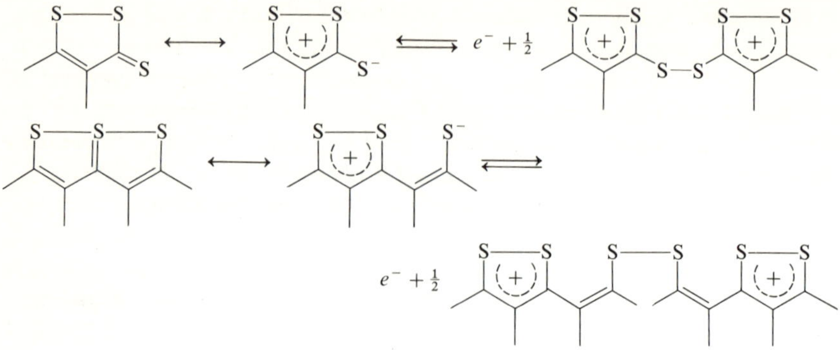

SCHEME 7

c. *Peroxide Attack of 1,2-Dithiole-3-thiones.* As indicated in a preceding review (Ref. 1: p. 49), peracetic acid oxidizes the thione sulfur to sulfate and generally gives the corresponding 1,2-dithiolium salt in good yield. 1,2-Dithiole-3-thiones being the most easily available 1,2-dithiole derivatives, this reaction is particularly useful for the synthesis of 3- or 5-unsubstituted 1,2-dithiolium cations. This type of reaction has been recently reviewed.[51] This reaction had been described as a "reduction" of 1,2-dithiole-3-thiones, which is difficult to reconcile with the fact that peracetic acid is evidently reduced. According to the present knowledge of the chemistry of 1,2-dithiole compounds, it seems most likely that the reaction of peracetic acid with 1,2-dithiole-3-thiones proceeds first through an S-oxidation of the thione sulfur to a sulfinate which afterward loses SO_2 to yield a carbene which in turn is protonated (Scheme 8).

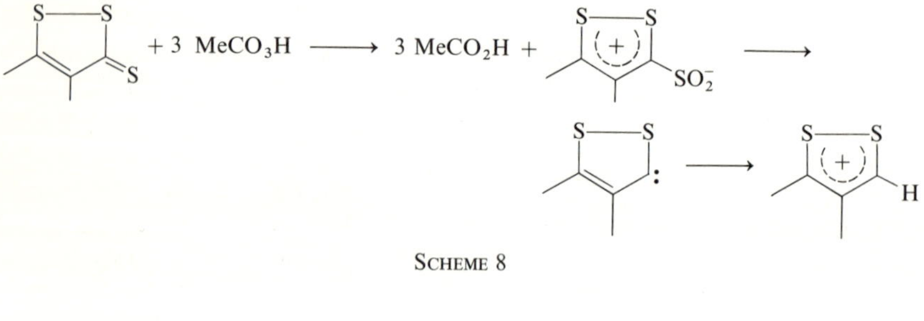

SCHEME 8

Although this reaction gives good results with aryl-1,2-dithiole-3-thiones, it does not work well with aromatic fused rings such as 1,2-benzodithiole-3-thione.[52] Some examples of this reaction are given in the preparations of the

---

[51] S. M. Loosmore and D. M. McKinnon, *Phosphorus Sulfur* **1**, 185 (1976).
[52] E. Klingsberg, *J. Org. Chem.* **37**, 3226 (1972).

unsubstituted,[10] 4-aryl,[53] 3-carboxy,[54] 3-aryl,[53] 3-styryl and substituted styryl[55] compounds, and the 5-methylnaphtho[1,2-c]dithiolium salt and its 4,5-dihydro derivative.[52]

Peracetic acid or hydrogen peroxide are generally used in this preparation. If the 1,2-dithiole-3-thione has bulky aliphatic substituents, nitric acid at 60–80°C may also be used as oxidizing agent, as in the preparation of 3-*tert*-butyl-4-neopentyl-1,2-dithiolium cation,[56] which can also be obtained using hydrogen peroxide.[27]

Chlorine in hot acetic acid gives very poor yields (0–5%) of 3-(*p*-methoxyphenyl)-1,2-dithiolium hydrogen sulfate from 5-(*p*-methoxyphenyl)-1,2-dithiole-3-thione.[57]

d. *Halogenation.* Under controlled conditions, halogens give, with 1,2-dithiole-3-thiones, adducts that may be considered as 3-halogenothio-1,2-dithiolium halides (Eq. 6).[57–59] These adducts are useful intermediates for synthesis[32,60] and easily lose sulfur, with formation of 3-halogeno-1,2-dithiolium halides.[61,62] As a side reaction, chlorination of the 1,2-dithiole ring may occur.[57,62]

$$\begin{array}{c}\text{[structure]} + X_2 \longrightarrow \text{[structure]}\ X^- \end{array} \quad (6)$$

A more selective reagent for preparing 3-chloro-1,2-dithiolium cations from 1,2-dithiol-3-ones or from 1,2-dithiole-3-thiones is oxalyl chloride (Scheme 9).[59,61,63–65]

Similarly, oxalyl bromide leads to 3-bromo-1,2-dithiolium bromides.[66] Thiophosgene or *N*-trichloromethyldichloromethanimine may also be used for converting either 1,2-dithiol-3-ones or 1,2-dithiole-3-thiones into the

---

[53] M. G. Voronkov and T. Lapina, *Khim. Geterotsikl. Soedin.*, 452 (1970).
[54] E. Klingsberg, U.S. Patent 3,186,995 (1965) [*CA* **63**, 7014 (1965)].
[55] C. Metayer and G. Duguay, *C. R. Hebd. Seances Acad. Sci., Ser. C* **273**, 1457 (1971).
[56] B. A. M. Oude-Alink and D. Redmore, U.S. Patent 3,959,313 (1976) [*CA* **85**, 94346 (1976)].
[57] H. Quiniou and N. Lozac'h, *Bull. Soc. Chim. Fr.*, 1167 (1963).
[58] N. Lozac'h and O. Gaudin, *C. R. Hebd. Seances Acad. Sci.* **225**, 1162 (1947).
[59] F. Boberg, *Justus Liebigs Ann. Chem.* **678**, 67 (1964).
[60] J. L. Adelfang, *J. Org. Chem.* **31**, 2388 (1966).
[61] J. Faust and R. Mayer, *Justus Liebigs Ann. Chem.* **688**, 150 (1965).
[62] Hercules Powder, British Patent 900,805 (1962) [*CA* **61**, 3113 (1964)].
[63] J. Bader, *Helv. Chim. Acta* **51**, 1409 (1968).
[64] R. Wiedermann, W. von Gentzkow, and F. Boberg, *Justus Liebigs Ann. Chem.* **742**, 103 (1970).
[65] J. Faust and R. Mayer, *Angew. Chem.* **75**, 573 (1963).
[66] J. Faust, H. Spies, and R. Mayer, *Naturwissenschaften* **54**, 537 (1967).

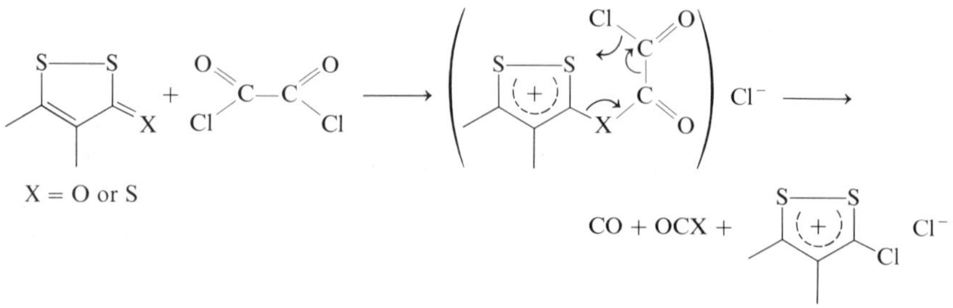

SCHEME 9

corresponding 3-chloro-1,2-dithiolium cation.[67] For the same purpose, sulfur dichloride has been used with 1,2-dithiole-3-thiones.[52,68]

With phosphorus oxytrichloride and 1,2-dithiol-3-ones, the first reaction step is probably the formation of a 3-(dichlorophosphoryloxy)-1,2-dithiolium chloride (18), which isomerizes to a 3-chloro-1,2-dithiolium dichlorophosphate (19).

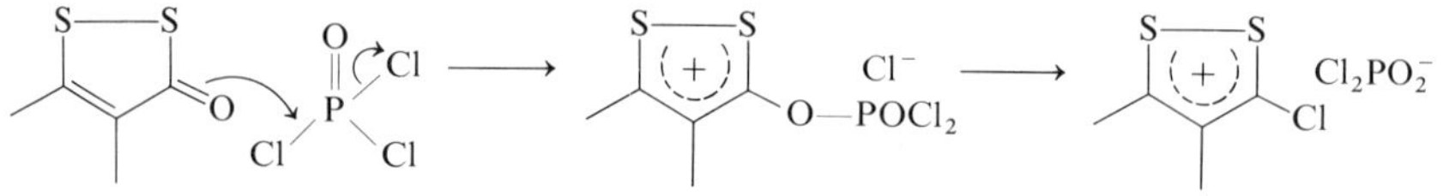

From 19, a crystalline 3-chloro-1,2-dithiolium perchlorate may be obtained with perchloric acid.[69,70] However, as perchlorates present explosion dangers, it has been suggested that for purposes of small-scale characterization the perchlorate may be replaced by the perrhenate anion.[61]

With 2 moles of phosphorus pentachloride reacting with 1 mole of 1,2-dithiol-3-one, a 3-chloro-1,2-dithiolium hexachlorophosphate (20) is obtained.[61]

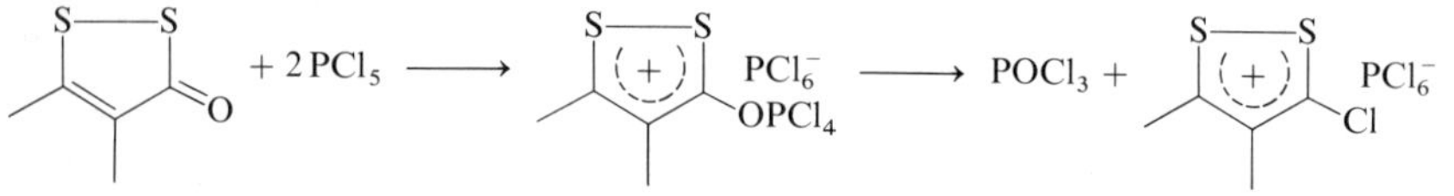

In several instances, reactions involving 1,2-dithiolium ions have been performed directly with a mixture of 1,2-dithiol-3-one and phosphorus

[67] F. Boberg and W. von Gentzkow, *Justus Liebigs Ann. Chem.* **766**, 1 (1972).
[68] E. Klingsberg, *Chem. Ind. (London),* 1813 (1968).
[69] N. Lozac'h and C. T. Pedersen, *Acta Chem. Scand.* **24**, 3189 (1970).
[70] G. A. Reynolds, *J. Org. Chem.* **33**, 3352 (1968).

oxytrichloride. In such cases, the question of whether the reactive intermediate is **18** or **19** remains a matter of conjecture; for practical reasons we shall discuss these reactions later, together with the chemical properties of 3-chloro-1,2-dithiolium ions.

Exchange of the chlorine atoms of chloro-1,2-dithiolium ions is catalyzed by Lewis acids such as $AlCl_3$ and $SbCl_3$. This exchange is markedly easier in position 3 or 5 than in position 4.[71]

e. *Acylation.* 1,2-Dithiol-3-ones and 1,2-dithiole-3-thiones react with acetic anhydride in the presence of perchloric acid, giving 3-acetoxy- or 3-acetylthio-1,2-dithiolium perchlorates (Eq. 7).[61]

$$\underset{Ph}{\overset{S\text{---}S}{\diagdown}}\overset{H^+}{\underset{X}{\diagdown}}\overset{COMe}{\underset{COMe}{\diagdown O}} \longrightarrow MeCO_2H + \underset{Ph}{\overset{S\text{---}S}{\diagdown}}(+)\underset{X\text{---}COMe}{\diagdown} \quad (7)$$

X = O or S

f. *Alkylation.* As indicated in the previous review (Ref. 1: p. 54), 1,2-dithiole-3-thiones are easily alkylated by methyl iodide or dimethyl sulfate, while 1,2-dithiol-3-ones apparently cannot be alkylated under the same conditions. However, *O*-alkylation of various 1,2-dithiol-3-ones succeeds, with triethyloxonium tetrafluoroborate, to give various 4- or 5-aryl-3-ethoxyl-1,2-dithiolium tetrafluoroborates.[72,73] Methyl fluorosulfonate yields a dithiolium ion by *O*-methylation of an α-(1,2-dithiol-3-ylidene) ketone (**21**).[74] 1,6,6aλ^4-Trithiapentalenes have been similarly methylated by trimethyloxonium tetrafluoroborate.[75]

$$\underset{Ar}{\overset{S\text{---}S}{\diagdown}}\overset{O}{\underset{Ph}{\diagdown}} + FSO_2OMe \longrightarrow \underset{Ar}{\overset{S\text{---}S}{\diagdown}}(+)\underset{Ph}{\overset{OMe}{\diagdown}}$$

(**21**)

3-Methylthio-1,2-dithiolium salts are useful intermediates for synthesis and are generally obtained by reaction of methyl iodide or of dimethyl

---

[71] F. Boberg, R. Wiedermann, and J. Kresse, *J. Labelled Compd.* **10**, 297 (1974).
[72] J. Faust and J. Fabian, *Z. Naturforsch., Teil B* **246**, 577 (1969).
[73] C. Bouillon and J. Vialle, *Bull. Soc. Chim. Fr.*, 4560 (1968).
[74] D. Festal, J. Tison, Nguyen Kim Son, R. Pinel, and Y. Mollier, *Bull. Soc. Chim. Fr.*, 3339 (1973).
[75] M. M. Borel, A. Leclaire, G. Le Coustumer, and Y. Mollier, *J. Mol. Struct.* **48**, 227 (1978).

sulfate on 1,2-dithiole-3-thiones.[73,76-86] Alkylations of 1,2-dithiole-3-thiones have also been performed with ethyl or propyl iodide, iodoacetic acid, and ethyl iodoacetate,[87] bromoacetone and ω-bromoacetophenones,[88-92] bromoacetic acid, bromoacetamide, and ethyl bromoacetate.[92]

g. *Metallic Salts Adducts.* As indicated in the previous review (Ref. 1: p. 54), 1,2-dithiole-3-thiones give addition compounds with many salts of heavy metals. In most cases, only analytical data are available and the exact structure of the adducts has not been established, but it seems likely that in all cases the thione sulfur plays the role of electron donor toward the metal, thus conferring a positive character to the dithiole ring.

Adducts of the 1,2-dithiole-3-thione with silver nitrate and copper dichloride have been described,[80] and for various 1,2-dithiole-3-thiones, adducts containing iron, cobalt, nickel, and copper have been studied in more detail. The ligands (L) used are either unsubstituted 1,2-dithiole-3-thione or its methyl- or aryl-substituted derivatives. If X represents a halogen atom, the following types of coordination compounds have been characterized: $CuXL_2$ and $CuXL_3$,[93] FeX_2L_2,[93] CoX_2L_2,[94] and NiX_2L_2.[95] Complexes of titanium(III), tin(IV), antimony(III) and (V), and bismuth(III) have also been studied.[96]

Adducts with mercury(II) chloride have often been used to separate 1,2-dithiole-3-thiones from mixtures, and hydrosulfolysis of the adduct

---

[76] Y. Mollier and N. Lozac'h, *Bull. Soc. Chim. Fr.*, 700 (1960).
[77] G. Le Coustumer and Y. Mollier, *C. R. Hebd. Seances Acad. Sci., Ser. C* **270**, 433 (1970).
[78] J. P. Brown, *J. Chem. Soc. C*, 1077 (1968).
[79] E. Campaigne and R. D. Hamilton, *J. Org. Chem.* **29**, 2877 (1964).
[80] F. Challenger, E. A. Mason, E. C. Holdsworth, and R. Emmott, *J. Chem. Soc.*, 292 (1953).
[81] G. Le Coustumer and Y. Mollier, *Bull. Soc. Chim. Fr.*, 499 (1971).
[82] E. Klingsberg, *J. Heterocycl. Chem.* **3**, 243 (1966).
[83] A. Marei and M. M. A. El Sukkary, *U.A.R. J. Chem.* **14**, 101 (1971).
[84] C. Metayer, G. Duguay, and H. Quiniou, *Bull. Soc. Chim. Fr.*, 4576 (1972).
[85] R. M. Christie, A. S. Ingram, D. H. Reid, and R. G. Webster, *J. C. S., Perkin 1*, 722 (1974).
[86] Y. N'Guessan and J. Bignebat, *C. R. Hebd. Seances Acad. Sci., Ser. C* **280**, 1323 (1975).
[87] M. G. Voronkov and T. Lapina, *Khim. Geterotsikl. Soedin.*, 342 (1970).
[88] E. I. G. Brown, D. Leaver, and D. M. McKinnon, *J. Chem. Soc. C*, 1202 (1970).
[89] G. Caillaud and Y. Mollier, *Bull. Soc. Chim. Fr.*, 2018 (1970).
[90] G. Caillaud and Y. Mollier, *Bull. Soc. Chim. Fr.*, 331 (1971).
[91] G. Caillaud and Y. Mollier, *Bull. Soc. Chim. Fr.*, 2326 (1971).
[92] G. Caillaud and Y. Mollier, *Bull. Soc. Chim. Fr.*, 147 (1972).
[93] F. Petillon and J. E. Guerchais, *J. C. S., Dalton*, 1209 (1973).
[94] F. Petillon and J. E. Guerchais, *Bull. Soc. Chim. Fr.*, 2455 (1971).
[95] F. Y. Petillon and J. E. Guerchais, *Can. J. Chem.* **49**, 2598 (1971).
[96] F. Petillon and J. E. Guerchais, *J. Inorg. Nucl. Chem.* **37**, 1863 (1975).

regenerates the thione with excellent yields. Conversely, phenyl-1,2-dithiole-3-thiones have been suggested for gravimetric analysis of mercury.[97]

The adducts formed by mercury(II) acetate and 1,2-dithiole-3-thiones are of particular interest because they are often used for converting 1,2-dithiole-3-thiones into 1,2-dithiol-3-ones. The difference in the behavior of mercury(II) halides and acetate toward dithiolethiones suggests that a special mechanism operates in the latter case, probably a concerted cyclic reaction as indicated in Scheme 10. The 3-acetoxy-1,2-dithiolium cation thus obtained is rapidly converted to the corresponding dithiolone by reaction of the acetate anion or water.

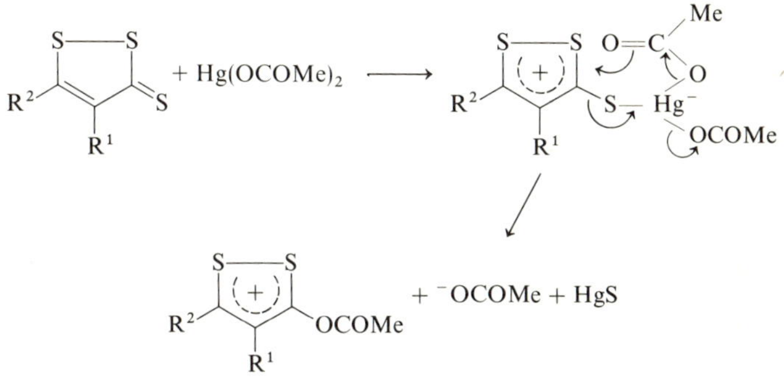

SCHEME 10

h. *Carbene and Nitrene Adducts.* Reaction of 1,2-benzodithiole-3-thione with ditosylcarbene was realized by heating the dithiole compound with ditosyldiazomethane in the presence of a catalytic amount of copper acetylacetonate.[98] The resulting compound may be considered either as a thiocarbonyl ylide (**22**) or as a dithiolium derivative (**23**).

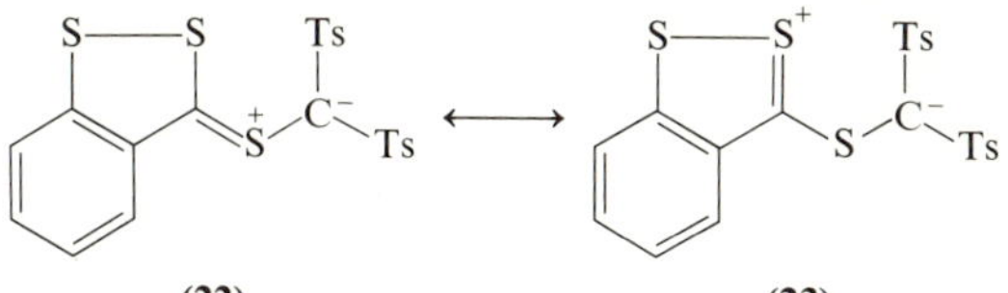

(22)   (23)

Similarly, nitrenes obtained by heating chloramines in methanol reacted with 1,2-dithiole-3-thiones[98] or 1,2-dithiole-3-selones.[99]

Ylides similar to **22** are probably formed in the deprotonation of 3-(acylmethylthio)-1,2-dithiolium ions which undergo an internal replacement reaction in the presence of bases (Section II,B,7,d).

---

[97] A. I. Busev and V. V. Evsikov, *Vestn. Mosk. Univ., Khim.* **13**, 81 (1965) [*CA* **77**, 28591 (1972)].
[98] S. Tamagaki and S. Oae, *Tetrahedron Lett.*, 1162 (1972).
[99] S. Tamagaki, K. Sakaki, and S. Oae, *Heterocycles* **2**, 39 (1974).

B. CHEMICAL PROPERTIES

1. General Remarks

Since the preceding review[1] many investigations into the chemical properties of 1,2-dithiolium cations have been conducted, particularly their reactions with nucleophiles. Such reactions cover a wide variety of types according to the nature of the reagents and of the substituents of the 1,2-dithiole ring.

As already indicated, the resonance hybrid corresponding to the 1,2-dithiolium cation is satisfactorily described by carbenium and sulfonium structures (Scheme 11). The fact that formal double bonds appear between atoms 1 and 5, 2 and 3, 3 and 4, and 4 and 5, but not between 1 and 2, is often highlighted by adopting the description of the charge delocalization **24**; this representation is used in the present chapter.

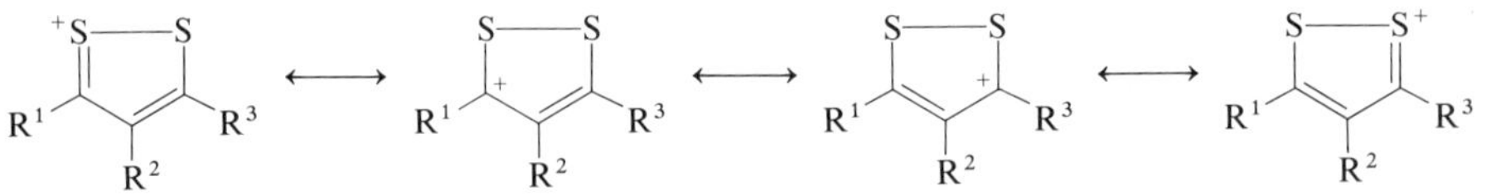

SCHEME 11

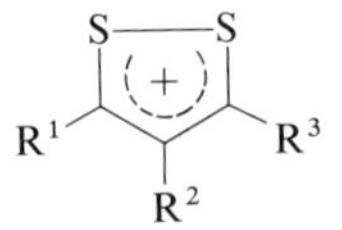

(24)

For substitutions in position 3 or 5 on the 1,2-dithiole ring various mechanisms may be considered. When R^1 (or R^3) is a good electrofuge such as H, carbenes (**25**) are likely intermediates.[100] Strong acidity slows hydrogen exchange, supporting such a mechanism. A similar mechanism is possible for the thiation of 1,2-dithiolium cations by elemental sulfur in the presence of pyridine (Eq. 8).

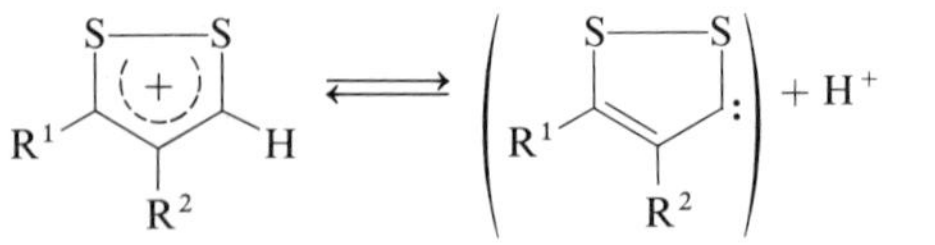

(25)

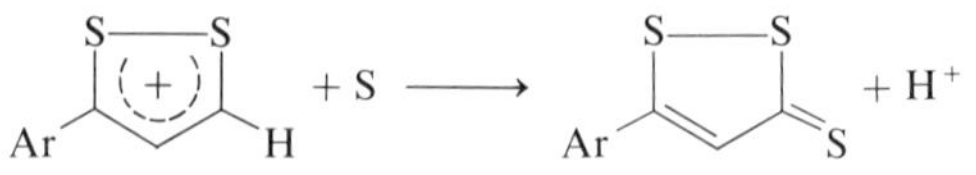

(8)

[100] H. Prinzbach, E. Futterer, and A. Lüttringhaus, *Angew. Chem., Int. Ed. Engl.* **5**, 513 (1966).

Sec. II.B]   THE 1,2- AND 1,3-DITHIOLIUM IONS   169

Attack by a nucleophile at position 3 (or 5) may lead either to substitution or to ring opening. The formation of an intermediate 3H-1,2-dithiole is generally assumed, although such intermediates have seldom been isolated (Scheme 12).

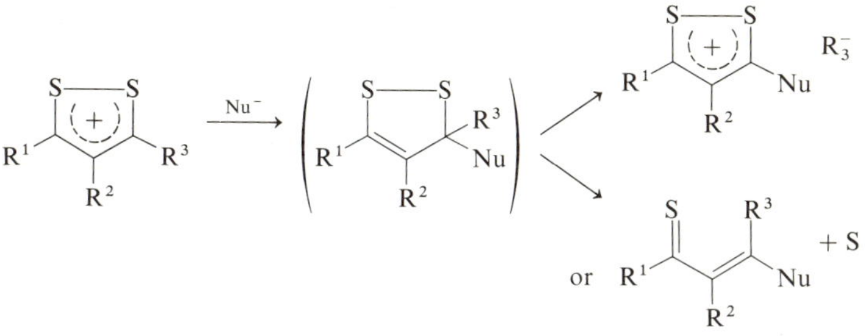

SCHEME 12

Nucleophilic attacks on 1,2-dithiolium cations may also result in substituent modification. This is the case when, for instance, pyridine reacts with a 3-methylthio-1,2-dithiolium cation (Eq. 9). Sulfur nucleophiles have also shown a demethylating action.[38]

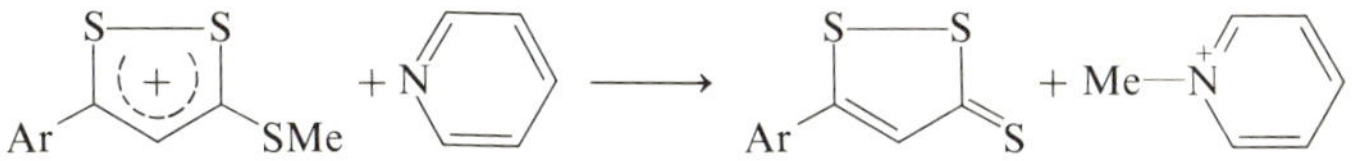

(9)

2. Hydrogen Exchange

When a 1,2-dithiolium cation has a hydrogen atom in position 3 or 5, deuteration can be realized either with a mixture of CF_3CO_2D and D_2O[100] or one of Me_2SO and D_2O.[101] As the deuteration rate diminishes when the acidity increases, a carbene mechanism has been suggested, as indicated in the preceding section.

A practical method consists in treating the 1,2-dithiolium perchlorate with a mixture of deuterium oxide and acetic anhydride. After reaction of deuterium oxide with the anhydride, practically complete deuteration at the free 3 and/or 5 positions is observed. Under the same conditions, methyl groups in positions 3 and/or 5 are also deuterated, but this inhibits deuteration on the ring.[102] This finding is in agreement with the carbene mechanism for ring deuteration. The presence of the 3-methyl group inhibits carbene formation because deprotonation gives preferentially a methylene dithiole (Scheme 13).

[101] H. Newman and R. B. Angier, *J. C. S., Chem. Commun.*, 353 (1967).
[102] G. Duguay and H. Quiniou, *C. R. Hebd. Seances Acad. Sci. Ser. C* **283**, 495 (1976).

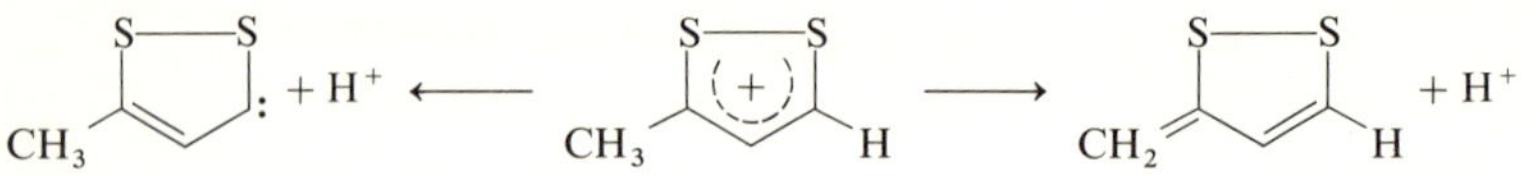

SCHEME 13

3. Reduction

1,2-Dithiolium ions, particularly those having a hydrogen atom in position 3 or 5, have oxidative properties that appear, for example, in dehydrogenation processes which follow condensation of 3-aryl-1,2-dithiolium ions with carbanionic reagents. No precise information is available on the nature of the reduction products arising from 1,2-dithiolium ions in these dehydrogenations.

On the other hand, electrochemical studies have given interesting information on the reduction products. Cathodic reduction of 1,2-dithiolium salts in acetonitrile, on a platinum electrode opposed to a calomel electrode, probably gives as the primary product a 1,2-dithiolyl radical, which dimerizes more or less rapidly. The dimer derived from 3,4-diphenyl-1,2-dithiolium cation is stable at ordinary temperature and may be isolated.[50,103–105] On the contrary, the free radical derived from the 3,5-diphenyl-1,2-dithiolium cation, which is stable at ordinary temperature in acetonitrile in an inert atmosphere, reversibly dimerizes at lower temperature: at −80°C the green color of the free radical disappears as does its ESR signal. Anodic oxidation of either the 1,2-dithiolyl radical or its dimer regenerates the 1,2-dithiolium cation.[50,103,104]

Electrochemical or chemical (Zn) reduction of phenaleno[1,9-cd]-1,2-dithiolium cation leads to the corresponding radical.[25]

The radical formed by cathodic reduction of 3,5-diaryl-1,2-dithiolium cations can undergo further reduction, giving a 1,3-diarylpropane-1,3-dithionate anion, which can be oxidized back to the corresponding radical and then to the 1,2-dithiolium cation.[103]

Flash photolysis of 1,2-dithiolium cations in ethanol also leads to the reduction of the cation at the expense of the solvent. Spectrometric evidence suggests the formation of a short-lived 1,2-dithiolyl radical accompanied by a more stable propane-1,3-dithionate anion.[106,107]

Reduction of 3-chloro-1,2-dithiolium ions by zinc has led to the formation of tetrathiafulvalene. (Eq. 10).[108]

[103] K. Bechgaard, V. D. Parker, and C. T. Pedersen, *J. Am. Chem. Soc.* **95**, 4373 (1973).
[104] C. T. Pedersen, K. Bechgaard, and V. D. Parker, *J. C. S., Chem. Commun.*, p. 430 (1972).
[105] C. T. Pedersen, *Angew. Chem. Int. Ed. Engl.* **13**, 349 (1974).
[106] C. T. Pedersen and C. Lohse, *Tetrahedron Lett.*, 5213 (1972).
[107] C. T. Pedersen and C. Lohse, *Acta Chem. Scand., Ser. B* **29**, 831 (1975).
[108] H. Behringer and E. Meinetsberger, *Tetrahedron Lett.*, 3473 (1975).

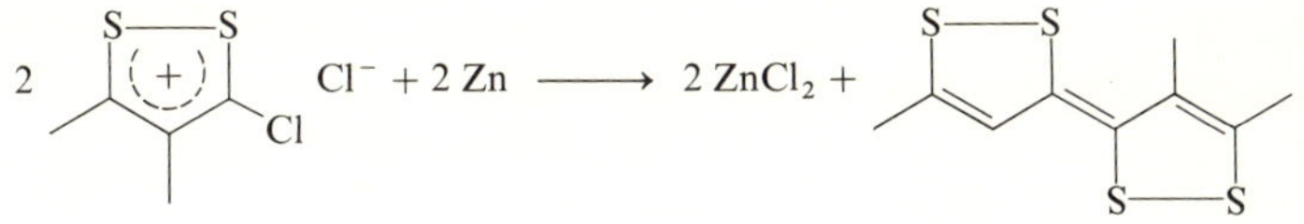

$$2 \underset{\text{Cl}}{\overset{S-S}{\underset{}{\bigcirc(+)}}} \text{Cl}^- + 2\,\text{Zn} \longrightarrow 2\,\text{ZnCl}_2 + \underset{S-S}{\overset{S-S}{\bigcirc=\bigcirc}} \qquad (10)$$

Alkyl or phenyl phosphites have been reported as having a reducing action on 1,2-dithiolium cations, with cleavage of the S—S bond (Eq. 11).[109]

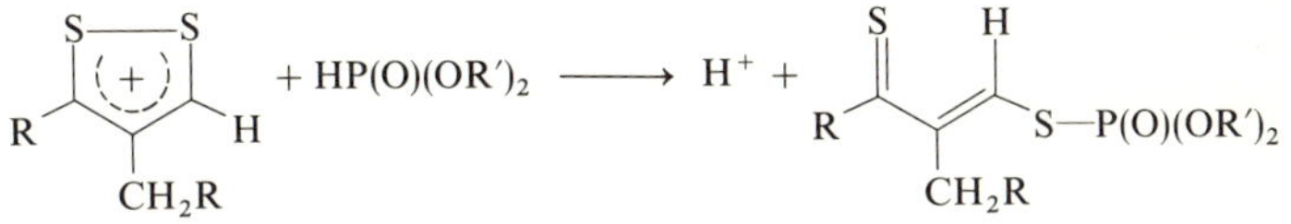

$$\underset{\text{CH}_2\text{R}}{\overset{S-S}{\underset{RH}{\bigcirc(+)}}} + \text{HP(O)(OR')}_2 \longrightarrow \text{H}^+ + \underset{\text{CH}_2\text{R}}{\overset{SH}{\underset{RS-\text{P(O)(OR')}_2}{\parallel}}} \qquad (11)$$

R = Me$_3$C
R' = Me, Et, Me$_2$CH, n-C$_4$H$_9$, Ph

4. Attack by Oxygen Nucleophiles

With strong bases such as HO⁻ and EtO⁻, 1,2-dithiolium ions undergo extensive destruction, and these reactions are of little preparative value. Under milder conditions the dithiole ring may be preserved. For instance, 4-phenyl-1,2-dithiolium sulfate and aqueous sodium bicarbonate give a yellow crystalline solid described as bis(4-phenyl-3H-1,2-dithiol-3-yl) ether.[101] This substance gives 4-phenyl-1,2-dithiolium perchlorate with perchloric acid.

3-Chloro- or 3-bromo-1,2-dithiolium ions are easily hydrolyzed to 1,2-dithiol-3-ones.[61,66,67] The same result is achieved in acetic acid, which is transformed into acetyl chloride (Eq. 12).[61,110]

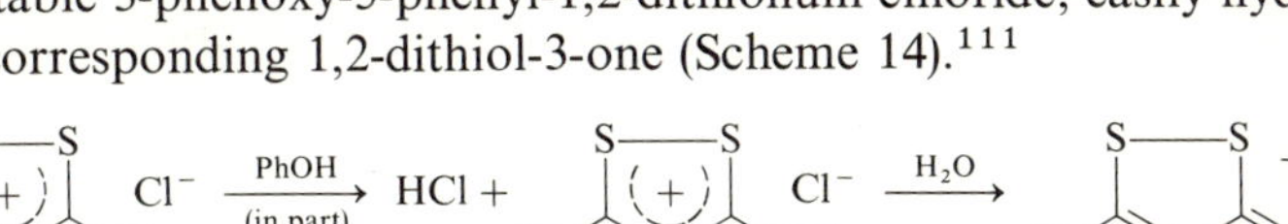

$$\underset{\text{Cl}}{\overset{S-S}{\bigcirc(+)}} + \text{MeCOOH} \longrightarrow \underset{O}{\overset{S-S}{\bigcirc}} + \text{H}^+ + \text{MeCOCl} \qquad (12)$$

3-Chloro-5-phenyl-1,2-dithiolium chloride reacts with phenol not only at the para carbon atom (see Section II,B,7,c) but also at the hydroxyl, to give the unstable 3-phenoxy-5-phenyl-1,2-dithiolium chloride, easily hydrolyzed to the corresponding 1,2-dithiol-3-one (Scheme 14).[111]

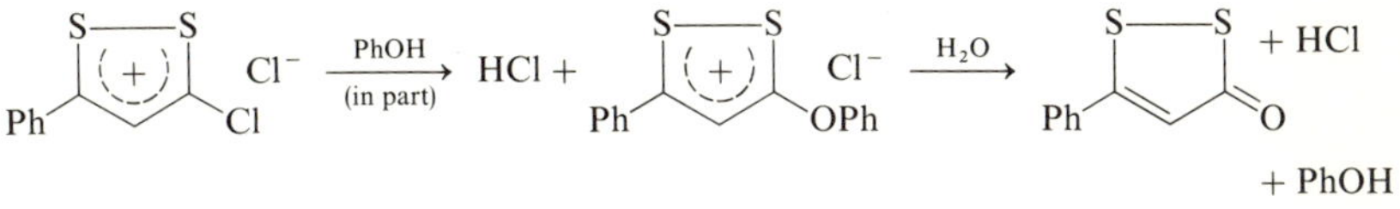

$$\underset{\text{Ph}\text{Cl}}{\overset{S-S}{\bigcirc(+)}}\text{Cl}^- \xrightarrow[\text{(in part)}]{\text{PhOH}} \text{HCl} + \underset{\text{Ph}\text{OPh}}{\overset{S-S}{\bigcirc(+)}}\text{Cl}^- \xrightarrow{\text{H}_2\text{O}} \underset{\text{Ph}\text{O}}{\overset{S-S}{\bigcirc}} + \text{HCl} + \text{PhOH}$$

SCHEME 14

[109] B. A. M. Oude-Alink, U.S. Patent 3,859,396 (1975) [*CA* **83**, 163,629 (1975)].
[110] J. Faust, H. Spies, and R. Mayer, *Z. Chem.* **7**, 275 (1967).
[111] J. Faust, B. Bartho, and R. Mayer, *Z. Chem.* **15**, 395 (1975).

Monocyclic 3-ethoxy-1,2-dithiolium ions are easily hydrolyzed to the corresponding dithiolone. The same compounds are obtained with anhydrous ethanol, which is transformed into diethyl ether.[111] With 3-ethoxy-1,2-benzodithiolium ion, hydrolysis gives only small quantities of 1,2-benzodithiol-3-one, the major product being 2,2′-dithiobis(benzoic) acid.[112]

Vilsmeier salts such as **26** are vinylogs of amino-1,2-dithiolium ions (**27**) and, accordingly, are attacked by nucleophiles. Hydrolysis of **26** is a good preparation of α-(1,2-dithiol-3-ylidene) aldehydes, as shown in the case of **28**.[85] Methanolysis may be also realized by heating in methanol.[113]

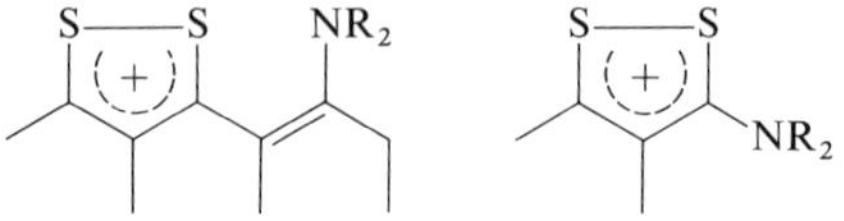

In an earlier review (Ref. 2: pp. 187 and 189), it was indicated that aqueous tertiary amines (triethylamine, pyridine) open the 1,2-dithiolium ring with

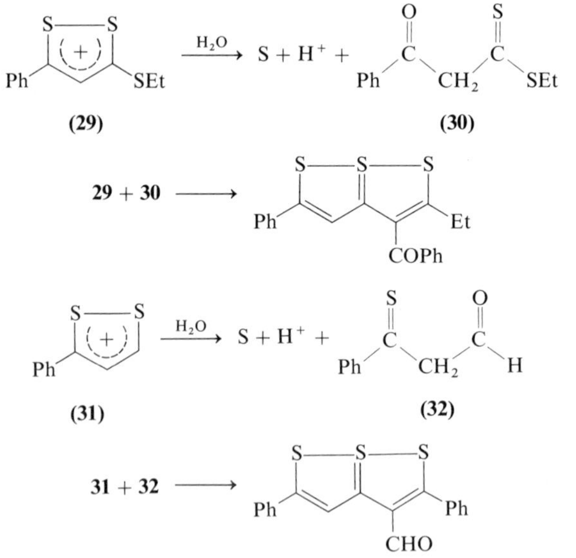

SCHEME 15

[112] J. Faust, *Z. Chem.* **15**, 478 (1975).
[113] R. Mayer and H. Hartmann, *Z. Chem.* **6**, 312 (1966).

Sec. II.B] THE 1,2- AND 1,3-DITHIOLIUM IONS

formation of a β-thioxoaldehyde or of a β-oxodithioester, according to the starting material. The β-thioxocarbonyl compound thus formed condenses with unchanged dithiolium ion to give a 1,6,6aλ⁴-trithiapentalene. Scheme 15 gives further examples of this reaction.[73,114]

5. *Attack by Sulfur or Selenium Nucleophiles*

3- or 5-Unsubstituted 1,2-dithiolium cations are converted to 1,2-dithiole-3-thiones by heating with sulfur in pyridine.[115] This reaction, which probably proceeds by attack of elemental sulfur on a carbene intermediate, has been extended to other cases (Eq. 13).[116]

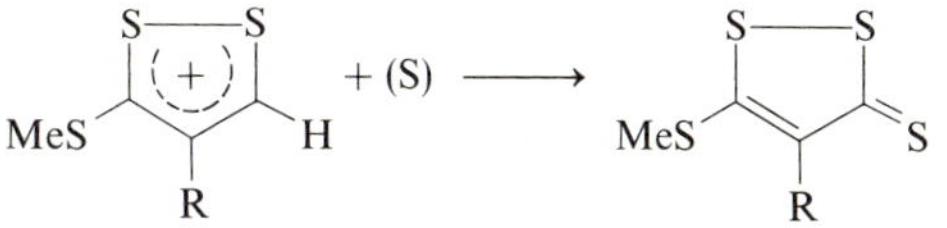

$$\text{MeS} \underset{R}{\overset{S-S}{\underset{H}{(+)}}} + (S) \longrightarrow \text{MeS} \underset{R}{\overset{S-S}{\underset{}{}}} S \qquad (13)$$

R	Yield (%)	R	Yield (%)
Ph	75	p-ClC$_6$H$_4$	55
p-MeC$_6$H$_4$	70	p-NO$_2$C$_6$H$_4$	70
p-EtC$_6$H$_4$	60	Et	45
p-Me$_3$C—C$_6$H$_4$	75	Me$_2$CH	10
p-EtMe$_2$C—C$_6$H$_4$	70	n-C$_5$H$_{11}$	45

4-Phenyl-1,2-dithiolium hydrogen sulfate, treated with aqueous hydrogen sulfide, gives a yellow solid that has been described as a thioether (**33**). With perchloric acid this compound regenerates the dithiolium cation in 70% yield.[101] Similar reactions have been described with mercaptans.[38]

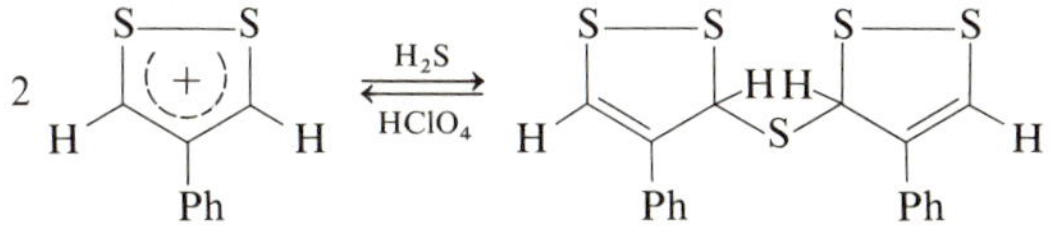

$$2 \underset{Ph}{\overset{S-S}{\underset{H\quad H}{(+)}}} \underset{HClO_4}{\overset{H_2S}{\rightleftharpoons}} \underset{Ph}{\overset{S-S}{\underset{H}{}}} \underset{S}{\overset{HH}{}} \underset{Ph}{\overset{S-S}{\underset{H}{}}}$$

(**33**)

3-Chloro- or 3-bromo-1,2-dithiolium ions react with hydrogen sulfide or methanethiolate anions giving 1,2-dithiole-3-thiones or 3-methylthio-1,2-dithiolium ions, respectively (Scheme 16).[61,66,110]

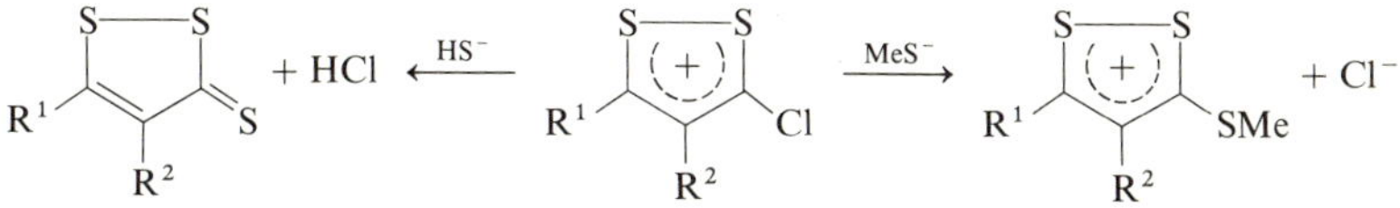

$$\underset{R^2}{\overset{S-S}{\underset{R^1}{}}}\!\!S + \text{HCl} \xleftarrow{\text{HS}^-} \underset{R^2}{\overset{S-S}{\underset{R^1}{(+)}}}\!\!\text{Cl} \xrightarrow{\text{MeS}^-} \underset{R^2}{\overset{S-S}{\underset{R^1}{(+)}}}\!\!\text{SMe} + \text{Cl}^-$$

SCHEME 16

---

[114] J. Bignebat and H. Quiniou, *Bull. Soc. Chim. Fr.*, 645 (1972).
[115] E. Klingsberg, *J. Org. Chem.* **28**, 529 (1963).
[116] A. Grandin, C. Bouillon, and J. Vialle, *Bull. Soc. Chim. Fr.*, 4555 (1968).

Thioacetic acid reacts with 3-chloro-1,2-dithiolium ions to give the corresponding 1,2-dithiole-3-thione. The mechanism shown in Scheme 17 has been suggested.[117] Hydroselenolysis of 3-methylthio-1,2-benzodithiolium ion gives the corresponding 1,2-dithiole-3-selone.[99,118]

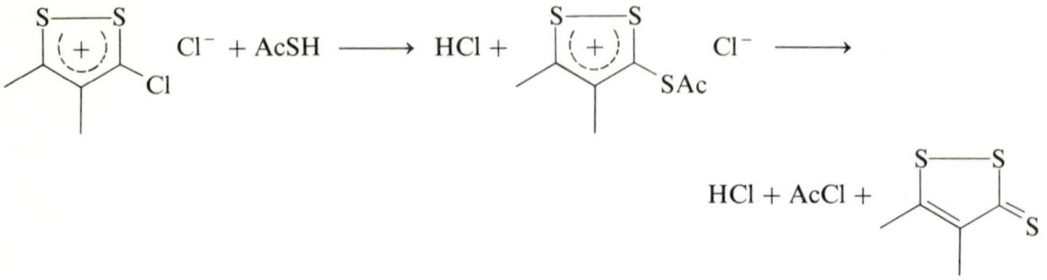

SCHEME 17

3-Ethoxy-1,2-dithiolium ions react in a similar way with aqueous sodium sulfide to give the corresponding 1,2-dithiole-3-thione.[112]

5-Phenyl-3-chloro-1,2-dithiolium chloride reacts with N-phenylselenourea to give 5-phenyl-1,2-dithiole-3-selone (**34**).[119]

(**34**)

6. Attack by Nitrogen Nucleophiles

a. *3- or 5-Unsubstituted 1,2-Dithiolium Ions.* These ions are known to react with ammonia to yield isothiazoles.[101,120–123] The mechanism shown in Scheme 18 has been suggested.[122] With 3-aryl-1,2-dithiolium salts,

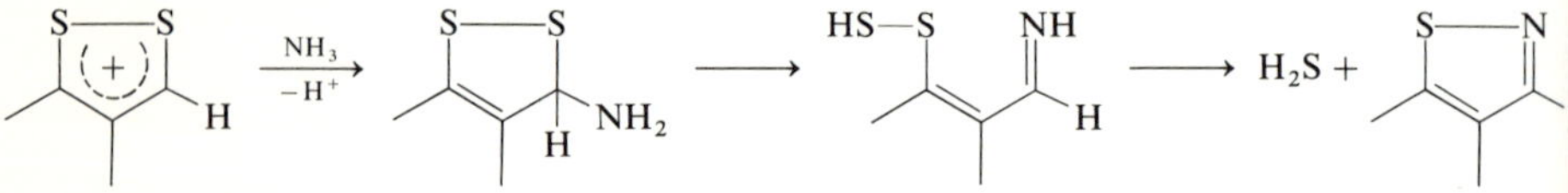

SCHEME 18

---

[117] G. J. Wentrup, M. Koepke, and F. Boberg, *Synthesis*, 525 (1975).
[118] S. Tamagaki, K. Sakaki, and S. Oae, *Heterocycles* **2**, 45 (1974).
[119] H. Spies, K. Gewald, and R. Mayer, *J. Prakt. Chem.* **313**, 804 (1971).
[120] G. Le Coustumer and Y. Mollier, *Bull. Soc. Chim. Fr.*, 3076 (1970).
[121] J. C. Poite, A. Perichaut, and J. Roggero, *C. R. Hebd. Seances Acad. Sci., Ser. C* **270**, 1677 (1970).
[122] R. A. Olofson, J. M. Landesberg, R. O. Berry, D. Leaver, W. A. H. Robertson, and D. M. McKinnon, *Tetrahedron* **22**, 2119 (1966).
[123] A. Perichaud, J. C. Poite, G. Mille, and J. P. Roggero, *Bull. Soc. Chim. Fr.*, 3830 (1972).

the main attack, as shown in Scheme 18, occurs at position 5. However, to a lesser extent, attack at position 3 may also occur (Eq. 14).[122]

$$\underset{Ph}{\overset{S\!-\!S}{\bigcirc_{(+)}}} + NH_3 \longrightarrow \underset{Ph}{\overset{S\!-\!N}{\bigcirc}} + \underset{Ph}{\overset{N\!-\!S}{\bigcirc}} \quad (14)$$
$$ 49\% 7\%$$

Reaction of a 3-aryl-1,2-dithiolium cation with phenylhydrazine gives a mixture of pyrazoles.[122,124] Similarly, with a 4-aryl-1,2-dithiolium ion, 1,2-dimethyl- or 1,2-diphenylhydrazine give a pyrazolium ion.[115] With hydroxylamine poor yields of mixtures of isoxazoles and isothiazoles are obtained.[122]

Reaction of 3-aryl-1,2-dithiolium ions with primary or secondary amines generally leads to a 3-amino-1-arylprop-2-enethione (**35**). This reaction has been performed with methylamine,[125] benzylamine,[38] 2-aminoethanethiol,[38] dimethylamine,[126] diethylamine,[127,128] pyrrolidine, piperidine, morpholine,[127] aniline,[125,129,130] N-methylaniline,[129,130] o-,p-toluidines, and o-,p-anisidines.[129] The same type of reaction is also observed with a styryl instead of an aryl in position 3.[55,131]

$$\underset{Ar}{\overset{S\!-\!S}{\bigcirc_{(+)}}} + RR'NH \longrightarrow \underset{Ar}{\overset{S}{\underset{H}{C}}}\!=\!\underset{H}{\overset{C}{\underset{NRR'}{C}}} + S + H^+$$

(**35**)

Two moles of amine may react with 1 mole of dithiolium ion (Eq. 15).[132] A similar reaction has been observed with 3-aryl-1,2-dithiolium ions and dimethylamine.[126]

$$\underset{\underset{Me}{Ph}}{\overset{S\!-\!S}{\bigcirc_{(+)}}}\!\!\!H + 2\,ArNH_2 \longrightarrow \underset{\underset{Me}{Ph}}{\overset{H\diagdown N\diagup Ar}{\bigcirc_{(+)}}}\!\!\!NHAr + H_2S + S \quad (15)$$

[124] M.-T. Bergeon, C. Metayer, and H. Quiniou, *Bull. Soc. Chim. Fr.*, 917 (1971).
[125] D. M. McKinnon and E. A. Robak, *Can. J. Chem.* **46**, 1855 (1968).
[126] F. Clesse, A. Reliquet, and H. Quiniou, *C. R. Hebd. Seances Acad. Sci., Ser. C* **272**, 1049 (1971).
[127] J. Bignebat, H. Quiniou, and N. Lozac'h, *Bull. Soc. Chim. Fr.*, 1699 (1966).
[128] J.-P. Pradere, G. Bouet, and H. Quiniou, *Tetrahedron Lett.*, 3471 (1972).
[129] J. Bignebat, H. Quiniou, and N. Lozac'h, *Bull. Soc. Chim. Fr.*, 127 (1969).
[130] D. Leaver, D. M. McKinnon, and W. A. H. Robertson, *J. Chem. Soc.*, 32 (1965).
[131] G. Duguay, C. Metayer, and H. Quiniou, *Bull. Soc. Chim. Fr.*, 2853 (1974).
[132] A. Reliquet and F. Reliquet-Clesse, *C. R. Hebd. Seances Acad. Sci., Ser. C* **275**, 689 (1972).

b. *1,2-Dithiolium Ions Substituted at Positions 3 and 5 by Hydrocarbon Groups.* 3,5-Dialkyl-1,2-dithiolium ions react with ammonia to give 3,5-dialkylisothiazoles.[13,14]

3,5-Diphenyl-1,2-dithiolium ion leads to 3-amino-1,3-diphenylprop-2-ene-1-thiones when reacted with methylamine,[125,133] ethylamine,[133] benzylamine, 2-aminoethanethiol,[38] piperidine,[126] aniline, toluidines, anisidines,[22] and chloranilines.[133] With dimethylamine a double condensation, analogous to the one shown in Eq. (15), occurs.[126]

With hydrazine, pyrazoles are obtained but generally in low yields. The relatively poor results obtained with 3,5-dialkyl-1,2-dithiolium ions and amines may be, at least in part, attributed to the deprotonation of alkyl substituents in the presence of a base (Eq. 16).[14]

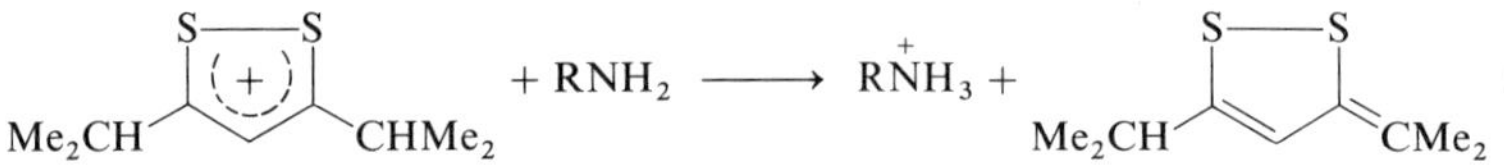

$$Me_2CH\underset{CHMe_2}{\overset{S-S}{(+)}} + RNH_2 \longrightarrow R\overset{+}{N}H_3 + Me_2CH\underset{CMe_2}{\overset{S-S}{\diagup\diagdown}} \quad (16)$$

As indicated in Section II,A,2,a, (5-aryl-1,2-dithiol-3-ylidene)malonic acids (**14**) give, by decarboxylation and protonation, 3-aryl-5-methyl-1,2-dithiolium cations (**36**). In the presence of aromatic amines, two types of compounds may be obtained[134]: **37**, resulting from the condensation of **36** with the amine, and **38**, resulting from the condensation of **36** with its conjugate base (cf. Ref. 1: p. 83).

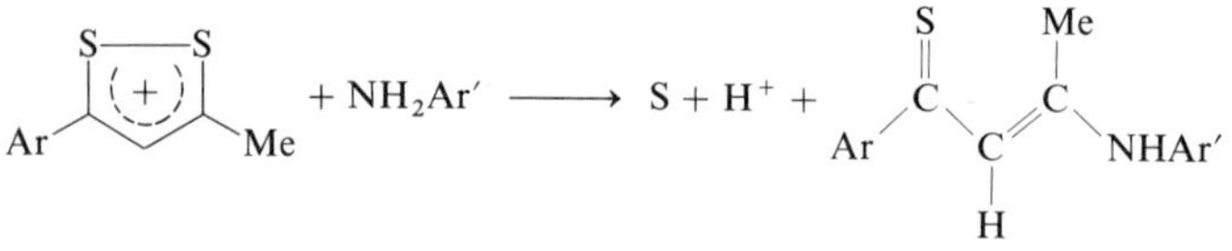

(**36**)   (**37**)

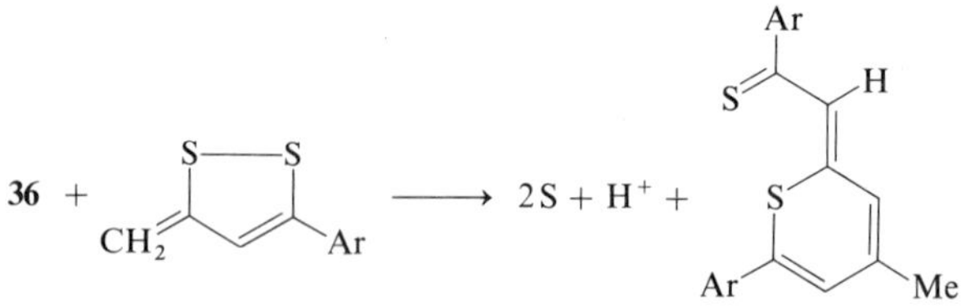

36 + $\longrightarrow$ 2S + H$^+$ +

(**38**)

A special type of nucleophilic attack on 1,2-dithiolium compounds is the pyridine-catalyzed rearrangement of 3-[(1,2-dithiol-3-ylidene)methyl]-1,2-

---

[133] E. Uhlemann and B. Zöllner, *Z. Chem.* **14**, 245 (1974).
[134] G. Duguay, H. Quiniou, and N. Lozac'h, *Bull. Soc. Chim. Fr.*, 4485 (1967).

Sec. II.B]   THE 1,2- AND 1,3-DITHIOLIUM IONS   177

dithiolium ions.[135,136] These reactions are pyridine-induced deprotonations. The more general type of nucleophilic attack leads to red compounds with a 1,6,6aλ^4-trithiapentalene structure whose mechanism is suggested in Scheme 19.[136] In some cases, together with this red compound, a blue isomer is also obtained. A structure of the diaryl thioketone type is likely for the latter (Scheme 20).

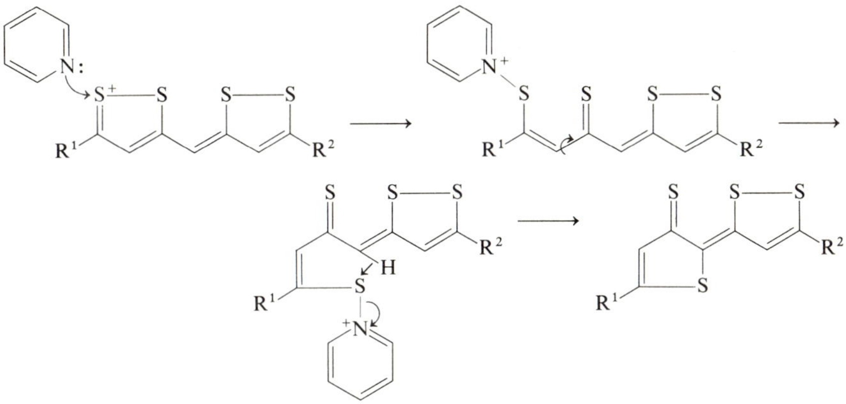

SCHEME 19

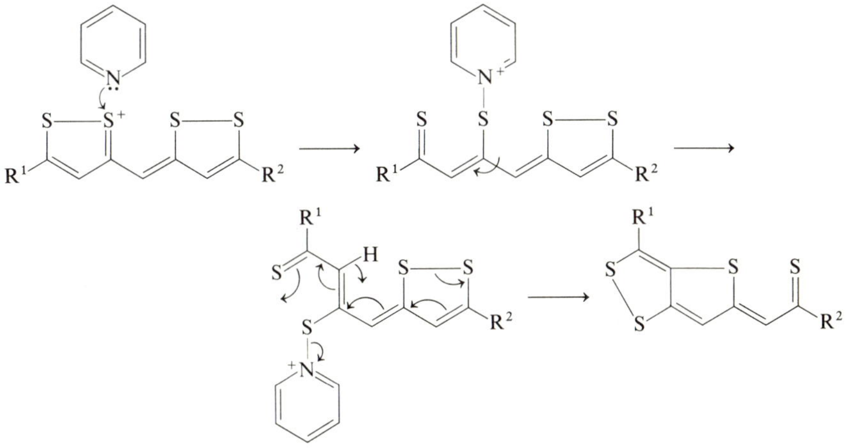

SCHEME 20

c. *3-Halogeno-1,2-dithiolium Ions.* 3-Chloro-5-phenyl-1,2-dithiolium ion reacts with methylamine, giving an aminothioamide (**39**),[110] and with

---

[135] C. Retour, M. Stavaux, and N. Lozac'h, *Bull. Soc. Chim. Fr.*, 3360 (1971).
[136] C. Lemarié-Retour, M. Stavaux, and N. Lozac'h, *Bull. Soc. Chim. Fr.*, 1659 (1973).

aniline, giving an imine (**40**).[61] 3-Bromo-5-phenyl-1,2-dithiolium ion reacts similarly with aniline.[66] In such reactions, the chlorine atom is more reactive than an alkylthio group (Eq. 17).[63] Similarly, 5-amino-1,2-dithiol-3-imines have been prepared from 5-amino-3-chloro-1,2-dithiolium ions and aromatic primary amines.[64]

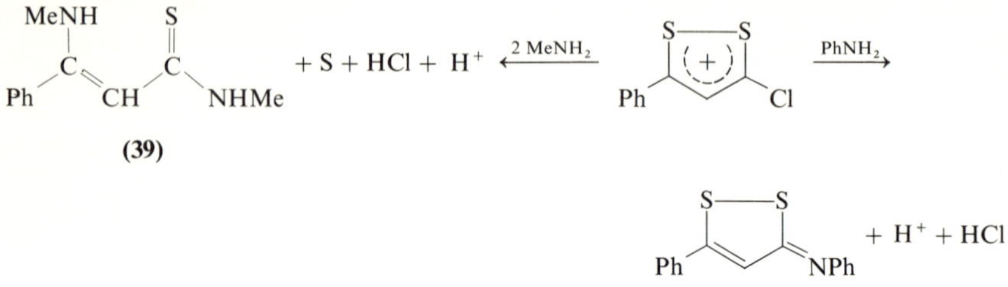

(**39**)

(**40**)

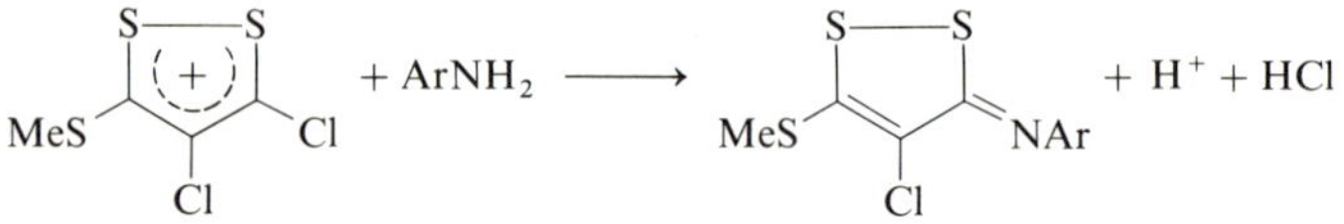

(17)

5-Aryl-3-chloro-1,2-dithiolium ions react with various secondary amines (morpholine, *N*-alkylanilines, diarylamines) to give 3-amino-5-aryl-1,2-dithiolium ions (**41**) which may also be considered as iminium derivatives (**42**).[137]

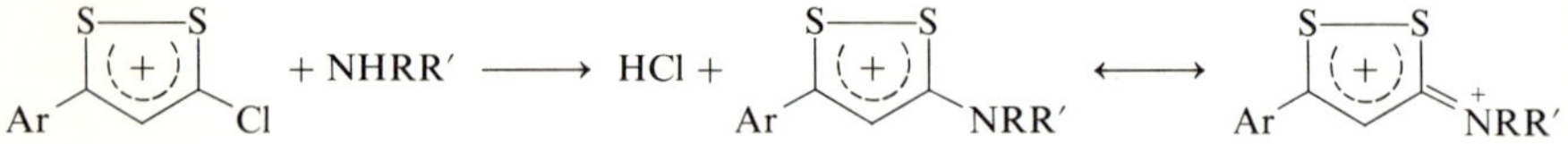

(**41**) (**42**)

Two moles of 3-chloro-1,2-benzodithiolium chloride, reacting with urea, give **43**, with partial bonding between oxygen and sulfur.[68]

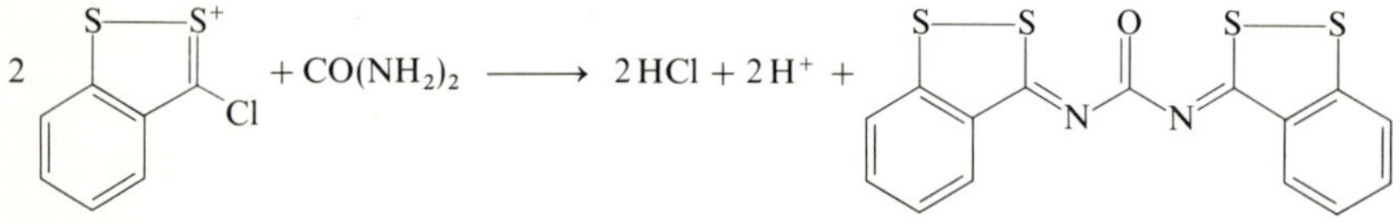

(**43**)

Reagents with "positive halogen" attack 3-chloro-1,2-dithiolium ions, with evolution of halogen (Eq. 18).[138–140]

---

[137] B. Bartho, J. Faust, R. Pohl, and R. Mayer, *J. Prakt. Chem.* **318**, 221 (1976).
[138] F. Boberg and R. Wiedermann, *Justus Liebigs Ann. Chem.* **728**, 36 (1969).
[139] F. Boberg and G. J. Wentrup, *Justus Liebigs Ann. Chem.*, 241 (1973).
[140] F. Boberg, G. J. Wentrup, and M. Koepke, *Synthesis*, 502 (1975).

(18)

X = ArCO, $ArSO_2$, MeOCO, EtOCO, Me_2NSO_2

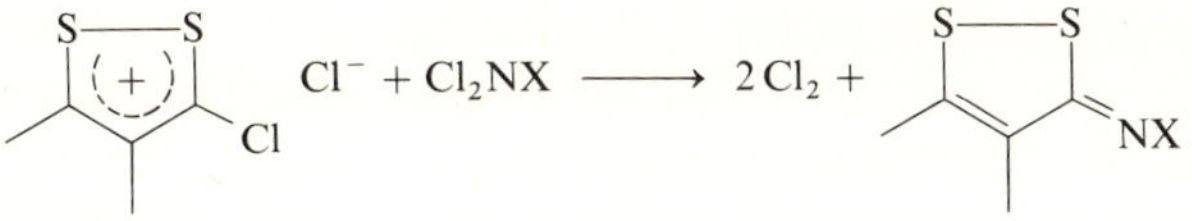

When there is a hydrogen atom at position 4 in the starting ion, this hydrogen may be substituted by chlorine during the condensation reaction.[138]

d. *3-Alkoxy and 3-Alkylthio-1,2-dithiolium Ions.* 3-Ethoxy-5-phenyl-1,2-dithiolium ion reacts with anhydrous ammonia in acetonitrile to give 3-ethoxy-5-phenylisothiazole (yield 69%).[112] With aniline, the same cation gives mainly (yield 54%) ethyl 3-mercapto-3-phenyl-*N*-phenylimidoacrylate (**44**), with a small quantity of imine (**45**) and dithiolone (**46**).[112]

(**44**) (**45**) (some) (**46**)

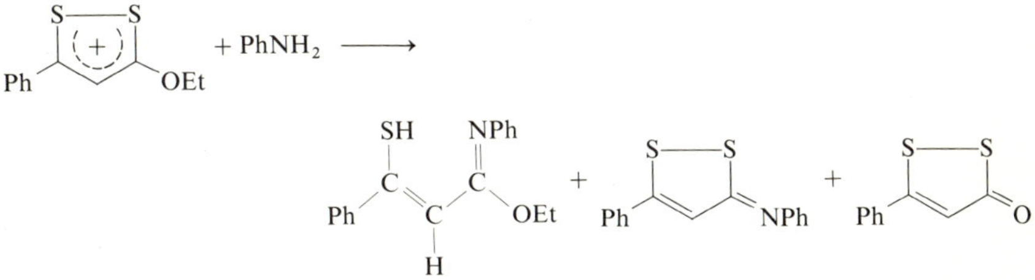

With aniline and 3-ethoxy-1,2-benzodithiolium ion, the major product is the *N*-phenylimine, while with other 3-ethoxy-1,2-dithiolium ions having no aryl substituent at position 5, the main product is an *O*-ethyl 3-anilinothioacrylate (Eq. 19).[112]

(19)

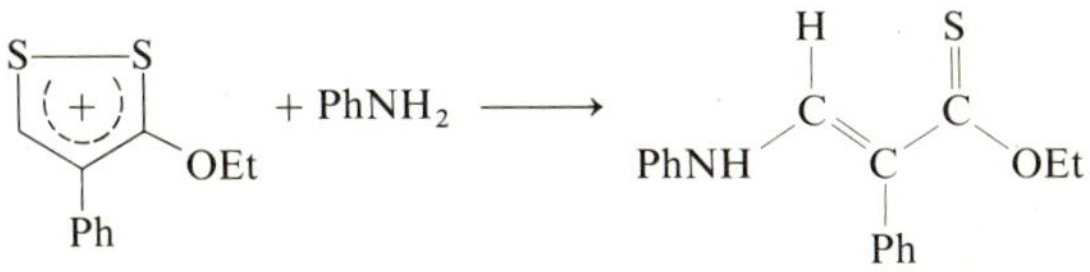

3-Methylthio-1,2-benzodithiolium ion reacting with formylhydrazine gives a formyl hydrazone (**47**) which gives easily the corresponding hydrazone (**48**) on hydrolysis.[141]

(**47**) (**48**)

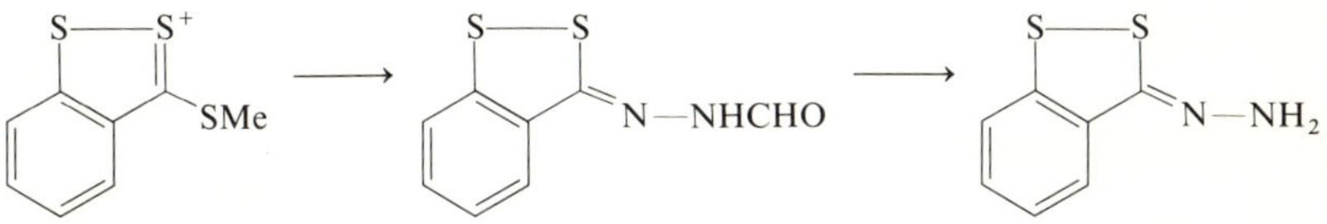

[141] S. Hünig, G. Kiesslich, K. H. Oette, and H. Quast, *Justus Liebigs Ann. Chem.* **754**, 46 (1971).

As shown in Scheme 21, aliphatic primary amines may react in three ways with 5-aryl-3-methylthio-1,2-dithiolium ions.[77,120] The products are: **49**, a 5-aryl-1,2-dithiole-3-thione, by sulfur demethylation; **50**, a methyl 3-alkylamino-3-aryldithioacrylate, by nucleophilic attack on carbon 5; and/or **51**, a 2-alkyl-5-arylisothiazoline-3-thione, by nucleophilic attack on carbon 3.

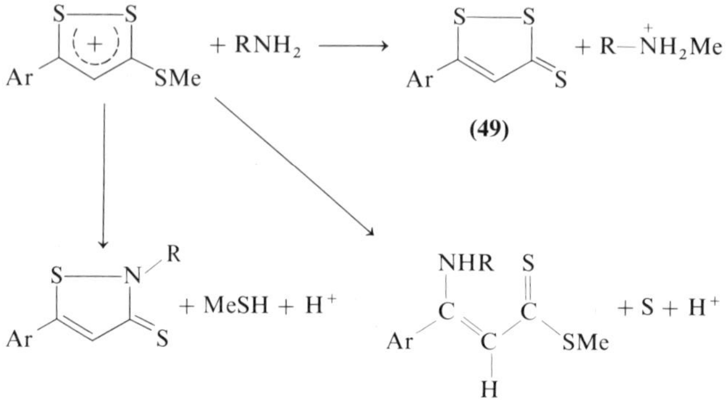

SCHEME 21

With aromatic primary amines, the above 5-aryl-3-methylthio cations give the corresponding N-arylimine (**52**).[61,75,86,142,143] Similarly, N-alkylarylamines lead to a 3-(N-alkylarylamino)-5-aryl-1,2-dithiolium ion (**53**), which can also be considered as an iminium ion (**54**).[142] In position 5, a 2-arylvinyl substituent has a similar effect (Eq. 20).[144]

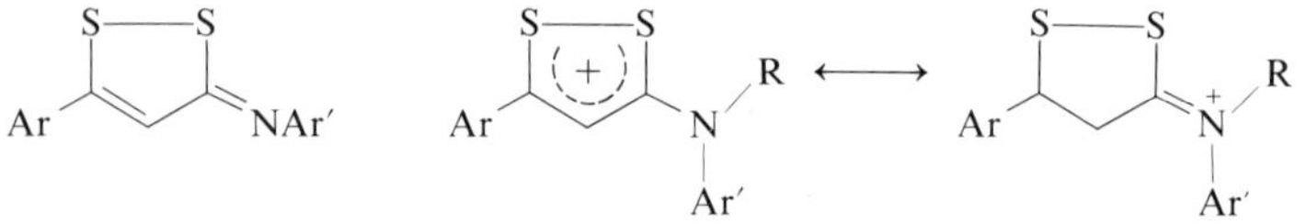

[142] C. Paulmier, Y. Mollier, and N. Lozac'h, *Bull. Soc. Chim. Fr.*, 2463 (1965).
[143] F. Boberg and W. von Gentzkow, *Justus Liebigs Ann. Chem.*, 247 (1973).
[144] R. Mayer and H. Hartmann, *Chem. Ber.* **97**, 1886 (1964).

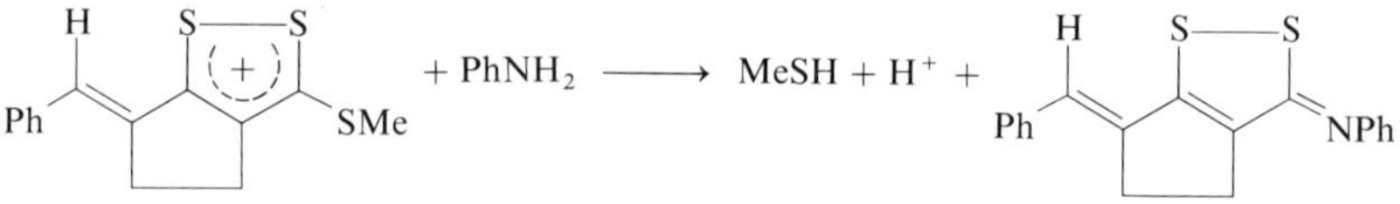

Vilsmeier salts (**55**) are also attacked by nitrogen nucleophiles such as aliphatic or aromatic primary amines.[16,145]

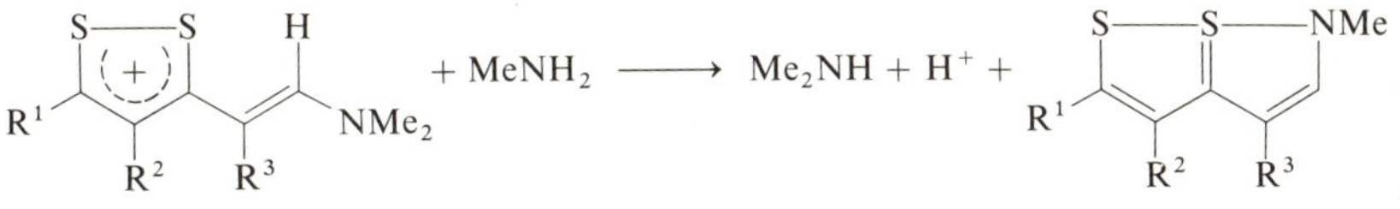

$$\underset{(\mathbf{55})}{\text{S-S structure with } R^1, R^2, R^3, H, NMe_2} + \text{MeNH}_2 \longrightarrow \text{Me}_2\text{NH} + \text{H}^+ + \text{S-S-NMe ring with } R^1, R^2, R^3$$

However, if in **55** R^1 is methylthio, nucleophilic attack by pyridine may give rise to S-demethylation.[84]

5-Aryl-3-methylthio-1,2-dithiolium ions react with secondary aliphatic amines.[81] As shown in Table I, the following compounds may be obtained in various yields: 5-aryl-1,2-dithiole-3-thione (**56**), by sulfur demethylation; methyl 3-aryl-3-(dialkylamino)dithioacrylate (**57**); and N,N-dialkyl-5-aryl-1,2-dithiol-3-iminium ion (**58**).

TABLE I
YIELDS OBTAINED IN REACTING 5-PHENYL-3-METHYLTHIO-1,2-DITHIOLIUM
ION WITH SOME SECONDARY ALIPHATIC AMINES

Amine	Yield (%)		
	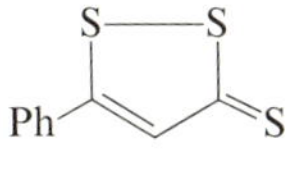 (**56**)	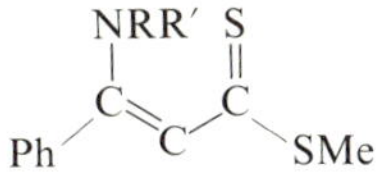 (**57**)	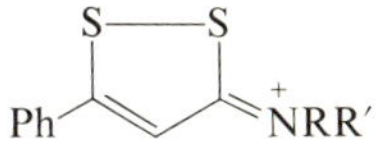 (**58**)
Me$_2$NH	6	48	35
Et$_2$NH	10	25	55
PhCH$_2$NHMe	8	10	66

3-Methylthio-1,2-dithiolium ions having no aryl substituent at position 5 react with primary and secondary amines to give 3-aminodithioacrylic esters (Eq. 21). Simultaneously, demethylation to the corresponding 1,2-dithiole-3-thione may occur to a large extent.[120,142,146–148] By reaction of two amine molecules instead of one, 3-aminothioacrylamides are sometimes obtained.[146]

---

[145] C. Metayer, G. Duguay, and H. Quiniou, *Bull. Soc. Chim. Fr.*, 163 (1974).
[146] E. J. Smutny, W. Turner, E. D. Morgan, and R. Robinson, *Tetrahedron* **23**, 3785 (1967).
[147] G. Le Coustumer and Y. Mollier, *Bull. Soc. Chim. Fr.*, 2958 (1971).
[148] G. Le Coustumer and Y. Mollier, *Bull. Soc. Chim. Fr.*, 3349 (1973).

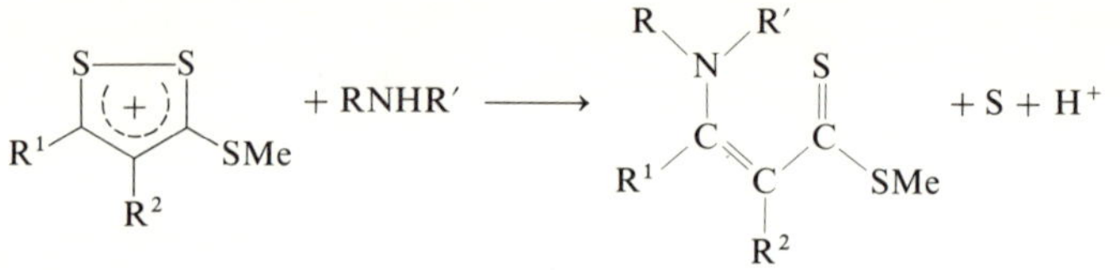

[diagram of equation (21): reaction of 1,2-dithiolium ion with R¹, R², SMe substituents with RNHR' giving open-chain product + S + H⁺]

$$R^1 \neq Ar \qquad (21)$$

R¹ ≠ Ar

e. *3-Halogenothio-1,2-dithiolium Ions.* In an earlier report it was stated that 4-aryl-3-bromothio-1,2-dithiolium ions react with primary amines to give 1,2-dithiol-3-imines.[60] It was shown later that the reaction product is in fact a 4-arylisothiazoline-5-thione (Scheme 22) in yields of ~50%.[149–151]

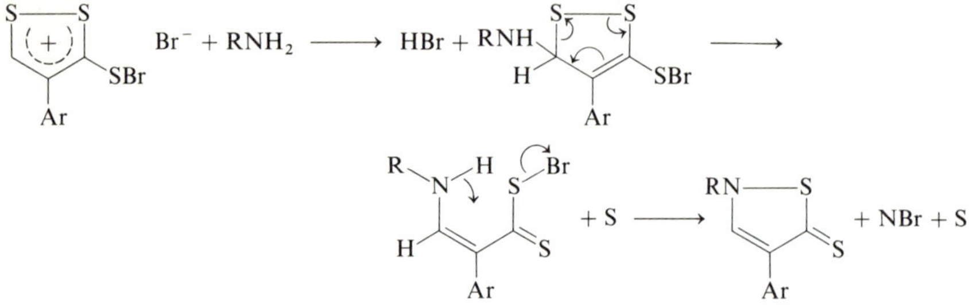

[Scheme 22 reaction diagram]

SCHEME 22

By contrast, 5-aryl-3-chlorothio(or 3-bromothio)-1,2-dithiolium ions react with primary arylamines to give 5-aryl-1,2-dithiol-3-imines.[32] This difference of behavior is clearly a result of the stabilizing effect of 5-aryl substituents on the 1,2-dithiole ring.

f. *3-Amino-1,2-dithiolium Ions.* These ions are resonance hybrids between dithiolium and iminium forms. Few examples of 5-aryl-3-amino-1,2-dithiolium ions have been studied, and it has been found that the course of the reaction depends markedly upon the nature of the nitrogen nucleophile.

With ammonia in anhydrous ethanol, 5-phenyl-3-amino-1,2-dithiolium cation gives 5-phenylisothiazole-3-thiol. With aniline or morpholine, however, the reaction involves loss of hydrogen sulfide and a 3-amino-5-phenylisothiazole is obtained (Scheme 23).[33]

Together with the reaction shown in Scheme 23, some of the 5-phenyl-3-amino-1,2-dithiolium ion reacts with the hydrogen sulfide produced according to Scheme 23 and some 5-phenyl-1,2-dithiole-3-thione is obtained.

---

[149] G. E. Bachers, D. M. McKinnon, and J. M. Buchshriber, *Can. J. Chem.* **50**, 2568 (1972).
[150] F. Boberg and W. von Gentzkow, *J. Prakt. Chem.* **315**, 965 (1973).
[151] M. S. Chauhan, M. E. Hassan, and D. M. McKinnon, *Can. J. Chem.* **52**, 1738 (1974).

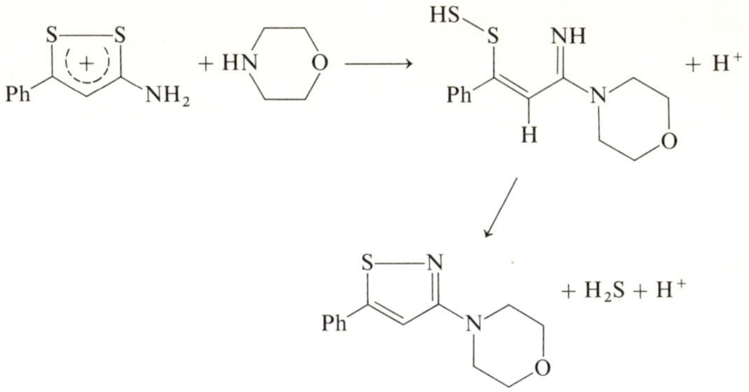

SCHEME 23

With hydrazine or phenylhydrazine, 5-aryl-3-amino-1,2-dithiolium cations give 5-aryl-3-aminopyrazoles.[33,152]

7. *Attack by Carbon Nucleophiles*

a. *3- or 5-Unsubstituted 1,2-Dithiolium Ions.* The literature prior to 1970 is cited in a previous review (see Ref. 2: pp. 173 and 188). In approximately neutral conditions, the reaction proceeds according to Eq. (22). Recent data concerning this reaction are given in Table II.[52,74,114,153–158]

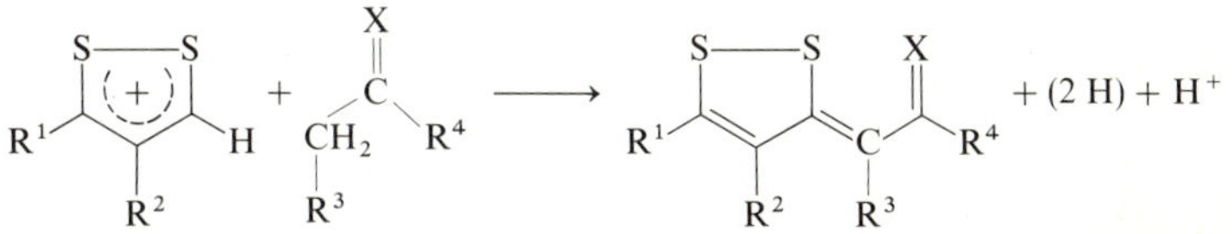

$$\text{dithiolium}^+ + \underset{R^3}{\underset{|}{CH_2}}\!\!-\!\!\underset{R^4}{\overset{X}{C}} \longrightarrow \text{product} + (2\,H) + H^+ \quad (22)$$

As indicated previously (Ref. 2: p. 188), if a 3- or 4-aryl-1,2-dithiolium ion is treated with a methyl acyldithioacetate in pyridine, the reaction takes a different course and gives a 3-acylthiopyran-2-thione.[156] Cyanothioacetamide and ethyl cyanoacetate react in a similar fashion.[159] These reactions are summarized in Scheme 24.

[152] A. Grandin and J. Vialle, *Bull. Soc. Chim. Fr.*, 4008 (1971).
[153] J. Bignebat and H. Quiniou, *Bull. Soc. Chim. Fr.*, 4181 (1972).
[154] E. Klingsberg, *Synthesis*, 213 (1971).
[155] Y. N'Guessan and J. Bignebat, *C. R. Hebd. Seances Acad. Sci., Ser. C* **272**, 1821 (1971).
[156] F. Clesse, J.-P. Pradere, and H. Quiniou, *Bull. Soc. Chim. Fr.*, 586 (1973).
[157] D. Festal, O. Coulibaly, R. Pinel, C. Andrieu, and Y. Mollier, *Bull. Soc. Chim. Fr.*, 2943 (1970).
[158] R. Pinel and Y. Mollier, *Bull. Soc. Chim. Fr.*, 1385 (1972).
[159] R. Pinel, N'Guyen Kim Son, and Y. Mollier, *C. R. Hebd. Seances Acad. Sci., Ser. C* **271**, 955 (1970).

TABLE II
REACTION OF 1,2-DITHIOLIUM CATIONS ON ACTIVE METHYLENE COMPOUNDS (EQ. 22)

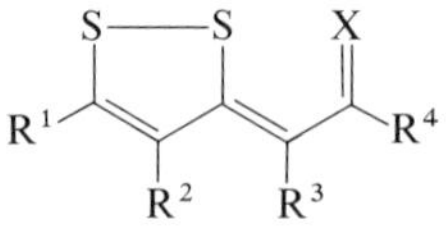

X	R^1	R^2	R^3	R^4	Reference
O	Ar	H	Me	Me, Et	74
O	t-Bu	H	H	p-$O_2NC_6H_4$	74
O	Ar	H	H^a	Ph	74
O	Ar	H	H	p-$CH_3C_6H_4$	153
O	H	Ph	H	Ph	154
O	Ph	Cl	H	Ar	155
O	Ph	H	H	2- or 4-Pyridyl	74
O	H	Ar	H	2- or 4-Pyridyl	74
O	t-Bu	H	CN	Ph	74
S	Ar	H	CHO	Ar	114
S	p-$MeOC_6H_4$	H	MeCO	MeS	156
S	Ar	H	ArCO	MeS	156
S	2-Thienyl	H	2-Thenoyl	MeS	156

X	R^1	R^2	R^3—R^4	Reference
^{18}O	Ph	H	—CH_2—CH_2—CH_2—	157
^{18}O	p-$MeOC_6H_4$	H	—CH_2—CH_2—CH_2—CH_2—	157
^{18}O	Ph	H	—CH=CH—CH=CH—^{b}	158
O	t-Bu	H	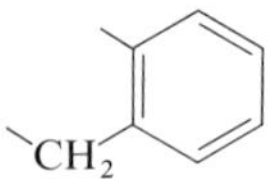 (o-tolyl-CH$_2$)	74
O	Ar	H	(o-tolyl-CH$_2$)	74
O	H	Ar	(o-tolyl-CH$_2$)	74

X	R^1—R^2	R^3—R^4	Reference
O	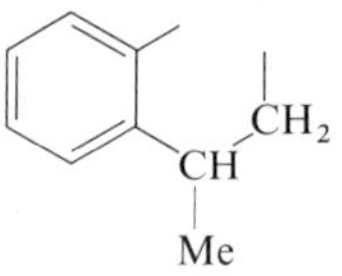 (o-tolyl-CH(Me)-CH$_2$)	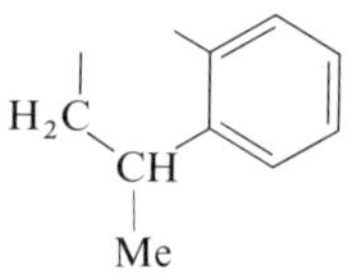 (o-tolyl-CH(Me)-CH$_2$)	52

^{a} In the product; in the starting methylene compound R^3 was CO_2H. Condensation is accompanied by decarboxylation.

^{b} In the product; in the starting ketone, [^{18}O]cyclohex-2-enone, R^3—R^4 was —CH_2—CH_2—CH=CH—. Here condensation is accompanied by dehydrogenation of the cyclohexenone ring.

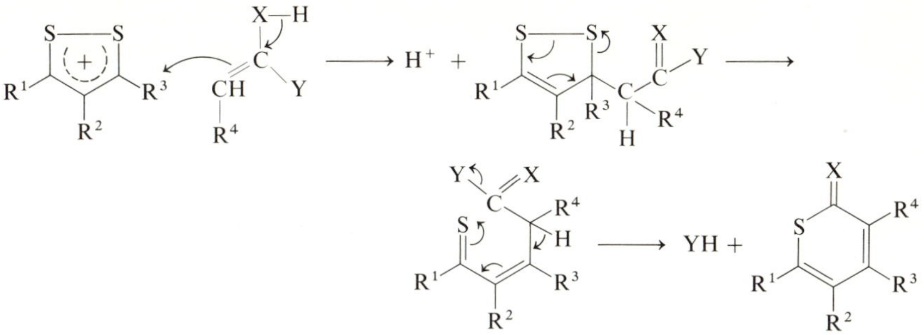

Sec. II.B] THE 1,2- AND 1,3-DITHIOLIUM IONS

R⁴	X	Y	Ref.
COAr	S	SMe	156
CN	S	NH₂	159
CN	O	OEt	159

SCHEME 24

Whereas the former condensation with retention of the 1,2-dithiole ring required the presence of a hydrogen atom as R^3, this is not necessary in the formation of thiopyran-2-thiones, as shown in Table III.

TABLE III
SYNTHESIS OF 3-ACYLTHIOPYRAN-2-THIONES[156]

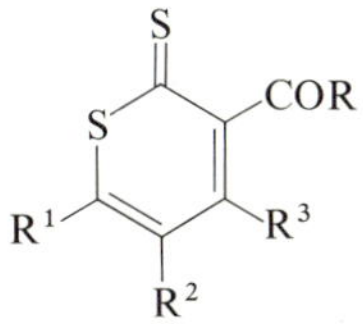

R	R^1	R^2	R^3	Yield (%)
Ar	Ar'	H	H	34–64
Ar	2-Thienyl	H	H	25–28
2-Thienyl	p-MeOC₆H₄	H	H	57
Me	Ar	H	H	38–57
Me	2-Thienyl	H	H	16
Me	H	Ph	H	49
Ar	H	Ph	H	54–69
2-Thienyl	H	Ph	H	67
Me	Ar	H	Ar	36–42
Me	2-Thienyl	H	2-Thienyl	50
Ar	Ar'	H	Ar'	27–39
p-MeC₆H₄	Ph	Ph	H	80
p-MeC₆H₄	Ph	Me	Ph	3

b. *1,2-Dithiolium Ions Substituted in Positions 3 and 5 by Hydrocarbon Groups.* As stated in the preceding section, condensation with active methylene groups in moderately basic medium (pyridine) leads to thiopyran-2-thiones.[156,159] A number of these compounds are listed in Table III. Similar results are obtained with sodium methylate as catalyst.[160]

When there is a methyl in position 3 or 5 in a 1,2-dithiolium cation, bases attack the cation giving the conjugate base which, in turn, reacts with another 1,2-dithiolium ion, probably according to the mechanism indicated in the previous review (Ref. 1: p. 83). Some new results concerning this autocondensation have since been published.[40,134]

c. *3-Halogeno-1,2-dithiolium Ions.* Publications prior to 1970 have been cited previously (see Ref. 2: pp. 175–176). 3-Chloro-1,2-benzodithiolium chloride reacting with cyclopentanone and various cyclohexanones gave α-(1,2-benzodithiol-3-ylidene) cycloalkanones. The same cation reacted with 3,4,5-trimethylphenol to give a dithiolium ion which can be deprotonated by pyridine (Scheme 25).[52]

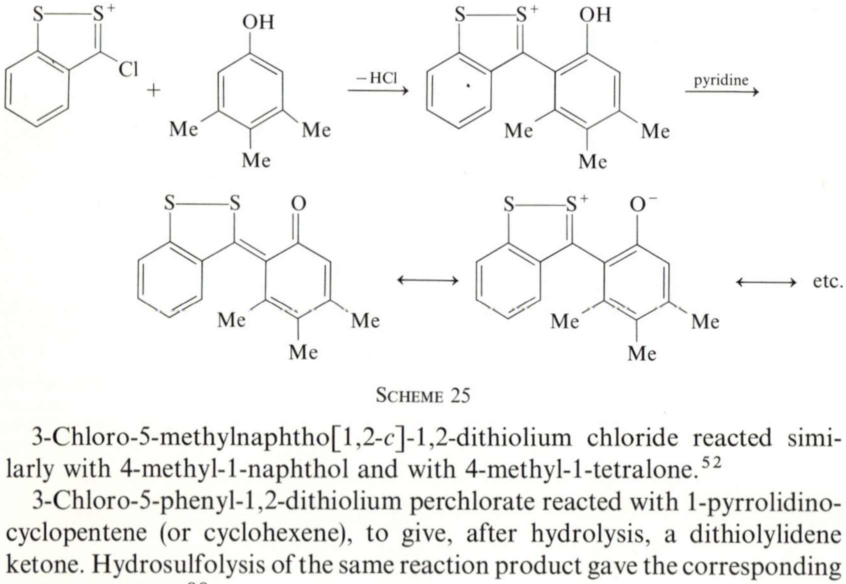

SCHEME 25

3-Chloro-5-methylnaphtho[1,2-*c*]-1,2-dithiolium chloride reacted similarly with 4-methyl-1-naphthol and with 4-methyl-1-tetralone.[52]

3-Chloro-5-phenyl-1,2-dithiolium perchlorate reacted with 1-pyrrolidinocyclopentene (or cyclohexene), to give, after hydrolysis, a dithiolylidene ketone. Hydrosulfolysis of the same reaction product gave the corresponding trithiapentalene.[88]

Other reactions of 3-chloro-1,2-dithiolium salts with carbon nucleophiles facilitated by electron-releasing nitrogen groups include 5-aryl-3-chloro-1,2-

---

[160] I. Shibuya, Japan. Kokai 75 130,765 (1975) [*CA* **85**, 21110 (1976)].

dithiolium ions and 3-chloro-1,2-benzodithiolium ions with N,N-dialkylarylamines and with diphenylamine which give 3-(aminoaryl)-1,2-dithiolium ions.[161]

d. *3-Alkylthio-1,2-dithiolium Ions.* Publications prior to 1970 are cited in a previous review (Ref. 2: pp. 176, 177, 189, 190).

In the presence of acetic acid and pyridine, 5-phenyl- or 5-*tert*-butyl-3-methylthio-1,2-dithiolium ions have been reacted with cyanoacetone to give, in the usual way, an α-(1,2-dithiol-3-ylidene) ketone.[74] A similar reaction has been observed with 4-hydroxy-6-methylpyran-2-one and with 4-hydroxycoumarin (Eq. 23).[162,163] In the same way, 4-hydroxy-6-methylchromene-2-thione and 3-methylthio-5-phenyl-1,2-dithiolium cation gave the fused trithiapentalene **59**.[162]

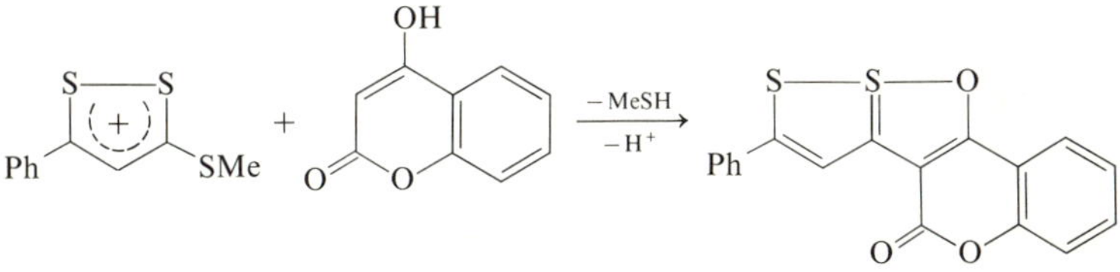

(23)

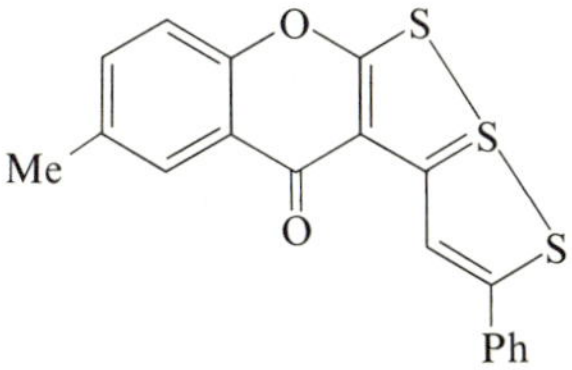

(59)

With 3-oxoglutaric anhydride, 5-aryl-3-methylthio-1,2-dithiolium cations may lead to a double condensation.[164,165]

When a 3-methylthio-1,2-dithiolium cation reacts with a β-oxoalkanoate anion, the condensation product easily loses carbon dioxide to give α-(1,2-dithiol-3-ylidene) ketones. This occurs, for instance, in the reactions of 3-methylthio-5-phenyl-1,2-dithiolium iodide with sodium benzoylacetate,[88]

---

[161] B. Bartho, J. Faust, and R. Mayer, *Z. Chem.* **15**, 440 (1975).
[162] F. M. Dean, J. Goodchild, and A. W. Hill, *J. C. S., Perkin 1*, 1022 (1973).
[163] E. G. Frandsen, *Tetrahedron* **33**, 869 (1977).
[164] E. G. Frandsen, *J. C. S., Chem. Commun.*, 851 (1977).
[165] E. G. Frandsen, *Tetrahedron* **34**, 2175 (1978).

and 3-methylthio-5-*tert*-butyl-1,2-dithiolium perchlorate with sodium 3-oxocyclohexanecarboxylate.[166]

Similarly, sodium enolates of 3-oxo-3-phenylpropanal react with 5-aryl-3-methylthio-1,2-dithiolium ions.[153,155] Sometimes only one product can be characterized[155]; in other cases a mixture of the two possible geometric isomers (**60** and **61**) is obtained.[153]

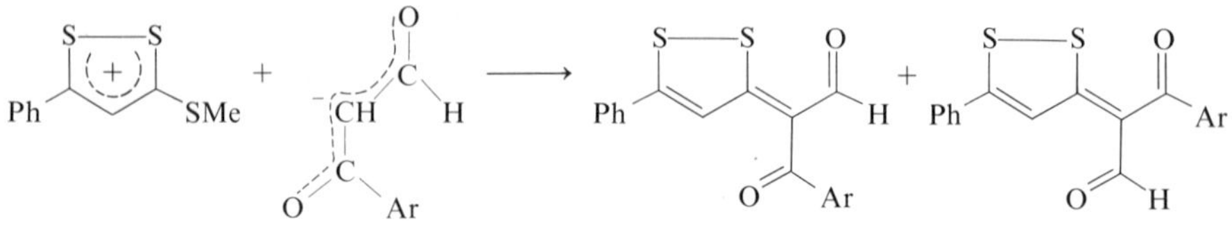

(**60**) (**61**)

5-Phenyl-3-methylthio-1,2-dithiolium methosulfate reacts with methyl acetyldithioacetate to give **62**.[167] A similar reaction has been realized with methyl pivaloyldithioacetate.[85]

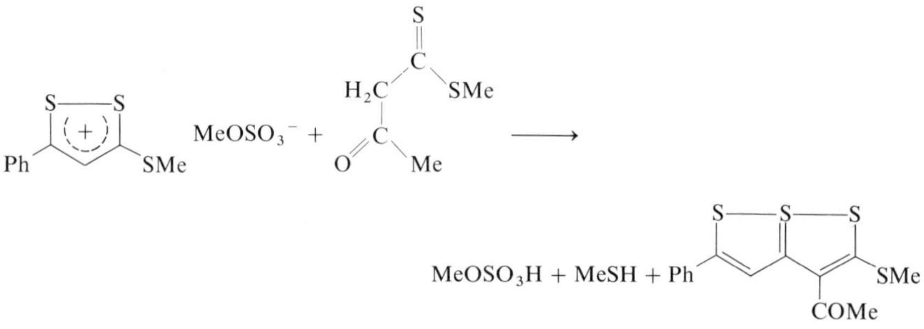

(**62**)

With 4-hydroxy-6-methylpyran-2-one and with 4-hydroxycoumarin, 4-aryl-3-methylthio-1,2-dithiolium cations react quite differently from their 5-aryl isomers, undergoing opening of the dithiole ring with subsequent recyclization into a thiopyrone.[163]

5-Aryl- or 5-*tert*-butyl-3-methylthio-1,2-dithiolium ions react with anthrone in the presence of pyridine to give the quinonoid structure **63**.[168] A similar reaction, leading to **64**, is observed with sodium 2,6-dimethylphenate.[169]

---

[166] R. M. Christie and D. H. Reid, *J. C. S., Perkin 1*, 880 (1976).
[167] R. J. S. Beer, D. Cartwright, R. J. Gait, and D. Harris, *J. Chem. Soc. C*, 963 (1971).
[168] R. Pinel and Y. Mollier, *Bull. Soc. Chim. Fr.*, 608 (1973).
[169] R. Pinel and Y. Mollier, *Bull. Soc. Chim. Fr.*, 1032 (1973).

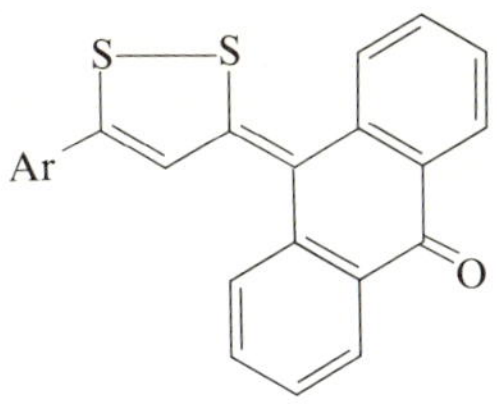

Sec. II.B] THE 1,2- AND 1,3-DITHIOLIUM IONS 189

(63) 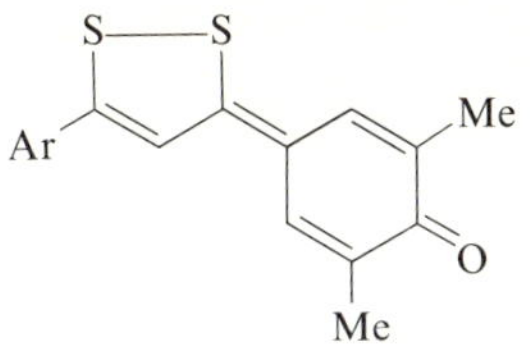   (64)

Similarly, both the 2-phenyl- and the 2,4-diphenylimidazolate anions give, with 3-methylthio-5-phenyl-1,2-dithiolium ion, dithiadiazafulvalenes.[170]

4-Aryl-3-methylthio-1,2-dithiolium ions react with N-alkylarylamines, to give methyl 2-aryl-3-(4-alkylaminophenyl)dithioacrylates (Eq. 24).[171]

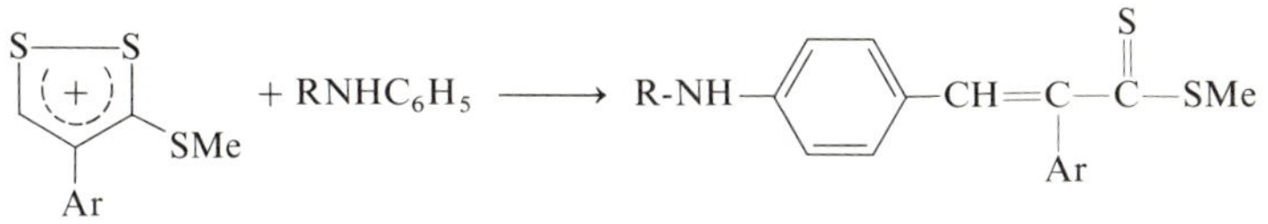

$$\text{[dithiolium-SMe, Ar]} + \text{RNHC}_6\text{H}_5 \longrightarrow \text{R-NH-C}_6\text{H}_4\text{-CH=C(Ar)-C(=S)-SMe} \quad (24)$$

Ar	R
Ph	Me
Ph	Et
p-MeC$_6$H$_4$	Me

3-Methylthio-1,2-dithiolium ions react with ethylidenemalononitrile in an acetic acid–pyridine mixture (Eq. 25).[172]

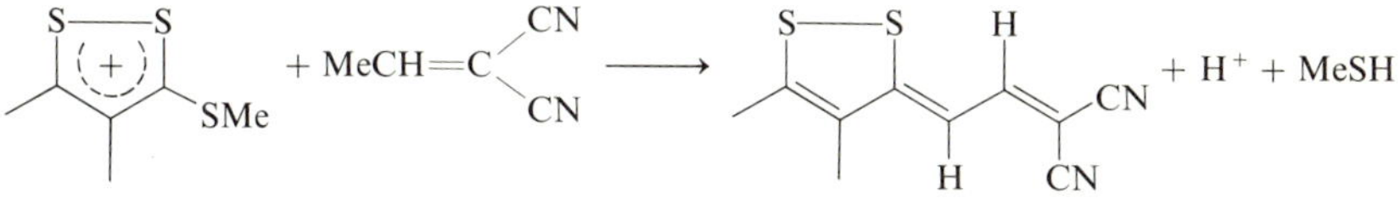

$$\text{[dithiolium-SMe]} + \text{MeCH=C(CN)}_2 \longrightarrow \text{product} + \text{H}^+ + \text{MeSH}$$

(25)

3-(Acylmethylthio)-1,2-dithiolium ions, in the presence of a basic catalyst, undergo internal replacement.[88–91,173] The first intermediate (65) is probably formed by deprotonation of the methylene group between the carbonyl and the sulfur linked to the dithiole ring. This zwitterion may cyclize, and the resulting thiirane then decompose, giving either a dithiolylidene ketone (66) or a disulfide (67) (Scheme 26).

---

[170] R. Weiss and R. Gompper, *Tetrahedron Lett.*, 481 (1970).
[171] G. Le Coustumer and Y. Mollier, *C. R. Hebd. Seances Acad. Sci., Ser. C* **274**, 1215 (1972).
[172] J. M. Catel and Y. Mollier, *C. R. Hebd. Seances Acad. Sci., Ser. C* **282**, 361 (1976).
[173] G. Caillaud and Y. Mollier, *Bull. Soc. Chim. Fr.*, 151 (1972).

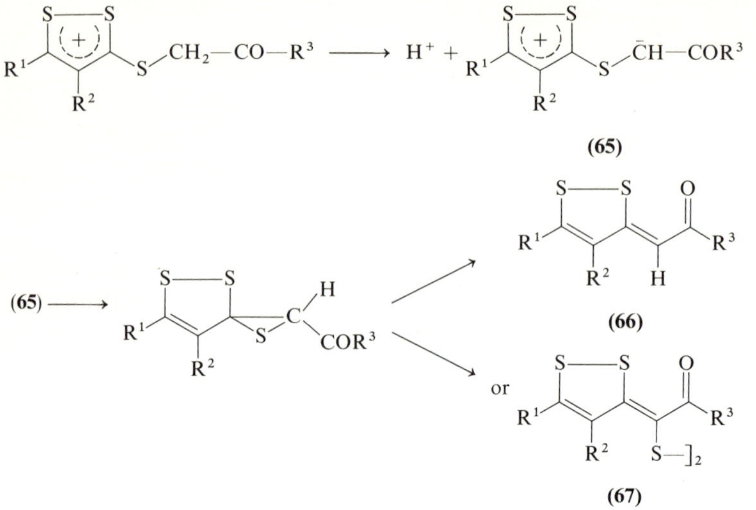

SCHEME 26

e. *3-Amino-1,2-dithiolium Ions.* In basic medium (sodium methylate) nitromethane attacks 3-amino-5-phenyl-1,2-dithiolium ions to give a nitrothiophene derivative (Eq. 26).[137]

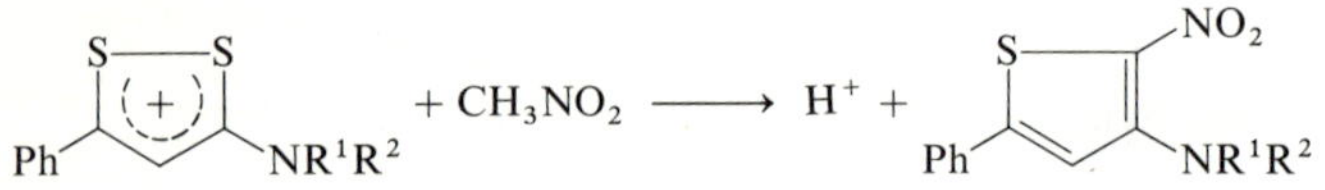

$$\text{[dithiolium]} + CH_3NO_2 \longrightarrow H^+ + \text{[nitrothiophene]} \qquad (26)$$

8. Electrophilic Attack

Owing to their positive charge, 1,2-dithiolium cations are, of course, almost unreactive to electrophiles. On the other hand, their conjugate bases, such as 3-alkylidene-1,2-dithioles, react easily with electrophilic reagents such as aromatic aldehydes, dimethylformamide, and nitrous acid. Some of these reactions are useful for the synthesis of partially bonded compounds and have been described, for papers prior to 1970, in a previous review (Ref. 2: pp. 181, 187, 193).

Aldehydes condense readily with 3-methyl-1,2-dithiolium ions, giving styryl derivatives (**68**).[55,174] Vilsmeier salts (**69**) are obtained by reacting dimethylformamide,[113] dimethylthioformamide,[16,84] *N*-methyl-*N*-phenylthioformamide, or dimethylformamide diethylacetal[145] with dithiolium compounds having α-methylene groups.

---

[174] R. Mayer, K. Fabian, H. Kröber, and H. Hartmann, *J. Prakt. Chem.* **314**, 240 (1972).

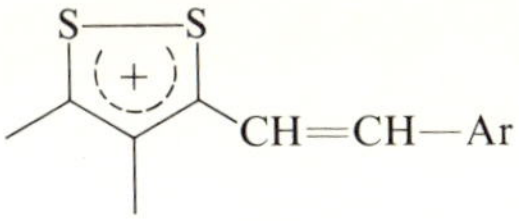

Sec. II.B]   THE 1,2- AND 1,3-DITHIOLIUM IONS   191

(68)

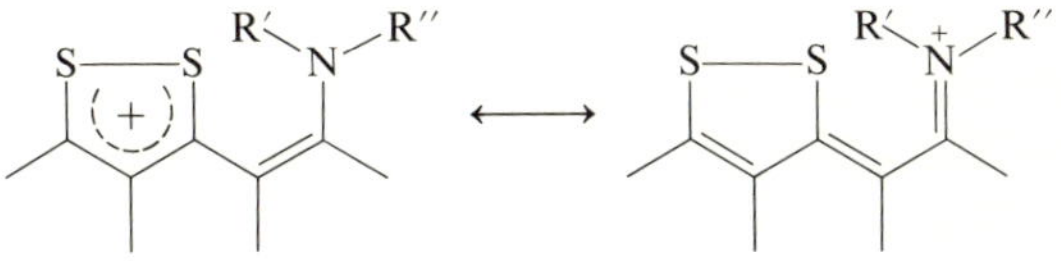

(69)

Similarly, nitrosation leads to 70,[85,175] whereas aryl diazonium ions give 71.[176,177]

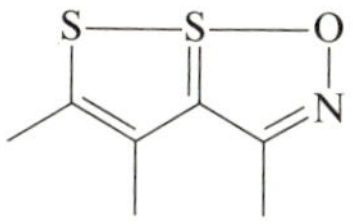

(70) 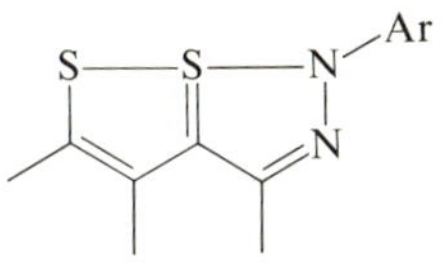   (71)

The electrophilic reagent may also be a 1,2- or 1,3-dithiolium ion.[178-180] An example of this reaction is given in (Eq. 27).

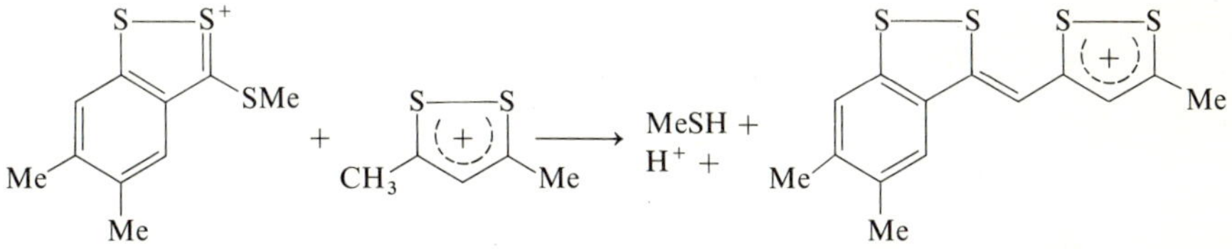

(27)

3-Ethoxy-2-cyanoacrylonitrile has also been used as electrophilic reagent attacking the conjugate base of a 1,2-dithiolium cation (Eq. 28).[172]

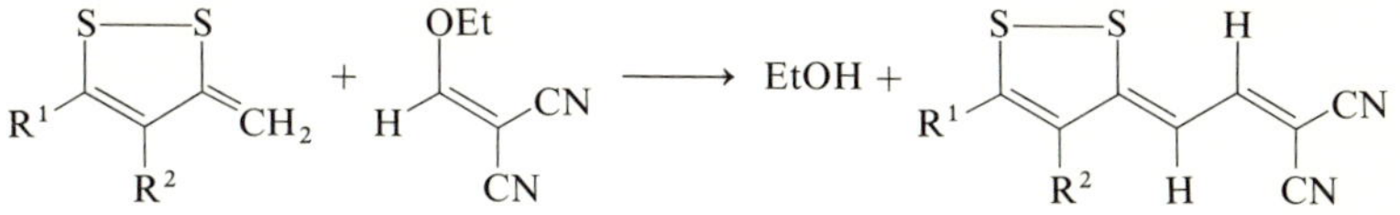

(28)

[175] J. G. Dingwall, A. R. Dunn, D. H. Reid, and K. O. Wade, *J. C. S., Perkin 2*, 1360 (1972).
[176] R. M. Christie, A. S. Ingram, D. H. Reid, and R. G. Webster, *J. C. S., Chem. Commun.*, 92 (1973).
[177] R. M. Christie and D. H. Reid, *J. C. S., Perkin 1*, 228 (1976).
[178] E. Klingsberg, *Synthesis*, 29 (1972).
[179] E. Klingsberg, U.S. Patent 3,299,055 (1967) [*CA* **66**, 105,899 (1967)].
[180] E. I. G. Brown, D. Leaver, and D. M. McKinnon, *J. C. S., Perkin 1*, 1511 (1977).

C. Physical Properties

1. *Quantum Mechanical Calculations*

Since the earlier review[1] reactivity calculations have been made for the 3- and 4-phenyl-1,2-dithiolium cations. The π-electron density, superdelocalizability, atom localization energy, and Z^0 factor have been given for the unsubstituted positions of the heterocycle and phenyl group.[181]

CNDO/2 calculations involving various parametrization methods and performed either with only s- and p-orbitals of sulfur, or including d-orbitals, have been applied to the 3-mercapto-1,2-dithiolium cation; and the modification of electronic spectra on 1,2-dithiole-3-thione protonation has been given theoretical interpretation.[182]

The π-electron distributions for 3- and 4-phenyl-1,2-dithiolium cations, derived from an improved Hückel method, have been compared with bond length data and UV spectra. The phenyl substituents appear to have only a small influence on the π-electron structure of the 1,2-dithiolium ring, but the π-bond order of the linkage connecting the phenyl and dithiolium rings is more pronounced with the 3-phenyl isomer.[183]

A question which has aroused much controversy is the contribution of sulfur d-orbitals to the ground-state bonding of various sulfur heterocycles including, among others, 1,2- and 1,3-dithiolium cations. A calculation using a linear combination of Gaussian orbitals (LCGO) led to the conclusion that the intervention of d-orbitals is not significant, and a satisfactory agreement of these calculations with polarographic half-wave reduction potentials has been found.[184]

On the other hand, a CNDO theoretical study of the electronic spectra of the 1,2-dithiolium cation and of its 3-methyl-, 4-methyl-, 3-phenyl-, and 4-phenyl derivatives shows that extension of the basis to sulfur d-orbitals leads to a limited but significant improvement of the correlation between theoretical and experimental results.[185]

A CNDO/2 method extending the atomic orbitals basis to sulfur d-orbitals has been applied to the study of the cathodic reduction of 1,2-dithiolium cations.[186]

---

[181] A. Mehlhorn, J. Fabian, and R. Mayer, *Z. Chem.* **5**, 23 (1965).
[182] D. Gonbeau, C. Guimon, and G. Pfister-Guillouzo, *Tetrahedron* **29**, 3599 (1973).
[183] A. Hordvik and E. Sletten, *Acta Chem. Scand.* **20**, 1938 (1966).
[184] M. H. Palmer and R. H. Findlay, *Tetrahedron Lett.*, 4165 (1972).
[185] C. Guimon, D. Gonbeau, and G. Pfister-Guillouzo, *Tetrahedron* **29**, 3399 (1973).
[186] C. Guimon, D. Gonbeau, G. Pfister-Guillouzo, K. Bechgaard, V. D. Parker, and C. T. Pedersen, *Tetrahedron* **29**, 3695 (1973).

2. Molecular Structure

Improved precision of bond-length measurements has been achieved, and in some cases it was possible to show that two or four slightly different 1,2-dithiolium units can exist in the same crystal lattice.[187,188] Compounds for which recent measurements are reported are indicated in Table IV.[37,187–195] Some unit cell dimensions and space-group data for a number of other 1,2-dithiolium salts have been recorded by Hordvik et al.[191]

TABLE IV
RECENT BOND-LENGTH DETERMINATIONS OF 1,2-DITHIOLIUM SALTS

$$\begin{array}{c} S\!\!-\!\!S \\ \diagdown\;\;\diagup \\ R^1\!\!-\!\!\!\!-\!\!R^3 \\ R^2 \end{array}\quad X^-$$

R^1	R^2	R^3	X^-	Reference
Me	H	Me	$\frac{1}{2}(Fe^{II}Cl_4)^{2-}$	187, 189
Me	H	Me	$\frac{1}{2}(Co^{II}Cl_4)^{2-}$	190
H	Ph	H	Cl^- (monohydrate)	191, 192
H	Ph	H	I^-	193
Ph	H	Ph	$\frac{1}{2}(Hg^{II}Cl_4)^{2-}$	188
Ph	H	Ph	$\frac{1}{2}(Fe^{II}Cl_4)^{2-}$	188
Ph	H	Ph	$\frac{1}{2}[(Fe^{III}Cl_4)^-, Cl^-]$	194
MeS	MeS	MeS	I^-	37
MeCONH	H	MeCONH	Br^-	195

A linear bond-length/bond-order correlation has been proposed for the S—S bond in 1,2-dithiolium cations. Calculated π-bond orders and experimental sulfur–sulfur bond lengths for 3- and 4-phenyl-1,2-dithiolium cations are in good agreement with this correlation.[196]

---

[187] H. C. Freeman, G. H. W. Milburn, C. E. Nockolds, R. Mason, G. B. Robertson, and G. A. Rusholme, *Acta Crystallogr., Sect. B* **30**, 886 (1974).
[188] R. Mason, G. B. Robertson, and G. A. Rusholme, *Acta Crystallogr., Sect. B* **30**, 894 (1974).
[189] H. C. Freeman, G. H. W. Milburn, C. E. Nockolds, P. Hemmerich, and K. H. Knauer, *J. C. S., Chem. Commun.*, 55 (1969).
[190] G. A. Heath, P. Murray-Rust, and J. Murray-Rust, *Acta Crystallogr., Sect. B* **33**, 1299 (1977).
[191] A. Hordvik, E. Sletten, and J. Sletten, *Acta Chem. Scand.* **20**, 1172 (1966).
[192] F. Grundtvig and A. Hordvik, *Acta Chem. Scand.* **25**, 1567 (1971).
[193] A. Hordvik and E. Sletten, *Acta Chem. Scand.* **20**, 1874 (1966).
[194] R. Mason, G. B. Robertson, and G. A. Rusholme, *Acta Crystallogr., Sect. B* **30**, 906 (1974).
[195] A. Hordvik and H. M. Kjoege, *Acta Chem. Scand.* **20**, 1923 (1966).
[196] A. Hordvik, *Acta Chem. Scand.* **20**, 1885 (1966).

Studies of potential organic metals have prompted structure determination of tetrathiotetracene (TTT) (**15**) derivatives. The crystal structure of the (TTT)I$_x$ anion-radical salts ($x = 1$ or 1.5) have been examined. They indicate that iodine chains are arranged in channels between the TTT stacks.[44]

3. Electronic Absorption Spectra

The electronic spectra of 1,2-dithiolium, 1,2-benzodithiolium, 3-mercapto-1,2-dithiolium, and 3-mercapto-1,2-benzodithiolium cations, dissolved in aqueous perchloric acid, have been measured and were interpreted by Pariser–Parr–Pople (PPP) calculations.[197] The spectrum of 3-phenyl-5-morpholino-1,2-dithiolium perchlorate in methylene chloride has been measured.[198]

Protonation of 4-(5-phenyl-1,2-dithiol-3-ylidene)cyclohexadien-1-one to 5-phenyl-3-(p-hydroxyphenyl)-1,2-dithiolium cation results in a strong hypsochromic effect on the longest wavelength absorption band (530 to 467 nm).[199]

Various derivatives of the 3-[(1,2-dithiol-3-ylidene)methyl]-1,2-dithiolium cations have been investigated. Their absorption spectra are quite different from those of the corresponding protonated species, which are quite similar to the spectrum of the 3-methyl-5-phenyl-1,2-dithiolium cation. These facts have been interpreted as resulting from a S—S partial bonding between the two dithiole rings of the nonprotonated cations. This interaction disappears with protonation of the methylidyne group linking the two dithiole rings.[200–202]

1,2-Dithiolium-1,1,3,3-tetracyanopropenides, diversely substituted, have an absorption band around 460–530 nm which results from charge transfer and whose position and intensity depend markedly upon the solvent.[203]

The effect of alkyl groups on the UV spectra of 1,2-dithiolium cations and the charge-transfer spectrum with iodide as donor have been compared with the results of SCF–LCI calculations.[204]

4. NMR, ESR, and ESCA Data

Few general studies have appeared since the previous review.[1] For various alkyl-1,2-dithiolium cations, the chemical shifts of the ring and methyl

---

[197] J. Fabian, K. Fabian, and H. Hartmann, *Theor. Chim. Acta* **12**, 319 (1968).
[198] H. Hartmann, *J. Prakt. Chem.* **313**, 1113 (1971).
[199] R. Gompper and H. U. Wagner, *Tetrahedron Lett.*, 165 (1968).
[200] J. Fabian and H. Hartmann, *Z. Chem.* **12**, 349 (1972).
[201] J. Fabian and H. Hartmann, *Tetrahedron* **29**, 2597 (1973).
[202] J. Fabian, H. Hartmann, and K. Fabian, *Tetrahedron* **29**, 2609 (1973).
[203] J.-M. Catel and Y. Mollier, *C. R. Hebd. Seances Acad. Sci., Ser. C* **280**, 673 (1975).
[204] K. Fabian, H. Hartmann, J. Fabian, and R. Mayer, *Tetrahedron* **27**, 4705 (1971).

protons determined by NMR were correlated with calculated π-charges. A linear relationship exists between the respective chemical shifts of the ring and methyl protons at the same positions of dithiolium rings otherwise similarly constituted.[204]

Restricted rotation around C—N bonds has been studied for 3-dialkylamino-1,2-dithiolium cations[205] and for 3-(2-dimethylaminovinyl)-1,2-dithiolium cations (Vilsmeier salts).[16]

ESR spectra have been obtained with the rather stable free radicals obtained by reduction of 1,2-dithiolium cations such as 3,5-diaryl-1,2-dithiolium cations,[103,104] and the phenaleno[1,9-cd]dithiolium cation.[25] ESR spectra have also been used in connection with conductivity studies of complexes containing dithiole rings.[46] (see also Section II,C,5.)

ESCA spectra of various 1,2-dithiolium salts show only one S(2p) signal for both sulfur atoms of the ring, a fact which shows a high degree of charge delocalization in the ring. Experimental values may be found in Lindberg et al.[206] and Schneller and Swartz.[207]

5. Conductivity

Some derivatives of 1,2-dithiole have shown interesting electrical properties arising from the electron-donor/electron-acceptor properties of the 1,2-dithiole/1,2-dithiolium systems. These properties have been particularly studied in ring systems containing two 1,2-dithiole rings such as **72** or **75**, to which correspond, respectively, the cation radicals **73** or **76** and the

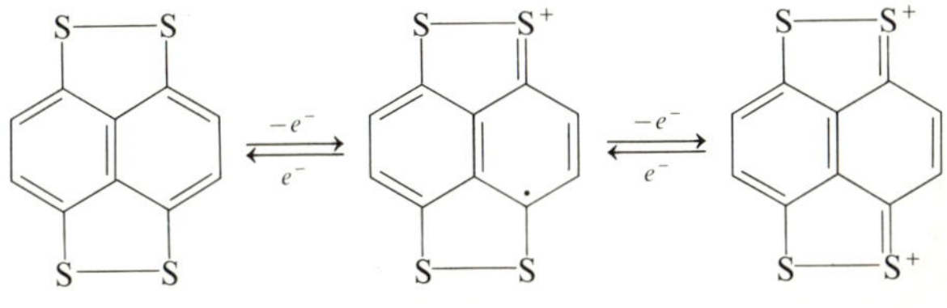

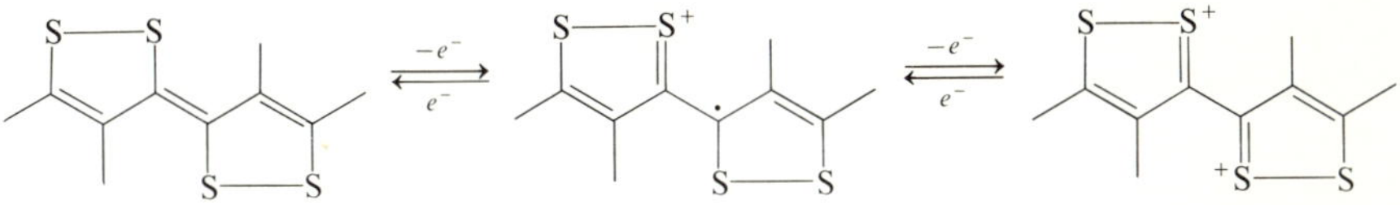

[205] M. -L. Filleux-Blanchard, G. Le Coustumer, and Y. Mollier, *Bull. Soc. Chim. Fr.*, 2607 (1971).
[206] B. J. Lindberg, S. Högberg, G. Malmsten, J. E. Bergmark, O. Nilsson, S. E. Karlsson, A. Fahlman, U. Gelius, R. Pinel, M. Stavaux, Y. Mollier, and N. Lozac'h, *Chem. Scr.* **1**, 183 (1971).
[207] S. W. Schneller and W. E. Swartz, *J. Heterocycl. Chem.* **11**, 105 (1974).

TABLE V
ABBREVIATIONS USED FOR CONSTITUENTS OF ORGANIC METALS

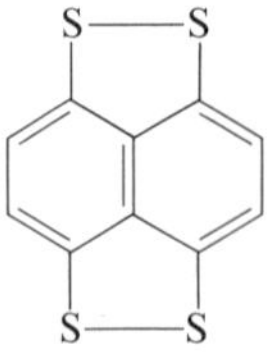

TTN: Tetrathionaphthalene
Systematic name: Naphtho[1,8-cd: 4,5-c'd']-bis[1,2]dithiole

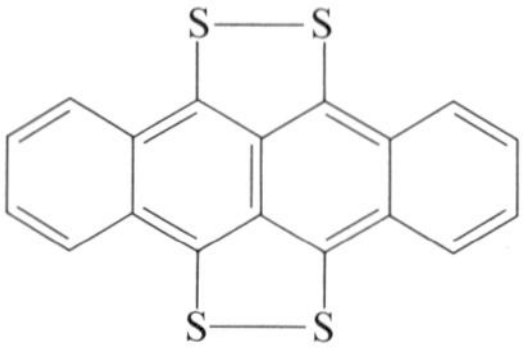

TTT: Tetrathiotetracene [a]
Systematic name: Naphthaceno[5,6-cd: 11,12-c'd']-bis[1,2]dithiole

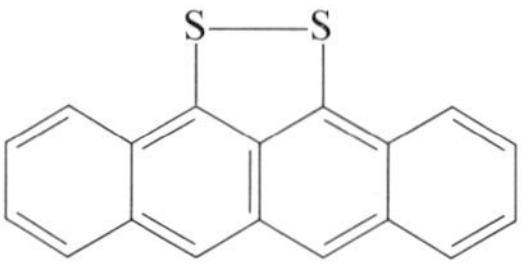

DTT: Dithiotetracene
Systematic name: Naphthaceno[5,6,cd]-1,2-dithiole

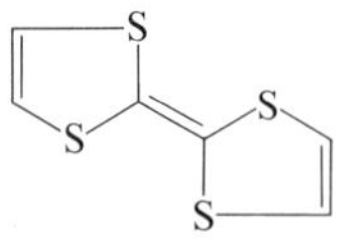

TTF: Tetrathiafulvalene
Systematic name: Bis(1,3-dithiol-2-ylidene)

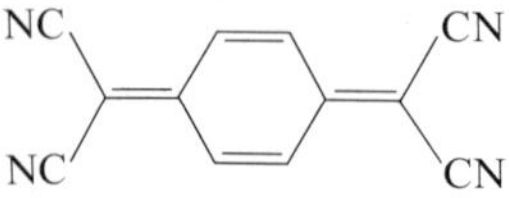

TCNQ: Tetracyanoquinodimethane
Systematic name: Cyclohexa-2,5-diene-1,4-diylidene-bis(malononitrile)

[a] Care should be taken in that the abbreviation TTN (for tetrathionaphthacene) has been used also for this compound.[45]

dications **74** or **77**. For simplicity's sake, only one of the resonance forms is given in these formulas.

Electron-donor properties of systems **72, 75,** and their derivatives are shown in the formation of complexes with electron acceptors such as tetracyanoethylene or chloranil. We shall designate the constituents of the complexes studied here by the most commonly used abbreviations which are summarized in Table V along with the corresponding trivial and systematic names.

Among the various 1,2-dithiole derivatives, by far the most studied is tetrathiotetracene (TTT), described in 1948 by Marschalk and Stumm.[208]

[208] C. Marschalk and C. Stumm, *Bull. Soc. Chim. Fr.*, 418 (1948).

TABLE VI
Conductivities[a] at Room Temperature of Various Tetrathiotetracene (TTT) Salts or Complexes

Derivative	Conductivities σ (Ω^{-1} cm^{-1})	Reference
(TTT)$_3$(o-chloranil)	25–50 × 10^{-2}	46
(TTT)$_3$(o-bromanil)	12–17 × 10^{-2}	46
(TTT)$_3$(tetracyanoethylene)$_2$	7 × 10^{-2}	46
(TTT)Cl, H$_2$O	4.3 × 10^{-4}	43
(TTT)Br	2 × 10^{-2}	43
(TTT)I	140 × 10^{-2}	43
(TTT)(SCN)	23 × 10^{-2}	43
(TTT)(HSO$_4$)	1 × 10^{-2}	42
(TTT)(HSO$_4$)$_2$	3 × 10^{-8}	42

[a] Measured from compressed powder pellets.

The preparation of mono- and dications from TTT was described in 1952.[41] However, the electrical properties of TTT salts or charge-transfer complexes were only disclosed much later.

Conductivities of various derivatives of TTT have been measured on compressed-powder pellets. Some results are summarized in Table VI. It appears, for example, that dicationic salts of TTT are much less conducting than monocationic salts.[42] The selenium analog of TTT shows better conducting properties, and the corresponding monocation hydrogen sulfate has a conductivity about twice that of the corresponding TTT salt.[42]

Single crystals of the (TTT)(TCNQ) complex have been studied between 2°K and room temperature.[45] At room temperature, the conductivity of this complex is 1 Ω^{-1} cm^{-1}.

A complex that arouses particular interest is the triodide (TTT)$_2$I$_3$ which has properties quite different from those of the iodide (TTT)I. The interaction of TTT with iodine in nitrobenzene with different proportions of the reagents leads to single crystals of the two phases.[44] For (TTT)I, at room temperature, the single crystal conductivity is about 30–70 Ω^{-1} cm^{-1} and does not change very much on cooling until 200°K, when it falls rapidly. In contrast, (TTT)$_2$I$_3$ single crystal conductivity at room temperature is rather high (about 600–1200 Ω^{-1} cm^{-1}), and it increases considerably on cooling to 50°K, but then drops again as the temperature is reduced still further. These conductivities are comparable with those of (TTF)(TCNQ) complexes. The temperature dependence of the electrical properties of (TTT)$_2$I$_3$ has been studied in detail.[209,210]

[209] L. C. Isett and E. A. Perez-Albuerne, *Solid State Commun.* **21**, 433 (1977).
[210] B. Hilti and C. W. Mayer, *Helv. Chim. Acta* **61**, 501 (1978).

Dithiotetracene (DTT) has been found as a by-product of some TTT syntheses,[41,42] and its most likely structure is given in Table V although quinonoid structures may also perhaps be considered.[42] A better synthesis of DTT has been given more recently and the (DTT)(TCNQ) complex has a single crystal conductivity of $3 \cdot 10^{-6}\ \Omega^{-1}\ cm^{-1}$.[47]

Difficulty in the synthesis of TTN has retarded conductivity studies, although it seems to be a promising component for organic metals. The single crystal conductivity of (TTN)(TCNQ) at room temperature is $40\ \Omega^{-1}\ cm^{-1}$, compared with $1\ \Omega^{-1}\ cm^{-1}$ found for (TTT)(TCNQ).[48]

A quite different group of semiconductors is constituted by 1,2-dithiolium 5-thioxo-1,2-dithiole-3-thiolates which are charge-transfer salts (Eq. 29).[211]

$$\text{(29)}$$

For Ar = phenyl, the single crystal conductivity of the charge-transfer salt is about $0.17 \cdot 10^{-12}\ \Omega^{-1}\ cm^{-1}$, much less than for the (TTF)(TCNQ) complex, a fact which has been ascribed to a different type of crystal packing.[212]

6. Mass Spectra

Mass spectra have been obtained from various 3-alkylthio- and 3-arylthio-1,2-dithiolium iodides thermolyzed in the ion source of the mass spectrometer. The spectra from alkylthio salts can be rationalized by assuming the primary formation of a 1,2-dithiolyl radical. The 3-arylthio compounds give a bis(1,2-dithiol-3-ylidene), probably through formation of a carbene intermediate.[213]

A study of alkyl- or aryl-substituted 1,2-dithiolium salts leads to similar conclusions. One group of these salts undergoes thermolysis to a dithiolyl radical, and in the mass spectrum the parent ion is the dithiolyl ion, which often loses a hydrogen atom. However, sometimes the parent ion is stable enough to be observed in the spectrum. This is the case for 3,5-diphenyl-1,2-dithiolium bromide. Other salts, such as 3- or 4-phenyl-1,2-dithiolium bromide, probably first expel a proton, giving a carbene intermediate.[214]

---

[211] N. Loyaza and C. T. Pedersen, *J. C. S., Chem. Commun.*, 496 (1975).
[212] O. Simonsen, N. Loyaza, and C. T. Pedersen, *Acta Chem. Scand., Ser. B* **31**, 281 (1977).
[213] C. T. Pedersen, N. Loyaza Huaman, and J. Moeller, *Acta Chem. Scand., Ser. B* **28**, 1185 (1974).
[214] C. T. Pedersen and J. Moeller, *Tetrahedron* **30**, 553 (1974).

III. The 1,3-Dithiolium Ion

A. Methods of Preparation

1. With Formation of the 1,3-Dithiole Ring

a. *From α-Dithiols.* Preparation of 1,3-benzodithiolium salts **80** from benzene-1,2-dithiols (**78**) has been widely used and was already cited in a previous review (Ref. 1: p. 103). Thus, compounds **80** are obtained by acid-catalyzed condensation of **78** with aldehydes followed by oxidation of the intermediate **79** by various oxidizing agents. The reaction can also be carried out with dialdehydes.

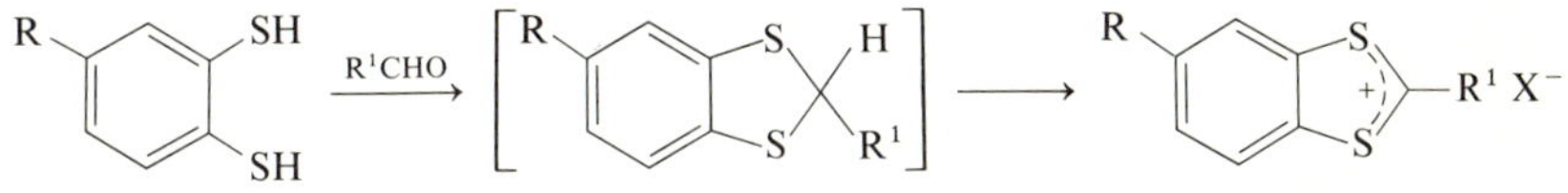

Another route to compounds **80** is the condensation of **78** with carboxylic acids in the presence of phosphorus oxychloride: moderate to good yields have been reported[215-217] for some 2-phenyl- and 2-unsubstituted examples (see also Table XIV in Ref. 1: p. 105). Similarly, 4,5-diphenyl-1,3-dithiolium perchlorate is obtained in 45% yield from 1,2-diphenylethylene-1,2-dithiol and formic acid (Eq. 30).[218]

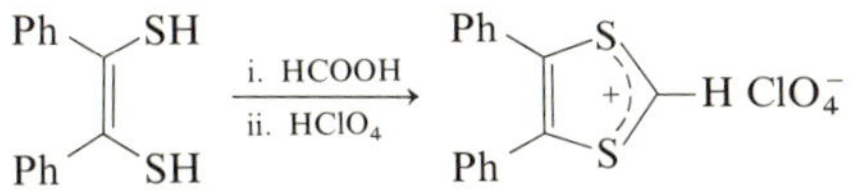

Acyl chlorides also react with various benzene-1,2-dithiols, leading to 1,3-benzodithiolium salts. Moderate yields (11-70%) are reported for a range of 2-alkyl and 2-aralkyl examples.[219]

By reaction of acetic or benzoic esters with benzene-1,2-dithiol in tetrafluoroboric acid/ether complex, 2-methyl- or 2-phenyl-1,3-benzodithiolium salts are easily prepared in excellent yields.[220]

4-Methylbenzene-1,2-dithiol, in methyl chlorodithioformate gives 2-methylthio-5-methyl-1,3-benzodithiolium perchlorate (**81**) (33% yield) on treatment with 70% perchloric acid.[79]

---

[215] S. Hünig, G. Kiesslich, H. Quast, and D. Scheutzow, *Justus Liebigs Ann. Chem.*, 310 (1973).
[216] G. Scherowsky and J. Weiland, *Justus Liebigs Ann. Chem.*, 403 (1974).
[217] I. Degani, R. Fochi, and P. Tundo, *J. Heterocycl. Chem.* **11**, 507 (1974).
[218] K. M. Pazdro, *Rocz. Chem.* **43**, 1089 (1969).
[219] P. Appriou and R. Guglielmetti, *Bull. Soc. Chim. Fr.*, 510 (1974).
[220] I. Degani and R. Fochi, *Synthesis*, 263 (1977).

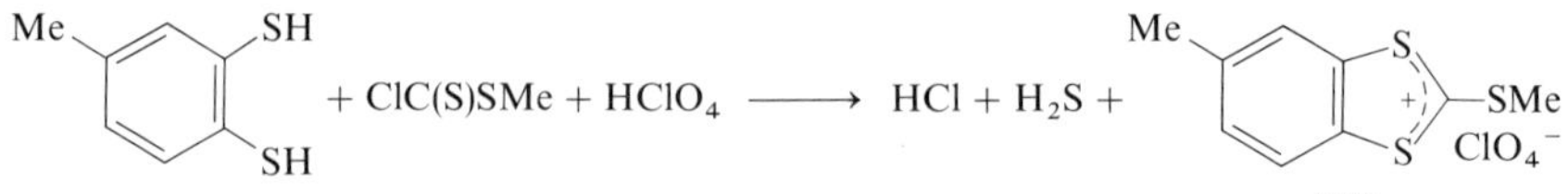

[Scheme showing: Me-benzene-1,2-dithiol + ClC(S)SMe + HClO₄ → HCl + H₂S + Me-benzodithiolium-SMe ClO₄⁻]

(81)

These methods are now easy to apply. 4-Methylbenzene-1,2-dithiol is a commercial product and benzene-1,2-dithiol is conveniently prepared by reacting anthranilic acid with isopentyl nitrite in the presence of carbon disulfide.[221,222] 2-Isopentyloxy-1,3-benzodithiole thus obtained, treated by the tetrafluoroboric/ether complex gives, in 65–70% yield, 1,3-benzodithiolium tetrafluoroborate, which is easily cleaved by sodium in liquid ammonia, giving benzene-1,2-dithiol in 80–85% yield (Scheme 27).[223]

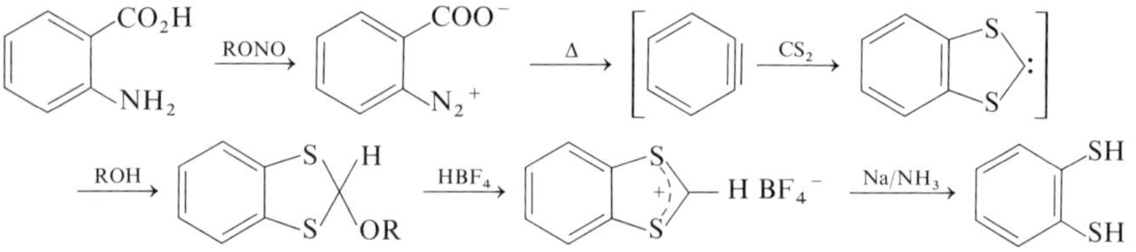

SCHEME 27

b. *From α-Oxoalkyl Dithiocarboxylates.* α-Oxoalkyl dithiocarboxylates **82** can be cyclized to the salts **83** by warming either in 70% perchloric acid or concentrated sulfuric acid. This acid-catalyzed cyclodehydration has sometimes been performed in the presence of H_2S or of the complex H_2S/BF_3 [224] (Table VII).[10,79,88,225–230]

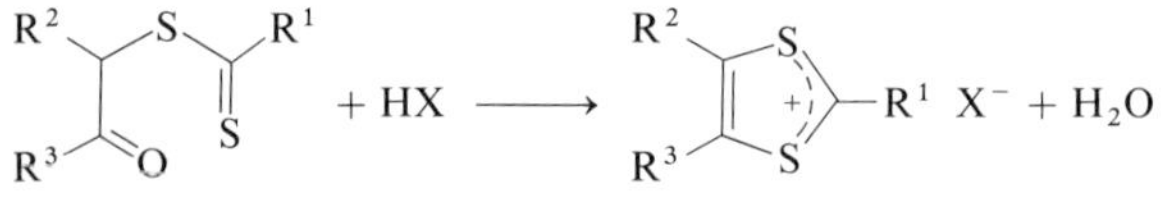

[Reaction scheme: **82** + HX → **83** + H_2O, with R¹, R², R³ substituents]

(82) R¹: aliphatic, aromatic,   (83)
alkylthio, or dialkylamino

---

[221] J. Nakayama, *Synthesis* 38 (1975).
[222] J. Nakayama, E. Seki, and M. Hoshino, *J. C. S., Perkin 1*, 468 (1978).
[223] I. Degani and R. Fochi, *Synthesis*, 471 (1976).
[224] E. Campaigne and R. D. Hamilton, *Q. Rep. Sulfur Chem.* **5**, 275 (1970).
[225] E. Campaigne and N. W. Jacobsen, *J. Org. Chem.* **29**, 1703 (1964).
[226] D. Leaver and W. A. H. Robertson, *Proc. Chem. Soc., London* 252 (1960).
[227] E. Campaigne, R. D. Hamilton, and N. W. Jacobsen, *J. Org. Chem.* **29**, 1708 (1964).
[228] M. Ahmed, J. M. Buchshriber, and D. M. McKinnon, *Can. J. Chem.* **48**, 1991 (1970).
[229] A. Mas, J. -M. Fabre, E. Torreilles, L. Giral, and G. Brun, *Tetrahedron Lett.*, 2579 (1977).
[230] A. Takamizawa and K. Hirai, *Chem. Pharm. Bull.* **17**, 1924 (1969).

TABLE VII
1,3-DITHIOLIUM SALTS (83) FROM α-OXOALKYL DITHIOCARBOXYLATES

R^1	R^2	R^3	Reference
Ph	H	Ph	10, 225
Ph	H	Me	10, 226
Me	H	Ph	10, 226
Ph	Ph	Ph	10
RS	H	Ar	88, 227
EtS	Ph	Ph	228
R_2N	H	H	229
R_2N	H	Me, Ar	225, 230
R_2N	R'	R'	225, 229
R_2N	Ph	Ph	230
Ph_2N, PhMeN	H	Ph	79

The diester **84** with cold concentrated sulfuric acid yields the 2,2'-(*p*-phenylene)bis(1,3-dithiolium) salt (**85**).[231]

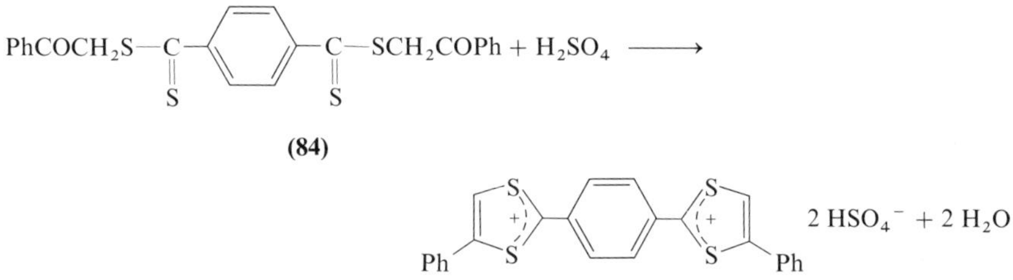

$$PhCOCH_2S-\underset{S}{\overset{\parallel}{C}}-\underset{}{\bigcirc}-\underset{S}{\overset{\parallel}{C}}-SCH_2COPh + H_2SO_4 \longrightarrow$$

(**84**)

[structure **85**] $2\ HSO_4^- + 2\ H_2O$

(**85**)

Compounds **86**, with two 1,3-dithiolium nuclei in the same molecule, are also obtained by acidic cyclization of diesters. By the action of H_2S or NaHS, compounds **86** give the thiones **87** which, treated with peracetic and then perchloric acid, lead to the salts **88**.[232,233] These last can be obtained from **86** in another way, consisting first of a reduction by $NaBH_4$ leading to **89**, which gives **88** by loss of an anion.[233,234] In both cases, yields are excellent. The explosive perchlorate **90** is prepared similarly.[235]

---

[231] Y. Ueno, M. Bahry, and M. Okawara, *Tetrahedron Lett.*, 4607 (1977).
[232] Y. Ueno, Y. Masuyama, and M. Okawara, *Chem. Lett.*, 603 (1975).
[233] G. Schukat, L. van Hinh, and E. Fanghänel, *J. Prakt. Chem.* **320**, 404 (1978).
[234] M. L. Kaplan, R. C. Haddon, and F. Wudl, *J. C. S., Chem. Commun.*, 388 (1977).
[235] J. R. Andersen, V. V. Patel, and E. M. Engler, *Tetrahedron Lett.*, 239 (1978).

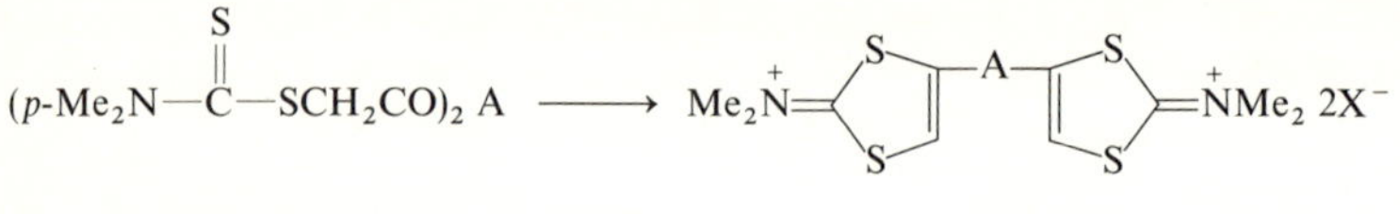

$(p\text{-Me}_2\text{N}-\overset{\text{S}}{\underset{\|}{\text{C}}}-\text{SCH}_2\text{CO})_2 \text{ A} \longrightarrow \text{Me}_2\overset{+}{\text{N}}=\underset{\text{S}}{\overset{\text{S}}{\diagup\!\!\!\diagdown}}-\text{A}-\underset{\text{S}}{\overset{\text{S}}{\diagup\!\!\!\diagdown}}=\overset{+}{\text{N}}\text{Me}_2 \text{ 2X}^-$

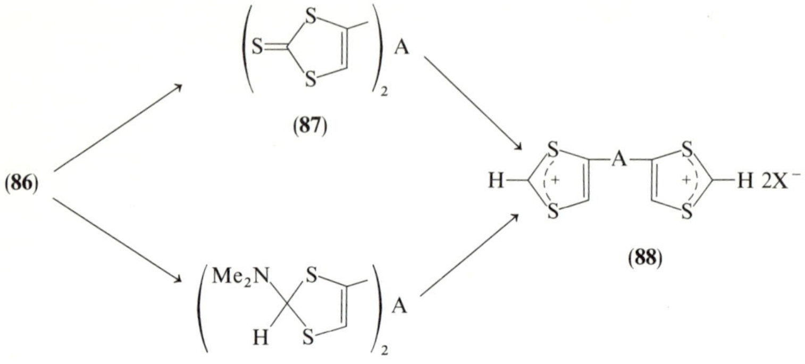

(86)

(86) → (87), (88), (89)

A = —(CH₂)₃—, —(CH₂)₄—, —C₆H₄—, —C₆H₄—C₆H₄—

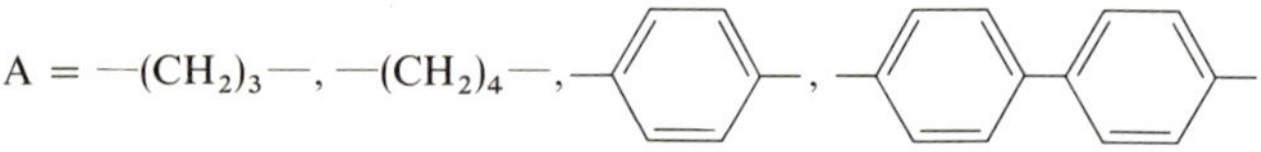

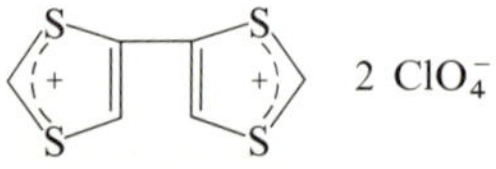

(90)

c. *From Cyanoalkyl Dithiocarboxylates.* Cyclization of cyanoalkyl dithiobenzoates **(91)** under action of strong acids HCl, H₂SO₄, or HNO₃ is a route to 4-amino-1,3-dithiolium salts **(92)**. By the action of acyl chlorides upon **91** the *N*-acyl derivatives **93** are obtained. Their IR spectra indicate that they exist, at least in part, in the imidol form.[236,237]

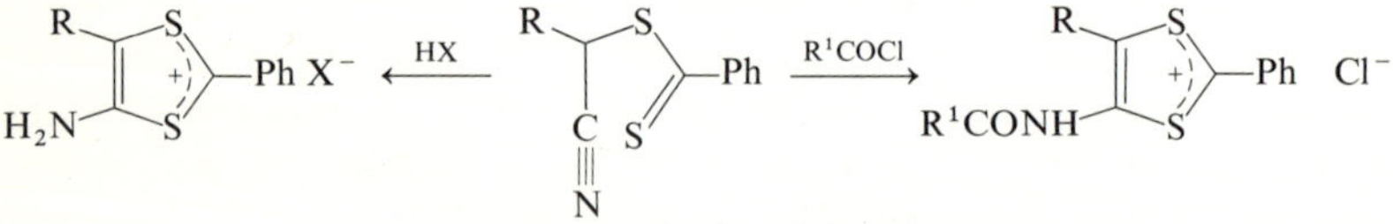

(92) ←[HX]— **(91)** —[R¹COCl]→ **(93)**

92: R = H, Ph; X = Cl, HSO₄, NO₃
93: R = H, R¹ = Ph, Me, Et, MeOCO(CH₂)₂, MeOCO(CH₂)₄

---

[236] M. Ohta and M. Sugiyama, *Bull. Chem. Soc. Jpn.* **36**, 1437 (1963).
[237] M. Ohta and M. Sugiyama, *Bull. Chem. Soc. Jpn.* **38**, 596 (1965).

d. From α-(Thioacylthio) carboxylic Acids.

To have a hydroxy group substituted on C-4 in the 1,3-dithiolium nucleus, cyclizations of α-(thioacylthio)carboxylic acids (94) by perchloric acid were investigated. The corresponding 1,3-dithiolium perchlorates (95) are thus obtained. On the basis of NMR and UV data, the 2-phenyl derivative exists in the 4-hydroxy form whereas the 2-dimethylamino and 2-methylthio derivatives exist in the 4-oxo form.[227]

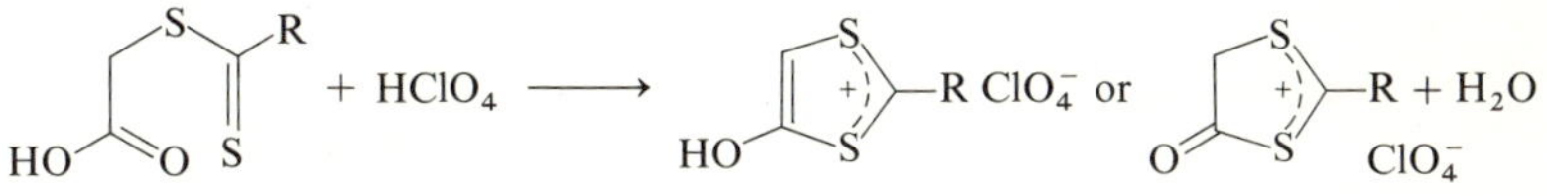

(94) R = NMe$_2$, SMe, Ph   (95)

Cyclodehydration of α-(thioaroylthio)- or α-(thiocarbamoylthio)carboxylic acids (96) with acetic anhydride in the presence of catalytic amounts of boron trifluoride etherate, or with a mixture of acetic anhydride and triethylamine (1:1), leads to the 2-aryl- or 2-amino-1,3-dithiolium-4-olates (97) listed in Table VIII.[237–243] Chemical properties, dipole moments, NMR, IR, and UV spectra, and charge distribution calculations are consistent with a mesoionic structure.[238,239,241]

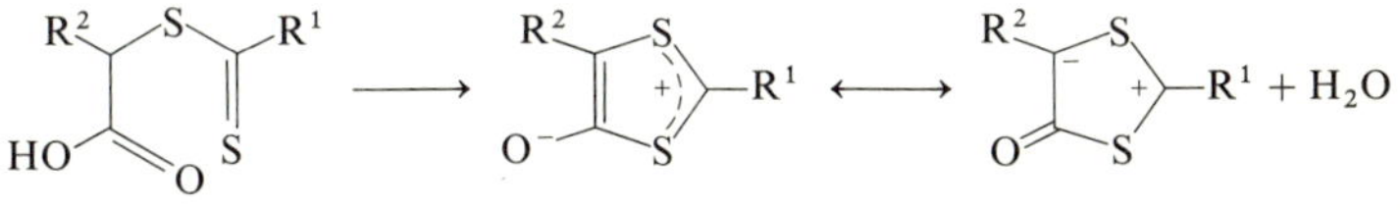

(96) R^1 = Ar or NR$_2$ (97)

TABLE VIII
1,3-DITHIOLIUM-4-OLATES (97)

R^1	R^2	Yield (%)	Reference
Ar	H	33–80	237–240
Ar	Ar'	29–93	239, 241, 242
p-MeOC$_6$H$_4$	COOEt	45	242
Ph	Me	34	239
R$_2$N[a]	Ph	72–92	243

[a] Piperidino, morpholino, pyrrolidino.

[238] K. T. Potts, D. R. Choudhury, A. J. Elliott, and U. P. Singh, *J. Org. Chem.* **41**, 1724 (1976).
[239] H. Gotthardt, M. C. Weisshuhn, and B. Christl, *Chem. Ber.* **109**, 740 (1976).
[240] K. T. Potts and U. P. Singh, *J. C. S. Chem. Commun.*, 569 (1969).
[241] H. Gotthardt and B. Christl, *Tetrahedron Lett.*, 4743 (1968).
[242] K. T. Potts, S. J. Chen, J. Kane, and J. L. Marshall, *J. Org. Chem.* **42**, 1633 (1977).
[243] H. Gotthardt and C. M. Weisshuhn, *Chem. Ber.* **111**, 2021 (1978).

Compounds of type **97** also result from *p*-methoxydithiobenzoic acid and α-bromoacyl chlorides in benzene containing triethylamine.[242]

e. *From* N,N-*Dialkyldithiocarbamates.* *S*-Vinyl *N,N*-dialkyldithiocarbamates (**98**) are obtained from sodium dithiocarbamates and 1,2-dibromoethane. They can readily be converted to 2-dialkylamino-1,3-dithiolium salts (**99**) by a brominative cyclization followed by thermal elimination of hydrogen bromide.[244] The dication **100** is obtained similarly.[245]

(**98**) (**99**)

(**100**) X = ClO_4, HSO_4, Cl

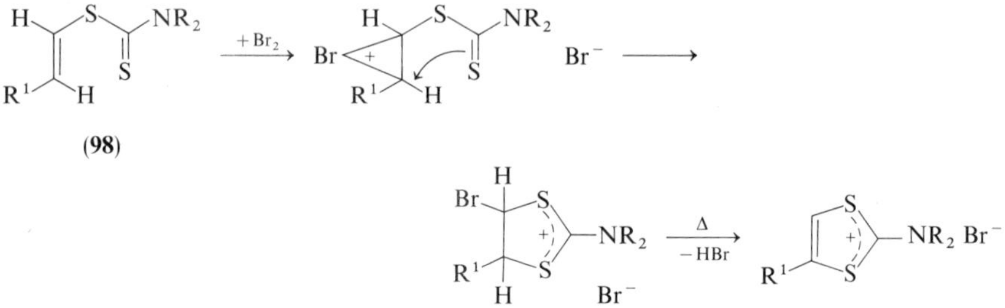

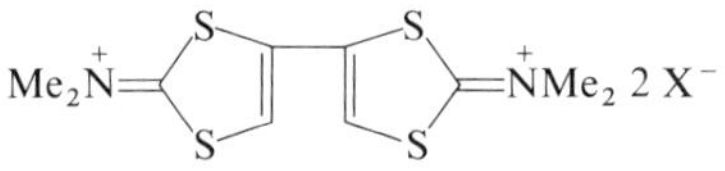

When the dithiocarbamates **101** are treated with bromine, 2-amino-4,7-dihydroxy-1,3-benzodithiolium salts (**103**) are obtained. Their formation is explained by initial addition of bromine at a sulfur atom, leading to **102**, which spontaneously loses hydrogen bromide.[246]

(**101**) (**102**) (**103**)

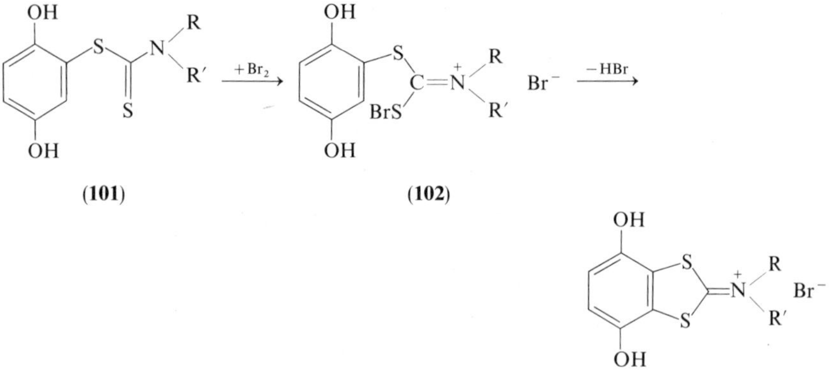

[244] K. Hiratani, H. Shiono, and M. Okawara, *Chem. Lett.*, 867 (1973).
[245] T. Haga, Japan. Kokai 77 83,561 [*CA* **88**, 37778 (1978)].
[246] R. L. N. Harris and L. T. Oswald, *Aust. J. Chem.* **27**, 1309 (1974).

Another advantageous route to **107** is the pyrolysis of 2-dialkylamino-4-(α-bromoalkyl)-1,3-dithiolan-2-ylium obtained in almost quantitative yield by bromination of allyl dithiocarbamates (**104**). Intermediates **105** have been isolated and characterized.[247] Their transformation probably involves an initial 1,2-dehydrobromination giving **106**, which undergoes 1,3-hydrogen shift.[248]

(104) **(105)**

(106) **(107)**

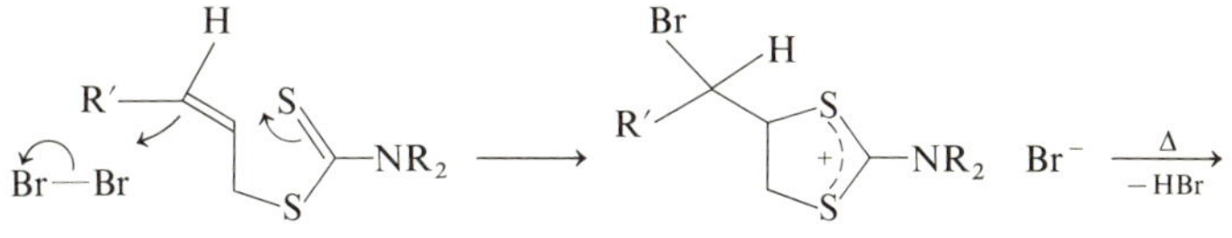

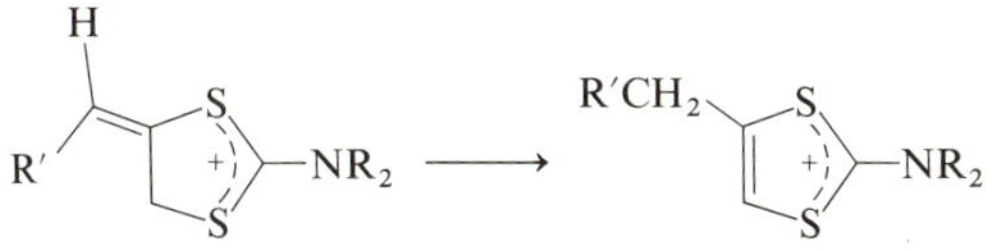

With α,β-dichloroketones such as chloranil or 2,3-dichloro-1,4-naphthoquinone, dithiocarbamic acid salts yield compounds **108** and **109**.[249,250]

(108) $R^1 = R^2 =$ Me or $—R^1—R^2— = (CH_2)_5$

(109)

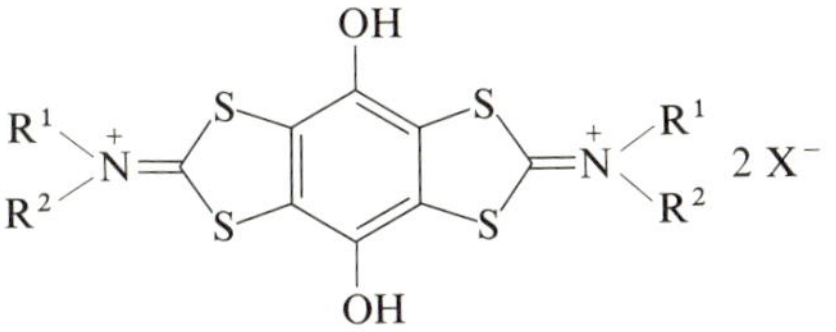

Reaction of sodium dithiocarbamate with an equimolar amount of the chloroacetaldehyde/sodium hydrogen sulfite adduct (**110**) in dimethylformamide at room temperature yields **111** quantitatively. The latter, treated first with concentrated sulfuric acid and then with sodium tetraphenylborate, gives the salt **112** in 86% yield.[251,252] Under the same conditions, piperidinium dithiobenzoate reacts with **110** providing in 64% yield 2-phenyl-1,3-dithiolium tetraphenylborate.[252]

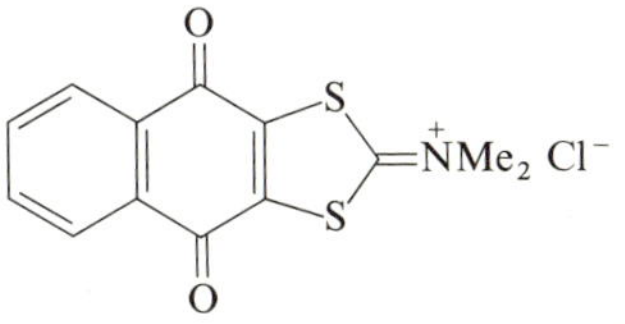

[247] K. Hiratani, T. Nakai, and M. Okawara, *Chem. Lett.*, 1041 (1974).
[248] T. Nakai, K. Hiratani, and M. Okawara, *Bull. Chem. Soc. Jpn.* **49**, 827 (1976).
[249] N. G. Demetriadis, S. J. Huang, and E. T. Samulski, *Tetrahedron Lett.*, 2223 (1977).
[250] T. Haga, Japan. Kokai 77 83,594 [*CA* **88**, 22876 (1978)].
[251] M. Okawara and Y. Ueno, Japan. Kokai 76 100,076 [*CA* **86**, 189908 (1977)].
[252] Y. Ueno, A. Nakayama, and M. Okawara, *Synthesis*, 277 (1975).

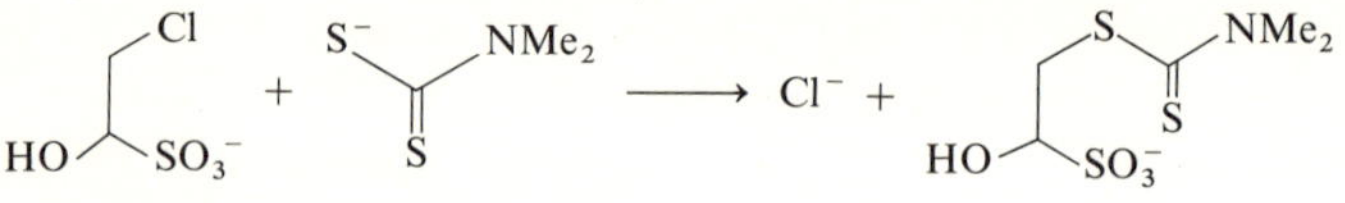

(110) + [S⁻-C(NMe₂)=S] → Cl⁻ + (111)

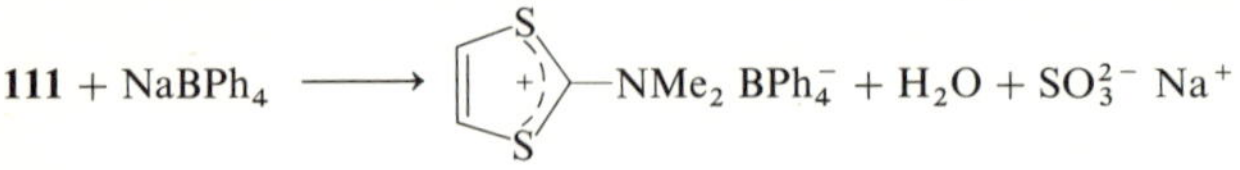

(110): HO-CH(Cl)-CH-SO₃⁻ type structure

(111): HO-CH-CH(SO₃⁻)-S-C(=S)-NMe₂

111 + NaBPh₄ ⟶ (112) + H₂O + SO₃²⁻ Na⁺

(112): 2-dimethylamino-1,3-dithiolium tetraphenylborate —NMe₂ BPh₄⁻

f. *From Thioacids or Trithiocarbonates and α-Substituted Ketones.* Excess thioacids or S-alkyl thioesters react with α-halogeno-, α-hydroxy-, α-mercapto-, or α-diazoketones in the presence of hydrogen iodide or perchloric acid, leading to various 1,3-dithiolium salts (Eq. 31).[253]

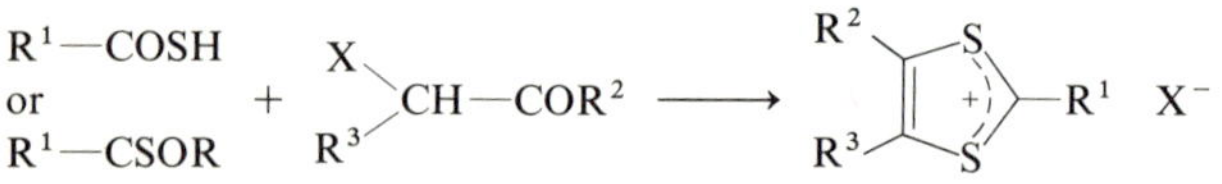

$$R^1-COSH \text{ or } R^1-CSOR + \underset{R^3}{\overset{X}{>}}CH-COR^2 \longrightarrow [\text{1,3-dithiolium}]-R^1 \quad X^- \quad (31)$$

X = Halogen, OH, SH

1,2-Diphenyl-2-chloroethanone with sodium methyl trithiocarbonate in alcohol, then with perchloric acid, gives 2-methylthio-4,5-diphenyl-1,3-dithiolium perchlorate in 35% yield (Eq. 32).[254]

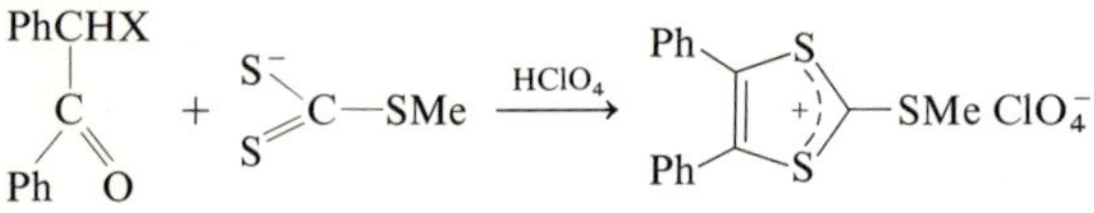

$$\underset{Ph}{\overset{PhCHX}{\underset{O}{>}C}} + \overset{S^-}{\underset{S}{>}}C-SMe \xrightarrow{HClO_4} [\text{4,5-Ph}_2\text{-1,3-dithiolium}]-SMe \ ClO_4^- \quad (32)$$

g. *By Degradation of Heterocyclic Compounds.* By the action of hydrogen iodide in ethanol at 40°C, 3-methylthio-1,4,2-dithiazines undergo ring contraction, yielding 2-amino-1,3-dithiolium salts (Eq. 33).[255]

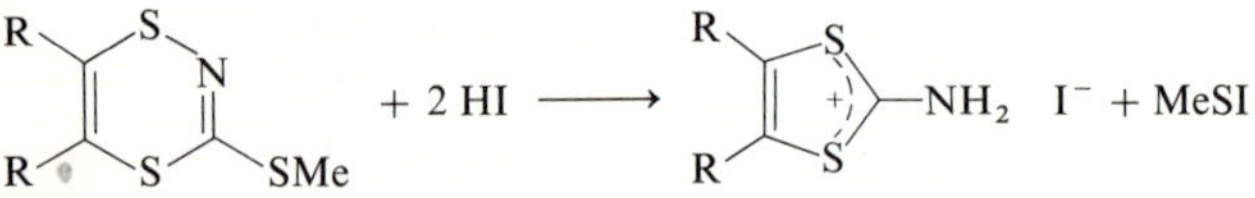

$$[\text{R,R-1,4,2-dithiazine-SMe}] + 2 HI \longrightarrow [\text{R,R-1,3-dithiolium}]-NH_2 \ I^- + MeSI \quad (33)$$

2. From 1,3-Dithiole Derivatives

a. *Protonation.* Like 1,2-dithiole derivatives, 1,3-dithiole derivatives **113** may be considered as conjugate bases of dithiolium cations. They are soluble and stable in strong acids. The protonated species **114** may be

---
[253] K. Fabian and H. Hartmann, *J. Prakt. Chem.* **313**, 722 (1971).
[254] E. Campaigne and F. Haaf, *J. Org. Chem.* **30**, 732 (1965).
[255] E. Fanghänel, *J. Prakt. Chem.* **318**, 127 (1976).

represented by various limiting formulas. Their importance in the equilibrium is discussed in the previous review (Ref. 1: pp. 115 and 118).

$$\left(\underset{S}{\overset{S}{\bigcirc}}=X \longleftrightarrow \underset{S}{\overset{S}{\bigcirc}}-X^-\right) + H^+ \rightleftharpoons \left(\underset{S}{\overset{S}{\bigcirc}}-XH \longleftrightarrow \underset{S}{\overset{S}{\bigcirc}}=\overset{+}{X}H\right)$$

(113) X = O, S (114)

Some 2-alkylidene-1,3-dithioles are converted to 1,3-dithiolium salts by protonation with perchloric acid. These salts tend to lose a proton, and weak bases are sufficient to displace the equilibrium completely toward the neutral species. This is given as proof of the low resonance energy of the dithiolium cation (Eq. 34).[1]

$$\underset{R^2}{\overset{R^1}{\bigcirc}}\underset{S}{\overset{S}{\diagdown}}C\underset{R^4}{\overset{R^3}{\diagup}} \xrightleftharpoons[-H^+]{HClO_4} \underset{R^2}{\overset{R^1}{\bigcirc}}\underset{S}{\overset{S}{\diagdown}}-CHR^3R^4 \; ClO_4^- \qquad (34)$$

b. *Formation by Anion Loss or Electron Loss.* The preparation of 1,3-dithiolium salts by hydride removal from 1,3-dithioles is sometimes implied in the synthesis from benzene-1,2-dithiol and carbonyl compounds, but in these cases the intermediate 1,3-dithioles are not isolated (Section III,A,1).

Unlike 3H-1,2-dithioles, numerous 2H-1,3-dithioles (115) have been isolated and are rather stable. By reaction with trityl salts, they can lose a hydride ion, giving 1,3-dithiolium salts; representative examples are listed in Table IX.[256-260] Yields are fair to good. This elimination is favored by electron-releasing substituents on C-2, and 4,5-dicyano-1,3-dithiole is stable toward oxidizing agents.

TABLE IX
1,3-DITHIOLIUM SALTS FROM 1,3-DITHIOLES (115)

R^1	R^2	R^3	Yield (%)	Reference
—CH=CH—CH=CH—		H	75	256
—CH=CH—CH=CH—		Ar	94–97	257, 258
—CH=CH—CH=CH—		Cyclohexyl	90	257
Ph	H	Ar	23–77	259
H, Ph	Ph	Ph	46–87	260
H	Ph	n-C$_3$H$_7$	66	260

[256] J. Nakayama, K. Fujiwara, and M. Hoshino, *Chem. Lett.*, 1099 (1975).
[257] I. Degani and R. Fochi, *Synthesis*, 759 (1976).
[258] I. Degani and R. Fochi, *J. C. S., Perkin 1*, 1886 (1976).
[259] K. Hirai, *Tetrahedron* 27, 4003 (1971).
[260] A. Takamizawa and K. Hirai, *Chem. Pharm. Bull.* 17, 1931 (1969).

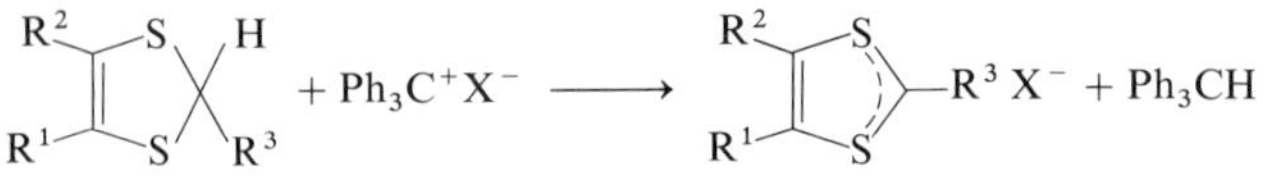

(115)

2,2'-(p-Phenylene)- and 2,2'-(m-phenylene)bis(1,3-benzodithiolium) ions have been prepared analogously, starting from the corresponding bis(1,3-benzodithioles). With the ortho derivatives a monocation is obtained.[261] The dication of the para derivatives have also been obtained by reaction of N-chlorosuccinimide.[262]

With one molar equivalent of trityl hexachloroantimonate, compound **116** is converted to a monocation **117** in 88% yield. The latter is transformed by triethylamine in almost quantitative yield into the quinonoid compound **118**. The relationship between this system and the dibenzotetrathiafulvalene is expected to be similar to that between tetracyanoethylene and TCNQ, and the donor properties of **118** were demonstrated by charge-transfer complex formation.[263]

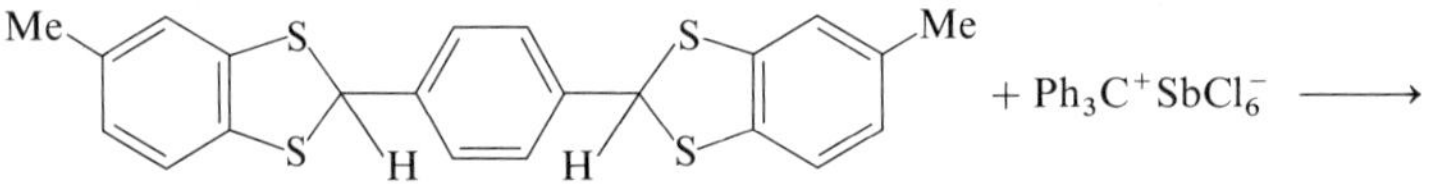

(116)

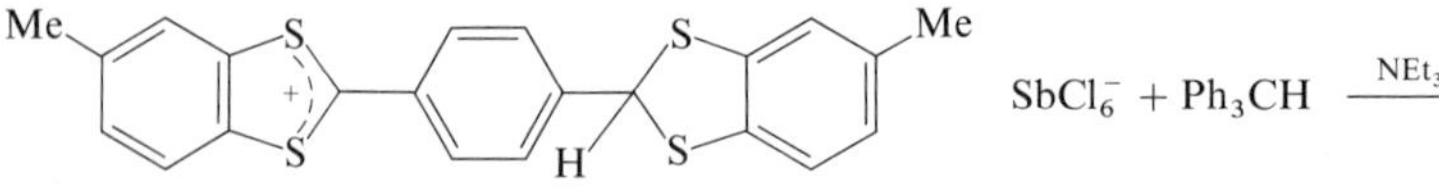

(117)

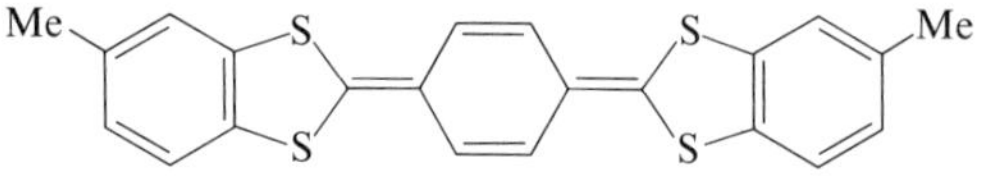

(118)

2-Alkoxy- or 2-alkylthio-1,3-dithioles (**119**) are easily prepared.[221,222,264,265] They react with trityl tetrafluoroborate in acetonitrile[266] or tetrafluoroboric acid in acetic anhydride[222,256,265,267] giving the 2-unsubstituted

[261] J. Nakayama, K. Ueda, M. Hoshino, and T. Takemasa, *Synthesis*, 770 (1977).
[262] M. V. Lakshmikantham and M. P. Cava, *J. Org. Chem.* **43**, 82 (1978).
[263] Y. Ueno, A. Nakayama, and M. Okawara, *J. C. S., Chem. Commun.*, 74 (1978).
[264] J. Nakayama, *Synthesis*, 436 (1975).
[265] F. Wudl and M. L. Kaplan, *J. Org. Chem.* **39**, 3608 (1974).
[266] J. Nakayama, K. Fujiwara, M. Imura, and M. Hoshino, *Chem. Lett.*, 127 (1977).
[267] J. Nakayama, K. Fujiwara, and M. Hoshino, *Bull. Chem. Soc. Jpn.* **49**, 3567 (1976).

1,3-dithiolium salts **120** in yields higher than 80%. Other salts of the 1,3-benzodithiolium cation can be obtained using sulfuric, perchloric, hydriodic, or hydrobromic acid instead of trityl tetrafluoroborate or tetrafluoroboric acid.[267]

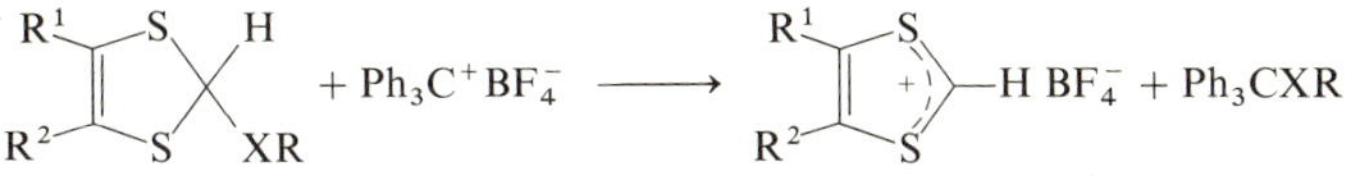

$$R^1\diagdown S\diagdown H + Ph_3C^+BF_4^- \longrightarrow R^1\diagdown S\diagup^+ - H\ BF_4^- + Ph_3CXR$$
$$R^2\diagup S\diagup XR \qquad\qquad\qquad R^2\diagup S$$

(119) X = O, S (120)

A similar and convenient route to 2-unsubstituted-4-aryl-1,3-dithiolium salts (**122**) is by the action of acids in ethanol upon 2-amino-4-aryl-1,3-dithioles (**121**). Yields are generally higher than 90%.[230]

Various 1,3-dithiolium salts prepared from **119** and **121** are given in Table X.

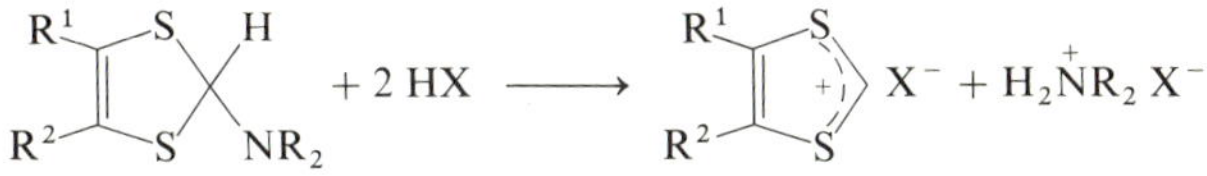

$$R^1\diagdown S\diagdown H + 2\ HX \longrightarrow R^1\diagdown S\diagup^+ X^- + H_2\overset{+}{N}R_2\ X^-$$
$$R^2\diagup S\diagup NR_2 \qquad\qquad\qquad R^2\diagup S$$

(121) (122) X = ClO$_4^-$, HSO$_4^-$

TABLE X
SYNTHESIS OF 1,3-DITHIOLIUM SALTS BY ANION LOSS

R^1	R^2	Leaving group	Reference
—CH=CH—CH=CH—		RO	222, 256, 267
—CMe=CH—CH=CH—		iso-C$_5$H$_{11}$O	222
—CH=CR—CH=CH—[a]		iso-C$_5$H$_{11}$O	222
—CH=(C$_6$H$_4$)=CH—[b]		iso-C$_5$H$_{11}$O	222
H	H	MeS	265
—CH=CH—CH=CH—		RS	256, 267
H	Ar	R$_2$N	230
Ph	Ph	R$_2$N	230

[a] R = Me, Cl, I, NO$_2$.
[b] Naphtho[2,3-d]-fusion.

The electron-donor properties of tetrathiafulvalene [TTF, 2,2′-bis(1,3-dithiol-2-ylidene)] and of its substitution derivatives **123** have recently been reviewed.[268] Accordingly, only papers subsequent to this review will be covered here.

[268] M. Narita and C. U. Pittman, *Synthesis*, 489 (1976).

Oxidation, carried out electrochemically or with halogens or thiocyanogen, proceeds in two steps leading to radical cations **124** and then dications **125** (Scheme 28).[215,233,269-272] Starting from **123a** and thiocyanogen, or **123d** or **123e** and halogens, the oxidation stops at the first step.

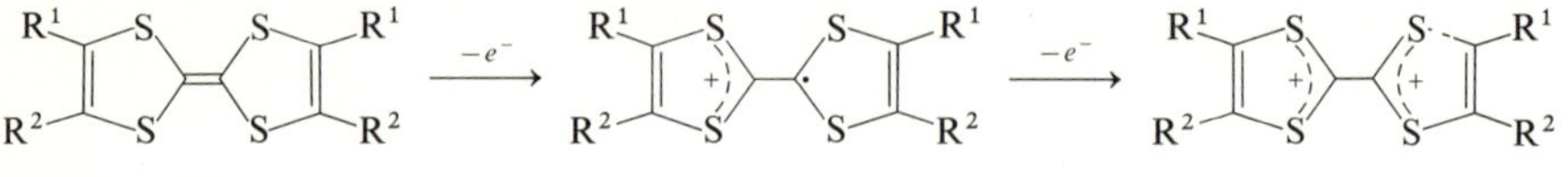

$$(123) \xrightarrow{-e^-} (124) \xrightarrow{-e^-} (125)$$

	R^1	R^2	Ref.		R^1	R^2	Ref.
a	H	H	215, 269–271	e	H	p-Ph—C$_6$H$_4$	233
b	Me	Me	271	f	—CH$_2$—CH$_2$—CH$_2$—CH$_2$—		215
c	MeS	MeS	272	g	—CH=CH—CH=CH—		215
d	H	Ph	233	h	—CH=C(Me)—CH=CH—		215

SCHEME 28

The radical-cation salts can be obtained by direct oxidation of **123** by halogens or thiocyanogen, or by exchange between (TTF)$_3$(BF$_4$)$_2$ and an anion, but these methods give compounds with variable stoicheiometries. It was recently found that electrochemical oxidation of **123** afforded controlled conditions leading to salts of predictable stoicheiometry.[273]

TTF is a powerful electron donor which combines with a variety of acceptors to produce charge-transfer complexes in which the ratio of each component can vary. With tetracyanoquinodimethane (TCNQ) as the acceptor the complex has the formula **126**. Chemical modifications in either the donor or the acceptor entail variations of stoicheiometry.[268,274]

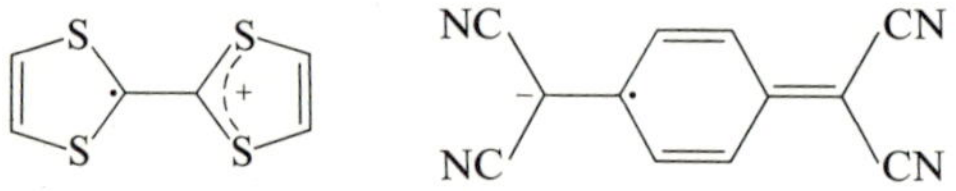

(126)

Polymers of TTF also react with bromine and iodine providing radical-cation halides.[233]

---

[269] F. Wudl, G. M. Smith, and E. J. Hufnagel, *J. C. S., Chem. Commun.*, 1453 (1970).
[270] D. L. Coffen, J. Q. Chambers, D. R. Williams, P. E. Garrett, and N. D. Canfield, *J. Am. Chem. Soc.* **93**, 2258 (1971).
[271] H. Strzelecka, L. Giral, J. M. Fabre, E. Torreilles, and G. Brun, *C. R. Hebd. Seances Acad. Sci., Ser. C* **284**, 463 (1977).
[272] P. R. Moses and J. Q. Chambers, *J. Am. Chem. Soc.* **96**, 945 (1974).
[273] F. B. Kaufman, E. M. Engler, D. C. Green, and J. Q. Chambers, *J. Am. Chem. Soc.* **98**, 1596 (1976).
[274] R. C. Wheland, *J. Am. Chem. Soc.* **99**, 291 (1977).

Sec. III.A]   THE 1,2- AND 1,3-DITHIOLIUM IONS   211

c. *Peroxide Attack of 1,3-Dithiole-2-thiones.* The unsubstituted 1,3-dithiolium bisulfate is obtained in 87% yield by treatment of 1,3-dithiole-2-thione by peracetic acid in acetone at $-40°C$.[275] Its 4-phenyl derivative and some 4,4′-polymethylenebis(1,3-dithiolium) salts are prepared similarly from the corresponding 1,3-dithiole-2-thiones.[232] However, unlike 1,2-dithiole derivatives, this method has not been developed widely, probably owing to difficulties in the preparation of starting materials.

d. *Halogenation.* Iodine has a double role in the transformation of **127** to **128**. It acts as oxidizing agent of the hydroquinone and adds to the sulfur atoms as in 1,2-dithiole-3-thiones (Section II,A,2).[276]

(127)   (128)

e. *Alkylation.* As indicated in the previous review (Ref. 1: p. 117), 1,3-dithiole-2-thiones, like 1,2-dithiole-3-thiones, are very sensitive to alkylating agents, giving 2-alkylthio-1,3-dithiolium salts, in high yields (Eq. 35).

$$R^1\text{-}S\text{-}C\text{=}S + R^3X \longrightarrow R^1\text{-}S\text{-}C\text{-}SR^3 \; X^- \quad (35)$$

Alkylating agents commonly used are methyl iodide,[277] dimethyl sulfate,[178,216,278] trimethyloxonium tetrafluoroborate,[229] and triethyloxonium tetrafluoroborate.[272] When $R^1 = R^2 = CO_2Me$ or CN, the $-M$ effect of these substituents hinders formation of a positive charge in the ring and alkylation is possible only with methyl fluorosulfonate.[279] When $R^1 = R^2 = CO_2H$, decarboxylation occurs during the reaction with methyl iodide and 2-methylthio-1,3-dithiolium iodide is obtained.[280]

Methyl iodide also reacts with 1,3-dithiol-2-imines (**129**) leading to **130**[119,255]; in one of the examples[255] an *N*-methylthio group is lost.

[275] L. R. Melby, H. D. Hartzler, and W. A. Sheppard, *J. Org. Chem.* **39**, 2456 (1974).
[276] T. Haga, Japan. Kokai 77 95,695 [*CA* **88**, 62375 (1978)].
[277] H. Behringer, D. Bender, J. Falkenberg, and R. Wiedenmann, *Chem. Ber.* **101**, 1428 (1968).
[278] R. Gompper and E. Kutter, *Chem. Ber.* **98**, 1365 (1965).
[279] N. F. Haley, *J. Org. Chem.* **43**, 678 (1978).
[280] E. Klingsberg, U.S. Patent 3,187,009 [*CA* **63**, 4431 (1965)].

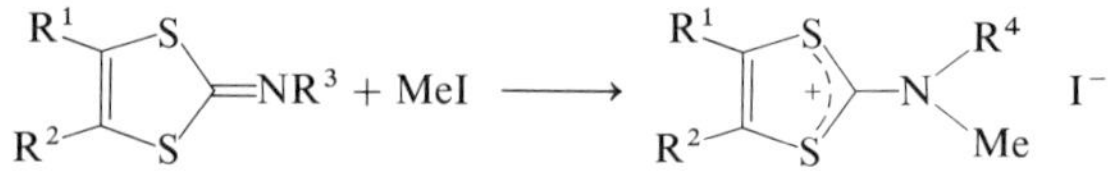

(129) → (130)

R^1—R^2 = —$(CH_2)_4$—, R^3 = MeS, R^4 = Me, yield 8%
R^1 = H, R^2 = R^3 = R^4 = Ph; yield 83%.

f. *Side-Chain Modifications.* The 2-methyl substituent of a 1,3-dithiolium cation has strong acidic properties. The conjugate base undergoes electrophilic attack by various compounds, leading to diversely 2-substituted 1,3-dithiolium salts (Section III,B,9).

When **131** is treated with primary aromatic amines and perchloric acid in ethanol, salts **132** are obtained.[281]

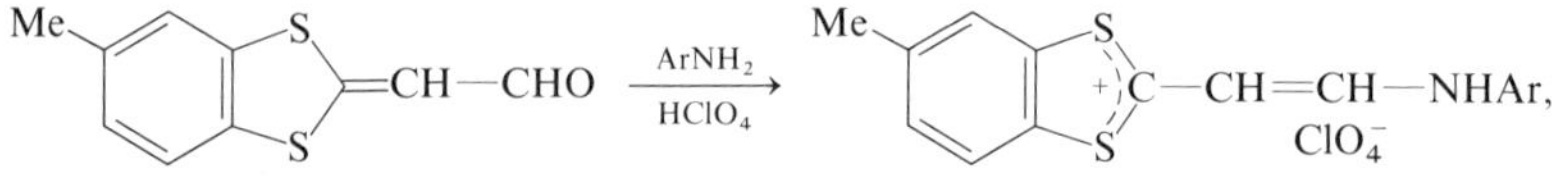

(131) → (132)

B. Chemical Properties

1. *General Remarks*

The discovery of the unusually high solid-state electrical conductivity of the charge-transfer complex of TTF with TCNQ has prompted extensive investigations of 1,3-dithiolium salts, an important class of intermediates for the synthesis of tetrathiafulvalene derivatives.

Many convenient routes to 1,3-dithiolium salts have been developed. Chemical properties of these salts have been actively studied, especially their behavior toward nucleophilic reagents.

The 1,3-dithiolium cation is a six π-electron system in a positively charged five-membered ring. Although the positive charge is often shown implicitly as delocalized equally over all the ring atoms, 1,3-dithiolium salts react exclusively at the 2-position with nucleophilic reagents. Moreover the much smaller delocalization of the positive charge on C-4 than on C-2 is shown by the difference in behavior of the salts **133** and **134**. Whereas the methyl on C-2 condenses easily with benzaldehyde, the methyl on C-4 does not

[281] R. Wizinger-Aust, *Q. Rep. Sulfur. Chem.* **5**, 191 (1970).

react with aromatic aldehydes, even in the presence of pyridine or triethylamine. However, ^{13}C-NMR chemical shifts do indicate a certain extent of delocalization of charge to C-4 and C-5 (see Section III,C,4).

(133) (134)

Among the nucleophilic reagents that react with 1,3-dithiolium salts, tertiary aliphatic amines have given interesting results. Numerous TTF derivatives have been prepared by treatment of 2-unsubstituted 1,3-dithiolium salts by triethylamine or *N*-ethyldiisopropylamine (Section III,B,6,a). The 1,3-dithiolium cation probably first undergoes a deprotonation leading to a cyclic carbene which can also be represented by dipolar structures (Eq. 36).

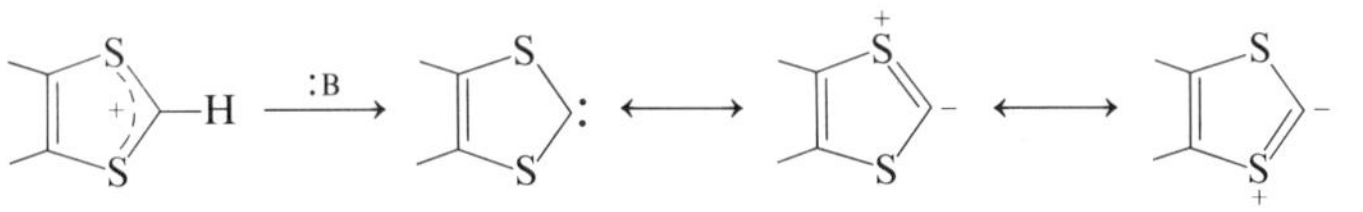

$$\qquad\qquad\qquad\qquad\qquad\qquad\qquad\qquad\qquad\qquad (36)$$

Carbenes have also been considered as intermediates in the thermal dissociation of 2-alkoxy-1,3-dithioles (Section III,B,4) and in the reactions of carbon disulfide with activated alkynes[282–284] and with benzyne.[285,286]

Another type of deprotonation may occur with 1,3-dithiolium cations containing phenolic hydroxyls. Thus, treatment of 4-aryl-2-(4-hydroxyphenyl)-1,3-dithiolium perchlorate with triethylamine gives **135**, stabilized by its zwitterionic structure **136**.[259]

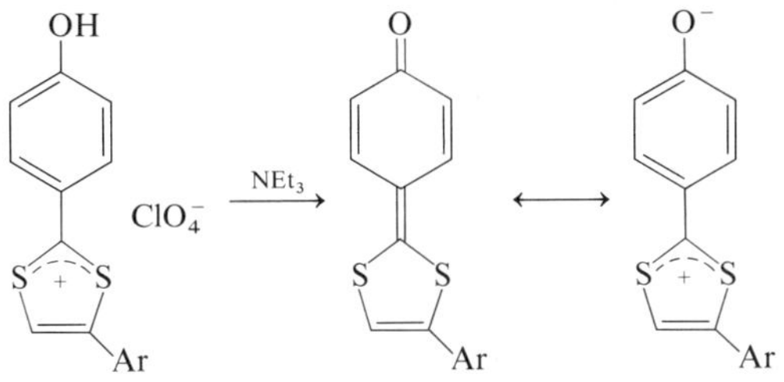

(135) (136)

[282] H. D. Hartzler, *J. Am. Chem. Soc.* **92**, 1412 (1970).
[283] H. D. Hartzler, *J. Am. Chem. Soc.* **95**, 4379 (1973).
[284] D. L. Coffen, *Tetrahedron Lett.*, 2633 (1970).
[285] J. Nakayama, *J. C. S., Chem. Commun.*, 166 (1974).
[286] E. K. Fields and S. Meyerson, *Tetrahedron Lett.*, 629 (1970).

2-Amino-4,7-dihydroxy-1,3-benzodithiolium salts treated with sodium acetate also lose a proton to give **137**.[246]

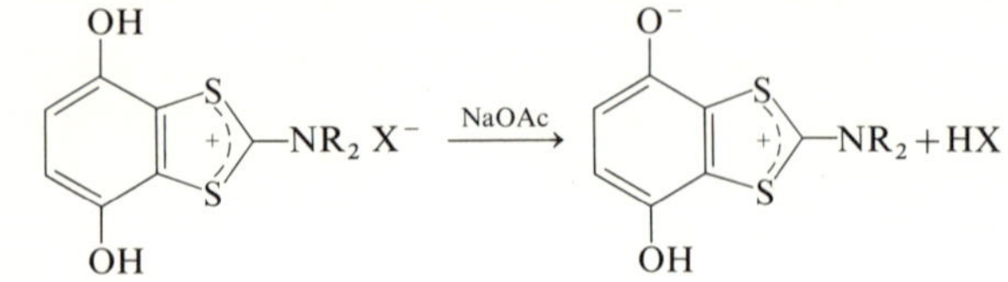

$$\text{[structure with OH, OH groups]} \xrightarrow{\text{NaOAc}} \text{[structure 137]} + HX$$

(137)

2. Hydrogen Exchange

The acidity of H-2 in 1,3-dithiolium cations has been observed by NMR. In CF_3CO_2D/D_2O, H/D exchange occurs between at the 2-position, the speed being much greater for 4-aryl than for 4-methyl derivatives. A carbene intermediate is proposed (Eq. 37).[287]

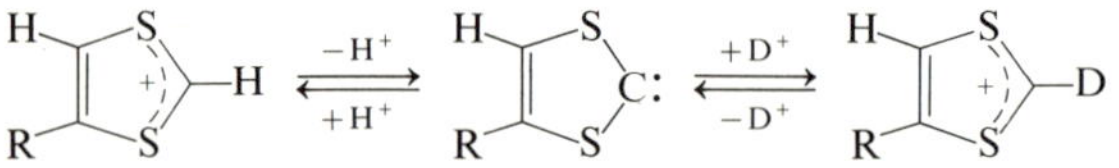

$$\text{[equation 37: dithiolium H exchange via carbene]} \tag{37}$$

3. Reduction

The ability of 1,3-dithiolium salts to add hydride ions has been widely studied. 2-Unsubstituted salts[260,267] as well as 2-alkyl or 2-aryl,[130,257] 2-amino,[230,234] and 2-methylthio[229,265] derivatives are easily reduced to give 1,3-dithioles in excellent yields (Eq. 38). Reducing agents are sodium hydrosulfide in ethanol[130]; sodium borohydride in methanol, ethanol, or tetrahydrofuran[230,234,265,267]; lithium aluminium hydride[260] or deuteride[207] in ether; or sodium borodeuteride in acetonitrile.[257]

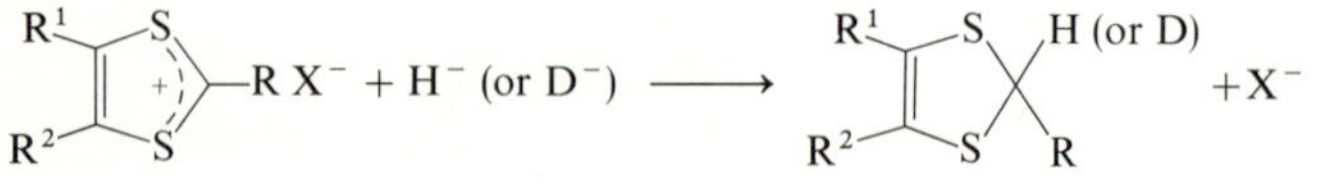

$$\text{[equation 38]} \tag{38}$$

R = H, SMe, NR_2, aryl, cyclohexyl

With sodium in liquid ammonia, 1,3-benzodithiolium tetrafluoroborate is cleaved to benzene-1,2-dithiol.[223]

---

[287] H. Prinzbach, H. Berger, and A. Lüttringhaus, *Angew. Chem.* **77**, 453 (1965).

4-Phenyl-1,3-dithiolium or 1,3-benzodithiolium cations abstract hydride from cycloheptatriene, giving the thermodynamically more stable tropylium salt and 4-phenyl-1,3-dithiole or 1,3-benzodithiole.[256,259,267]

Reduction of the dication **138** by excess lithium iodide gives a complex which seems to exist as an equilibrium mixture of a neutral charge-transfer complex **139** and ion-radical salt **140**. The ratio of iodine, $n = 2.8$, is estimated by elemental analysis.[231]

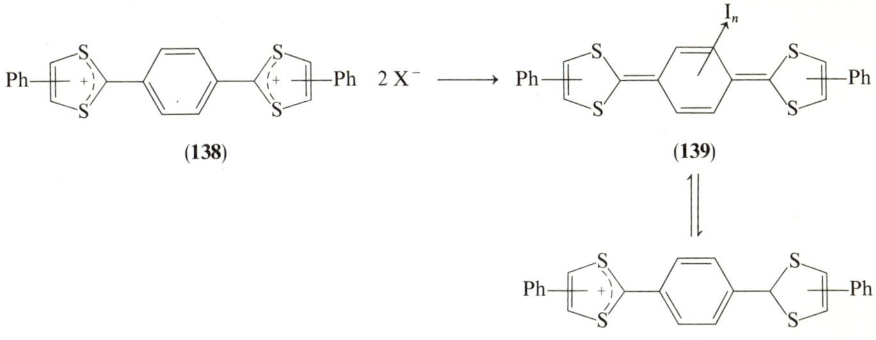

(138)   (139)

(140)

Differential pulse polarography measurements of 2,2'-(*p*-phenylene)bis-(1,3-benzodithiolium) bis(tetrafluoroborate) in acetonitrile show two reductions corresponding to the monocation-radical and the neutral compound **141**; these species are highly unstable.[288]

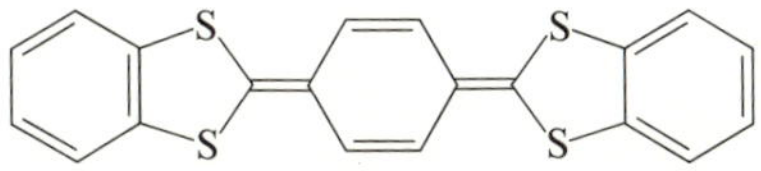

(141)

Coupling reactions are sometimes observed in the reduction of 1,3-dithiolium salts. Thus, use of sodium bis-diglyme-hexacarbonylvanadate(1−) as reducing agent of the parent compound **142** leads to the 2,2'-bis(1,3-dithiolyl).[289] Better yields (up to 92%) are obtained by use of zinc dust in a mixture acetic acid–benzene–water.[290]

---

[288] M. Sato, M. V. Lakshmikantham, M. P. Cava, and A. F. Garito, *J. Org. Chem.* **43**, 2084 (1978).
[289] A. R. Siedle and R. B. Johannesen, *J. Org. Chem.* **40**, 2002 (1975).
[290] A. Kruger and F. Wudl, *J. Org. Chem.* **42**, 2778 (1977).

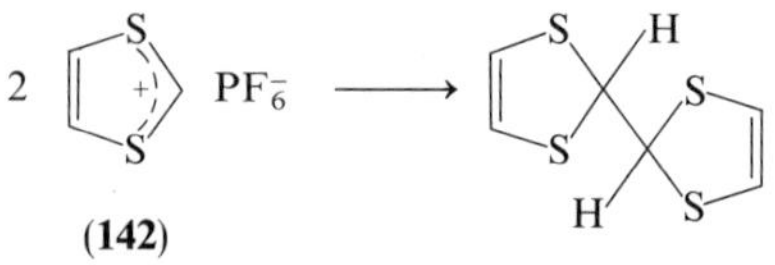

(142)

Coupling reactions are observed when treating 2-methylthio-1,3-dithiolium iodides with zinc. The intermediates **143** are not isolated; they spontaneously lose dimethyl disulfide to give TTF derivatives.[291]

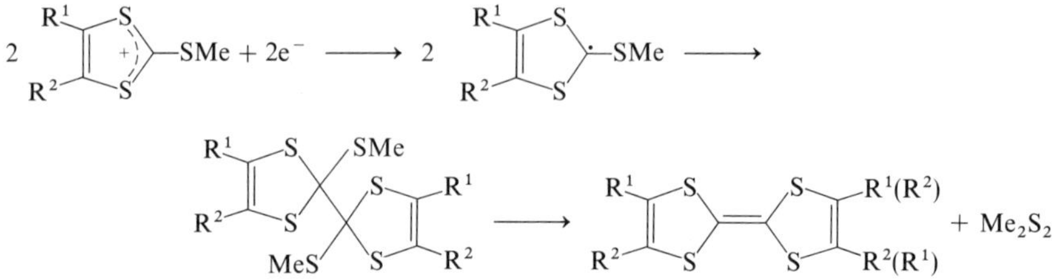

(143)

R^1	R^2	Yield (%)
$-CH_2-CH_2-CH_2-CH_2-$		57
$p\text{-MeC}_6H_4$	H	60
$p\text{-BrC}_6H_4$	H	33
$p\text{-ClC}_6H_4$	H	50
Ph	Ph	50
$-CH=CH-CH=CH-$		30

By electrolytic reduction of 2-ethylthio-4,5-bis(methylthio)-1,3-dithiolium ion, a dimer **144** is obtained. It gives a TTF derivative on pyrolysis.[272]

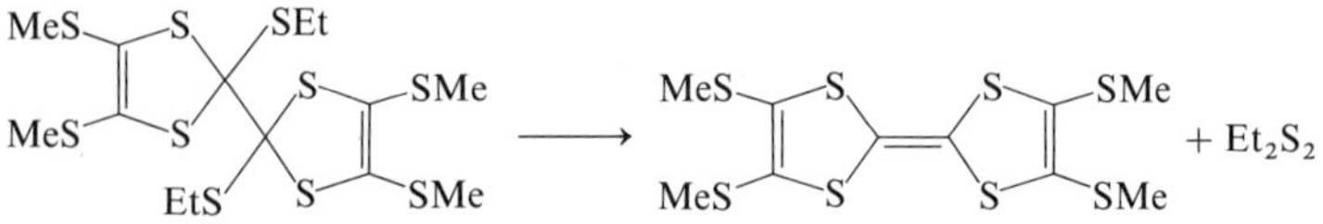

(144)

4. Attack by Oxygen Nucleophiles

Unlike the 1,2-dithiolium ring, the 1,3-dithiolium ring is stable in strong basic medium. Thus, salts **145** react with alcoholates giving 2-alkoxy-1,3-dithioles.[292] Even under milder conditions, by dissolution in methanol, ethanol, or 1- or 2-propanol, 1,3-dithiolium salts form 2-alkoxy-1,3-dithioles

---

[291] E. Fanghänel, Le van Hinh, and G. Schukat, *Z. Chem.* **16**, 317 (1976).
[292] E. Fanghänel and R. Mayer, *Z. Chem.* **4**, 384 (1964).

Sec. III.B] THE 1,2- AND 1,3-DITHIOLIUM IONS

in excellent yields.[216,217,260,267] The structures of these compounds were deduced from the NMR and UV spectra which indicated that they are covalent compounds. 2-Phenyl-1,3-benzodithiolium perchlorate reacts similarly with ethyl alcohol to give 2-ethoxy-2-phenyl-1,3-benzodithiole in 92% yield.[217]

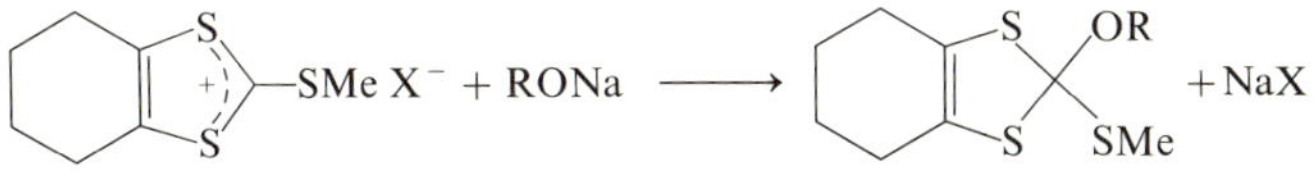

(145)

R = Me, Et, n-C$_3$H$_7$, Me$_2$CH

4-Aryl-2-methoxy-1,3-dithioles regenerate the starting salts in quantitative yield by addition of acids. Therefore purification of the salts through their 2-methyl ethers is very effective and 2-methyl ethers can be used instead of salts in acidic medium.[260]

2-Alkoxy-1,3-benzodithioles (146) react in acetic acid at room temperature with aromatic compounds having strong electron-donor substituents to give 2-aryl derivatives in good yields. The dithiole probably dissociates into alkoxide ion and 1,3-benzodithiolium which acts as electrophilic reagent with aromatic compounds.[293]

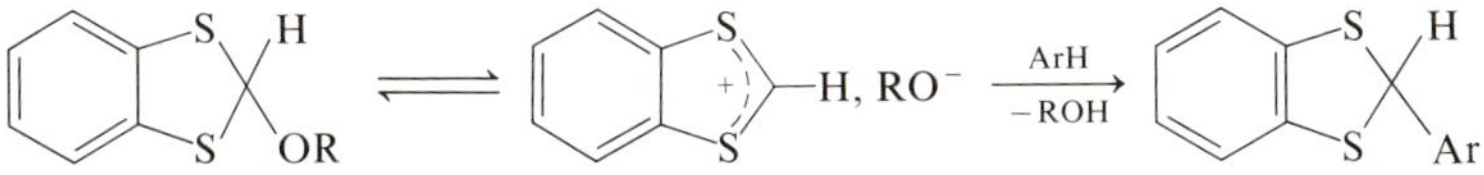

(146)   (147)

Heating 146 at 200°C for 2 hours gives dibenzotetrathiafulvalene (149) as main product. This is explained by the thermal dissociation of 146 to 147 which undergoes a deprotonation to give the carbene 148. Reaction of 148 upon 147 can lead to an adduct which would lose a proton giving 149.[294]

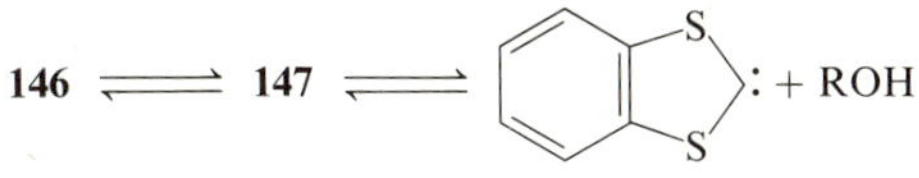

146 ⇌ 147 ⇌

(148)

147 + 148 ⟶

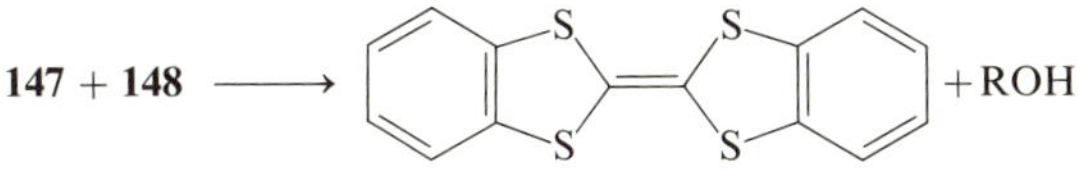

(149)

[293] J. Nakayama, *Synthesis*, 170 (1975).
[294] J. Nakayama, *Synthesis*, 168 (1975).

The behavior of 1,3-dithiolium salts toward water is discussed in the previous review (Ref. 1: p. 125). Accordingly, we shall describe mainly results obtained since 1966.

Action of water on the 1,3-dithiolium cation **150** gives **151**, which regenerates the original cation on treatment with acid.[260]

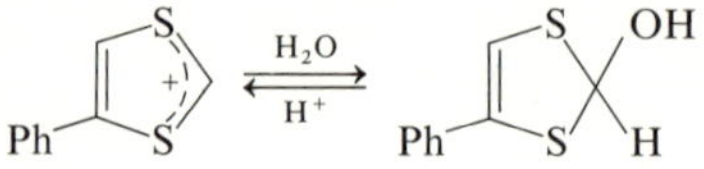

(150)   (151)

The products of hydrolysis of the 1,3-benzodithiolium cation **152** depend upon the experimental conditions. In acetonitrile–water (3:1) or acetone–water (4:1), after 1 hour, **153** is obtained.[295,296] In an acetonitrile–water mixture (4:1), after several hours, or in triethylamine–water (1:15), **154** is obtained in excellent yields.[267,296] The mechanism probably involves nucleophilic attack of **152** by water and tautomerism of the 2-hydroxy compound **155** with the open-chain derivative **156**. Reaction of **156** with another mole of **152** affords **153**, which can lose carbon monoxide to give **157**. This last can, in turn, react with **152**, giving **154**. Intermediate **155** has not been isolated but was identified by NMR in the reaction mixture (Scheme 29).[296]

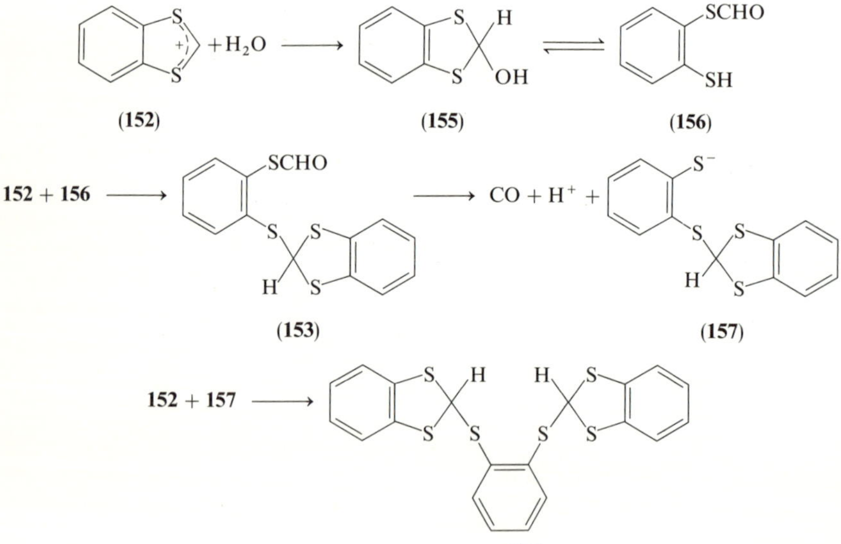

(152)   (155)   (156)

152 + 156 ⟶   (153)   ⟶ CO + H⁺ +   (157)

152 + 157 ⟶   (154)

SCHEME 29

[295] D. Buza and S. Szymanski, *Rocz. Chem.* **45**, 501 (1971).
[296] D. Buza and J. Szymoniak, *Rocz. Chem.* **48**, 765 (1974).

When **152** is treated with triethylamine in dimethylformamide containing traces of water, **154** is obtained in poor yield, but dibenzotetrathiafulvalene is isolated in good yield. Increasing the quantity of water favors formation of **154** at the expense of the dibenzotetrathiafulvalene formation, owing to partial hydrolysis of **152**.[297]

5. *Attack by Sulfur Nucleophiles*

2-Dialkylamino and 2-ethylthio-1,3-dithiolium salts are cleaved by hydrogen sulfide,[229,249,298] or sodium hydrosulfide,[228,232,252,299] giving 1,3-dithiole-2-thiones in excellent yields. Sodium sulfide and sodium hydrosulfide react with 1,3-benzodithiolium salts in methanol giving the sulfide **158** and the dithiole **159**.

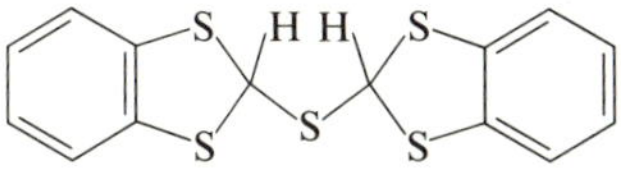

(**158**) 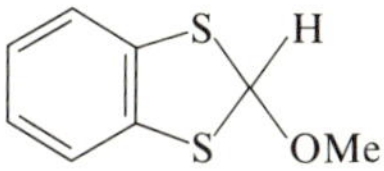   (**159**)

Like alcoholate anions, thiolate,[260] dithiocarbamate,[260,300] and *O*-ethyl dithiocarbonate anions[300] add easily in boiling ethanol or acetonitrile at C-2 of 1,3-dithiolium salts (Eq. 39).

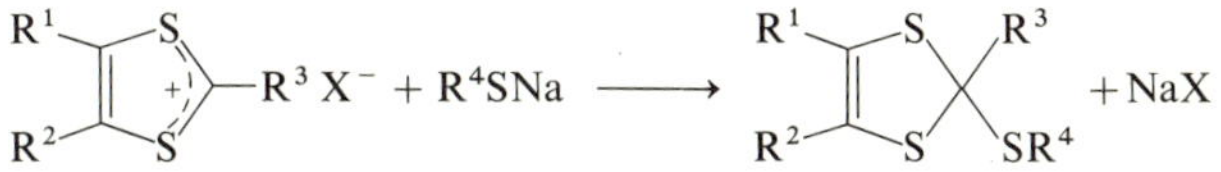

$$R^3 = H, MeS$$
$$R^4 = Me, Ar, R_2NCS, EtOCS$$

Reaction of 4-aryl-1,3-dithiolium salts with 2 moles of potassium *O*-ethyl dithiocarbonate in acetone affords **160** and not the expected adduct. This result has been explained by the solvation effect of the nucleophile which depends on the dielectric constants of acetone and acetonitrile.[300] Treatment of **160** with perchloric acid yields the starting cation with evolution of hydrogen sulfide (Eq. 40).

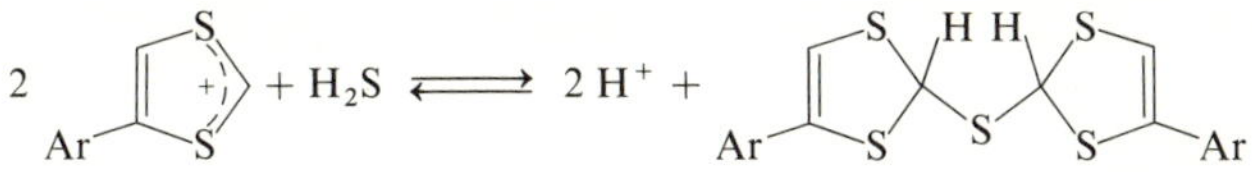

(**160**)

[297] D. Buza, A. Gryff-Keller, and S. Szymanski, *Rocz. Chem.* **44**, 2319 (1970).
[298] E. Fanghänel, *Z. Chem.* **7**, 58 (1967).
[299] T. Haga, Japan. Kokai 77 95,694 [*CA* **88**, 62376 (1978)].
[300] K. Hirai, H. Sugimoto, and T. Ishiba, *J. Org. Chem.* **42**, 1543 (1977).

The 1,3-benzodithiolium cation, in equilibrium with the corresponding alkoxydithioles (**146**) in methanol or ethanol, reacts with ethanethiol to give 2-ethylthio-1,3-benzodithiole.[267] In a similar way the reaction of 2-isobutoxy-1,3-benzodithiole with thiols of all kinds in acetic acid at room temperature yields the corresponding 2-alkylthio- or 2-arylthio-1,3-benzodithioles, presumably via the 1,3-benzodithiolium ion.[264]

6. *Attack by Nitrogen Nucleophiles*

a. *2-Unsubstituted 1,3-Dithiolium Ions.* Ammonium acetate or chloride reacts with 1,3-benzodithiolium tetrafluoroborate in acetonitrile in the presence of pyridine to give **161** in 86% yield.[267]

Ammonia and aromatic and aliphatic primary amines react similarly with a double molar amount of a 4,5-diphenyl-1,3-dithiolium or 1,3-benzodithiolium salt. Both amino hydrogens are replaceable by 1,3-benzodithiol-2-yl groups, leading to compounds **162**.[267,301,302] With secondary aliphatic amines, a similar reaction occurs which leads to 2-amino-1,3-dithioles (**163**) in good yields.[260,267,301,302]

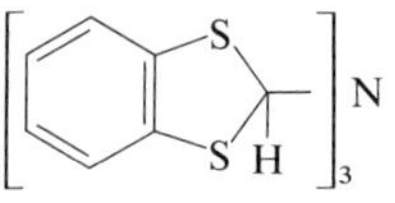

(**161**) 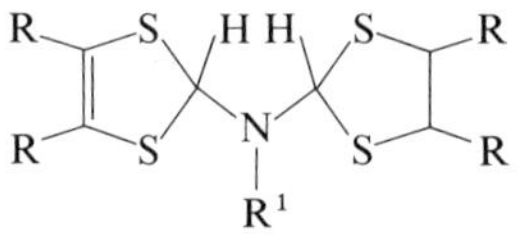   (**162**) 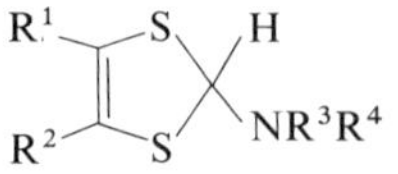   (**163**)

R = Ph or R–R = —CH=CH—CH=CH—
R^1 = H, Et, Ph, $PhCH_2$

1,3-Benzodithiolium cation (**152**) reacts with piperidine, dibenzylamine, and diisopropylamine, giving two products: the expected 2-amino-1,3-dithiole (**163**) and **154**. The latter is the major product, if not the only one, when **152** is treated with morpholine or pyrrolidine.[303] The mechanism leading to **154** is probably very similar to that of Scheme 29.

b. *2-Aryl-1,3-dithiolium Ions.* Adduct **164** is obtained from 2,4,5-triphenyl-1,3-dithiolium perchlorate and dry ammonia in benzene. This product is unstable and easily regenerates the starting material.[130]

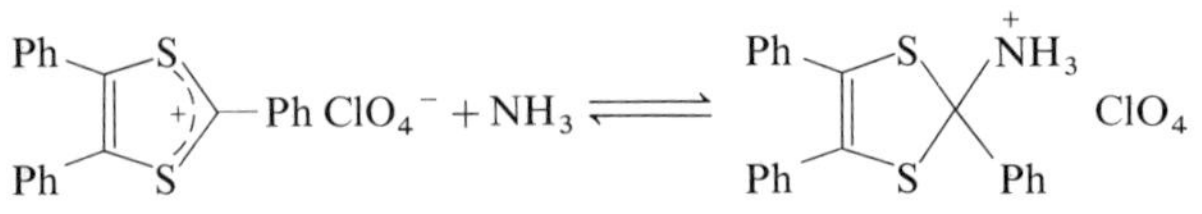

(**164**)

[301] K. M. Pazdro and W. Polaczkowa, *Rocz. Chem.* **44**, 1823 (1970).
[302] K. M. Pazdro and W. Polaczkowa, *Rocz. Chem.* **45**, 811 (1971).
[303] D. Buza and H. Adamowicz, *Rocz. Chem.* **50**, 1823 (1976).

c. *2-Alkylthio-1,3-dithiolium Ions.* Aromatic or aliphatic primary amines react with 2-methylthio-1,3-dithiolium cations in equimolar amounts to produce 2-amino-1,3-dithiolium cations. With an excess of primary amine, the corresponding 2-imino-1,3-dithiole is obtained.[141,304]

Secondary amines and 2-methylthio-1,3-dithiolium cations give, in good yield, mesomeric 2-amino-1,3-dithiolium cations (Eq. 41).[79] Two moles of 2-methylthio-4-phenyl-1,3-dithiolium perchlorate with piperazine form **165**.[79]

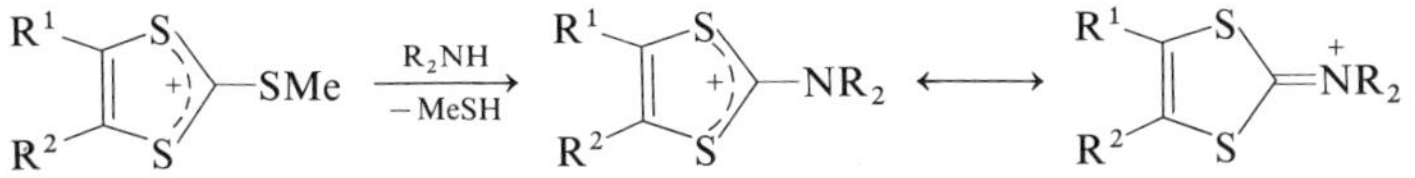

$$\underset{R^2}{\overset{R^1}{\diagup}}\hspace{-2mm}\underset{S}{\overset{S}{\diagdown}}\hspace{-2mm}\text{SMe} \quad \xrightarrow[-\text{MeSH}]{R_2\text{NH}} \quad \underset{R^2}{\overset{R^1}{\diagup}}\hspace{-2mm}\underset{S}{\overset{S}{\diagdown}}\hspace{-2mm}\text{NR}_2 \longleftrightarrow \underset{R^2}{\overset{R^1}{\diagup}}\hspace{-2mm}\underset{S}{\overset{S}{\diagdown}}\hspace{-2mm}=\overset{+}{\text{NR}}_2 \quad (41)$$

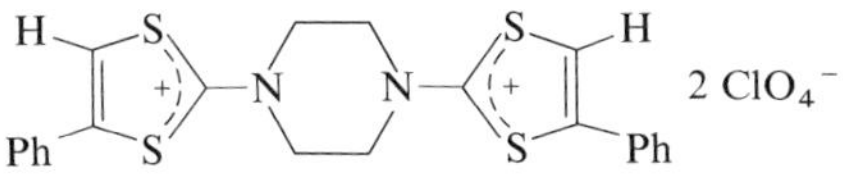

(165)

In acetonitrile, 2-ethylthio-4-phenyl-1,3-dithiolium cation and 2,4-diphenylpyrrole give a red 1:1 adduct which has been described as a charge-transfer complex **166**. The latter slowly decomposes when recrystallized from acetonitrile to give the heterofulvalene **167**. 2,5-Diphenylpyrrole reacts similarly.[170] No complex has been obtained in the imidazole series, but the anion **168** reacts with 2-ethylthio-4-phenyl-1,3-dithiolium cation to form the heterofulvalene **169**.[170]

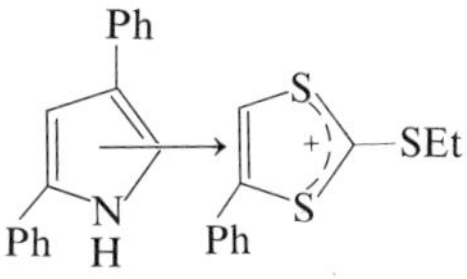

(166) 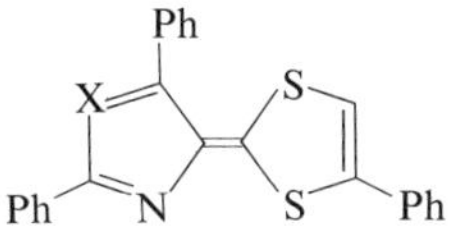   (167) X = CH   (168)
        (169) X = N

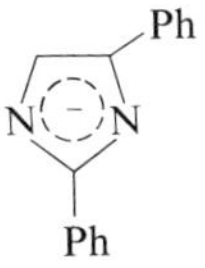

d. *2-Amino-1,3-dithiolium Ions.* 4-*p*-Chlorophenyl-2-dimethylamino-1,3-dithiolium perchlorate with ammonia in ethanol gives the 4-*p*-chlorophenyl-1,3-dithiole-2-thione (43%).[224] The mechanism is unknown, but decomposition of some starting material must provide the extra sulfur atom.

7. Attack by Carbon Nucleophiles

a. *2-Unsubstituted 1,3-Dithiolium Ions.* Nucleophilic addition of Grignard reagents to 1,3-benzodithiolium perchlorate in dry ether is immediate and leads to 2-substituted 1,3-benzodithioles which can be hydrolyzed in high yields to aldehydes.[258]

[304] E. Campaigne, T. Bosin, and R. D. Hamilton, *J. Org. Chem.* **30**, 1677 (1965).

TABLE XI
REACTION OF 1,3-DITHIOLIUM CATIONS WITH ACTIVE METHYLENE COMPOUNDS (SCHEME 30)

Starting material				Product type	Reference
R^1	R^2	R^3	R^4		
H	H	—CO—CH=CH—CO—		172	305
—CH=CH—CH=CH—		CN	CN	172	267
—CH=CH—CH=CH—		Ac	Ac	170	267
—CH=CH—CH=CH—		Ac	EtOCO	170	267
—CH=CH—CH=CH—		—CO—CH$_2$—CMe$_2$—CH$_2$—CO—		170	267
—CH=CH—CH=CH—		H	Ac	170	216
—CH=CMe—CH=CH—		H	Ac	170	216
—CH=CH—CH=CH—		H	Ac	171	216
—CH=CMe—CH=CH—		H	Ac	171	216

Activated methylenes may react with one or two 1,3-dithiolium ions (Scheme 30). When the methylene compound condenses with an equimolar amount of 1,3-dithiolium ion, the adduct **170** is easily oxidized and the final product is a 1,3-dithiol-2-ylidene derivative of the starting methylene compound. Some results are given in Table XI.[216,267,305]

Starting from 4-phenyl-1,3-dithiolium cation and tropolone, a compound of type **170** is obtained, but halogenotropolones give mixtures of compounds **170** and **171**.[306]

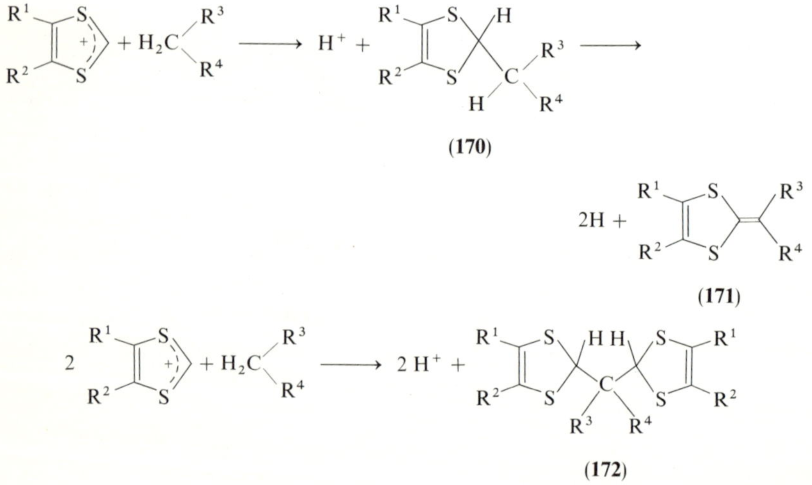

SCHEME 30

[305] J. Nakayama, M. Ishihara, and M. Hoshino, *Chem. Lett.*, 77 (1977).
[306] K. Takahashi, K. Morita, and K. Takase, *Chem. Lett.*, 1505 (1977).

Nucleophilic attack by a negative carbon atom at C-2 of 4-*p*-bromophenyl-1,3-dithiolium ion has also been realized by the sulfur ylide 1-dimethyl-sulfonio-1-benzoylmethanide.[307]

Para positions of arylamines, phenols, and phenol ethers are electron-rich and react at C-2 of 1,3-dithiolium ions, leading to compounds **173**.[267,302] When both para positions are occupied, as in N,N,N',N'-tetramethyl-*p*-phenylenediamine, ortho condensations occur.[261]

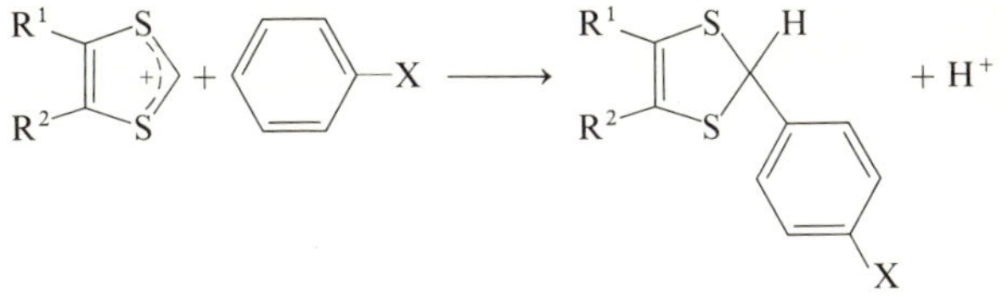

$$X = NR_2, OR, OH \qquad (173)$$

Reaction of 2,4- and 2,6-disubstituted phenols with 2 moles of 1,3-benzodithiolium tetrafluoroborate, followed by treatment with triethyl-amine, leads to compounds **174** and **175**, respectively.[259,308]

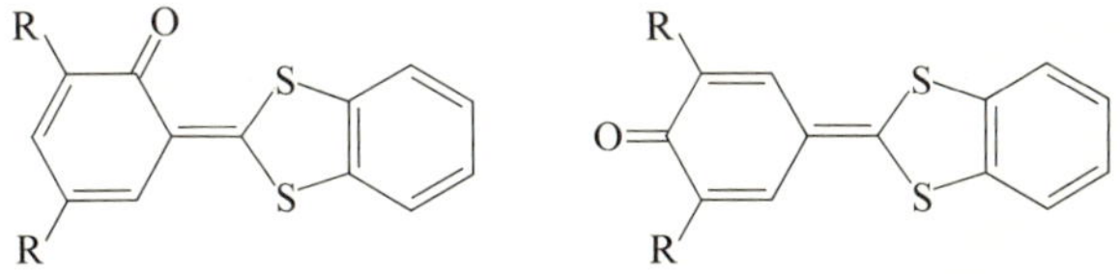

(**174**) R = Me, *t*-Bu, Cl   (**175**) R = *i*-Pr, *t*-Bu, Cl

After the discovery of the interesting electrical properties of tetrathia-fulvalene (TTF)–tetracyanoquinodimethane (TCNQ) complexes, many TTF derivatives have been prepared by reaction of 1,3-dithiolium salts with tertiary amines. This reaction has been interpreted as proceding by deprotonation of a 1,3-dithiolium cation to the corresponding carbene which in turn reacts as a nucleophile on the C-2 of another 1,3-dithiolium cation. This topic having been recently reviewed, we refer in Table XII[220,222,229,267,296,309,310] only to papers subsequent to this review.[268]

1,3-Dithiolium salts have been used with various anions for this coupling reaction. Hydrogen sulfates are hygroscopic, and difficulties in their purification hinder their use. Perchlorates entail explosion hazards but hexafluorophosphates and tetrafluoroborates are convenient starting materials.[268,311]

---

[307] K. Hirai, H. Sugimoto, and T. Ishiba, *Tetrahedron* **33**, 1595 (1977).
[308] J. Nakayama, K. Yamashita, M. Hoshino, and T. Takemasa, *Chem. Lett.*, 789 (1977).
[309] J. M. Fabre, E. Torreilles, J. P. Gilbert, M. Chanaa, and L. Giral, *Tetrahedron Lett.*, 4033 (1977).
[310] F. Wudl, A. A. Kruger, M. L. Kaplan, and R. S. Hutton, *J. Org. Chem.* **42**, 768 (1977).
[311] J. P. Ferraris, T. O. Poehler, A. N. Bloch, and D. O. Cowan, *Tetrahedron Lett.*, 2553 (1973).

TABLE XII
TETRATHIAFULVALENE DERIVATIVES PREPARED BY DEPROTONATION OF 1,3-DITHIOLIUM SALTS[a]

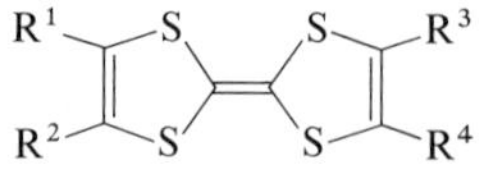

$$R^1 \underset{S}{\overset{S}{\diagdown}} \underset{S}{\overset{S}{\diagup}} R^3$$
$$R^2 \underset{S}{\overset{}{\diagup}} \underset{S}{\overset{}{\diagdown}} R^4$$

R^1	R^2	R^3	R^4	Yield (%)	Reference
R^b	R^b	R^b	R^b	—[c]	229, 309, 310
H	Me	Me	H	—[c]	310
Me	Me	H	H	26	310
Me	Me	Me	Me	20	310
Me	Et	Me[d]	Et[d]	—[c]	229, 309
—(CH$_2$)$_3$—		H	H	22	309
—(CH$_2$)$_3$—		Me	Et	12	309
—(CH$_2$)$_3$—		—(CH$_2$)$_3$—		—[c]	309
—CH=CH—CH=CH—		—CH=CH—CH=CH—		84	267, 296
—CMe=CH—CH=CH—		—CMe=CH—CH=CH—		84, 43	222
—CH=CR—CH=CH—[e]		—CH=CR—CH=CH—[e]		67, 95[f]	222
—CH=(C$_6$H$_4$)=CH—[g]		—CH=(C$_6$H$_4$)=CH—[g]		99	220

[a] Complement to Table I, p. 499 of ref. 268.
[b] R = H, Et, n-propyl.
[c] Yield not indicated.
[d] Stereochemistry not indicated.
[e] R = Me, Cl, I, NO$_2$.
[f] The formation of both Z and E isomers is observed.
[g] Naphtho[2,3-d]-fused.

Until 1976, only trialkylamines were used in this reaction although N,N-dimethylformamide and 1,8-diazabicyclo[5.4.0]undec-7-ene can also give satisfactory results.[222,267]

As shown in Eq. (42), polymers containing the TTF structure are synthesized by this method. Their charge-transfer complex formation with some acceptors and oxidation by halogens have been examined.[232,233]

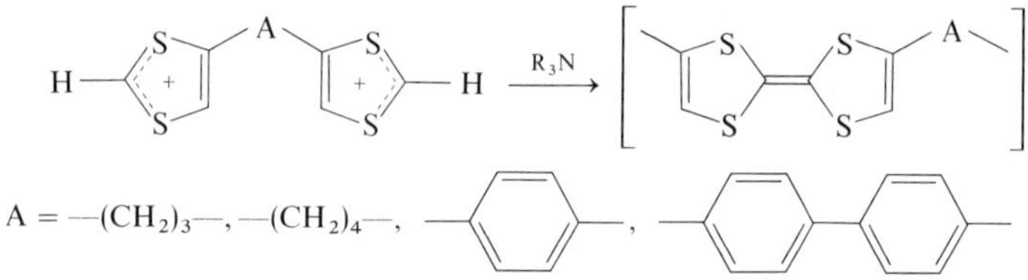

$$H-\underset{S}{\overset{S}{\diagup\!\diagdown}}^{+}\!\!-A-\underset{S}{\overset{S}{\diagup\!\diagdown}}^{+}\!\!-H \xrightarrow{R_3N} \left[\underset{S}{\overset{S}{\diagup\!\diagdown}}=\underset{S}{\overset{S}{\diagup\!\diagdown}}-A\right]_n \quad (42)$$

A = —(CH$_2$)$_3$—, —(CH$_2$)$_4$—, —C$_6$H$_4$—, —C$_6$H$_4$—C$_6$H$_4$—

Preparation of **177** is also carried out in this way with a large excess of 1,3-dithiolium tetrafluoroborate in order to avoid polymerization of **176**. The expected product is obtained in 21% yield together with a high yield of TTF and polymers.[234]

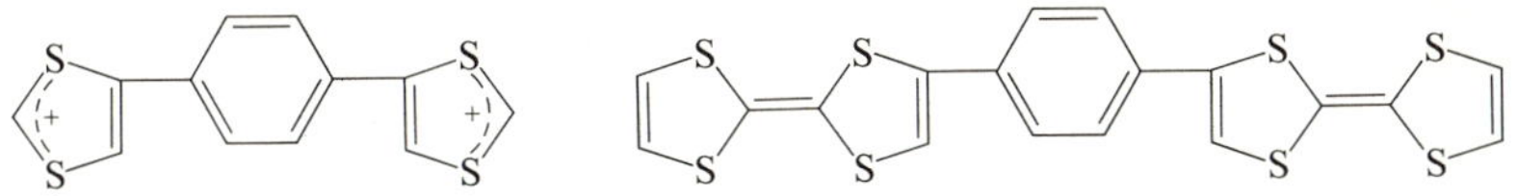

(176)   (177)

Cross reactions are observed between 1,3-dithiolium salts and electron-rich olefins such as bis(1,3-diphenylimidazolidin-2-ylidene), bis(2,3-dihydro-*N*-methylbenzothiazol-2-ylidene), 2-(1,3-benzodithiol-2-ylidene)-*N*-methyl-2,3-dihydrobenzothiazole, or tetramethyl TTF. An example of these reactions is shown in Eq. (43).[312]

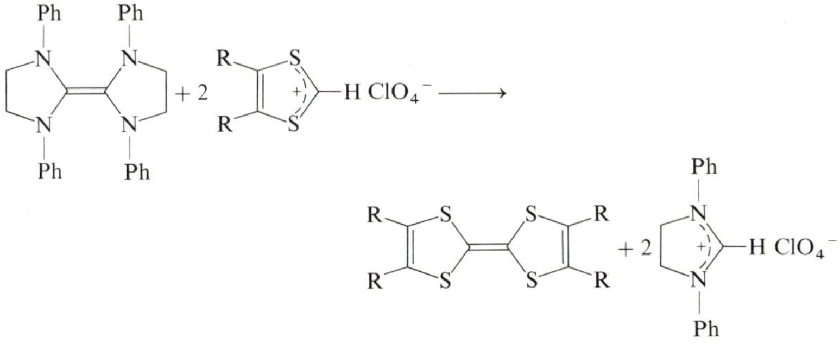

$$R = Me, Ph, -(CH_2)_4-$$

1,3-Dithiolium salts react with the phosphoranes **178**. When excess triethylamine is added in the reaction mixture, a TTF derivative is obtained. In a first step, **178** undergoes nucleophilic addition to the 1,3-dithiolium cation. Then triethylamine brings about elimination of triphenylphosphine and formation of TTF (Scheme 31). This synthesis allows the preparation of

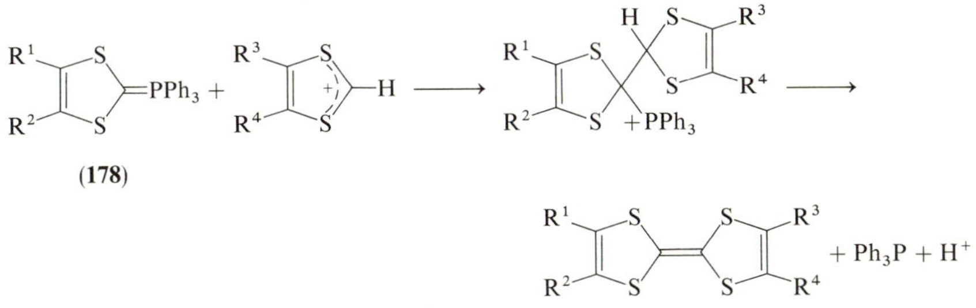

SCHEME 31

[312] D. Buza and W. Krasuski, *Rocz. Chem.* **50**, 2007 (1976).

a variety of TTF derivatives from two different 1,3-dithiolium cations without the simultaneous formation of symmetrical products.[313]

1,3-Dithiolium perchlorates also react as electrophilic reagents with 2-piperidino-1,3-dithioles or N-methyl-2-piperidino-2,3-dihydrobenzothiazole. In each case, TTF derivatives are simultaneously obtained.[314,315]

b. *1,3-Dithiolium Ions Substituted at the 2-Position by Hydrocarbon Groups.* Like 2-unsubstituted 1,3-benzodithiolium salts, corresponding compounds substituted at the 2-position undergo nucleophilic addition of Grignard reagents. The resulting dithioles can be hydrolyzed to ketones, and this seems a good route to these compounds (Scheme 32).[258]

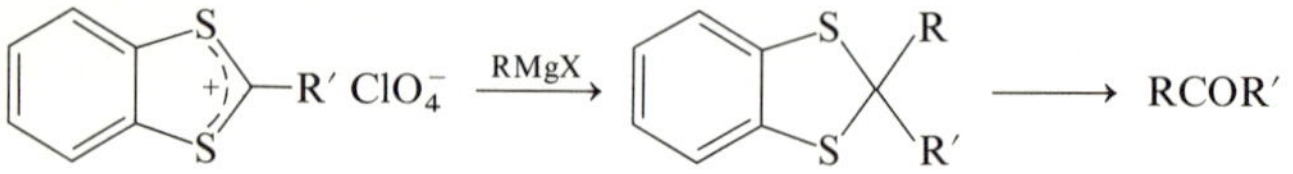

SCHEME 32

c. *2-Alkylthio-1,3-dithiolium Ions.* Grignard reagents derived from primary halides condense with 2-methylthio-1,3-benzodithiolium cation giving **180**. The initial adduct **179** has not been isolated, even when an excess of dithiolium salt is used.[292]

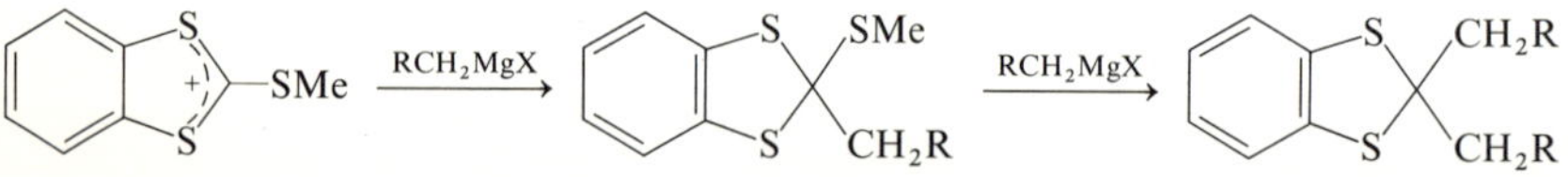

(179) (180)

Reactions of 2-methylthio-1,3-dithiolium salts with active methylene compounds have been extensively reported and reviewed.[1] They are performed in refluxing ethanol, acetic acid, or acetonitrile, pyridine being used as catalyst. Reagents frequently used include ethyl cyanoacetate, acetyl-

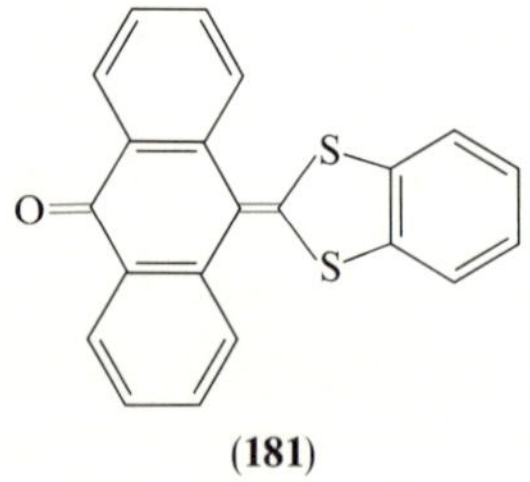

(181)

---

[313] N. C. Gonnella and M. P. Cava, *J. Org. Chem.* **43**, 369 (1978).
[314] D. Buza and W. Krasuski, *Rocz. Chem.* **46**, 2377 (1972).
[315] D. Buza and W. Krasuski, *Rocz. Chem.* **49**, 2007 (1975).

acetate, or benzoylacetate, cyanothioacetamide, malononitrile, indane-1,3-dione, and 2-hydroxytropone.[1,173,277,305,316] When 10H-9-anthrone reacts with 2-methylthio-1,3-benzodithiolium perchlorate, **181** is obtained.[278]

The nucleophilic reagent may also be a 3-methyl-1,2-dithiolium cation which can lose a proton (Section II,B,8). Accordingly, compounds **182** are formed in acetic acid containing pyridine[317] or in boiling ethanol.[178,179]

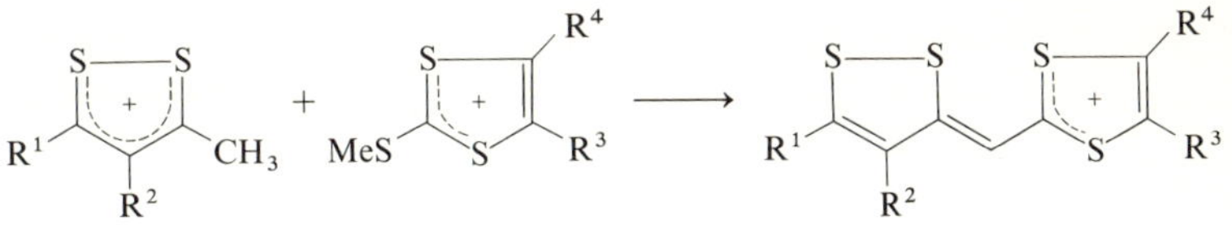

(**182**)

In chloroform, **183** reacts with 2-methylthio-4,5,6,7-tetrahydro-1,3-benzodithiolium cation to give **184**, which is demethylated by triethylamine to **185**.[318]

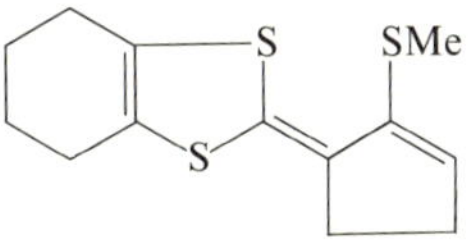

(**183**) 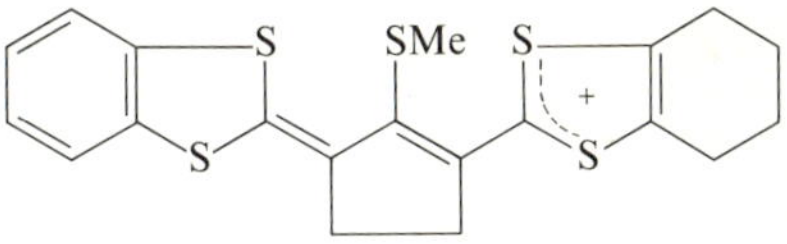   (**184**)

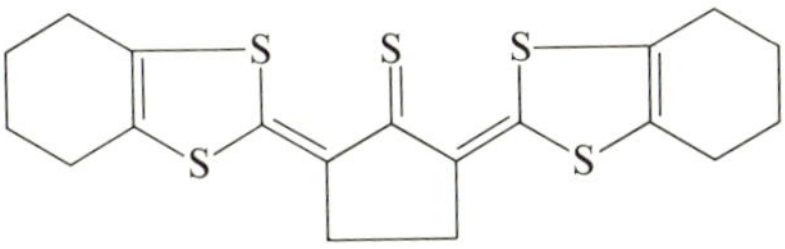

(**185**)

The para positions of N,N-dimethylaniline, diphenylamine, phenol, and anisole are electron-rich and accordingly attack 2-alkylthio-1,3-dithiolium cations at C-2, yielding various 2-aryl-1,3-dithiolium cations.[1,79,278,280] A similar reaction occurs with 2,6-bis(*tert*-butyl)phenol and with 2-naphthol, and by subsequent deprotonation ketones such as **174** and **175** are obtained.[278]

Enamines **186** also react with 2-methylthio-1,3-dithiolium cations **187**. In chloroform, in the presence of triethylamine, an addition product is formed which leads to the ketone **188** by hydrolysis or to the thioketone **189** by hydrosulfolysis.[318]

---

[316] J. Nakayama, M. Ishihara, and M. Hoshino, *Chem. Lett.*, 287 (1977).
[317] D. B. J. Easton, D. Leaver, and D. M. McKinnon, *J. Chem. Soc. C*, 642 (1968).
[318] E. Fanghänel, *J. Prakt. Chem.* **317**, 137 (1975).

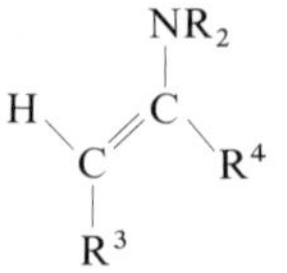

(186) 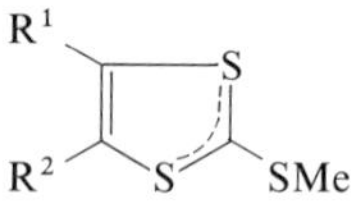 (187) 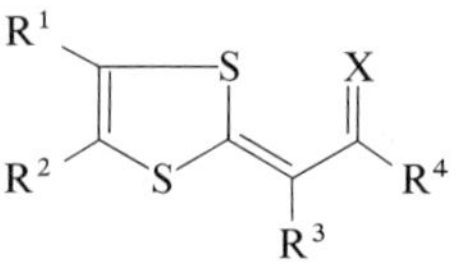 (188) X = O (189) X = S

Like the corresponding 1,2-dithiole derivatives (Section II,B,7,d), 2-phenacylthio-1,3-dithiolium ions **190**, treated by triethylamine in acetic acid undergo an internal rearrangement, yielding first the disulfide **191** and then **192**.[88]

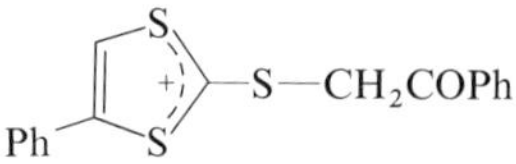

(190) 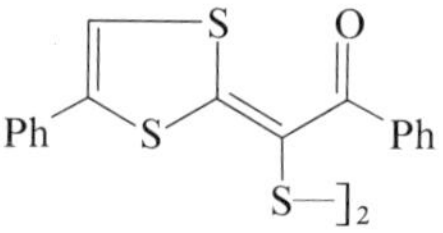 (191) 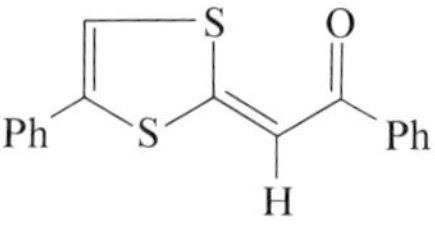 (192)

d. *2-Amino-1,3-dithiolium Ions.* 4-Phenyl-2-piperidino-1,3-dithiolium hydrogen sulfate is also very reactive toward active methylene compounds and leads, with elimination of piperidinium salt, to 2-alkylidene derivatives.[319,320] The reaction proceeds at room temperature or on brief refluxing in methylene chloride in the presence of triethylamine with compounds such as barbituric acid, ethyl acetoacetate, ethyl cyanoacetate, cyanoacetamide, malononitrile, indane-1,3-dione, benzoylacetone, dimedone, acetylacetone, deoxybenzoin, nitromethane, and nitroethane.

8. Attack by Phosphorus Nucleophiles

Trialkyl- and triarylphosphines react with 1,3-benzodithiolium cations, giving a phosphonium salt which is deprotonated by butyllithium to give **193**. Similarly, trialkylphosphites, in the presence of sodium iodide, lead to a dialkyl phosphonate which is deprotonated by butyllithium yielding **194**. Either **193** or **194** can react with carbonyl compounds R^1COR^2 to give the 2-alkylidene-1,3-dithioles **195**.[321]

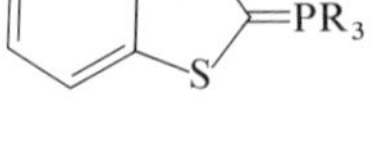

(193) 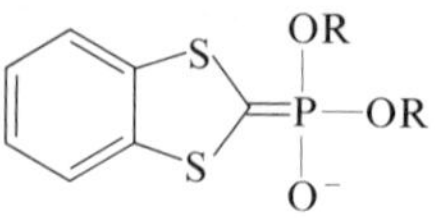 (194) 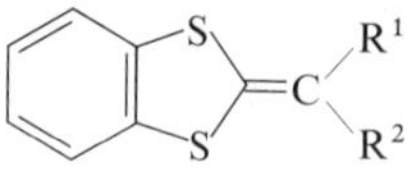 (195)

[319] K. Hirai, T. Ishiba, and H. Sugimoto, *Chem. Pharm. Bull.* **20**, 1711 (1972).
[320] K. Hirai, *Tetrahedron Lett.*, 1137 (1971).
[321] K. Ishikawa, Kin-ya Akiba, and N. Inamoto. *Tetrahedron Lett.*, 3695 (1976).

9. Electrophilic Attack

α-Hydrogens on 2-alkyl-1,3-dithiolium ions **196** exhibit acidic properties explained by the resonance stabilization of the conjugate base **197**.

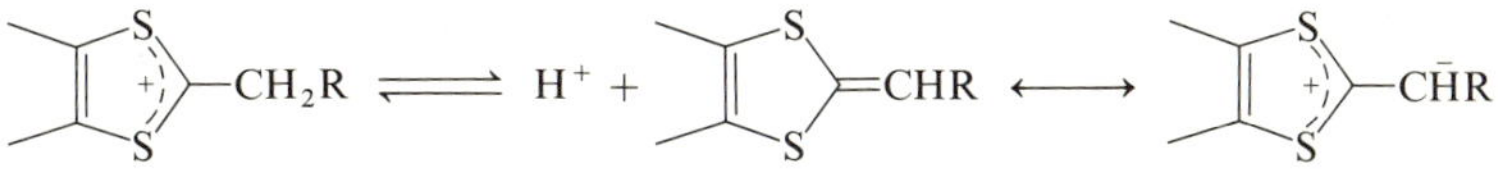

(**196**) R = H, alkyl or aryl   (**197**)

Aromatic aldehydes react with **196**, probably through the intermediate **197** in acetic acid or anhydride, without addition of base. This reaction, which leads to cations **198**, has been performed with various aldehydes: benzaldehydes,[1,74] cinnamaldehydes,[174,322] salicylaldehydes,[219,323,324] and 2-phenyl-4H-1-benzopyran-3-carbaldehyde.[325] Similar reactions occur with ketones and pyrones but need a strong base as catalyst.

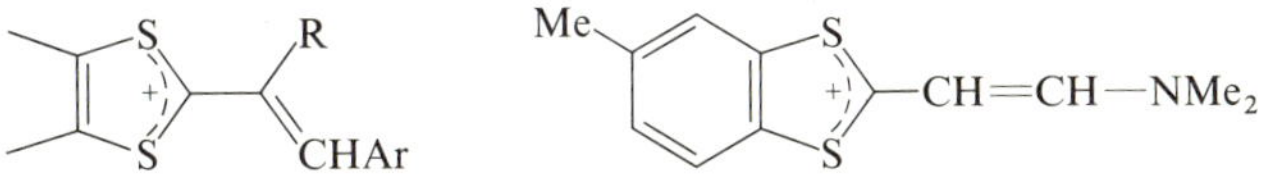

(**198**)   (**199**)

Condensation of dimethylformamide and 2,5-dimethyl-1,3-benzodithiolium perchlorate in phosphorus oxychloride gives a 2-(2-dimethylaminovinyl) derivative **199** in excellent yield. A 1:1 condensation product is also obtained from 2,5-dimethyl-1,3-benzodithiolium perchlorate and malonodialdehydedianilide hydrochloride in hot acetic anhydride.[281] A double condensation between 2-methyl-1,3-dithiolium cations and ethyl orthoformate, leads to **200**.[174,326]

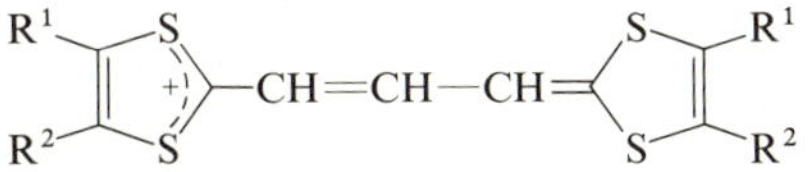

(**200**) R^1 = H, R^2 = Ph
R^1R^2 = —CH=CMe—CH=CH—

The electrophilic reagent may also be a 1,3-dithiolium ion which adds upon the conjugate base **197** in acetic acid. The intermediate loses methanethiol or is oxidized giving **201** (Scheme 33).[317,326]

---

[322] L. Soder and R. Wizinger, *Helv. Chim. Acta* **42**, 1779 (1959).
[323] L. E. Nivorozhkin, N. S. Loseva, and V. I. Minkin, *Khim. Geterotsikl. Soedin.*, 318 (1972).
[324] P. Appriou and R. Guglielmetti, *C. R. Hebd. Seances Acad. Sci., Ser. C* **275**, 1549 (1972).
[325] G. A. Reynolds and J. A. VanAllan, *J. Org. Chem.* **36**, 600 (1971).
[326] L. Soder and R. Wizinger, *Helv. Chim. Acta* **42**, 1733 (1959).

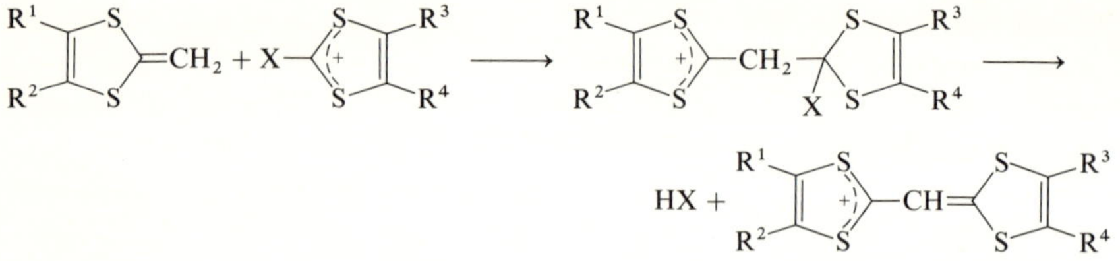

SCHEME 33

(201) X = SMe or H

10. Addition Reactions of 2,5-Disubstituted 1,3-Dithiolium-4-olates

These compounds, which can be represented by mesomeric structures such as **202** and **203**, behave as 1,3-dipoles, undergoing ready cycloadditions with acetylenic dipolarophiles. The bicyclic compound shown in Scheme 34 is a likely intermediate, but elimination of carbon oxysulfide occurs so easily that only the thiophene is obtained. This is a useful synthesis for a variety of thiophenes available with difficulty by other routes.[238,240,327-329]

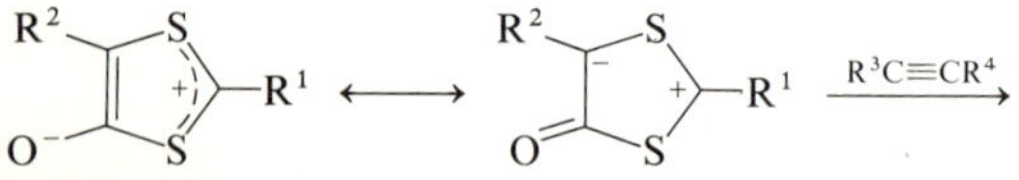

(202)　(203)

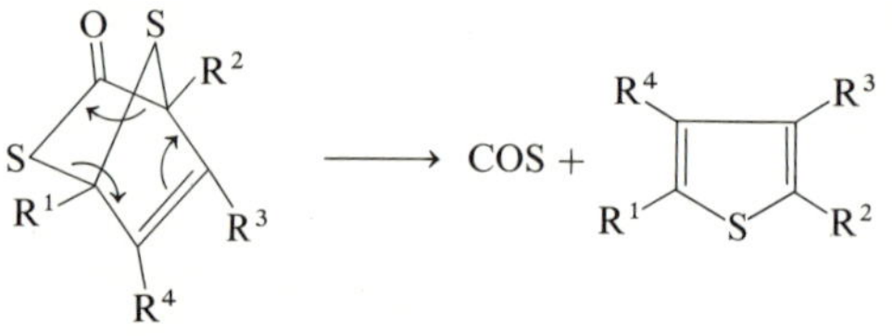

SCHEME 34

With benzyne, formed *in situ* from 1-aminobenzotriazole, the 2,5-diphenyl-1,3-dithiolium-4-olate (Section III,A,1,d) gives 1,3-diphenylbenzo[c]thiophene in 36% yield.[330,331]

Both with olefinic dipolarophiles such as *N*-arylmaleimide, acrylonitrile, or ethyl acrylate and with azirines, stable 1:1 adducts are formed.[238]

---

[327] H. Gotthardt, M. C. Weisshuhn, and B. Christl, *Chem. Ber.* **109**, 753 (1976).
[328] H. Gotthardt and B. Christl, *Tetrahedron Lett.*, 4747 (1968).
[329] H. Gotthardt and C. M. Weisshuhn, *Chem. Ber.* **111**, 2028 (1978).
[330] S. Nakazawa, T. Kiyosawa, K. Hirakawa, and N. Kato, *J. C. S., Chem. Commun.*, 621 (1974).
[331] H. Kato, S. Nakazawa, T. Kiyosawa, and K. Hirakawa, *J. C. S., Perkin 1*, 672 (1976).

Dimethyl maleate, dimethyl fumarate, (Z)- and (E)-dibenzoylethylene, maleic anhydride, methyl cinnamate, styrene, methylstyrene, norbornenes and norbornadiene, cyclopentene, tetramethyl- and tetracyanoethylene,[332] and dimethyl tricyclo[4.2.2.0^{2,5}]deca-3,7,9-triene-7,8-dicarboxylate[333] react similarly with 2,5-diaryl-1,3-dithiolium-4-olates. In many cases the stereochemistry of the products has been elucidated.

Trifluoroacetic anhydride dimerizes 2-alkylthio-5-phenyl-1,3-dithiolium-4-olates to **204**. They are the first reported dimers derived from five-membered mesoionic compounds.[334] In solution these dimers show temperature- and solvent-dependent equilibrium with the monomers. Thus, the reaction of the dimer or of the monomer with dimethyl acetylenedicarboxylate proceeds with elimination of carbon oxysulfide and formation of the thiophene derivative **205**.

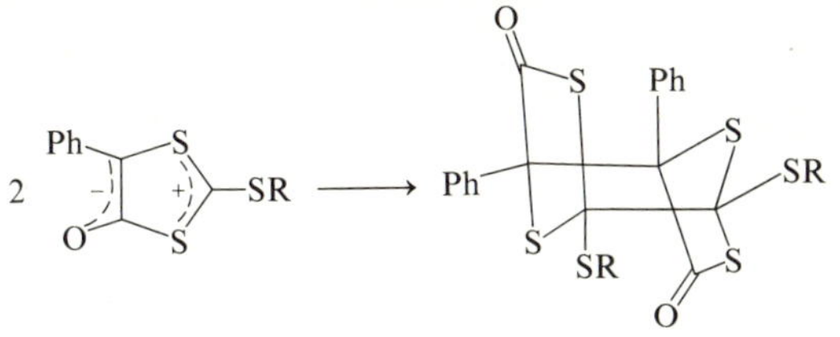

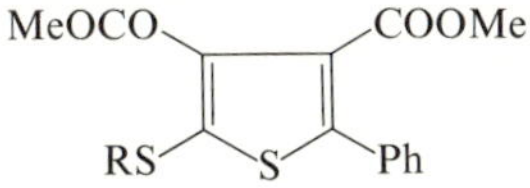

(204)

(205)

C. PHYSICAL PROPERTIES

1. Quantum Mechanical Calculations

Since the previous review,[1] some calculations on the 1,3-dithiolium system have been performed using different parameters.

The Pariser–Parr–Pople method provides a satisfactory interpretation of the absorptions in the UV and visible regions.[197,204,335,336] Transition

---

[332] H. Gotthardt and B. Christl, *Tetrahedron Lett.*, 4751 (1968).
[333] H. Matsukubo and H. Kato, *J. C. S., Perkin 1*, 2562 (1976).
[334] H. Gotthardt, C. M. Weisshuhn, O. M. Huss, and D. J. Brauer, *Tetrahedron Lett.*, 671 (1978).
[335] J. Fabian, A. Mehlhorn, and R. Zahradnik, *J. Phys. Chem.* **72**, 3975 (1968).
[336] R. Mayer, J. Sühnel, H. Hartmann, and J. Fabian, *Z. Phys. Chem. (Leipzig)* **256**, 792 (1975).

energies calculated by a variable-integrals method are in excellent agreement with experiment.[337]

π-Electron densities and reactivity indexes of 4-aryl-1,3-dithiolium ion have been calculated by the HMO method.[181,338] π-Electron densities at the C-2 positions of different cations showed a linear correlation with experimental pK_{R^+} values, and an excellent correlation between pK_{R^+} and the Hammett σ_p constants was established.[338]

The total and orbital energies of 1,3-dithiolium cation were calculated with and without d-orbitals, using a linear combination of Gaussian orbitals.[184] According to these calculations, C-2 is negatively charged, although nucleophilic substitutions occur exclusively at C-2. These results were reconciled by assuming that the presence of the reagent may induce a change of polarization on C-2.

Charge densities and energies calculated by Pople's method for 1,3-dithiolium cation and for the corresponding carbene were compared with those of thiazolium and imidazolium species. It is concluded that sulfur tends to facilitate ionization of hydrogen on C-2 and to increase the electrophilicity of the resulting carbene, while nitrogen has a weak ionization effect on C-2 and favors nucleophilicity of an adjacent carbene. This is confirmed by comparison of the chemical properties of these cations.[339]

Theoretical studies of the redox system formed by TTF and its radical-cation, using repulsion integrals calculated by two different methods, showed the possibilities and limitations of the Coulomb repulsion integrals to estimate formation constants of radicals.[340]

2. Molecular Structure

Few X-ray crystal studies on 1,3-dithiolium system have been performed. The structures of 2,4-diphenyl- and 2,4,5-triphenyl-1,3-dithiolium perchlorates have been reported.[341]

The interatomic distances and angles of TTF and of its radical-cation in the TTF–TCNQ complex were measured. There are no significant differences in the angles between the neutral and ionized species. However, in the cation the sulfur–carbon distances are nearly all equivalent, while in TTF the sulfur to bridging-carbon distances are longer than the other S—C bonds, and C—C distances are both longer in the cation.[342]

[337] Z. Yoshida and T. Kobayashi, *Theor. Chim. Acta* **20**, 216 (1971).
[338] A. Takamizawa and K. Hirai, *Chem. Pharm. Bull.* **18**, 865 (1970).
[339] H. C. Sorensen and L. L. Ingraham, *J. Heterocycl. Chem.* **8**, 551 (1971).
[340] P. Carsky, S. Hünig, D. Scheutzow, and R. Zahradnik, *Tetrahedron* **25**, 4781 (1969).
[341] A. Hordvik, *Acta Chem. Scand.* **17**, 1809 (1963).
[342] T. J. Kistenmacher, T. E. Phillips, and D. O. Cowan, *Acta Crystallogr.* **30**, 763 (1974).

3. IR and Electronic Absorption Data

Few IR data of 1,3-dithiolium salts have been reported.[1,79,224] 1,3-Dithiolium-4-olates have been studied more in relation to their mesoionic and resonance hybrid structure 97. They show a strong absorption band whose position varies between 1575 and 1610 cm^{-1} according to the substituents.[237-243] This is also observed in the spectra of other mesoionic compounds and seems characteristic of this structure.[343]

More recent UV measurements on 2-unsubstituted 1,3-dithiolium salts were carried out in acidic medium and were found to be quite different from previous results[10] in ethanol solution. It was shown that UV spectra taken in ethanol were not the spectra of 1,3-dithiolium salts but were those of 2-ethoxy-1,3-dithioles formed by nucleophilic attack of ethanol on C-2 in the cation.[230]

The parent cation, dissolved in 70% perchloric acid, absorbs with a maximum intensity at 242 and 264 nm (log ε = 3.80 and 3.55). Introduction of alkyl substituents displaces the longest wavelength peak up to 306 nm in the 4-ethyl-2,5-dimethyl-1,3-dithiolium cation.[204]

Many UV spectra of 4-phenyl-2-substituted 1,3-dithiolium salts have been measured.[259,260,306] In 2-(p-substituted phenyl) derivatives, a marked bathochromic shift is induced by electron-donor groups, in the order H, Me, OPh, OH, OMe.[259]

The 4-aryl-1,3-dithiolium salts absorb at lower wavelengths than the 2-aryl analogs. This has been explained by the possibility, in the latter, of delocalization of the positive charge on the aryl group.[224]

A linear relationship exists between the wavenumber of the longest wavelength absorbed by the 2-(p-substituted phenyl)-1,3-dithiolium salts and Hammett σ_p values.[259]

Triphenyl-1,3-dithiolium perchlorate in 50% H_2SO_4 shows intense absorption at 231, 253, 310, and 403 nm (log ε = 4.35, 4.14, 4.00, 4.19).[260]

Many spectra of 2-substituted 1,3-benzodithiolium salts have been measured.[199,217,220] 2-Phenyl-1,3-benzodithiolium tetrafluoroborate in acetonitrile containing 2% H_2SO_4 absorbs at 247, 290, and 389 nm (log ε = 3.67, 3.66, 4.39). Introduction of electron-donor substituents in the para position induces a bathochromic shift.[220]

4-Alkyl-2-dialkylamino-1,3-dithiolium salts in ethanol or acetonitrile have a long wavelength absorption in the range 294–308 nm, and another between 241 and 249 nm.[244,248,252] Replacement of the alkyl by a phenyl induces a slight bathochromic shift of the longer wavelength band. Substitution on the para position of this phenyl causes a further bathochromic displacement which varies according to the substituent.[79,230,244] 2-(p-

---

[343] M. Ohta and C. Shin, *Bull. Chem. Soc. Jpn.* **38**, 704 (1965).

Dimethylaminophenylazo)-1,3-benzodithiolium perchlorate in methanol absorbs intensely at 566 nm (log ε = 4.59).[141]

Ultraviolet spectra of 2-aryl-1,3-dithiolium-4-olates have been measured. The longest wavelength band is found at 508 nm in the 5-methyl-2-phenyl derivative.[239] In 2,5-diaryl derivatives the maximum falls in the range 533–599 nm and is attributed to a $\pi \to \pi^*$ transition.[239,241] In the 2-amino-5-aryl-1,3-dithiolium-4-olates, a band between 456 and 485 nm is similarly assigned.[243]

The UV–visible spectra of compounds **206** have been measured,[200,281,292,317] and the maxima of the longest wavelengths are compared and discussed.[201,202,281] They lie in the range 503–536 nm when $n = 0$ and increase by 150 nm when $n = 1$ and again by 90 nm for $n = 2$. The influence of R seems weak, whereas bathochromic shifts are observed by substitution on the condensed ring in benzo derivatives.[281]

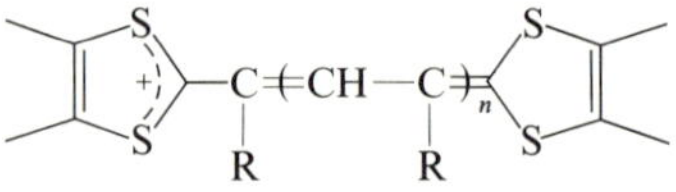

(**206**) $n = 0, 1, 2$

The UV spectra of TTF and of its radical-cation and dication have been measured. Absorption intensities are lower in oxidized forms between 250 and 350 nm, but intense absorptions appear between 400 and 600 nm in the radical-cations and between 400 and 700 nm in dications. In the unsubstituted radical-cation, an absorption at 435 nm is attributed to a $\pi \to \pi^*$ transition, and one at 580 nm to a $\pi \to \sigma^*$ transition.[270]

In the UV–visible spectra of components of the redox system formed by **207**, **208**, and **209**, it is remarkable that the longest wavelength maximum for the dication appears between 520 and 590 nm (log ε = 4.4) whereas the radical-cation (Ar = Ph) shows intense absorption at 596, 705, and 815 nm.[344] Wavelengths of maximum absorption (λ_{max}) for various dications are listed in Table XIII.

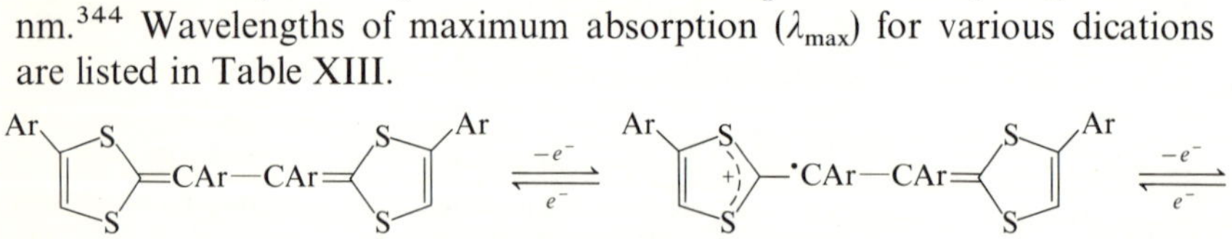

(**207**)   (**208**)

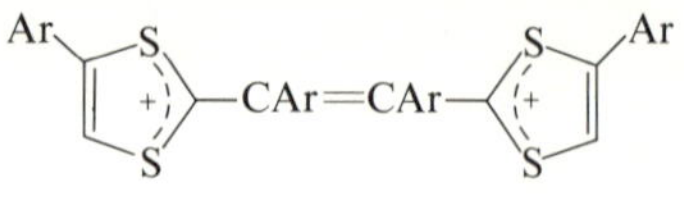

(**209**)

[344] R. Mayer and H. Kröber, *J. Prakt. Chem.* **316**, 907 (1974).

TABLE XIII
UV Spectra of Some 1,3-Dithiolium Dications

Compound	Solvent	$\lambda_{max}(\varepsilon)$ (nm)	Reference
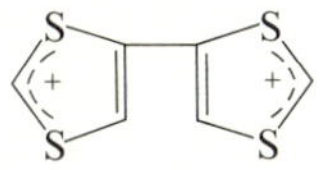 [structure] 2 ClO$_4^-$	CF$_3$CO$_2$H	252	235
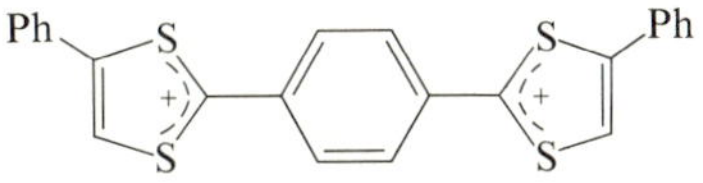 [structure with Ph groups] 2 HSO$_4^-$	MeCN	245.5, 281, 318, 412	231
 [benzodithiolium structure, para] 2 BF$_4^-$	MeCN	243s(12,000), 288(7,500), 412(24,300)	261
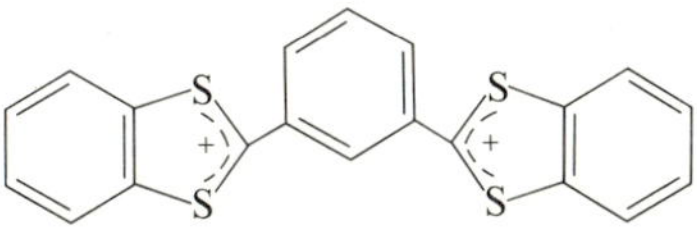 [benzodithiolium structure, meta] 2 BF$_4^-$	MeCN	248s(10,400), 283(8,300), 377(30,300)	261
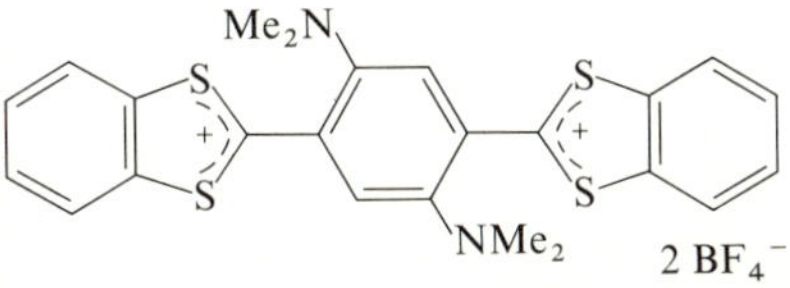 [benzodithiolium structure with Me$_2$N, NMe$_2$] 2 BF$_4^-$	MeCN	274s(17,200), 282(14,300), 434(35,200), 600(6,000)	261

4. NMR Data

A considerable number of papers report ^{1}H-NMR data of various types of 1,3-dithiolium cations: the unsubstituted cation,[204,265,345] 2-unsubstituted cations,[204,230,287,302] 2-amino cations,[1,79,230,235,248] other 2-substituted cations,[173,204,231,259,260,345–347] 1,3-benzodithiolium cations,[216,222,256,267,297,345,347] and 1,3-dithiolium-4-olates.[238–240]

---

[345] K. Sakamoto, N. Nakamura, M. Oki, J. Nakayama, and M. Hoshino, *Chem. Lett.*, 1133 (1977).
[346] H. Hartmann, *J. Prakt. Chem.* **313**, 730 (1971).
[347] G. A. Olah and J. L. Grant, *J. Org. Chem.* **42**, 2237 (1977).

TABLE XIV
^{13}C-NMR Data for Some 1,3-Dithiole Derivatives[a]

No.	Compound	δ C-2	δ C-4	Solvent	Ref.
(210)	 [1,3-dithiolium] BF_4^-	221.2	46.4	CF_3CO_2D	345
(211)	 [1,3-dithiolium] BF_4^-	179.5	146.2	CF_3CO_2D	345
(212)	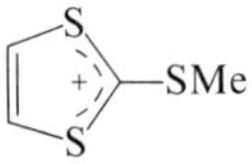 [1,3-dithiolium]—SMe I^-	166.4	139.7	CF_3CO_2D	345
(213)	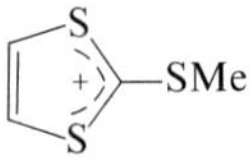 [1,3-dithiolium]—SMe I^-	149.5	137.7	$SO_2\,(-40°C)$	347
(214)	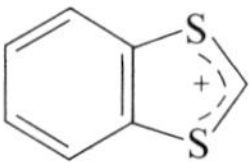 [benzo-1,3-dithiolium] BF_4^-	182.4	146.0	CF_3CO_2D	345
(215)	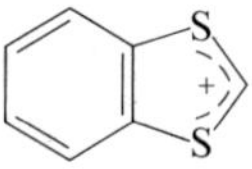 [benzo-1,3-dithiolium] ClO_4^-	184.9	145.5	CD_3CN	347
(216)	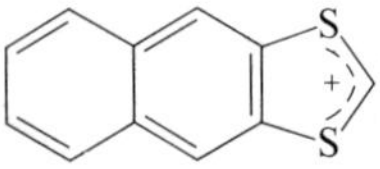 [naphtho-1,3-dithiolium] BF_4^-	187.1	139.6 or 135.4	CF_3CO_2D	345
(217)	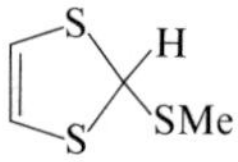 [1,3-dithiole-H,SMe]	61.2	115.4	$CDCl_3$	345
(218)	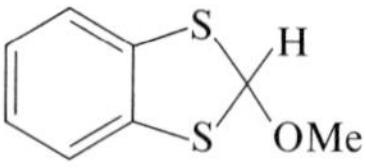 [benzo-1,3-dithiole-H,OMe]	90.9	136.1	$CDCl_3$	345
(219)	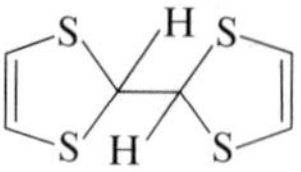 [bis-1,3-dithiole-H,H]	60.3	115.6	CCl_4	289

[a] Chemical shifts given in ppm downfield from TMS, internal reference.

Sec. III.C]   THE 1,2- AND 1,3-DITHIOLIUM IONS   237

In agreement with theoretical findings, the H-2 signals of 1,3-dithiolium cations appear at lower field (~ 2 ppm) than H-5 signals. The chemical shifts of the ring protons and methyl protons of methyl- and dimethyl-1,3-dithiolium salts were correlated with calculated π-charges, and confirm the nonuniform charge distribution in the ring.[204] A linear relationship exists between the H-5 chemical shifts of 2-(p-substituted phenyl)-4-phenyl-1,3-dithiolium salts and Hammett σ_p values.[259]

Restricted rotation around C—N bonds have been studied for 2-dialkylamino-1,3-dithiolium cations.[1,224]

The H-5 signals of 2-aryl-1,3-dithiolium-4-olates in $CDCl_3$ appear at higher fields than do H-5 signals of other salts. This reflects an increased shielding at C-5 due to delocalization of the negative charge on O-4 and C-5 and is in agreement with structural assignments. NMR and UV measurements indicate that the cations of these compounds exist in a 4-hydroxy form.[224]

In 1977, the first ^{13}C-NMR spectra of 1,3-dithiolium cations were reported. They are considered as reflecting the charge densities on carbon atoms and proving the aromatic character of the system. If we consider Table XIV, the C-2 signal in the 1,3-dithiolanylium cation appears at lower field than in 1,3-dithiolium cations, whereas the C-4 signal appears at a much higher field. This suggests that the positive charge is more delocalized in the latter ions.

The C-2 and C-4 signals of compounds **217–219** are typical of sp^3 and sp^2 carbons respectively, which is in good agreement with the assigned structures.

The observed ^{13}C chemical shifts of **214** were roughly correlated with the calculated electron densities.[345,348]

The NMR spectrum of 1,3-dithiole-2-thione in 1:1 FSO_3H-SbF_5 in SO_2 indicates that protonation occurs on the thiocarbonyl sulfur. This ^{13}C spectrum consists of two peaks at 145.2 and 140.0 ppm, attributed to C-2 and C-4 respectively; and these shifts, very similar to those of **213**, show the formation of a 2-mercapto-1,3-dithiolium cation.

5. *Conductivity*

The charge-transfer complexes of TTF with various acceptors exhibit interesting electrical conductivities (listed in Ref. 268: p. 512). For example, the room-temperature electrical conductivity of $(TTF)^{\ddag}(TCNQ)^{\overline{\cdot}}$ is 652 Ω^{-1} cm^{-1} and increases considerably with decreasing temperature with a

---

[348] J. Koutecky, J. Paldus, and R. Zahradnik, *Collect. Czech. Chem. Commun.* **25**, 617 (1960).

maximum of $1.47 \times 10^4 \, \Omega^{-1} \, cm^{-1}$ at 58°K. For these reasons electronic properties and crystal structures of the complexes have been extensively studied in relation to their conductivity.[349-353] There are many publications on the physics of this subject which are outside the scope of this review. The conductivity of radical-cation halides of polymers of TTF depends on stoichiometric composition and is lower than with monomer salts.[233] The room-temperature compressed-pellet electrical conductivity of **220** was found to be 10^5 larger than that of $Na_2Ni[S\!-\!C(CN)\!=\!C(CN)\!-\!S]_2$, indicating an important effect of TTF on the conductivity.[354]

$$\text{TTF}^{2+} \; 2\,Cl^- \;+\; Ni[S_2C_2(CN)_2]_2^{2-} \; 2\,Na^+ \longrightarrow TTFNi[S_2C_2(CN)_2]_2 \;+\; 2\,NaCl$$

(220)

6. Miscellaneous Physical Properties

The ESR spectra of some radical-cations formed by oxidation of TTF derivatives are reported.[215,272,310]

Mass spectra of 2-aryl-, 2-amino-5-aryl- and 2,5-diaryl-1,3-dithiolium-4-olates have been recorded.[239,243,355] They show relatively low-intensity molecular ions which all decomposed by initial loss of CO. Ions R^1CS^+ and R^2CS^+ are present and R^2CS^+ is often the base peak for **202**.

Photolysis of 2,5-diphenyl-1,3-dithiolium-4-olates afforded 2,3,5,6-tetraphenyl-1,4-dithiin, sulfur, and diphenylacetylene. The last suggests the possible intermediacy of diphenylthiirene and formation of such a species in decomposition of mesoionic compounds is considered to be general.[356]

Dipole moments of 2,4-diaryl-1,3-dithiolium-4-olates lie between 4.96 and 5.98 D and are in good agreement with the accepted structure.[239,241]

The temperature dependence of the Mössbauer spectra of bis(2-phenyl-1,3-dithiolium) or bis(4,5-dimethyl-1,3-dithiolium) tetrachloroferrate(II) between 4 and 293°K has been reported.[357]

---

[349] R. C. Wheland and J. L. Gillson, *J. Am. Chem. Soc.* **98**, 3916 (1976).
[350] M. Weger and J. Friedel, *Phys. Abstr.* **38**, 241 (1977).
[351] M. J. Cohen, L. B. Coleman, A. F. Garito, and A. J. Heeger, *Phys. Rev. B* **13**, 5111 (1976).
[352] J. H. Perlstein, *Angew. Chem., Int. Edn Engl.* **16**, 519 (1977).
[353] M. Weger and H. Gutfreund, *Comments Solid State Phys.* **8**, 135 (1978).
[354] F. Wudl, C. H. Ho, and A. Nagel, *J. C. S., Chem. Commun.*, 923 (1973).
[355] K. T. Potts, R. Armbruster, E. Houghton, and J. Kane, *Org. Mass Spectrom.* **7**, 203 (1973).
[356] H. Kato, M. Kawamura, and T. Shiba, *J. C. S., Chem. Commun.*, 959 (1970).
[357] R. L. Martin and I. A. G. Roos, *Aust. J. Chem.* **24**, 2231 (1971).

Addition of 0.1–10% of 1,2- and 1,3-dithiolium salts with various alkyl or aryl substituents to photoconductors causes improvement in photoconduction.[358,359]

Some 1,3-dithiolium cations have been electrochemically oxidized. Peak potentials of the oxidation waves and the reduction waves which result from oxidation products were measured, but no well-defined product, could be identified.[360]

---

[358] G. A. Reynolds, C. V. Wilson, and B. C. Cossar, Ger. Offen. 1,807,359 [*CA* **72**, 95321 (1970)].
[359] G. A. Reynolds and J. A. VanAllan, French Patent 2,055,690 [*CA* **76**, 87190 (1972)].
[360] P. R. Moses, J. Q. Chambers, J. O. Sutherland, and D. R. Williams, *J. Electrochem. Soc.* **122**, 608 (1975).

Advances in Imidazole Chemistry

M. R. GRIMMETT

Chemistry Department, University of Otago, Dunedin, New Zealand

I. Introduction 242
II. Formation of the Imidazole Ring 242
   A. Formation of the 2,3-, 3,4-, and 1,5- (and Sometimes 1,2-) Bonds 243
   B. Formation of the 1,2- and 1,5-Bonds 245
   C. Formation of the 1,2- or the 1,5-Bond 246
   D. Formation of the 1,5- and 3,4-Bonds 250
   E. Formation of the 1,2- and 2,3-Bonds 252
   F. Formation of the 1,5- and 2,3-Bonds 253
   G. Formation of the 4,5-Bond or Both 1,2- and 4,5-Bonds 257
   H. From Other Heterocycles 260
      1. Ring Expansions 260
      2. From Other Five-Membered Heterocycles 262
      3. Ring Contractions 264
   I. Miscellaneous Methods 269
III. Physical Properties 270
   A. Dipole Moments 270
   B. Acid and Basic Strength 270
   C. Ultraviolet and Visible Spectra 273
   D. Infrared Spectra 274
   E. NMR Spectra 275
   F. ESR Spectra 277
   G. Mass Spectrometry 277
   H. Quantum-Mechanical Studies 280
   I. Hammett Studies 281
IV. Chemical Properties 282
   A. Structure and Tautomerism 282
   B. Electrophilic Substitution Reactions 288
      1. At Ring Nitrogen Atoms 288
      2. At Ring Carbon Atoms 297
   C. Nucleophilic Substitution Reactions 305
   D. Reactions Involving Radicals: Photochemical Reactions 309
   E. Thermal Reactions 315
      1. Thermal Decomposition Reactions 315
      2. Thermal Rearrangements 316
   F. Reactions of Substituent Groups 317
      1. Acyl (and Aroyl) Groups 317
      2. Alkyl, Aryl, and Unsaturated Hydrocarbon Substituents 318

3. Amino and Diazo Groups 320
4. Cyano and Cyanoalkyl Groups. 321
5. Hydroxy and N-Oxide Functions 321
6. Quaternary Salts. 322
7. Sulfur Derivatives 322
8. Miscellaneous Substituents 323
V. Appendix. 323

I. Introduction

Since the earlier review[1] in 1970 there have been considerable advances in the field, prompted—at least in part—by the recognition of the importance of the imidazole nucleus in biological processes and by the growing list of applications of imidazoles as pharmaceuticals and in industrial processes. In the intervening years a number of publications[2-8] have dealt with aspects of imidazole chemistry, usually in relation to other azoles. The present article aims to survey developments in imidazole chemistry to mid-1979, limiting the discussion to the simple heteroaromatic ring (condensed or reduced imidazoles are not covered). The extensive literature dealing with metal complexes, pharmacology, and biology of imidazoles has not been reviewed.

II. Formation of the Imidazole Ring

As was mentioned in the earlier review,[1] considerable difficulties arise in the classification of imidazole syntheses. There is a great variety of methods, few of which have wide general application, and in many instances the mechanism is poorly understood. For example, some syntheses almost certainly involve concerted cycloaddition reactions, but sound evidence for this process is lacking. In addition, numerous examples have appeared of conversions of other heterocyclic compounds into imidazoles. While the

---

[1] M. R. Grimmett, *Adv. Heterocycl. Chem.* **12**, 103 (1970).
[2] C. A. Matuszac and A. J. Matuszac, *J. Chem. Educ.* **53**, 280 (1976).
[3] M. R. Grimmett, in "Methodicum Chimicum" (J. Falbe, ed.), Vol. 4, Sect. 6.2.1, p. 242. Thieme, Stuttgart, 1980.
[4] M. R. Grimmett, *Encicl. Chim.* **4**, 413 (1979).
[5] K. Schofield, M. R. Grimmett, and B. R. T. Keene, "Heteroaromatic Nitrogen Compounds, the Azoles." Cambridge Univ. Press, London and New York, 1976.
[6] M. R. Grimmett, *Compr. Org. Chem.* **4**, 357 (1979).
[7] M. R. Grimmett, *Org. Chem., Ser. One* **4**, 55 (1973); *Ser. Two* **4**, 51 (1975).
[8] Various authors, in "Specialist Periodical Reports, Aromatic and Heteroaromatic Chemistry" (C. W. Bird and G. W. H. Cheeseman, eds.), Vols. 1–6. Chemical Society, London, 1973–1978.

majority of these have only limited synthetic utility, they do possess intrinsic chemical interest and thereby merit inclusion in this section.

A general subdivision of synthetic procedures similar to that of Volume 12[1] has been followed.

A. FORMATION OF THE 2,3-, 3,4-, AND 1,5- (AND SOMETIMES 1,2-) BONDS

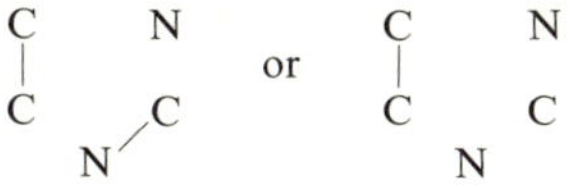

The use of α-dicarbonyl compounds with ammonia and an aldehyde continues as a common synthetic method.[9-16] Thus a series of 2-aryl-5-trifluoromethyl-4-phenylimidazoles (2) have been prepared[11] from the previously unknown 3,3,3-trifluoro-1-phenylpropane-1,2-dione monohydrate (1) as shown in Eq. (1). As an extension of earlier studies of imidazole formation by the action of ammonia on reducing oligosaccharides, Richards[17] has developed what appears to be a simple, sensitive method for linkage identification in polysaccharides. The action of ammonia on periodate-oxidized polysaccharides gives rise to 2-imidazolon-4-ylethanol from 1→4-linked sugars, imidazole and 4-methylimidazole from 1→6-linked sugars, and no imidazoles from 1→3-linked sugars.[17] A variety of vinyl-, hydroxy-, and methoxy-substituted 2-arylimidazole-4,5-dicarboxylic acids

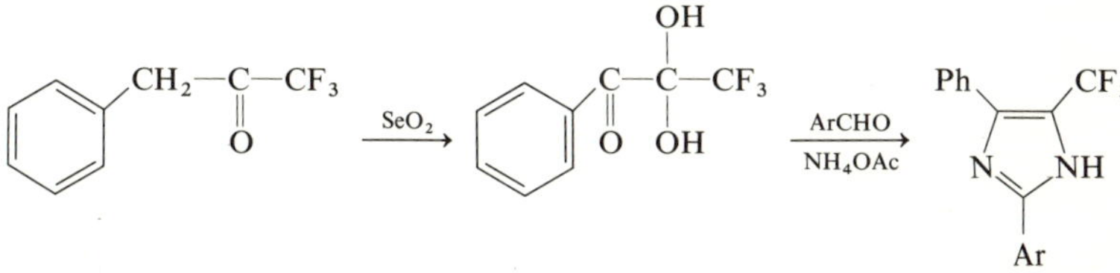

---

[9] D. M. White, *J. Org. Chem.* **35**, 2452 (1970).
[10] A. A. Bardina, B. S. Tanaseichuk, and A. A. Khomenko, *J. Org. Chem. USSR* (*Engl. Transl.*) **7**, 1307 (1971).
[11] J. G. Lombardino, *J. Heterocycl. Chem.* **10**, 697 (1973).
[12] P. Schneiders, J. Heinze, and H. Baumgärtel, *Chem. Ber.* **106**, 2415 (1973).
[13] M. Hoffer, and A. McDonald U.S. Patent 3,652,579 (1972) [*CA* **76**, 153741 (1972)].
[14] B. S. Tanaseichuk and S. V. Yartseva, *J. Org. Chem. USSR* (*Engl. Transl.*) **7**, 1299 (1971).
[15] S. V. Yartseva and B. S. Tanaseichuck, *J. Org. Chem. USSR* (*Engl. Transl.*) **7**, 1558 (1971).
[16] U. Lang and H. Baumgärtel, *Chem. Ber.* **106**, 2079 (1973).
[17] E. L. Richards, *Aust. J. Chem.* **23**, 1033 (1970).

have been prepared using the classical approach which involves reaction of tartaric acid dinitrate with ammonia and aldehydes.[18]

The Bredereck[19] modification which uses an α-halo- (or otherwise functionalized) carbonyl compound with formamide has not received the attention that might have been expected. The synthesis of the imidazolepropanol **3** from 3-bromo-2-methoxytetrahydropyran provides one of the only examples[20] (Eq. 2).

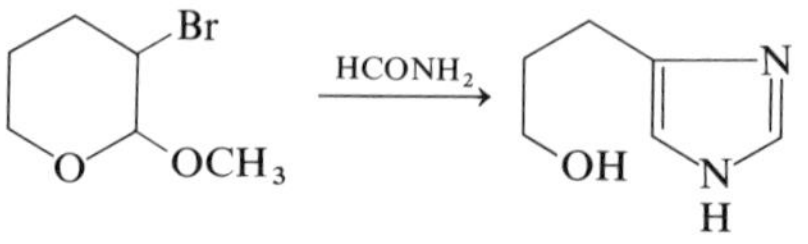

$$\text{(tetrahydropyran-Br, OCH}_3\text{)} \xrightarrow{\text{HCONH}_2} \text{imidazole-CH}_2\text{CH}_2\text{OH} \quad (2)$$

(**3**)

The interactions of 2-hydroxy-[21] or α-bromo-[22,23] ketones and an imidic ester are exemplified by the synthesis[23] of a new imidazole alkaloid (**4**) which had been isolated from the Urticaceae family (see Scheme 1). The reaction of α-bromoketone, ammonia, and an aldehyde gives imidazoles.[24]

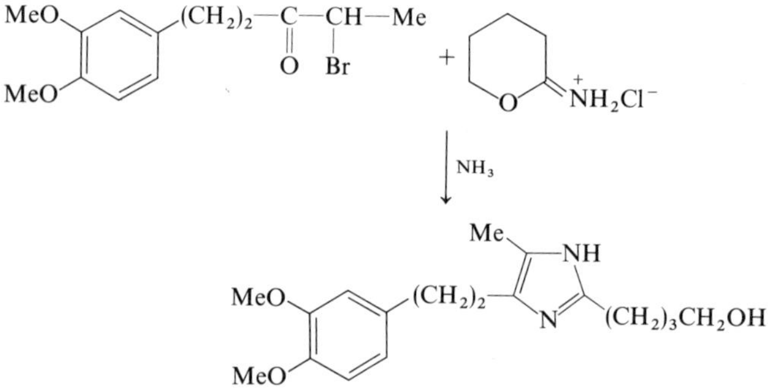

(**4**)

SCHEME 1

---

[18] B. Krieg, R. Schlegel, and G. Manecke, *Chem. Ber.* **107**, 168 (1974).
[19] H. Bredereck, R. Gompper, H. G. Schuh, and G. Theilig, *Angew. Chem.* **71**, 753 (1959).
[20] G. A. Kivits and J. Hora, *J. Heterocycl. Chem.* **12**, 377 (1975).
[21] P. Dziuron and W. Schunack, *Arch. Pharm. (Weinheim, Ger.)* **306**, 347 (1973).
[22] H.-J. Sattler, H.-G. Lennartz, and W. Schunack, *Arch. Pharm. (Weinheim, Ger.)* **312**, 107 (1979).
[23] N. K. Hart, S. R. Johns, J. A. Lamberton, J. W. Loder, and R. H. Nearn, *Aust. J. Chem.* **24**, 857 (1971).
[24] J. J. Baldwin and F. C. Novello, French Patent 2,081,360 (1972)[*CA* **77**, 61994 (1972)].

Imidazole N-oxides are accessible by reaction of an equimolar mixture of an α-diketone and an aldoxime or aldehyde in the presence of hydroxylamine.[25]

Functionalization of the α-position of an aryl ketone is achieved by sulfur in the reaction between ammonia, sulfur, and o-hydroxyacetophenone.[26] The product is a 2H-imidazoline-5-thione (5) as shown in Eq. (3). The rearrangement of cyclohexane-1,2-dione monophenylhydrazone in the presence of ammonium acetate in acetic acid gives an imidazole.[27]

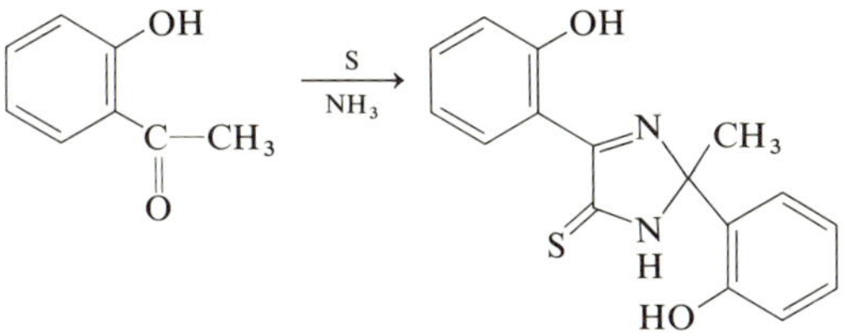

(3)

(5)

B. FORMATION OF THE 1,2- AND 1,5-BONDS

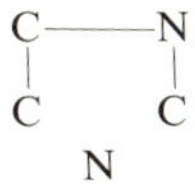

A number of 1,2,5-triarylimidazoles with luminescent properties have been prepared by the cyclization of phenacylbenzamides with arylamines in the presence of phosphorus trichloride.[28,29]

The condensation products of formate esters in the presence of base with N-substituted N-formylglycine esters cyclize with ammonia to yield 1-substituted imidazole-4,5-dicarboxylic esters. These compounds may be reduced to the dialdehydes en route to imidazo[4,5-d]pyridazines.[30]

An extension of the known reaction of amines with N-cyanoalkylimidates (6) gives rise to 5-aminoimidazole nucleosides (7).[31,32]

---

[25] K. Akagane and G. G. Allan, *Chem. Ind. (London)*, 38 (1974).
[26] W. Ried and E. Nyiondi-Bonguen, *Justus Liebigs Ann. Chem.*, 134 (1973).
[27] V. I. Shvedov, L. B. Altukhova, and A. N. Grinev, *Khim. Geterotsikl. Soedin.* 7, 131 (1971) [*CA* 76, 153710 (1972)].
[28] O. N. Poplin and V. G. Tishchenko, *Chem. Heterocycl. Compd. (Engl. Transl.)* 8, 1142 (1972).
[29] V. G. Tishchenko and O. N. Poplin, *Stsintill. Org. Lyuminofory*, 93 (1972) [*CA* 79, 115494 (1973)].
[30] H. Schubert and H. D. Rudorf, *Z. Chem.* 11, 175 (1971).
[31] G. Mackenzie and G. Shaw, *J. Chem. Res. (S)*, 184 (1977).
[32] C. G. Beddows and D. V. Wilson, *J. C. S. Perkin I*, 1773 (1972).

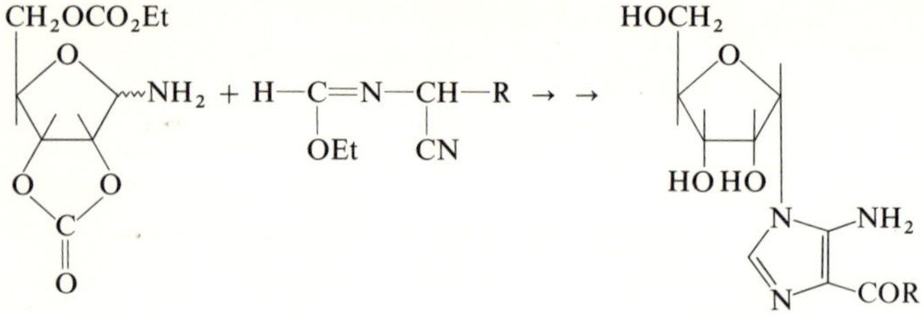

C. FORMATION OF THE 1,2- OR THE 1,5-BOND

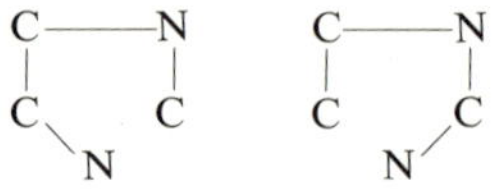

Diphenylimidazoles, along with 2-imidazoline-5-thiones and 5-imino-2-thiazolines, are formed when α,α-diphenylglycine thioamides (RCONHCPh$_2$·CSNH$_2$) are cyclized by aluminum chloride in boiling toluene.[33] A reaction which displays some similarities is the acid-induced cyclization of an N-acetyl-β-glucopyranosylamine.[34] Dry hydrogen chloride induces ring-closure of PhCH=NC(CN)=C(R)SCH$_3$ to chloroimidazoles.[35]

Oxidation of Schiff bases of the monoamide **8** derived from diaminomaleonitrile gives 4-cyanoimidazole-5-carboxamides[36] (Eq. 4). A new synthesis of 2-formyl-4-methyl-1-phenylimidazole 3-oxide (**10**) involves the fluoride-ion-promoted cyclization of the trimethylsilyl ether of N-dichloroacetyl-N-phenylaminopropanone oxime (**9**).[37]

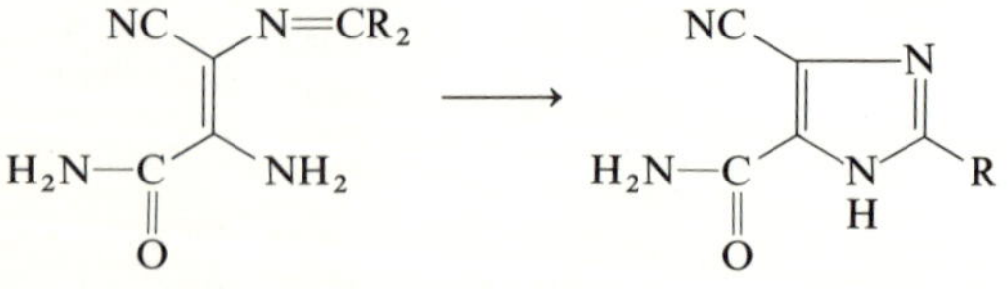

[33] J. Nyitrai and K. Lempert, *Acta Chim. Acad. Sci. Hung.* **73**, 43 (1972) [*CA* **77**, 164600 (1972)].
[34] S. Hirano and R. Yamasaki, *Carbohydr. Res.* **43**, 377 (1975).
[35] K. Hartke and B. Seib, *Pharmazie* **25**, 517 (1970) [*CA* **75**, 5398 (1971)].
[36] Y. Ohtsuka, *A.C.S. Congr. Abstr.* No. 393 (1979).
[37] Y. Mizuno and Y. Inoue, *Chem. Commun.*, 124 (1978).

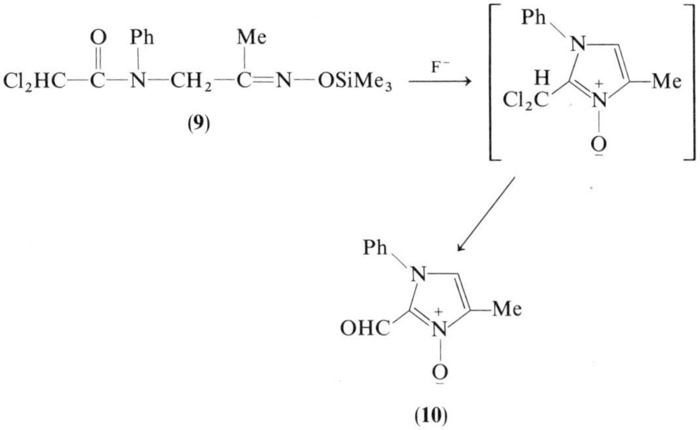

Nucleophilic attack by thiophenate on N-(1-cyanoalkyl) alkylideneamine N-oxides (11) induces ring closure to 4-phenylthioimidazoles by a mechanism such as that shown in Scheme 2. The reaction is accelerated by small amounts of piperidine, but inhibited at temperatures above the melting point of the N-oxide.[38]

When α-amidinonitriles are heated under reflux in acetic anhydride in the presence of one equivalent of sodium acetate, they cyclize to amino- or amidinoimidazoles.[39] An electron-deficient trifluoromethyl carbon becomes part of the imidazole ring when $(CF_3)_2C{=}NC(R^1){=}NR^2$ are heated in the

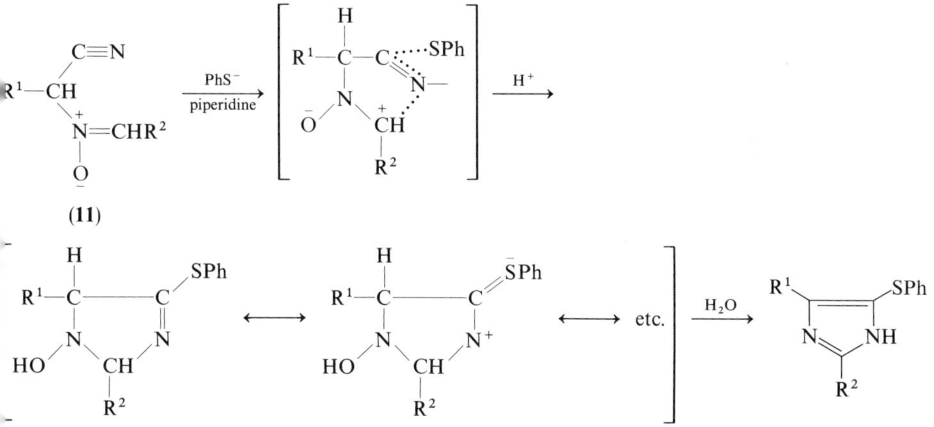

SCHEME 2

[38] M. Masui, K. Suda, M. Yamauchi, and C. Yijima, *J. C. S. Perkin I*, 1955 (1972).
[39] M. Julia and T. Huynh-Dinnh, *Bull. Soc. Chim. Fr.*, 1303 (1971).

presence of stannous chloride; 5-fluoro-4-trifluoromethylimidazoles are formed.[40] 2-Aryloxymethylimidazoles have been prepared[41] from suitable amidines by an adaptation of an earlier method.[42]

The reaction sequence of Scheme 3 might well be classified as an imidazole synthesis from other heterocycles, but it seems more logical to treat this as a cyclization involving formation of the 1,5-bond. The enaminoketone condensation products (12) of 3-amino-1,2,4-oxadiazoles and 1,3-dicarbonyl compounds are cyclized by base to imidazoles (13) in 60–80% yields.[43] Such a reaction makes use of the well-established general attack of a nucleophilic center in the side chain on N-2 of the oxadiazole ring. Benzamidine combines with 2-amino-3-phenacyl-1,3,4-oxadiazolium bromides to produce 1-acyl-amino-2-benzimidoylamino-4-arylimidazoles.[44]

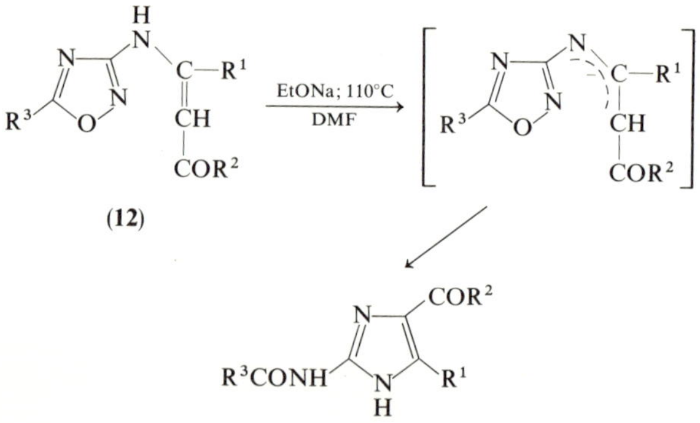

SCHEME 3

Mesoionic imidazoles (15) are accessible from the reaction of N-methyl-N-(N'-phenylbenzimidoyl)aminoacetonitrile (14) with an acid chloride, followed by sodium bicarbonate treatment. Dry hydrogen chloride converts 14 into the imidazolium salt 16 which can also be transformed into 15, or which, with dilute potassium hydroxide, undergoes a Dimroth rearrangement to a 4-anilinoimidazole[45] (Scheme 4).

[40] R. Ottlinger, K. Burger, and H. Göth, *Tetrahedron Lett.*, 5003 (1978).
[41] E. R. Freiter, L. E. Begin, and A. H. Abdallah, *J. Heterocycl. Chem.* **10**, 391 (1973).
[42] J. K. Lawson, *J. Am. Chem. Soc.* **75**, 3398 (1953).
[43] M. Ruccia, N. Vivona, and G. Cusmano, *Tetrahedron* **30**, 3859 (1974).
[44] A. Hetzheim and G. Manthey, *Chem. Ber.* **103**, 2845 (1970).
[45] K. T. Potts and S. Husain, *J. Org. Chem.* **36**, 3368 (1971).

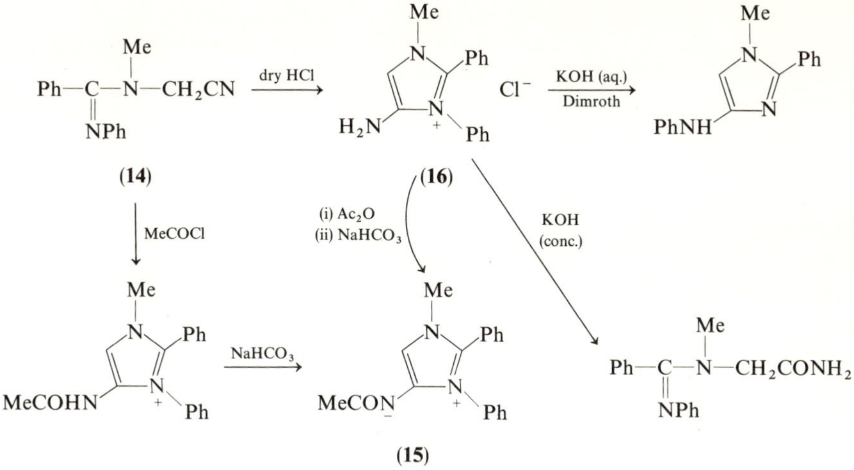

SCHEME 4

A Claisen rearrangement of the adduct (17) of an aromatic aldoxime and a propiolate ester provides a new and direct route to imidazole-4-carboxylates (see Scheme 5) in 61–72% yields.[46,47] This new synthesis apparently involves a [1,3]-sigmatropic shift.

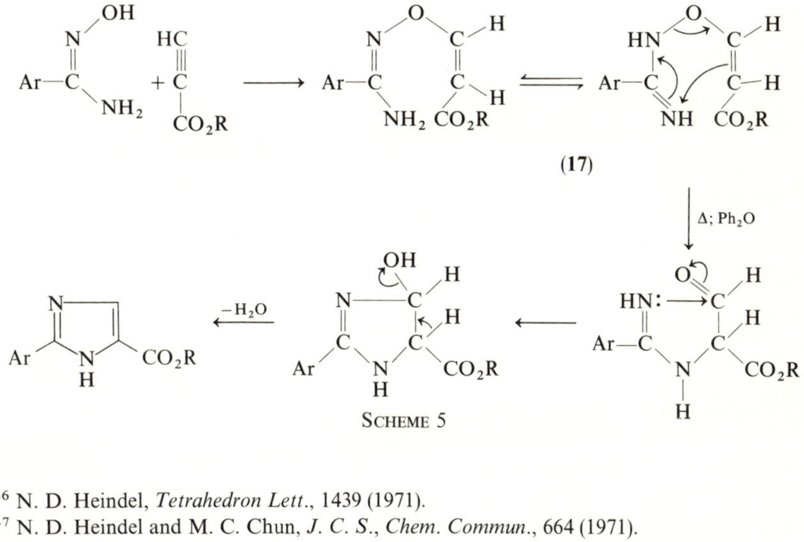

SCHEME 5

[46] N. D. Heindel, *Tetrahedron Lett.*, 1439 (1971).
[47] N. D. Heindel and M. C. Chun, *J. C. S., Chem. Commun.*, 664 (1971).

D. FORMATION OF THE 1,5- AND 3,4-BONDS

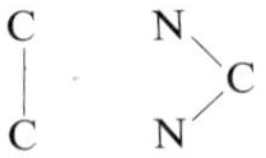

$$\begin{array}{cc} C & N \\ | & \diagdown C \\ C & N \diagup \end{array}$$

The reactions of amidines or guanidines with α-functionalized carbonyl compounds[48–51] continue to be utilized for the synthesis of imidazoles. Thus, the mixed anhydride of acetic and chloroacetic acids reacts with symmetrical diarylguanidines to give 1-aryl-2-arylaminoimidazolin-4-ones,[51] and there is competitive formation of imidazoles and pyrimidines in the reaction of benzamidine with 3-bromobenzo-4-pyrones (**18**).[50] Imidazoles are minor products, but are favored in nonpolar solvents. The use of α-dicarbonyl compounds with guanidine gives 2-amino-4-hydroxy-4-methyl-4H-imidazoles, which give excellent yields of 2-aminoimidazoles on catalytic hydrogenation.[52–54]

In addition to the above there are a number of processes which appear analogous. Salts of mesoionic 4-aminoimidazoles have been prepared by the

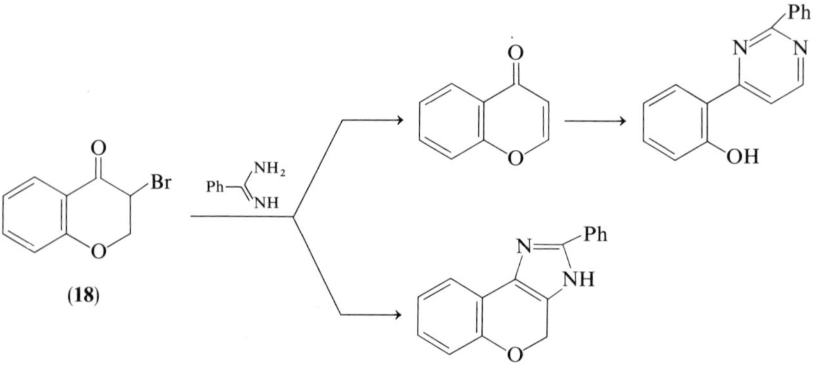

SCHEME 6

---

[48] H. Beyer and S. Schmidt, *Justus Liebigs Ann. Chem.* **748**, 109 (1971).
[49] M. Hamaguchi and T. Ibata, *Chem. Lett.*, 169 (1975).
[50] M.-C. Dubroencq, F. Rocquet, and F. Weiss, *Tetrahedron Lett.*, 4401 (1977).
[51] Yu. V. Svetkin and A. N. Minlibaeva, *J. Org. Chem. USSR (Engl. Transl.)* **7**, 1339 (1971).
[52] T. Nishimura, K. Nakano, S. Shibamoto, and K. Kitajima, *J. Heterocycl. Chem.* **12**, 471 (1975).
[53] T. Nishimura and K. Kitajima, *J. Org. Chem.* **41**, 1590 (1976).
[54] T. Nishimura and K. Kitajima, *J. Org. Chem.* **44**, 818 (1979).

reaction of amidines with haloacetonitriles,[55] and N-acetylamidrazones (**19**) give 1-aminoimidazoles with phenacyl bromide[56] (Eq. 5).

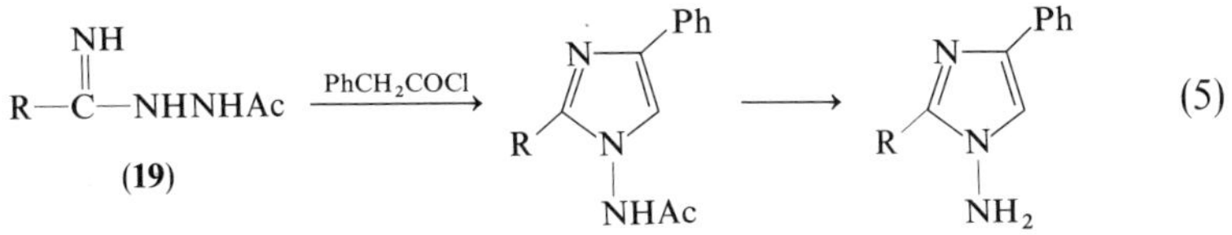

The interaction of amidines and suitably functionalized alkenes has proved a useful, recent route to imidazoles and imidazolines. In particular, N-chloro-N'-arylamidines react with enamines derived from aldehydes to give imidazolines which deaminate readily to imidazoles, especially when heated with pyridinium or triethylammonium chlorides[57,58] (see Scheme 7). Enamines derived from ketones only give chlorinated enamines, while enamines from aldehydes possessing two β-methyl groups rearrange to N-(2-amino-2,2-dialkyl)ethylidene amidines.[58] Further drawbacks to the method include the difficulty of synthesis of the starting vinylamines in a pure state, and the fact that only aryl enamines give good yields. Some of these problems can be overcome by the use of silyl enol ethers (**20**) in place of the enamines[59]

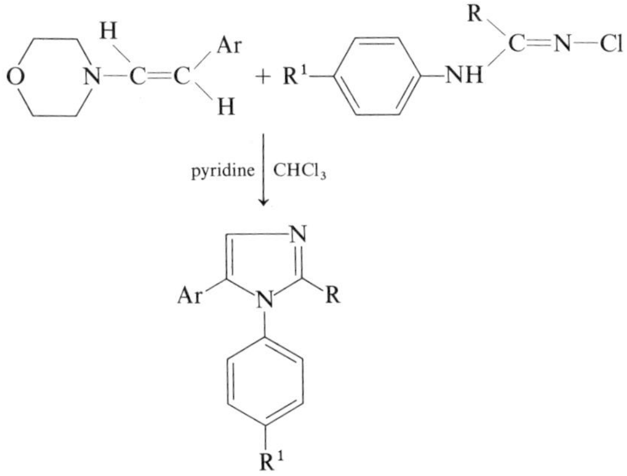

SCHEME 7

---

[55] A. Chinone, S. Sato, and M. Ohta, *Bull. Chem. Soc. Jpn.* **44**, 826 (1971).
[56] E. E. Glover, K. T. Rowbottom, and D. C. Bishop, *J. C. S. Perkin I*, 2927 (1972).
[57] D. Pocar, R. Stradi, and B. Gioia, *Tetrahedron Lett.*, 1839 (1976).
[58] L. Citerio, D. Pocar, R. Stradi, and B. Gioia, *J. C. S. Perkin I*, 309 (1978).
[59] L. Citerio and R. Stradi, *Tetrahedron Lett.*, 4227 (1977).

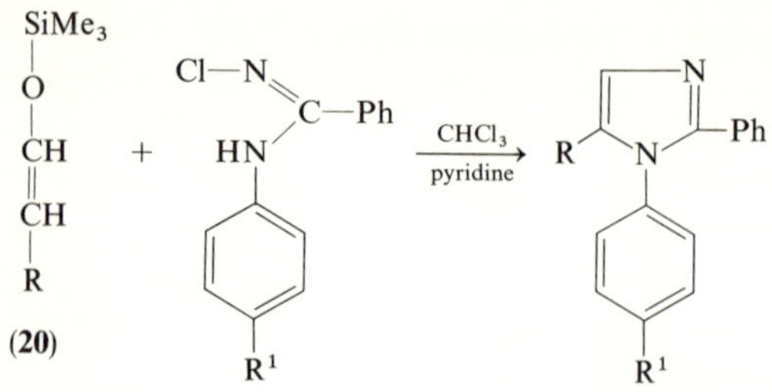

(Eq. 6), allowing a regiospecific synthesis of 5-alkylimidazoles. In a rather similar reaction, acyl vinyl phosphonium salts can be condensed with amidines to give imidazolyl phosphonium salts (22), presumably by way of 21 (Eq. 7).[60]

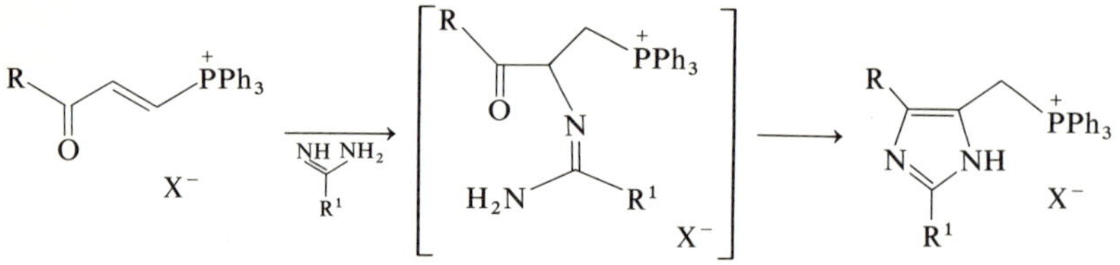

E. FORMATION OF THE 1,2- AND 2,3-BONDS

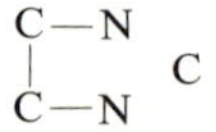

Reaction of α-aminooximes or α-dioximes with aldehydes leading to imidazole N-oxides will be discussed in Section II,F since all of these reactions have a common basis. The widely employed reaction of an α-diamine with an alcohol, aldehyde, or carboxylic acid has been restricted mainly to the synthesis of benzimidazoles.[61–63]

There has been considerable interest in the compound diaminomaleonitrile (23) (and also diaminofumaronitrile) as a source of imidazoles of

---

[60] R. L. Webb, C. S. Labaw, and G. R. Wellman, *A.C.S. Congr. Abstr.* No. 162 (1979).
[61] S. Weiss, H. Michaud, H. Prietzel, and H. Krommer, *Angew. Chem., Int. Ed. Engl.* **12**, 841 (1973).
[62] R. J. Hayward and O. Meth-Cohn, *J. C. S., Chem. Commun.*, 427 (1973).
[63] P. C. Unangst and P. L. Southwick, *J. Heterocycl. Chem.* **10**, 399 (1973).

potential biological interest. Thus, heating **23** in alcohol solution with an alkyl orthoformate[64] gives 4,5-dicyanoimidazole; with formic acid in refluxing xylene, 4-cyanoimidazole-5-carboxamide is formed[65]; with cyanogen chloride, 2-amino-4,5-dicyanoimidazole results[66]; with phosgene in the presence of triethylamine, the product is 4,5-dicyano-2-imidazolone.[67] High yields of 4-amino-5-cyanoimidazole are available on photolysis of **23**.[68,69] The reaction is believed to be unimolecular, proceeding via an electronically excited diaminofumaronitrile (**24**)[69] (Eq. 8). The reaction has been extended

$$\underset{(23)}{\underset{NC}{\overset{NC}{>}}\!\!\!=\!\!\!\underset{NH_2}{\overset{NH_2}{<}}} \;\underset{h\nu}{\rightleftharpoons}\; \underset{(24)}{\underset{NC}{\overset{H_2N}{>}}\!\!\!=\!\!\!\underset{NH_2}{\overset{CN}{<}}} \;\longrightarrow\; \left[\underset{NC}{\overset{H_2N}{\square}}\!\!\!\underset{H}{\overset{NH}{\square_N}}\right] \;\longrightarrow\; \underset{H_2N}{\overset{NC}{\square_N}}\!\!\!\underset{H}{\overset{N}{\square}} \qquad (8)$$

to the synthesis of 1-methyl-4,5-trimethylene-, -4,5-tetramethylene-, -4,5-pentamethyleneimidazoles, and some other derivatives.[70]

1-Methyl-2,5-diphenylimidazole can be prepared from the reaction of benzaldehyde and 2-methylaminophenylacetonitrile.[71]

F. FORMATION OF THE 1,5- AND 2,3-BONDS

$$\begin{array}{c} C\!\!-\!\!N \\ | \quad \; \\ C \quad C \\ \diagdown\!\!N\!\!\diagup \end{array}$$

Much of the synthetic activity involving formation of the 1,5- and 2,3-bonds has centered on the preparation of hydroxyimidazoles and their *N*-oxides. Since direct *N*-oxidation of the imidazole ring has not been accomplished, ring-synthetic methods provide the only means of access to compounds of this type. While not all of the reactions described below conform strictly to the classification of this section, the methods are sufficiently similar to permit

---

[64] Y. Yamada, T. Zama, and I. Kumashiro, Japanese Patent 04373 (1971) [*CA* **74**, 125693 (1971)].
[65] N. Asai, German Patent 2,160,673 (1972) [*CA* **77**, 114409 (1972)].
[66] W. A. Sheppard and O. W. Webster, *J. Am. Chem. Soc.* **95**, 2695 (1973).
[67] L. De Vries, *J. Org. Chem.* **36**, 3442 (1971).
[68] Y. Yamada, S. Nagashima, Y. Iwashita, A. Nakamura, and I. Kumashiro, Japanese Patent 35056 (1970) [*CA* **74**, 42357 (1971)].
[69] T. H. Koch and R. M. Rodehorst, *J. Am. Chem. Soc.* **96**, 6707 (1974).
[70] J. P. Ferris and R. W. Trimmer, *J. Org. Chem.* **41**, 19 (1976).
[71] E. Brunn, E. Funke, H. Gotthardt, and R. Huisgen, *Chem. Ber.* **104**, 1562 (1971).

such a deviation. The interactions of α-dioximes with aldehydes,[25,72–74] or α-oximinoketones[72,73,75–78] with aldimines,[76–78] aldehydes and ammonia[75] or amines,[75] or aldoximes[25,72,73] give 1-hydroxyimidazoles,[75] imidazole N-oxides,[74,77,78] and 1-hydroxyimidazole 3-oxides[25,72,73,76] (Eq. 9). The

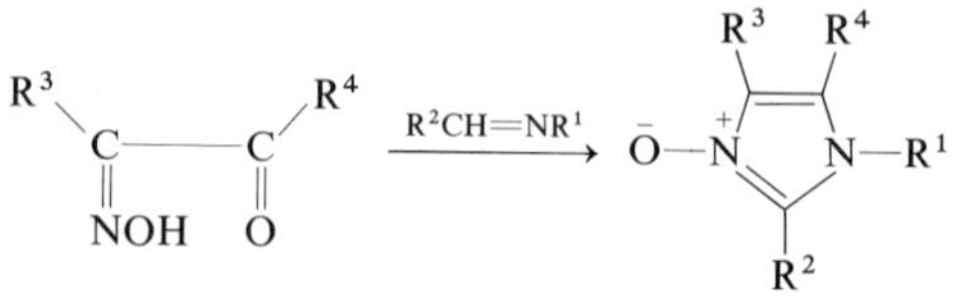

$$\underset{\substack{\| \\ \text{NOH} \quad \text{O}}}{\overset{R^3}{\text{C}} - \overset{R^4}{\text{C}}} \quad \xrightarrow{R^2\text{CH}=NR^1} \quad \bar{\text{O}} - \overset{R^3}{\underset{R^2}{\overset{+}{\text{N}}}} \overset{R^4}{\underset{}{\text{N}}} - R^1 \qquad (9)$$

(R^1 = H, OH, alkyl)

use of α-hydroxylaminooximes leads to 1-hydroxy-Δ^3-imidazoline 3-oxides which are readily dehydrated to imidazole N-oxides.[79]

Anti (E)-α-aminooximes (**25**) can be cyclized with phosgene [as well as by hydrolysis of the corresponding urethanes (**26**)] to 2-oxo-3-imidazoline

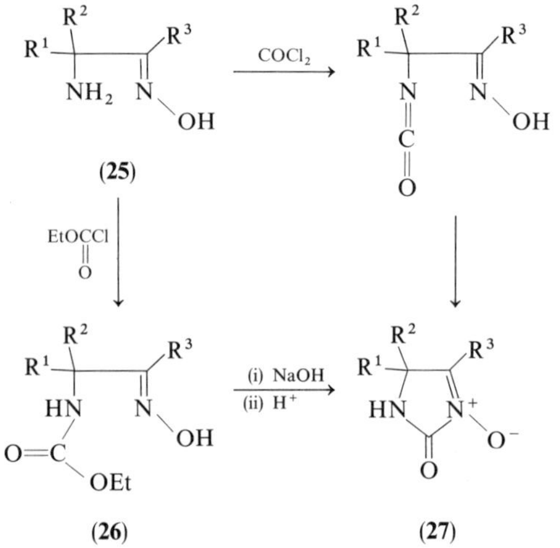

[72] P. Franchetti, M. Grifantini, C. Lucarelli, and M. L. Stein, *Farmaco, Ed. Sci.* **27**, 46 (1972) [*CA* **76**, 85750 (1972)].
[73] B. Krieg and W. Wohlleben, *Chem. Ber.* **108**, 3900 (1975).
[74] R. Bartnik, W. E. Hahn, and B. Orlowska, *Rocz. Chem.* **50**, 1875 (1976).
[75] K. Akagane, G. G. Allan, C. S. Chopra, T. Friberg, T. Mattila, S. O. Muircheartaigh, and J. B. Thomson, *Suom. Kemistil. B* **45**, 223 (1972) [*CA* **77**, 126496 (1972)].
[76] H. Towliati, *Chem. Ber.* **103**, 3952 (1970).
[77] I. J. Ferguson and K. Schofield, *J. C. S. Perkin I*, 275 (1975).
[78] H. Lettau, *Z. Chem.* **10**, 431 (1970).
[79] L. B. Volodarskii, *Chem. Heterocycl. Compd.* (*Engl. Transl.*) **9**, 1175 (1973).

3-oxides (**27**).[80] Oximes with α-hydrogen atoms lead to 3-hydroxy-4-imidazolin-2-ones under these conditions.[80] An investigation[81] of *syn*- and *anti*-α-aminooximes with acetimidic ester and orthoacetic acid ester has demonstrated that, whereas the former ester generates 2-methylimidazoles with both geometric isomers, orthoacetic acid reacts with *anti*-aminooximes to form the dimeric 4*H*-imidazole *N*-oxide, and *syn*-aminooximes are converted by it into pyrazine *N*-oxides. That not all such reactions give rise to *N*-oxides is evidenced by the formation of 4-acetyl-5-methyl-2-phenylimidazole from benzylamine and 3-oximinopentane-2,4-dione[82] and by the reactions of isonitrosoflavanones with Schiff bases.[83]

The reactions of α-aminocarbonyl compounds (and their acetals) with cyanamide continue to provide an attractive route to 2-aminoimidazoles.[84,85] A 40% yield of the 5-hydroxyimidazole derivative **29** was obtained from the interaction of norvaline methyl ester and *N*-isopropylacetonitrilium tetrachloroferrate (**28**).[86]

$$CH_3\text{—}C\equiv \overset{+}{N}\text{—}CH(CH_3)_2 \, FeCl_4^- + CH_3(CH_2)_2CH(NH_2)CO_2CH_3 \longrightarrow$$

(**28**)

(**29**)

An adaptation of the old Marckwald synthesis[87] has allowed an improved preparation of pilocarpine (**30**)[88] (see Scheme 8).

The exothermic reaction of α-oxothionamides with aldimines gives 4-mercaptoimidazoles.[89] Dimerization of ethyl α-amino-α-cyanoacetate gives

---

[80] H. Gnichtel and K.-E. Schuster, *Chem. Ber.* **111**, 1171 (1978).
[81] H. Gnichtel, W. Griebenow, and W. Loewe, *Chem. Ber.* **105**, 1865 (1972).
[82] A. C. Veronese, G. Zanotti, and A. Del Pra, *J. C. S., Chem. Commun.*, 443 (1977).
[83] A. R. Katritzky, M. Michalska, R. L. Harlow, and S. H. Simonsen, *Tetrahedron Lett.*, 4333 (1976).
[84] B. Cavalleri, R. Ballotta, and G. C. Lancini, *J. Heterocycl. Chem.* **9**, 979 (1972).
[85] A. Kreutzberger and R. Schuecker, *Arch. Pharm. (Weinheim, Ger.)* **306**, 169 (1973).
[86] R. Fuks, *Tetrahedron* **29**, 2153 (1973).
[87] W. Marckwald, *Ber. Dtsch. Chem. Ges.* **25**, 2354 (1892).
[88] J. I. DeGraw, *Tetrahedron* **28**, 967 (1972).
[89] F. Asinger, A. Sans, H. Offermanns, P. Krings, and H. Andree, *Justus Liebigs Ann. Chem.* **744**, 51 (1971).

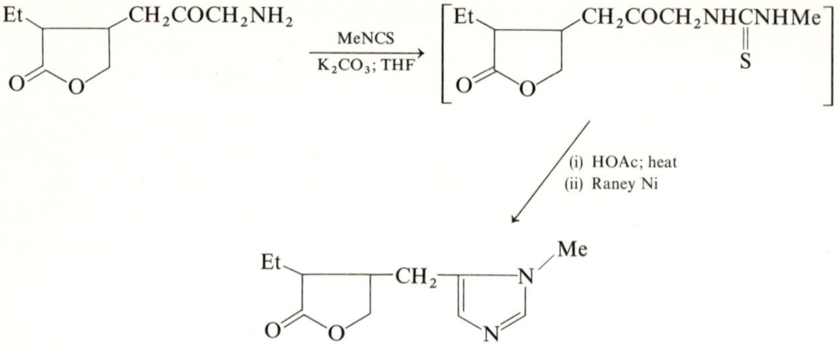

SCHEME 8

diethyl 5-aminoimidazole-2,4-dicarboxylate (31)[90] (Scheme 9). Another dissociation process is implicated in the conversion of 1,1-dimethyl-1-phenacylhydrazinium bromide into 2-benzoyl-4-phenylimidazole in refluxing propanol.[91]

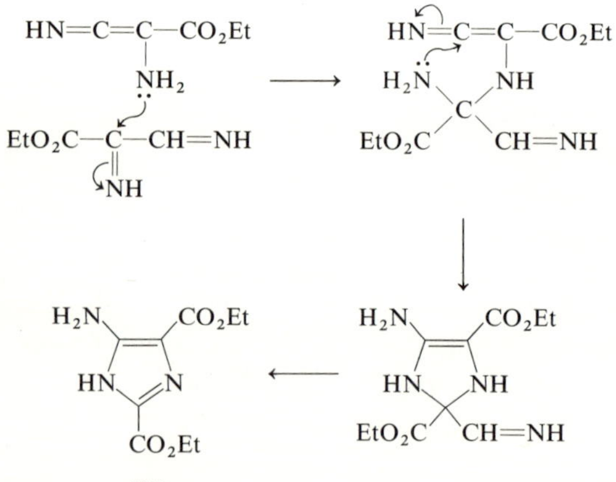

SCHEME 9

A low yield (~4%) of 4-t-butylamino-5-cyanoimidazole is obtained when 4 moles of HCN in HF reacts with 2,4,4-trimethylpent-2-ene. The reaction mechanism proposed[67] suggests that the cyclization process can be classified in this section.

[90] D. H. Robinson and G. Shaw, *J. C. S. Perkin I*, 1715 (1972).
[91] M. Koga and J. P. Anselme, *J. C. S., Chem. Commun.*, 53 (1973).

G. Formation of the 4,5-Bond or Both 1,2- and 4,5-Bonds

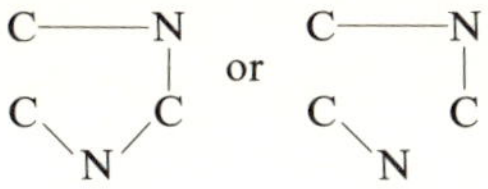

$$\begin{array}{cc} \text{C—N} & \text{C—N} \\ \text{C} \underset{\text{N}}{\diagdown} \text{C} \quad \text{or} \quad \text{C} \underset{\text{N}}{\diagdown} \text{C} \end{array}$$

In a reaction that bears similarities to an earlier synthesis of 4-amino-2-methylthioimidazoles,[92] N-cyanoimidic esters react with α-aminonitriles or α-aminoesters to give 4-aminoimidazoles[93] (Scheme 10).

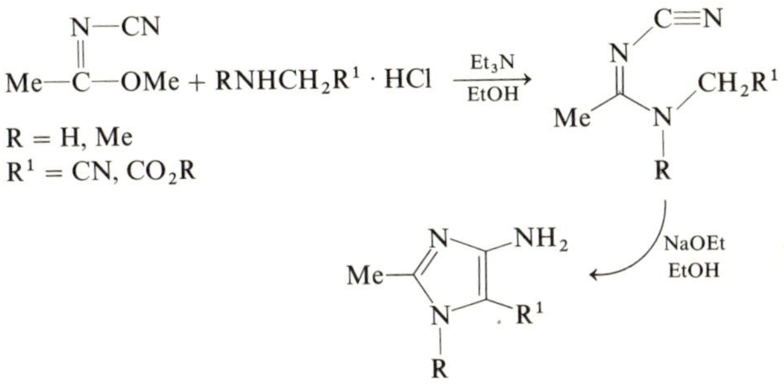

$$\text{Me—C(=N—CN)—OMe} + \text{RNHCH}_2\text{R}^1 \cdot \text{HCl} \xrightarrow[\text{EtOH}]{\text{Et}_3\text{N}}$$

R = H, Me
R^1 = CN, CO_2R

Scheme 10

Toluene-p-sulfonylmethyl isocyanide (**32**) ("Tosmic") reacts in a base-induced cycloaddition (in protic medium) with aldimines. Toluene-p-sulfinic acid is eliminated to yield 1,5-disubstituted imidazoles (**33**; R^3 = H); in reactions with imidoyl chlorides, hydrogen chloride is eliminated to give the toluene-p-sulfonyl derivatives (**33**: R^3 = p-tosyl)[94–96] (Eq. 10). Similarly,

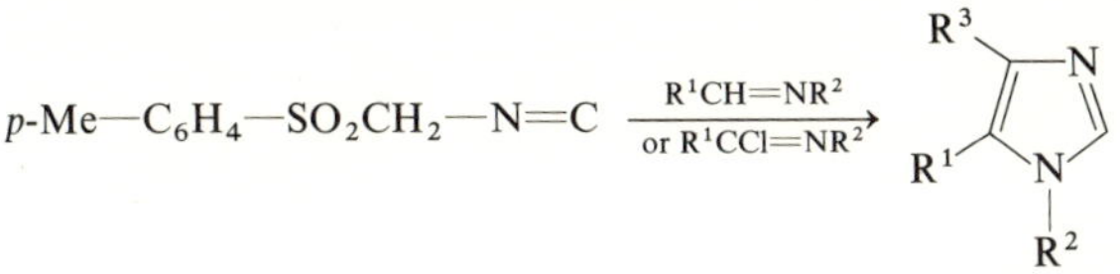

$$p\text{-Me—C}_6\text{H}_4\text{—SO}_2\text{CH}_2\text{—N=C} \xrightarrow[\text{or } R^1\text{CCl}=\text{NR}^2]{R^1\text{CH}=\text{NR}^2} \quad (10)$$

(**32**)　　　　　　　　　　(**33**)

---

[92] R. Gompper, M. Göng, and F. Saygin, *Tetrahedron Lett.*, 1885 (1966).
[93] A. Edenhofer, *Helv. Chim. Acta* **58**, 2192 (1975).
[94] A. M. van Leusen and O. H. Oldenziel, *Tetrahedron Lett.*, 2373 (1972).
[95] A. M. van Leusen, J. Wildeman, and O. H. Oldenziel, *J. Org. Chem.* **42**, 1153 (1977).
[96] O. H. Oldenziel and A. M. van Leusen, *Tetrahedron Lett.*, 2373 (1972).

α-tosylbenzyl and α-tosylethyl isocyanides give 1,4,5-trisubstituted imidazoles with aldimines.[95] Under aprotic conditions it is possible to isolate the primary cycloadducts, imidazolines, which ultimately form imidazoles. Since imidoyl chlorides are highly reactive toward nucleophiles, the natures of R^1 and R^2 are important; some aldimines (e.g., $R^1 = t$-butyl, $R^2 =$ cyclohexyl) are so prone to hydrolysis and the amide is formed so readily that even in very dry conditions there is no imidazole formation.[96] "Tosmic" also provides a source of N-tosylmethylimidic esters or thioesters which react with Schiff bases to give 1,2,5-trisubstituted imidazoles,[97] while high yields of 4-thioimidazoles result from the (base-induced) cyclization of thiomethyl isocyanides and nitriles[98] (Eq. 11). The regiospecific synthesis of 5-aminoimidazole-4-

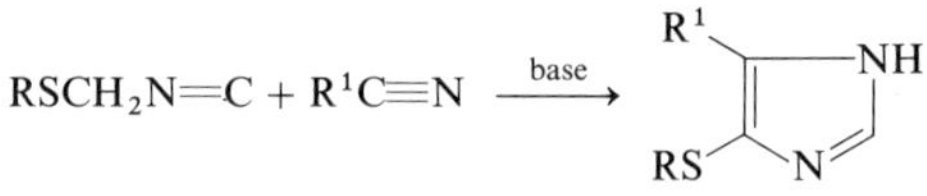

$$RSCH_2N=C + R^1C\equiv N \xrightarrow{\text{base}} \begin{array}{c} R^1 \\ \diagup \diagdown NH \\ RS \diagdown N \diagup \end{array} \qquad (11)$$

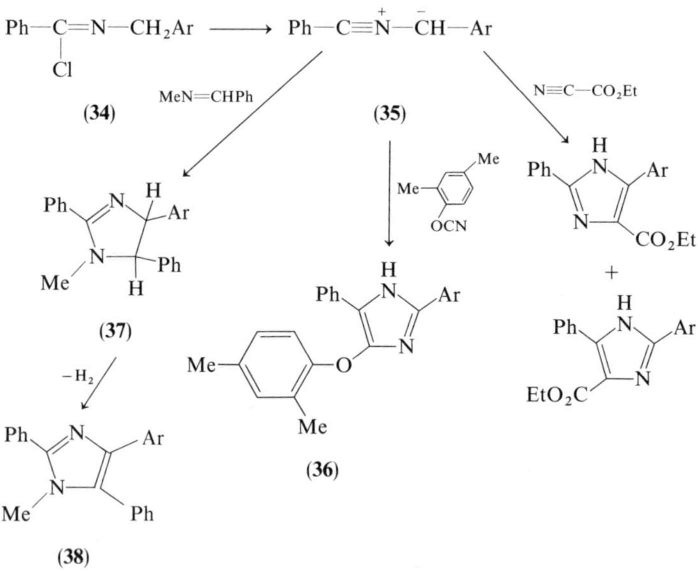

SCHEME 11

---

[97] H. A. Houwing, J. Wildeman, and A. M. van Leusen, *Tetrahedron Lett.*, 143 (1976).
[98] A. M. van Leusen and J. Schut, *Tetrahedron Lett.*, 285 (1976).

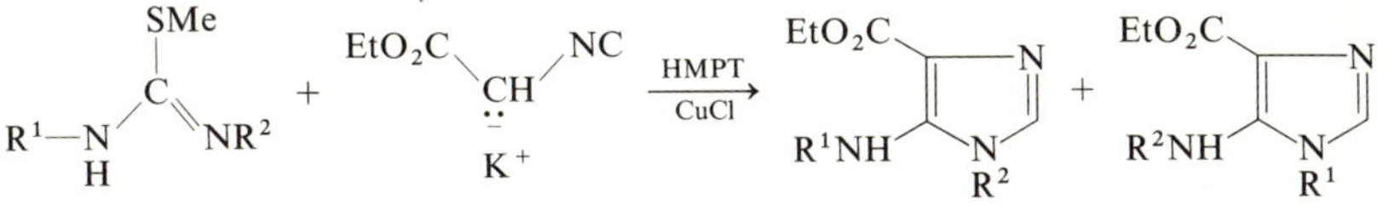

carboxylic esters via an isonitrile cycloaddition (Eq. 12) bears similarities to the above reactions,[99] as does the cycloaddition of the 1,3-dipolar ylide **35**, generated by the action of triethylamine on the imidoyl chloride **34**, to ethyl cyanoformate[100] (see Scheme 11). 2,4-Xylyl cyanate yields solely the adduct **36**; benzylidenemethylamine gave **38** via the intermediate **37**.[100] The 1,3-anionic cycloaddition of the 2-azaallyllithium compound **39** to aromatic nitriles leads to imidazoles, probably by elimination of lithium hydride[101] (Eq. 13). Electron-deficient nitriles add to mesoionic oxazoles (**40**) to give imidazoles[71] (Eq. 14), while yet a further novel synthesis involving cycloaddition is exemplified by the conversion of *N*-methyl-*C*-phenylnitrone (**41**) into **42** in the presence of potassium cyanide.[102]

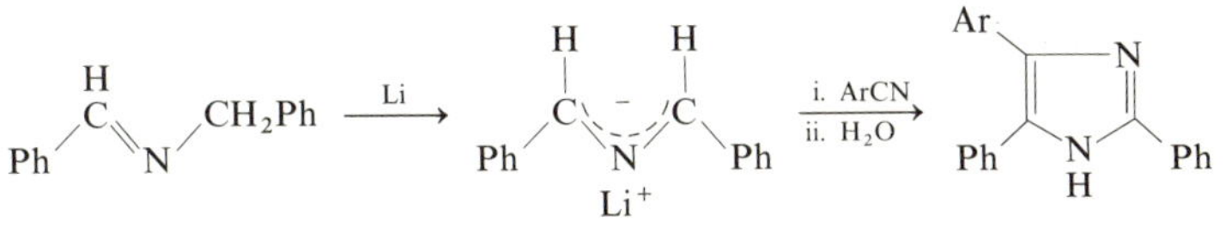

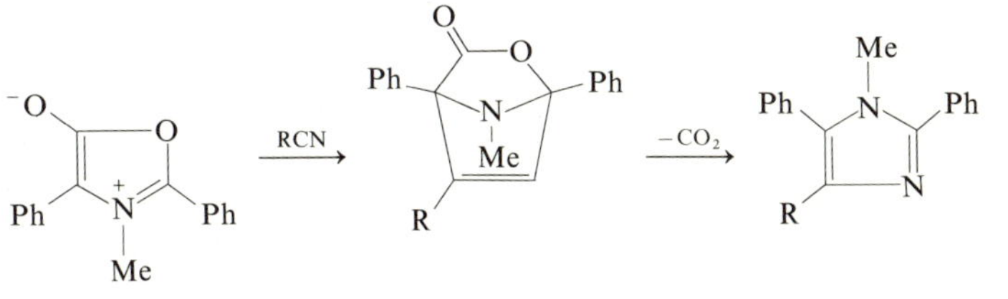

---

[99] J. T. Hunt and P. A. Bartlett, *Synthesis*, 741 (1978).
[100] K. Bünge, R. Huisgen, R. Raab, and H. J. Sturm, *Chem. Ber.* **105**, 1307 (1972).
[101] T. Kauffmann, A. Busch, K. Habersaat, and E. Köppelmann, *Angew. Chem., Int Ed. Engl.* **12**, 569 (1973).
[102] N. G. Clark and E. Cawgill, *Tetrahedron Lett.*, 2717 (1975).

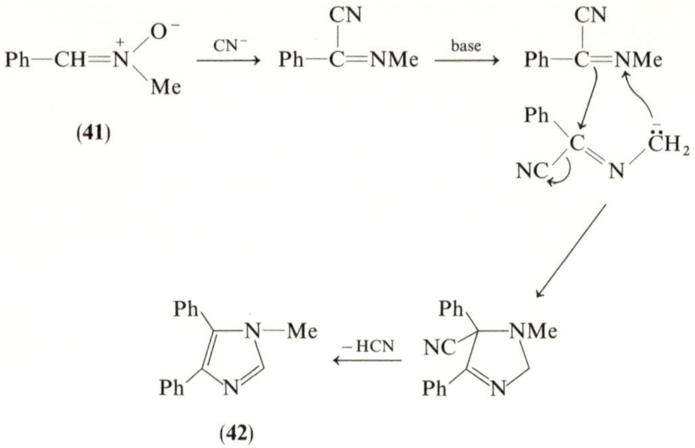

[Scheme showing:]

Ph—CH=N⁺(O⁻)(Me) →(CN⁻) Ph—C(CN)=NMe →(base) Ph—C(CN)=NMe ...
(41)

[leading to structures]

Ph—[imidazole ring]—N—Me with Ph ←(−HCN) Ph—[dihydroimidazole]—NMe with NC, Ph

(42)

H. FROM OTHER HETEROCYCLES

1. *Ring Expansions*

A number of ring expansions of 3- and 4-membered heterocycles have been reported. Among the former is the reaction of 2*H*-azirines (43) or their salts with nitriles, a reaction first reported by Leonard.[103] The reaction is promoted by perchloric acid[103] or boron trifluoride,[104,105] and is doubtless a 1,3-dipolar cycloaddition (Scheme 12) to the azaallyl intermediate cations 44. Photolysis, too, of such azirines (43) generates benzonitrile ylides (45); irradiation of 43 in the presence of ethyl cyanoformate gives a mixture of the oxazoline 46 and the imidazole 47 by 1,3-dipolar cycloaddition to the carbonyl and nitrile groups, respectively.[106] In the absence of a dipolarophile the geometrically isomeric 1,3-diazabicyclo[3.1.0]hex-3-enes (48) result from the addition of the ylide to the ground-state azirine. On further irradiation, 48 are transformed to a mixture of the imidazole 49 and the pyrimidine 50.[107] The photolysis of a formylazirine (which rearranges thermally to a pyrazole) gives a 1-phenylimidazole.[108]

An interesting example of a ring expansion of an azetine, involving a pericyclic mechanism, is the thermolysis of 4-cyano-1-*t*-octyl-3-*t*-octylamino-2-*t*-octyliminoazetine (51) into the imidazole[109] (Eq. 15). Compound 52

---

[103] N. J. Leonard and B. Zwanenberg, *J. Am. Chem. Soc.* **89**, 4456 (1967).
[104] H. Bader and H. J. Hansen, *Chimia* **29**, 264 (1975).
[105] H. Bader and H. J. Hansen, *Helv. Chim. Acta* **61**, 286 (1978).
[106] B. Jackson, H. Märky, H. J. Hansen, and H. Schmidt, *Helv. Chim. Acta* **55**, 919 (1972).
[107] A. Padwa, J. Smolanoff, and S. I. Wetmore, *J. C. S., Chem. Commun.*, 409 (1972).
[108] A. Padwa, J. Smolanoff, and A. Tremper, *Tetrahedron Lett.*, 29 (1974).
[109] L. De Vries, *J. Org. Chem.* **39**, 1707 (1974).

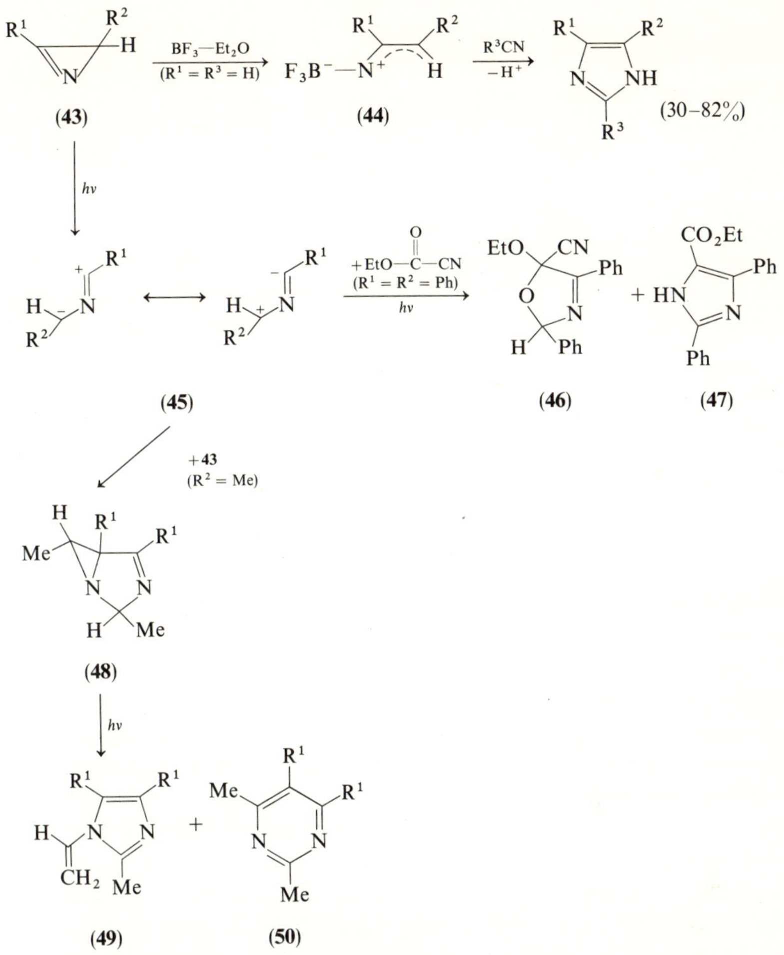

Sec. II.H]   ADVANCES IN IMIDAZOLE CHEMISTRY   261

SCHEME 12

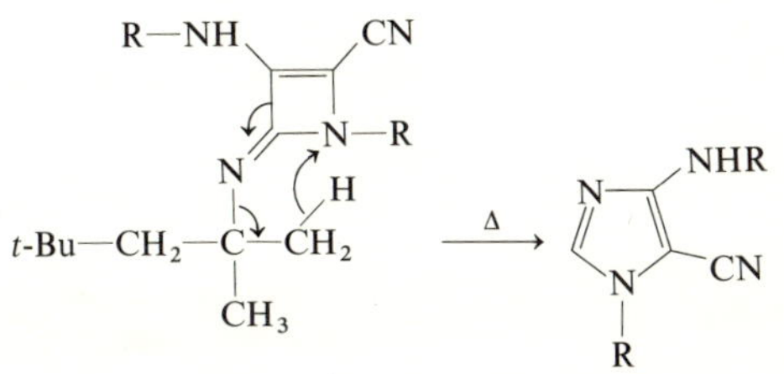

(15)

$$\underset{(52)}{\overset{H\ Ph}{\underset{O}{\overset{|}{C}}}\underset{N-Ph}{\overset{|}{\underset{|}{C}}}} \xrightarrow{-H^+} \left[\begin{array}{c} :C=N \diagdown \diagup Ph \\ \diagup \diagdown Ph \\ O \diagup NPh \end{array} \right] \longrightarrow \underset{Ph}{\overset{Ph\diagdown \diagup Ph}{\underset{O \diagdown N \diagup}{\bigg\langle}}_{N}} \quad (16)$$

(Eq. 16), which contains a β-lactam ring as well as an isocyanide group, rearranges in strongly basic medium, probably via the anion 53, to the imidazolinone.[110] Azirines have already been mentioned (Section II,E) as intermediates in the transformations of enaminonitriles into imidazoles.

2. From Other Five-Membered Heterocycles

Selenium has been found to be an efficient hydrogen acceptor in the dehydrogenation of 2-aryl-2-imidazolines to 2-arylimidazoles.[111] When 2,4,4,5-tetraphenyl-4H-imidazole is heated, it rearranges to the 1,2,4,5-tetraphenyl isomer.[112]

Oxazoles have long provided sources of imidazoles through oxygen–nitrogen exchange. Thus, 1-amino-4,5-dimethylimidazoles are formed when hydrazine is heated with the corresponding oxazoles in the presence of Brønsted acids.[113] There are many further examples[114–119] in which the oxazole is heated with ammonia, formamide, or amines. A novel oxazole-to-imidazole interconversion occurs when the oxazole 54 reacts with the imidoyl chloride 55 and phosphorus oxychloride.[119] The mechanism shown in Scheme 13 may involve the formation of an oxazolium salt which readily undergoes ring opening to a ketenimine (56). Internal nucleophilic attack of the imino function on the sp carbon atom of 56 induces ring closure.[119] Oxazolium salts, too, can be converted into imidazoles,[120,121] and in the transformation of 3-phenyloxazolium salts into 1-phenylimidazoles the remarkable stabilities of the nonaromatic intermediates and the forcing condi-

---

[110] I. Hoppe and U. Schöllkopf, *Chem. Ber.* **109**, 482 (1976).
[111] R. F. Klem, H. F. Skinner, and R. W. Isensee, *J. Heterocycl. Chem.* **7**, 403 (1970).
[112] G. Domany and J. Nyitrai, *Acta Chim. Acad. Sci. Hung* **90**, 109 (1976) [*CA* **86**, 72522 (1977)].
[113] W. Hafner and H. Prigge, German Patent 1,923,643 (1970) [*CA* **74**, 22838 (1971)].
[114] K. Fitzi and R. Pfister, Swiss Patent 554,872 (1974) [*CA* **82**, 43425 (1975)].
[115] K. Fitzi, Swiss Patent 561,718 (1975) [*CA* **83**, 131,598 (1975)].
[116] M. J. S. Dewar and I. J. Turchi, *J. Org. Chem.* **40**, 1521 (1975).
[117] K. Fitzi and R Pfister, U.S. Patent 3,901,908 (1975) [*CA* **83**, 206272 (1975)].
[118] B. Krieg and H. Lautenschläger, *Justus Liebigs Ann. Chem.*, 208 (1976).
[119] R. G. Harrison, M. R. J. Jolley, and J. C. Saunders, *Tetrahedron Lett.*, 293 (1976).
[120] Y. Kikugawa and L. A. Cohen, *Chem. Pharm. Bull.* **24**, 3205 (1976).
[121] G. N. Dorofeenko, L. V. Mezheritskaya, and V. I. Dulenko, *Khim. Geterotsikl. Soedin.*, 287 (1970) [*CA* **76**, 140618 (1972)].

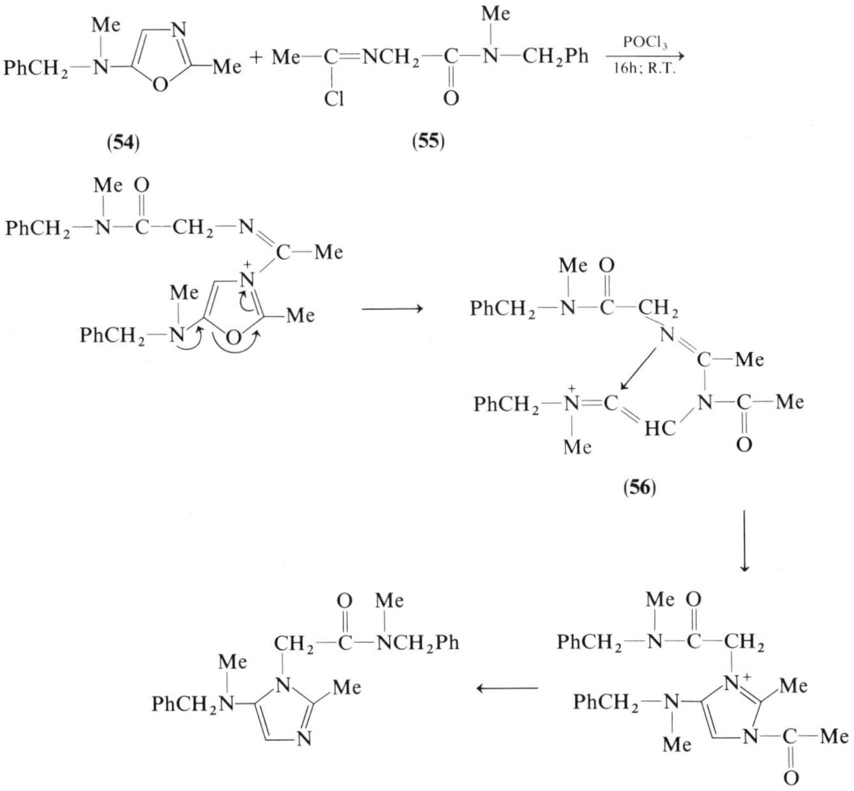

SCHEME 13

tions needed to dehydrate them have been noted.[121] Oxazolinones react with amines to form imidazolinones,[118,122] and an isoxazolin-5-onimine has been photolyzed to an imidazolin-2-one.[123] The ring transformations of oxazoles to imidazoles have been reviewed.[124] Phenyl isocyanate attacks C-2 of 2-substituted 4-methyloxazole N-oxides to cleave the ring and give, after recyclization, 5-hydroxy-4-methylene-1,2,5-trisubstituted 2-imidazolines.[125] This result contradicts an earlier suggestion[126] that the products were imidazole N-oxides. With ethyl α-amino-α-cyanoacetate a Δ²-isoxazoline derivative has been converted into a 1-(D-xylofuranosyl)imidazole.[127]

[122] A. M. Islam, A. M. Khalil, and M. S. El-Houseni, *Aust. J. Chem.* **26**, 1701 (1973).
[123] H. G. Aurich and G. Blinne, *Chem. Ber.* **107**, 31 (1974).
[124] T. Nishiwaki, *Synthesis*, 20 (1975).
[125] T. Goto, N. Honjo, and M. Yamazaki, *Chem. Pharm. Bull.* **18**, 2000 (1970).
[126] J. W. Cornforth and R. H. Cornforth, *J. Chem. Soc.*, 96 (1947).
[127] D. H. Robinson and G. Shaw, *J. C. S. Perkin I*, 774 (1974).

There are a number of examples of photochemically induced interconversions forming imidazoles. On irradiation, 3-methylpyrazole gives a mixture of 2- and 4-methylimidazoles,[128] and 4-aroyl-1-arylpyrazoles yield the diimidazol-4-yl derivative **57**[129] (Eq. 17).

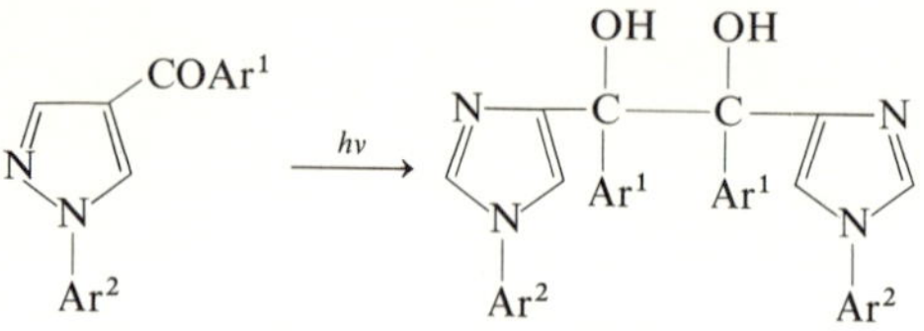

(17)

(57)

The dithiazole **58** is converted into the imidazodithiazole **59** and the imidazolethione **60** by a photochemical reaction of unknown mechanism[130] (Eq. 18). Thermolysis of the β-substituted α-(1-tetrazolyl)acrylamides **61** in the presence of copper has been reported[131] to give 2,4-disubstituted imidazole-5-carboxamides (Eq. 19).

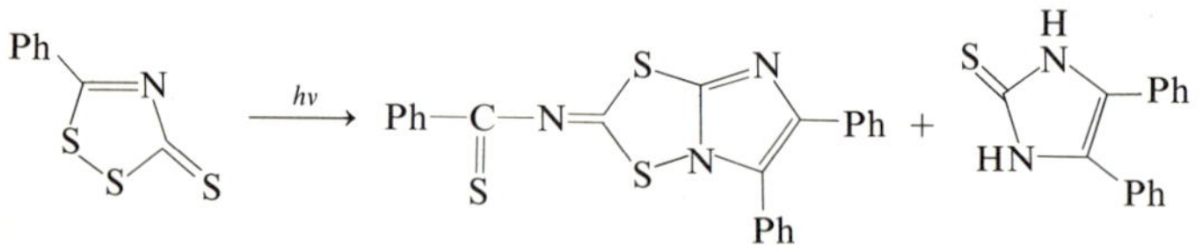

(18)

(58)   (59)   (60)

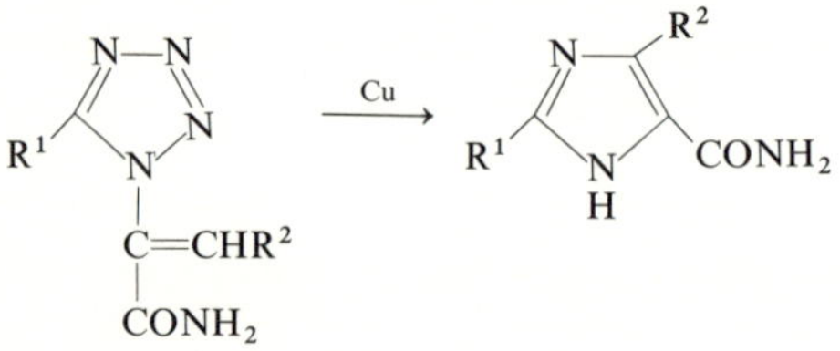

(19)

(61)

3. Ring Contractions

Much literature has appeared describing ring contractions of pyrimidines, pyrazines, and triazines into imidazoles, and the field has been reviewed.[132]

---

[128] R. M. Trimman, *Diss. Abstr. Int.* B **34**, 5917 (1974).
[129] R. Nishiwaki, F. Fujiyama, and E. Minamisono, *J. C. S. Perkin I*, 1871 (1974).
[130] R. Okazaki, F. Ishii, Ka. Okawa, and N. Inamoto, *J. Chem. Soc. Perkin I*, 270 (1975).
[131] J. Lykkeberg and B. Jerslev, *Acta Chem. Scand. Ser.* B **29**, 793 (1975).
[132] H. C. van der Plas, "Ring Transformations of Heterocycles," Vol. 2. Academic Press, New York, 1973.

5-Amino-4-chloro-2-phenylpyrimidine (62) reacts with potassium amide in liquid ammonia to form 4-cyano-2-phenylimidazole (63)[133]; 4-chloro-5-methyl-2-phenylpyrimidine (64) gives 4-ethynyl-2-phenylimidazole (65) under the same conditions.[133] These results are somewhat surprising in

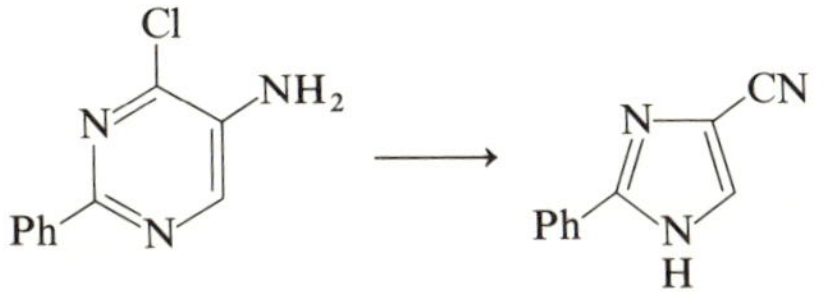

view of the observation that, while a pyridine (3-amino-2-bromopyridine) analogous to 62 rearranges to 3-cyanopyrrole under these reaction conditions, an analog of 64, 2-bromo-3-methylpyridine, does not give a pyrrole derivative.[134] Whereas in the formation of imidazoles 63 and 65 it is possible that either the 4,5- or 5,6-bonds in the pyrimidine could be cleaved, in the pyridine transformations the products can only be accounted for in terms of 2,3-bond fission. Clarification of the situation by ^{14}C labeling[135,136] demonstrated that the 5,6-bond in the pyrimidine must cleave and allowed the tentative mechanism depicted in Scheme 14 to be derived. It is not certain whether ring closure proceeds through the carbenoid species 66 or directly from 67. The conversion of pyrimidines into imidazoles has also been reported to occur during the pyrolysis of 4-azidopyrimidine giving 88% of the imidazole-1-carbonitrile,[137,138] a reaction which is believed to proceed through the intermediacy of a singlet nitrene. Similar work by Wentrup[139,140]

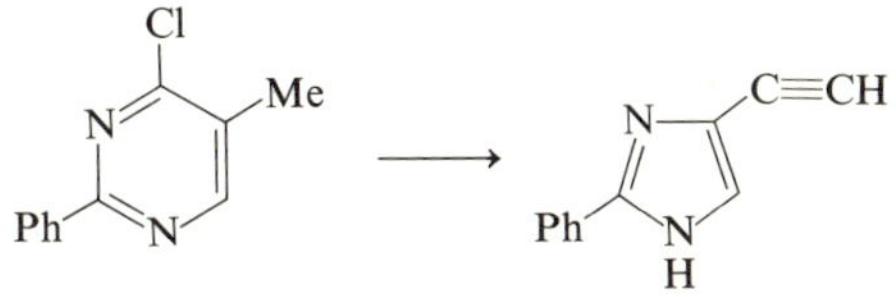

---

[133] H. W. van Meeteren and H. C. van der Plas, *Recl. Trav. Chim. Pays-Bas* **87**, 1089 (1968).
[134] H. J. den Hertog, R. J. Martens, H. C. van der Plas, and J. Bon, *Tetrahedron Lett.*, 4325 (1966).
[135] H. W. van Meeteren, H. C. van der Plas, and D. A. de Bie, *Recl. Trav. Chim. Pays-Bas* **88**, 728 (1969).
[136] H. W. van Meeteren and H. C. van der Plas. *Recl. Trav. Chim. Pays-Bas* **88**, 204 (1969).
[137] W. D. Crow and C. Wentrup, *Chem. Commun.*, 1026 (1968).
[138] W. D. Crow and C. Wentrup, *Chem. Commun.*, 1082 (1968).
[136] C. Wentrup and C. Thétaz, *Helv. Chim. Acta* **59**, 256 (1976).
[140] C. Wentrup, C. Thétaz, and R. Gleiter, *Helv. Chim. Acta* **55**, 2633 (1972).

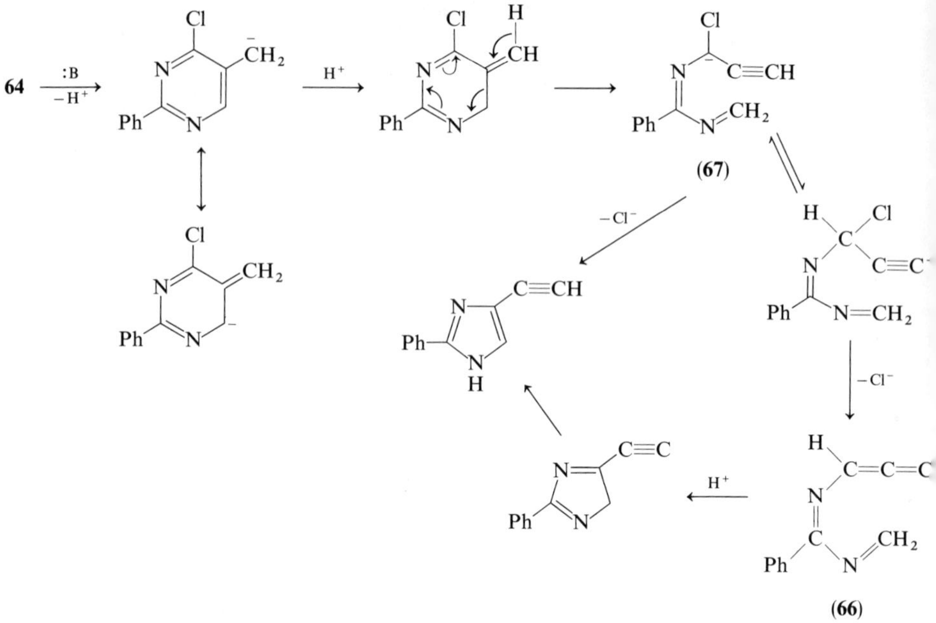

SCHEME 14

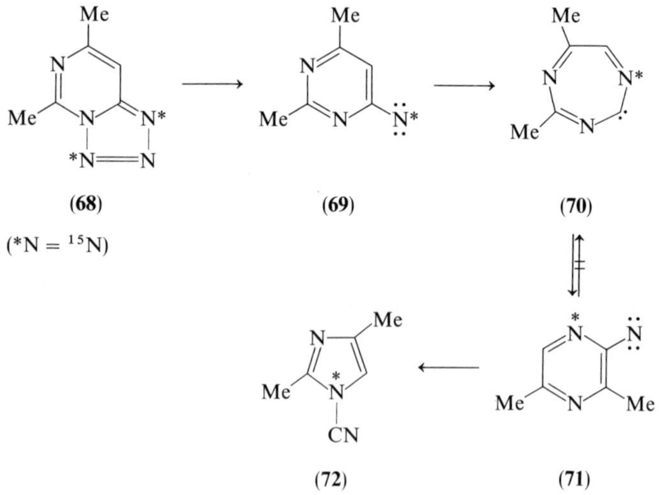

SCHEME 15

on the tetrazolopyrimidine **68** (tautomer of 6-azido-2,4-dimethylpyrimidine) labeled with ^{15}N as indicated by the asterisks in Scheme 15 has demonstrated that only the imidazole ring in the *N*-cyanoimidazole **72** is labeled. Such a result can be explained by the formation of the nitrene **69** which rearranges via the triazepine **70** irreversibly into the pyrazinylnitrene **71** which ring-contracts.[140]

Photochemical transformations of pyrimidines can also give rise to some imidazole products. For example, in benzene or methanol, photolysis of a number of substituted 1-oxides yielded small quantities of imidazoles[141–143] (Eq. 20). These reactions have been explained[143] by involving oxaziridines, 1,2,4-oxadiazepines, or zwitterions as possible intermediates.

(20)

Pyrazine *N*-oxides and 2,3-dihydropyrazines also rearrange photochemically into imidazoles. From 2,5-disubstituted pyrazine 1-oxides (**73**) two products, 2-acyl(or aroyl) 4-substituted imidazole (**74**) and 2,4-disubstituted imidazole (**75**), result.[144] The reaction (Scheme 16) presumably takes place through two isomeric oxaziridine derivatives.

2,3-Dihydropyrazine derivatives, on irradiation, give good yields of imidazoles.[145,146] The thermolysis of 2-azidopyrazine 1-oxide gives 2-cyano-1-hydroxyimidazole.[147,148]

The ring contraction of chloropyrazines to 2-cyanoimidazoles when treated with potassium amide in liquid ammonia has synthetic importance.[149–152] Labeling experiments have proved the origins of the nitrogen atoms in the products, and complex mechanisms have been proposed. It is

---

[141] R. A. F. Deeleman and H. C. van der Plas, *Recl. Trav. Chim. Pays-Bas* **92**, 317 (1973).
[142] F. Bellamy, P. Martz, and J. Streith, *Tetrahedron Lett.*, 3189 (1974).
[143] F. Roeterdink, H. C. van der Plas, and A. Koudijs, *Recl. Trav. Chim. Pays-Bas* **94**, 16 (1975).
[144] N. Ikekawa, Y. Honma, and R. Kenkyusho, *Tetrahedron Lett.*, 1197 (1967).
[145] P. Beak and J. L. Miesel, *J. Am. Chem. Soc.* **89**, 2375 (1967).
[146] T. Matsuura and Y. Ito, *Bull. Chem. Soc. Jpn.* **47**, 1724 (1974).
[147] R. A. Abramovich and B. W. Cue, U.S. Patent, 3,886,180 (1975) [*CA* **83**, 179058 (1975)].
[148] R. A. Abramovich and B. W. Cue, *J. Org. Chem.* **38**, 173 (1973).
[149] P. J. Lont, H. C. van der Plas, and A. Koudijs, *Recl. Trav. Chim. Pays-Bas* **90**, 207 (1971).
[150] P. J. Lont, H. C. van der Plas, and E. Bosma, *Recl. Trav. Chim. Pays-Bas* **91**, 1352 (1972).
[151] P. J. Lont and H. C. van der Plas, *Recl. Trav. Chim. Pays-Bas* **92**, 311 (1973).
[152] P. J. Lont and H. C. van der Plas, *Recl. Trav. Chim. Pays-Bas* **92**, 449 (1973).

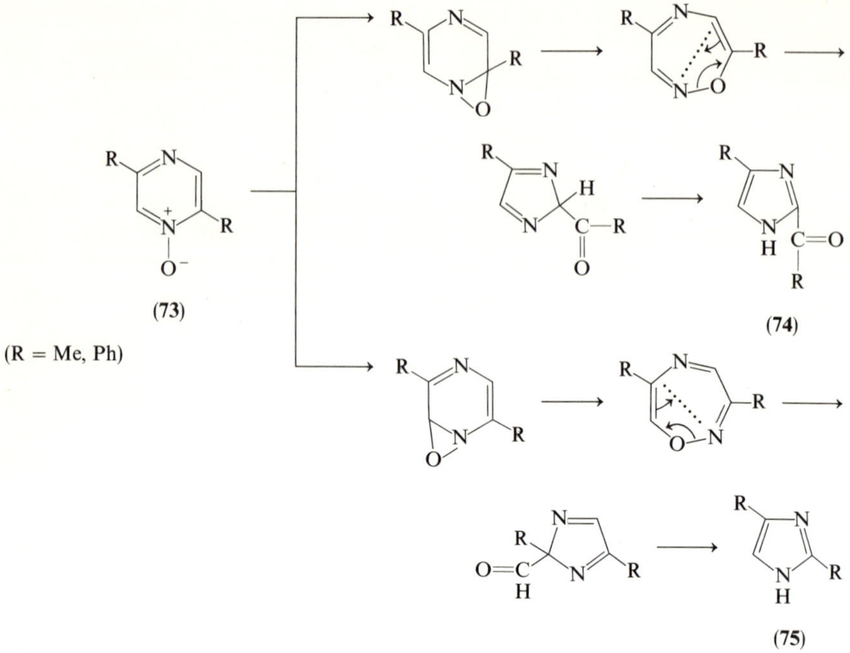

(73)

(R = Me, Ph)

(74)

(75)

SCHEME 16

apparent that the products seldom arise from simple nucleophilic attack at the carbon atom bearing the halogen.

When the dihydro-1,2,4-triazinone **76** is heated at 180°C with red phosphorus and hydroiodic acid, there is rupture of the 1,2-bond and subsequent ring closure to give 4,5-diphenylimidazolin-2-one (**77**)[153] (see Scheme 17); and reductive ring contractions to imidazoles (**81**) have also been reported[154] in the 1,3,5-triazine series. A further interesting ring contraction is that which occurs when 4H-1,3,5-thiadiazines (**78**) are treated with aliphatic amines at room temperature.[155] The triarylimidazoles (**81**) may be formed by initial formation of the 8π 1,3,5-thiadiazine anion (**79**) which probably loses sulfur via a thia-α-homoimidazole (**80**).

Only a few examples of ring contractions of seven-membered heterocycles have been reported. The acid-promoted contraction of 1,5-benzodiazepines into benzimidazoles is quite well known.[156,157] In the reaction of 5-benzyl-

[153] H. Biltz, *Justus Liebigs Ann. Chem.* **339**, 243 (1905).
[154] C. Grundmann and A. Kreutzberger, *J. Am. Chem. Soc.* **77**, 6659 (1955).
[155] G. Giordaro and A. Belli, *Synthesis*, 167 (1975).
[156] D. Lloyd, R. H. McDougall, and D. R. Marshall, *J. Chem. Soc.*, 3785 (1965).
[157] J. A. Barltrop, C. G. Richards, D. M. Russell, and G. Ryback, *J. Chem. Soc.*, 1132 (1959).

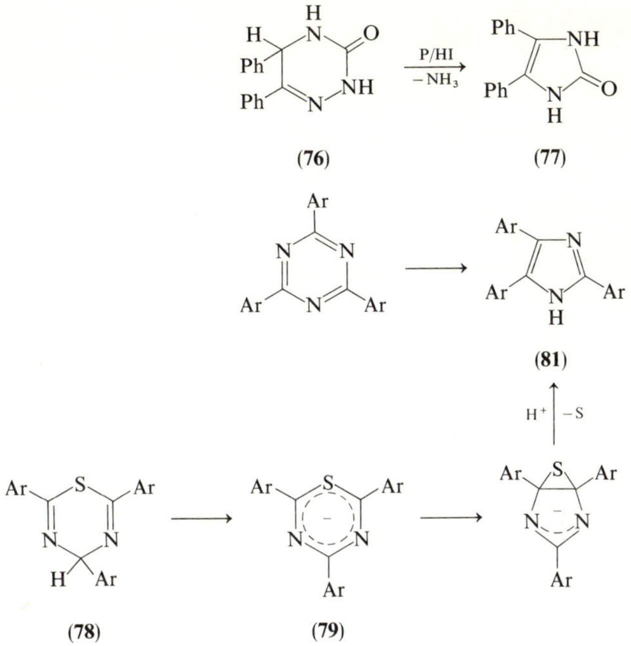

Sec. II.I] ADVANCES IN IMIDAZOLE CHEMISTRY 269

SCHEME 17

3,7-diaryl-4,6-dihydro-1,2,5-triazepines with bromine or N-bromosuccinimide, a halogenation–dehydrohalogenation process leads to imidazoles.[158] Imidazoles also result from the ring opening of such fused bicyclic systems as adenines.[159]

I. MISCELLANEOUS METHODS

There have been reports[160,161] of a novel synthesis of 2,4,5-trialkylimidazoles by the rhodium-catalyzed reactions of alkenes with carbon monoxide and ammonia. Yields are 50–60%. Benzylamine and derivatives react with carbon tetrachloride in the presence of a catalytic amount of metal carbonyls to yield 2,4,5-triarylimidazoles and -imidazolines.[162] The suggested reaction mechanism implicates an initially formed radical species which coordinates with the metal carbonyl.

[158] O. Tsuge and K. Kamata, *Heterocycles* **3**, 547 (1975).
[159] T. Fuji, T. Saito, and M. Kawanishi, *Tetrahedron Lett.*, 5007 (1978).
[160] Y. Iwashita and M. Sakuraba, *J. Org. Chem.* **36**, 3927 (1971).
[161] Y. Iwashita, M. Sakuraba, and A. Nakamura, Japanese Patent 19291 (1973) [*CA* **79**, 78800 (1973)].
[162] Y. Mori and J. Tsuji, *Tetrahedron* **27**, 4039 (1971).

III. Physical Properties

A. Dipole Moments

Values of molar Kerr constants and dipole moments of nitrogen azoles and their complexes with phenols have been obtained.[163-165] These complexes are formed by an intermolecular hydrogen bond between the pyridine-type nitrogen of the azole and the phenolic proton.[164,165] The use of dipole moments in conformational studies has shown that N-aryl- and C-aryl- and N-furyl- and C-furyl imidazoles (and benzimidazoles) are nonplanar, but 1-(α-furyl)-4,5-diphenylimidazoles do have a planar bicyclic fragment.[166] The dipole moments and conformations of azolides (N-acylazoles) have been studied.[167] In the 1-arylimidazoles the dipole is toward the aryl group.[168]

In 4,5-di-t-butylimidazole the molecule is essentially planar, but the C-4—C-5 bond is slightly stretched.[169] Among other imidazole derivatives which have been studied by X-ray are histidine hydrochloride,[170] 4-acetylamino-2-bromo-5-isopropyl-1-methylimidazole,[171] 4-acetyl-5-methyl-2-phenylimidazole,[82] and imidazole-4-acetic acid hydrochloride.[172]

B. Acid and Basic Strength

Comprehensive compilations of pK data for imidazoles have appeared[1,5] and more recent data have been collected in Table I.[168,173-181] The nucleophilic (or basic) nature of the multiply bonded ring nitrogen of imidazole

---

[163] S. B. Bulgarevich, V. S. Bolotnikov, V. N. Sheinker, O. A. Osipov, and A. D. Garnovskii, *J. Org. Chem. USSR* **12**, 191 (1976).
[164] S. B. Bulgarevich, V. N. Sheinker, A. D. Garnovskii, and A. S. Kuzharov, *J. Gen. Chem. USSR (Engl. Transl.)* **44**, 1957 (1974).
[165] S. B. Bulgarevich, V. N. Sheinker, O. A. Osipov, A. D. Garnovskii, R. I. Nikitina, and D. Y. Movshovich, *J. Gen. Chem. USSR (Engl. Transl.)* **44**, 1961 (1974).
[166] Yu. V. Kolodyazhnyi, A. D. Garnovskii, S. V. Serbina, O. A. Osipov, B. S. Tanaseichuk, L. T. Rezepova, and S. V. Yartseva, *Chem. Heterocycl. Compd. (Engl. Transl.)* **6**, 759 (1970).
[167] J. P. Fayet, M. C. Vertut, P. Mauret, R. M. Claramunt, and J. Elguero, *Rev. Roum. Chim.* **22**, 471 (1977).
[168] A. F. Pozharskii, L. M. Sitkina, A. M. Simonov, and T. N. Chegolya, *Chem. Heterocycl. Compd. (Engl. Transl.)* **6**, 194 (1970).
[169] G. J. Visser and A. Vos, *Acta Crystallogr., Sect. B* **27**, 1802 (1971).
[170] I. Bennett, A. G. H. Davidson, M. M. Harding, and I. Morelle, *Acta Crystallogr., Sect. B* **26**, 1722 (1970).
[171] F. P. van Remoortere and F. P. Boer, *J. Chem. Soc. B*, 976 (1971).
[172] G. P. Jones and P. J. Pauling, *J. C. Soc. Perkin II*, 34 (1976).
[173] A. C. M. Paiva, L. Juliano, and P. Boschcov, *J. Am. Chem. Soc.* **98**, 7645 (1976).
[174] T. Takeuchi, K. L. Kirk, and L. A. Cohen, *J. Org. Chem.* **43**, 3570 (1978).

TABLE I
pK_a Values for Imidazoles

Compound	Basic pK_a	Reference	Acid pK_a	Reference
Imidazole	7.25, 7.00	173, 174		
1-Me	7.30, 7.33	173, 175		
2-Me	8.16, 7.85	173, 174	15.10	174
4-Me	7.56	174	15.10	174
1,2-Me$_2$	8.00	174		
1,4-Me$_2$	7.20	174		
1,5-Me$_2$	7.70	174		
1-CH$_2$Ph	6.09	176		
1-Ph	5.10	168, 176*		
4-CF$_3$	2.28	174	10.6	174
4-Ph—2,5-(CF$_3$)$_2$			~8.1	11
1-Me—2-CH=NOH	5.92	177		
2-Br	3.85	178	11.03	178
4-Br	3.88	178	12.32	178
2-F	2.40	179	10.45, ~10.5	179, 180
4-F	2.44	179	11.92	174
2-F—1-Me	2.30	174		
4-F—1-Me	1.90	174		
5-F—1-Me	3.85	174		
2-F—4-Me	3.06	179	10.70	179
2-F—4-CH$_2$CH(NH$_2$)CO$_2$H	1.22	179	10.55	179
4-NO$_2$	−0.16	173		
1-(p-HO—C$_6$H$_4$)—	5.35	168, 176		
1-(m-HO—C$_6$H$_4$)—	5.23	168, 176		
1-(p-Me—C$_6$H$_4$)—	5.24	168, 176		
1-(m-Me—C$_6$H$_4$)—	5.24	168, 176		
1-(p-Br—C$_6$H$_4$)—	4.91	168, 176		
1-(p-COMe—C$_6$H$_4$)—	4.54	168, 176		
1-(p-NO$_2$—C$_6$H$_4$)—	3.96	168, 176		
2,4,5-Ph$_3$ 1-oxide	3.28	181	8.39	181
1-Me—2,4,5-Ph$_3$ 1-oxide	3.32	181		
1-OMe—2,4,5-Ph$_3$ 1-oxide	3.78	181		

---

[175] V. N. Sheinker, L. G. Tishchenko, A. D. Garnovskii, and O. A. Osipov, *Chem. Heterocycl. Compd. (Engl. Transl.)* **11**, 504 (1975).

[176] A. F. Pozharskii, T. N. Chegolya, and A. M. Simonov, *Khim. Geterotsikl. Soedin.* **4**, 503 (1968).

[177] I. N. Somin, N. I. Shapranova, and S. G. Kuznetsov, *Chem. Heterocycl. Compd. (Engl. Transl.)* **9**, 1074 (1973).

[178] G. B. Barlin, *J. Chem. Soc. B*, 641 (1967).

[179] H. J. C. Yeh, K. L. Kirk, L. A. Cohen, and J. S. Cohen, *J. C. S. Perkin II*, 928 (1975).

[180] K. L. Kirk, W. Nagai, and L. A. Cohen, *J. Am. Chem. Soc.* **95**, 8389 (1973).

[181] S. O. Chua, M. J. Cook, and A. R. Katritzky, *J. Chem. Soc. B*, 2350 (1971).

allows it to coordinate with metal ions such as Cd(II), Zn(II), and Mn(II), while coordination between a histidine unit and Fe(II) is involved in hemoglobin. One major consequence of the basicity of imidazole is that at physiological pH (about 7.4) substantial quantities of both the free base and the protonated imidazole species are present in a histidine unit of a protein. This means that histidine can act, according to the demands of its immediate environment, either as a proton acceptor or a proton donor. Histidine units in some enzymes (e.g., ribonuclease, aldolase, and some proteases) may have this role. The buffering action of histidine in the hemoglobin–oxyhemoglobin system is a further consequence. The comment has been made[2] that the imidazole groups of the histidine units in a polypeptide are the strongest bases present in any quantity at physiological pH; furthermore, the imidazolium cations are the strongest acids present in substantial amount, with variations in the pK_a induced by the local environment.

The pK_a values of some fluorinated imidazoles have been determined[174,179,180,182] for studies of relative reactivity in nucleophilic dehalogenation reactions. Although 2-fluoroimidazole dissociates to the anion with $pK = 10.45$,[179] the 4-fluoro isomer shows no sign of such dissociation up to $pH = 11.7$ ($pK = 11.92$[174]). This parallels the ionization constant data for the corresponding bromoimidazoles[178] (see Table II) and is attributed[180] to the more symmetrical anion of the 2-substituted compounds. That the fluoroimidazoles are both weaker bases and stronger acids than their bromo counterparts indicates that the inductive effect of halogen greatly overshadows its resonance effect.[179] Ionization constants have been measured for trifluoromethylimidazoles.[11,174]

The determination of the pK_a of histamine and its N-methyl derivatives has established that the N^τ form (**81**) predominates over the N^π form (**82**) of its cation by 4:1. This is probably due to the stability of the hydrogen-bonded form (**83**).[183] Further correlations [184,185] of pK_a values for 1- and 4(5)-substituted imidazoles by the Hammett equation show that for the

(81)   (82)   (83)

---

[182] T. Takeuchi, H. J. C. Yeh, K. L. Kirk, and L. A. Cohen, *J. Org. Chem.* **43**, 3565 (1978).
[183] G. R. Ganellin, *J. Pharm. Pharmacol.* **25**, 787 (1973).
[184] M. J. Collis and G. R. Edwards, *Chem. Ind. (London)*, 1097 (1971).
[185] N. Blazevic, F. Kajfez, and V. Sunjic, *J. Heterocycl. Chem.* **7**, 227 (1970).

former correlation is best with σ_I (though not significantly better than with σ_m), while for the latter σ_m is best (and σ_I is good). Such Hammett treatments have been critically reviewed recently.[5,186,187] In a study of m- and p-substituted 1-phenylimidazoles Pozharskii[168] has shown 1-phenylimidazole is a weaker base than 1-alkylimidazoles and that the phenyl substituents have their expected effects. The ρ value for correlation between these substituents on the phenyl ring and the ionization constants is $+0.753$, indicating the weakness of electron-transfer effects from the substituent R to the pyridine-type nitrogen in the imidazole ring. Strongly electron-attracting groups (p-NO_2, p-COMe) behave as though their σ values are greater than normal (viz., $+1.25$ and $+0.65$, respectively) i.e., close to the σ^- constants (1.27 and 0.87). This demonstrates the existence of a significant polar conjugation between the π electrons of the heterocycle and the substituent. The ionization constants of aminoimidazoles[188] and imidazole 3-oxides[181] have been studied.

C. ULTRAVIOLET AND VISIBLE SPECTRA

The spectra of nitroimidazoles show features that are altered characteristically by substituents and by pH, and which are useful for the purposes of orientation.[189] Chromatographic, spectrophotometric, and polarographic methods have been used[190] in the simultaneous determination of N-substituted and N-unsubstituted nitroimidazoles. The latter react with hydroxyl ions to form yellow nitroimidazole anions. The difficulty of reduction of these anions, together with the shift in the absorption maximum to longer wavelengths, makes the analysis possible.

A UV-spectroscopic study of 5-bromo-4-nitroimidazole is said to demonstrate that charge is localized on the peripheral heteroatom rather than on the heteroannular nitrogen.[191] Ultraviolet spectroscopy has been applied to studies of 2,4,5-triarylimidazolyl radicals,[192] 1-arylimidazoles,[168] and

[186] J. Elguero, C. Marzin, A. R. Katritzky, and P. Linda, *Adv. Heterocycl. Chem., Suppl.* **1**, 280 (1976).
[187] J. A. Zoltewicz and L. W. Deady, *Adv. Heterocycl. Chem.* **22**, 72 (1978).
[188] G. J. Litchfield and G. Shaw, *J. Chem. Soc. C*, 817 (1971).
[189] M. Hoffer, V. Toome, and A. Brossi, *J. Heterocycl. Chem.* **3**, 454 (1966).
[190] D. Dumanovic, S. Perkucin, and J. Volke, *Talanta* **18**, 675 (1971).
[191] V. P. Zhukov, V. A. Gubanov, V. S. Mokrushin, and A. K. Chirkov, *Ural. Konf. Spektrosk.*, 7th, 63 (1971) [*CA* **77**, 163734 (1972)].
[192] L. A. Cescon, G. R. Coraor, R. Dessauer, E. F. Silversmith, and E. J. Urban, *J. Org. Chem.* **36**, 2262 (1971).

acylated imidazoles.[193,194] Compared with 4(5)-acyl- or -aroylimidazoles, the 2-acyl compounds exhibit a bathochromic shift (20 nm).[193] In m- and p-substituted 2-aroyl-1-alkyl(-aryl or -aralkyl)imidazoles intense absorptions (log ε = 3.5–4.3) appear in the 295–309 nm regions, and a slightly less intense peak in the 233–266 nm region is also usually present.[194] The UV spectra of 2-alkylthioimidazole quaternary salts and 1,3-dialkylimidazole-2-thiones have been studied.[195,196]

D. Infrared Spectra

A study of the IR spectra of imidazole and 1-methylimidazole shows that the imidazole spectrum is most sensitive to temperature change between 100 and $-160°$C. This involves a shift in the band at 918 cm^{-1} corresponding to the nonplanar deformation vibrations of N—H and may be accounted for by a decrease in the N—H bond length with decrease in temperature causing an increase in bond strength. The observation that the band at 618 cm^{-1} is unaffected by temperature suggests that it has been incorrectly assigned[197,198] to a NH vibration.[199] A new matrix-isolation device developed to study hydrogen-bonded, high-boiling organic compounds allows the isolation of the monomeric species in an inert gas and has been applied to a study of imidazole, allowing some new band assignments.[200] Further studies of 4-methylimidazole[201] and salts,[202] and salts of imidazole and 1,3-dideuteroimidazole[203] have been carried out. The spectra of some 1,5-disubstituted 4-tosylimidazoles show bands in the regions 3120–3110 and 1620–1600 cm^{-1}.[94]

---

[193] S. Iwasaki, *Helv. Chim. Acta* **61**, 2843 (1978).
[194] M. R. Grimmett and K. H. R. Lim, unpublished; K. H. R. Lim, Ph.D. thesis, University of Otago, New Zealand, 1979.
[195] G. Assef, D. Bouin-Roubaud, J. Kister, and J. Metzger, *C. R. Hebd. Seances Acad. Sci., Ser. C* **283**, 143 (1976).
[196] J. Kister, D. Bouin-Roubaud, P. Hassanaly, H. Dou, and J. Metzger, *C. R. Hebd. Seances Acad. Sci., Ser. C* **287**, 201 (1978).
[197] M. Cordes and J. L. Walter, *Spectrochim. Acta, Part A* **24**, 24 (1968).
[198] M. Cordes and J. L. Walter, *Spectrochim. Acta, Part A* **24**, 237 (1968).
[199] Yu. D. Kanaskova, B. I. Sukhorukov, Yu. A. Pentin, and G. V. Komarovskaya, *Bull. Acad. Sci. USSR*, 1637 (1970).
[200] S. T. King, *J. Phys. Chem.* **74**, 2133 (1970).
[201] H. Wolff and E. Wolff, *Spectrochim. Acta, Part A* **27**, 2109 (1971).
[202] A.-M. Bellocq and C. Garrigou-Lagrange, *J. Chim. Phys.* **67**, 951 (1970).
[203] A.-M. Bellocq and C. Garrigou-Lagrange, *Spectrochim. Acta, Part A* **27**, 1091 (1971).

E. NMR Spectra

A detailed study of the spectra of 1-arylazoles and the corresponding compounds nitrated in the benzene ring has been applied to determination of the conformations of these molecules.[204,205]

In alkylimidazoles, ring protons adjacent to an N-methyl group can be differentiated from other ring protons by a characteristic shift in δ with variation in solvent[182]; in addition, H-5 appears at higher field than H-4 in nonpolar solvents. The other is reversed for polar solvents (see Table II). The orientation of N-methylation in 2-methyl-4(5)-phenylimidazole has been determined with lanthanide-shift reagents. For steric reasons the 1,2-dimethyl-5-phenyl isomer coordinates with the shift reagent while the 4-phenyl compound does not.[206]

TABLE II
NMR SOLVENT SHIFTS ($\Delta\delta$) FOR 1-METHYLIMIDAZOLE[182]

Position	δ (ppm)			$\Delta\delta$ (Hz)[a]	
	CDCl$_3$	Me$_2$SO$_4$-d_6	D$_2$O	Δ_1	Δ_2
H-2	7.41	7.55	7.57	−0.14	−0.16
H-4	7.03	6.88	6.99	+0.15	+0.04
H-5	6.87	7.08	7.07	−0.21	−0.20

[a] $\Delta_1 = \delta_{Me_2SO_4-d_6} - \delta_{CDCl_3}$; $\Delta_2 = \delta_{D_2O} - \delta_{CDCl_3}$.

A reinvestigation[207] of the spectra of simple quaternary salts of 1-methylimidazole shows that the process of quaternization exerts a pronounced deshielding effect on the ring protons (H-4, $\Delta\delta$ = +0.95 Hz; H-2, $\Delta\delta$ = +1.45 Hz). The spectrum of 1-methylimidazole methiodide showed a symmetrical structure with the equivalent 4- and 5-protons coupled (J = 1.8 Hz) to H-2, and the methyl signals appearing as a singlet.[207] Quaternization of imidazoles makes the proton at C-2 so acidic that it is often not visible in a D$_2$O spectrum of the quaternary salt[207,208]; in addition, the C-2 signal appears at very low field in the ^{13}C-NMR spectrum.[208]

Examination of coupling constants has allowed differentiation of 1,4- and 1,5-disubstituted imidazoles[209] and, in particular, suggests that when very

---

[204] J. Elguero, R. Jacquier, and G. Tarrago, *Bull. Soc. Chim. Fr.*, 1345 (1970).
[205] J. Elguero, R. Jacquier, and S. Mondon, *Bull. Soc. Chim. Fr.*, 1346 (1970).
[206] E. E. Glover and D. J. Pointer, *Chem. Ind. (London)*, 412 (1976).
[207] R. F. Borne, H. Y. Aboul-Enein, and J. K. Baker, *Spectrochim. Acta, Part A* **28**, 393 (1972).
[208] B. K. M. Chan, N. H. Chang, and M. R. Grimmett, *Aust. J. Chem.* **30**, 2005 (1977).
[209] H. R. Matthews and H. Rapoport, *J. Am. Chem. Soc.* **95**, 2297 (1973).

electronegative substituents are present, as in 4-nitroimidazoles, the H-2 signal does not necessarily appear at lowest field. This observation may necessitate reassignment of signals for some such compounds.

The concentration dependence of chemical shifts of NH protons of imidazole in various solvents has been studied.[210] In organic solvents the position and half-width of the NH signal depends on the concentration, and with increase in concentration the line shifts downfield. This shift ($\Delta = 1$ Hz for solution of imidazole in dimethyl sulfoxide) may be due to the formation of an ordinary hydrogen bond between imidazole molecules.[210]

Reports have appeared detailing spectra of complexes of organohalostannanes with 1-vinyl- and 1-ethylimidazoles,[211] of imidazole oximes,[212] and of fluoroimidazoles.[213]

The increasing interest in ^{13}C-NMR spectroscopy has resulted in a number of such studies in the imidazole field. Carbon-13 chemical shifts in a variety of solvents have been reported for imidazoles,[214–216] 4-methylimidazole and its cation,[217] phenylimidazoles,[218] 1-acetylimidazole,[219] and Fe(III) porphyrin–imidazole complexes.[220] The influence of Mn(II) and Cu(II) on ^{13}C nuclear relaxation rates in imidazole has been studied; ^{13}C-2 appears at lowest field (136.2), while ^{13}C-4 and ^{13}C-5 are at 122.3 Hz.[221] While ^{13}C-NMR studies have been of some value in conformational studies,[218,219] the shifts are of limited value only in ascertaining positions of tautomeric equilibrium for rapidly interconverting tautomers.[222]

Studies of NMR spectra involving ^{14}N,[223] ^{15}N,[224] and ^{19}F[213] nuclei have also appeared. In one of these[224] it has been shown that the ^{15}N shifts

[210] Yu. A. Teterin, A. G. Kiselev, and L. N. Nikolenko, *Chem. Heterocycl. Compd.* (*Engl. Transl.* **6**, 750 (1970).
[211] V. K. Voronov, Yu. N. Ivlev, E. S. Domnina, M. G. Voronkov, G. G. Skvortsova, and R. G. Mirskov, *Chem. Heterocycl. Compd.* (*Engl. Transl.*) **9**, 362 (1973).
[212] N. I. Shapranova, I. N. Somin, and S. G. Kuznetsov, *Chem. Heterocycl. Compd.* (*Engl. Transl.*) **9**, 1018 (1973).
[213] F. Fabra, C. Galvez, A. Gonzalez, P. Viladoms, and J. Vilarrasa, *J. Heterocycl. Chem.* **15**, 1227 (1978).
[214] M. C. Thorpe and W. C. Coburn, *J. Magn. Res.* **12**, 225 (1973).
[215] R. J. Pugmire and D. M. Grant, *J. Am. Chem. Soc.* **90**, 4232 (1968).
[216] F. J. Weigert and J. D. Roberts, *J. Am. Chem. Soc.* **90**, 3543 (1968).
[217] P. Haake, L. P. Bausher, and W. B. Miller, *J. Am. Chem. Soc.* **91**, 1113 (1969).
[218] M. Begtrup, *Acta Chem. Scand.* **27**, 3101 (1973).
[219] M. Begtrup, R. M. Claramunt, and J. Elguero, *J. C. S. Perkin II*, 99 (1978).
[220] H. Goff, *Chem. Commun.*, 777 (1978).
[221] R. E. Wasylishen and H. R. Graham, *Can. J. Chem.* **54**, 617 (1976).
[222] J. Elguero, C. Marzin, and J. D. Roberts, *J. Org. Chem.* **39**, 357 (1973).
[223] M. Witanowski, L. Stefaniak, H. Januszewski, Z. Grabowski, and G. A. Webb, *Tetrahedron* **28**, 637 (1972).
[224] M. Alei and W. E. Wageman, *Tetrahedron Lett.*, 667 (1979).

for imidazole and 1-methylimidazole in methylene chloride relative to aqueous solution are primarily due to strong hydrogen bonding between ^{15}N-3 of imidazole and water protons in solution.

F. ESR Spectra

Progress has been made in the interpretation of the ESR spectrum of 2,4,5-triphenylimidazolyl radicals by examining the spectra of the partially and completely deuterated species and by the application of simple molecular orbital (MO) calculations. It is evident that the radicals are not planar.[225] Some ambiguity is still present since INDO calculations indicate that the imidazyl radical should be close to a σ-radical, while ESR shows that the triphenylimidazyl radical is of the π type.[226] The results of ESR studies of imidazole anion-radicals have been thought to demonstrate that they exist in the tautomeric α-pyrrolenine form (**84**) rather than **85**.[227] However, a

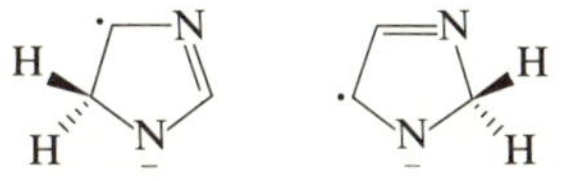

more recent CNDO study[228] has criticized this conclusion. The spectra of 1-hydroxy-2,4,5-triphenylimidazolyl and the corresponding 3-oxide radicals have been recorded and interpreted.[229]

G. Mass Spectrometry

The relative losses of unlabeled vs. labeled HCN from the $M^{+\cdot}$ and $[M - 1]^+$ ions of a number of specifically labeled 1-methylimidazoles (and 1-methylpyrazoles) show that hydrogen randomization in the molecular ions prior to fragmentation is insignificant.[230] The expulsion of HCN can follow two distinct pathways: elimination involving positions 2 and 3 (predominant in 1-methylimidazole), and elimination involving N(1) and its attached methyl group (predominant in 1-methylpyrazole). The molecular

---

[225] N. Cyr, M. A. J. Wilks, and M. R. Willis, *J. Chem. Soc. B*, 404 (1971).
[226] E. M. Evleth, R. M. Horowitz, and T. S. Lee, *J. Am. Chem. Soc.* **95**, 7948 (1973).
[227] P. H. Kasai and D. McLeod, *J. Am. Chem. Soc.* **95**, 27 (1973).
[228] E. Westhoff and W. Flossmann, *J. Am. Chem. Soc.* **97**, 6622 (1975).
[229] K. Volkamer and H. Zimmermann, *Chem. Ber.* **103**, 296 (1970).
[230] J. van Thuijl, K. J. Klebe, and J. J. van Houte, *Org. Mass Spectrom* **7**, 1165 (1973).

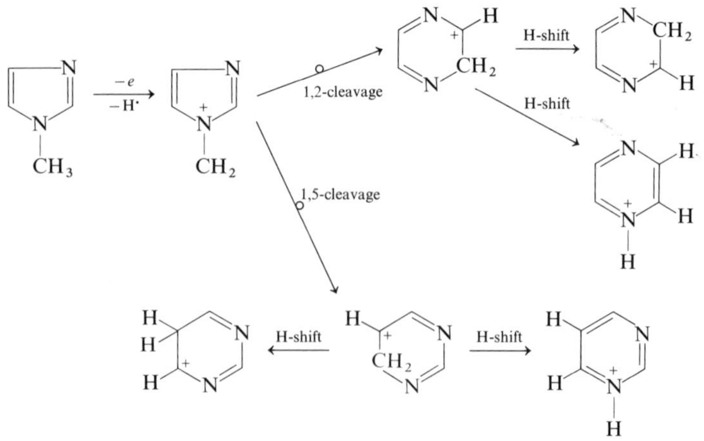

SCHEME 18

ions eject hydrogen radicals from the methyl groups specifically, although some minor contribution from position 5 is noted. The resultant $[M - 1]^+$ ions exhibit extensive, but incomplete, hydrogen randomization.[230] Loss of HCN from these ions is consistent with the intermediacy of ring-expanded ions[230] (see Scheme 18). From a similar study of the mass spectra of deuterated imidazoles, it has been shown that the loss of H˙ is specifically from the 4(5)-position, and that HCN loss is less specific, coming in order of preference from positions 2 > 4(5) > 1.[231]

The mass spectrum of 1-methyl-4-nitroimidazole-5-carboxylic acid amide shows the elimination of water by an electron impact as distinct from the usual thermal process.[232] In 1,4,5-trimethylimidazole-2-carboxylic acid the principal loss is that of CO_2. Subsequent loss of O˙, ˙OH, and HCN was also observed (Schofield et al.,[5] p. 173). Ferguson and Schofield[5] (pp. 172, 173) have reported considerable data for imidazole N-oxides. Thus 1-methoxy-4,5-dimethyl-2-phenylimidazole (which loses ˙OCH_3 and CH_3CN from the fragment m/e 171) is readily distinguished from 1,4,5-trimethyl-2-phenylimidazole 3-oxide (loss of ˙O and ˙OH). The spectra of 1-benzylimidazoles are dominated by the tropylium ion.[233]

2-Formyl-1,4,5-trimethyl- and 2-formyl-1,5-dimethylimidazole 3-oxides give very characteristic patterns: loss of CO occurs (as it does from 2-formyl-1-methylimidazole[234]). There appear to be two principal fragmentation schemes, one involving loss of the N-oxide group and then the formyl group, the other involving the alternative fragmentation order. The loss of ˙H, which

---

[231] K. J. Klebe, J. J. van Houte, and J. van Thuijl, Org. Mass Spectrom 1363 (1972).
[232] G. H. Lord, B. J. Millard, and J. Memel, J. C. S. Perkin I, 572 (1973).
[233] C. G. Begg, M. R. Grimmett, and P. D. Wethey, Aust. J. Chem. **26**, 2435 (1973).
[234] J. H. Bowie, R. G. Cooks, S. O. Lawesson, and G. Schroll, Aust. J. Chem. **20**, 1613 (1967).

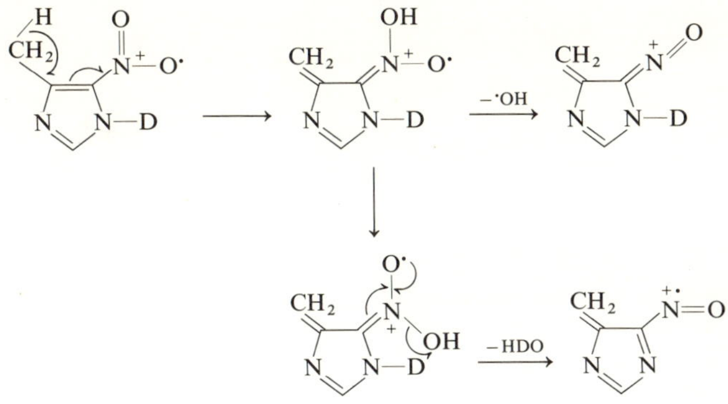

SCHEME 19

is observed with 2-formyl-1-methylimidazole, is not evident in the N-oxides, presumably because the hydrogen atom is lost in the ˙OH fragment. Fragment ions arising from loss of ˙O, ˙OH, ˙NO, and ˙NO$_2$ appear in the mass spectrum of 1,4,5-trimethyl-4-nitroimidazole; the spectrum, however, is complicated by loss of CHO, a previously unreported observation for nitroazoles (Schofield et al.,[5] pp. 172, 173), although present in some other nitroaromatics.

On electron impact, the 4-methyl-5-nitroimidazole molecular ion loses H$_2$O. This loss seems to require an NH adjacent to the nitro group; i.e., loss of H$_2$O comes from what is expected to be the least dominant tautomer[235] (Scheme 19). In 1-alkyl-5-nitroimidazoles (86) a similar hydrogen transfer is followed by an unusual rearrangement which results in loss of an aldehyde or ketone fragment from the molecular ion[236] (Scheme 20).

Molecular ions of 1-(2-hydroxy-2-phenylethyl) imidazoles eliminate benzaldehyde[237]. The parent ions of trans-1-styrylimidazoles cyclize in a manner similar to their photochemical cyclization[237].

Ketene elimination from azolides on electron impact gives rise to "molecular ions" of the corresponding unsubstituted azoles.[238] N-Acetylimidazoles lose CO, but the [M − CO]$^{+\cdot}$ ions do not have N- or C-methylazole structures. In a study of 2-aroylazoles Lim[194] has shown that a hydrogen radical is commonly lost from the molecular ion; CO loss from both the M$^{+\cdot}$ and [M − 1]$^{+}$ species is characteristic; p-methoxybenzoyl derivatives lose either formaldehyde or ˙CH$_3$ followed by CO from the [M − 29]$^{+}$ fragment.

[235] G. H. Lord and B. J. Millard, Org. Mass Spectrom. 2, 547 (1969).
[236] G. E. van Lear, Org. Mass Spectrom. 6, 1117 (1972).
[237] G. Cooper and W. J. Irwin, Org. Mass Spectrom. 10, 885 (1975).
[238] A. Maquestiau, Y. van Haverbeke, R. Flammang, R. M. Claramunt, and J. Elguero, Bull. Soc. Chim. Fr., 2693 (1975).

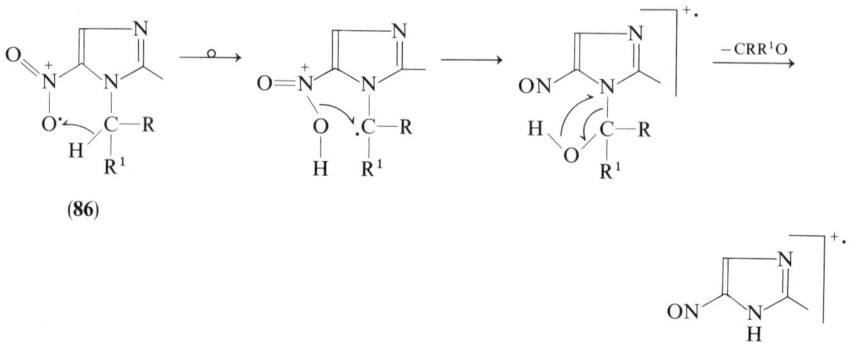

SCHEME 20

In the mass spectra of a number of methylated histamine derivatives the base peaks result from β-cleavage, usually with concomitant hydrogen rearrangement producing ions of mass $[M - CHR=NR]^+$, where R = H or CH_3.[239]

H. QUANTUM-MECHANICAL STUDIES

In the application of theoretical studies to the azole field many of these have attempted to achieve comparisons within the range of azole molecules. Thus, calculations of electron densities, dipole moments, and energies of formation give values that reflect the decrease in azole stability as the number of nitrogen atoms increase.[240] Good correlations between σ and total electron densities and ^1H and ^{13}C chemical shifts have been obtained.[241] Calculations (SCF) of π-electron distributions for the ground state of imidazole do not take into account the tautomeric equivalence of the 4- and 5-positions, but predict the order of electrophilic substitution as 5 > 2 > 4.[242,243] Various other quantum-mechanical calculations[243–247] have

[239] J. S. Dawborne, *Org. Mass Spectrom.* **6**, 211 (1972).
[240] M. Roche and L. Pujol, *J. Chim. Phys.* **68**, 465 (1971).
[241] G. Berthier, L. Praud, and J. Serre, *Jerusalem Symp. Quantum Chem. Biochem.* **2**, 40 (1970).
[242] M. Kamiya, *Bull. Chem. Soc. Jpn.* **43**, 3344 (1970).
[243] D. J. Evans, H. F. Thimm, and B. A. W. Coller, *J. C. S. Perkin II*, 865 (1978).
[244] A. D. Garnovskii, A. M. Simonov, and V. I. Minkin, *Chem. Heterocycl. Compd. (Engl. Transl.)* **9**, 88 (1973).
[245] B. Bartman, E. C. Gordon, M. Gonzalez-Kutas, D. S. Noyce, and B. B. Sandel, *J. Org. Chem.* **41**, 776 (1976).
[246] P. V. Schatnev, M. S. Shvartsberg, and I. Ya. Bernshtein, *Chem. Heterocycl. Compd. (Engl. Transl.)* **11**, 718 (1975).
[247] N. N. Zatsepina, I. F. Tupitsyn, A. A. Kane, and G. N. Sudakova, *Chem. Heterocycl. Compd. (Engl. Transl.)* **13**, 959 (1977).

been applied to reactions of imidazoles. The nucleophilic substitution reactions at C-2 in benzimidazoles and diazo coupling at C-2 in simple imidazoles have been discussed from the theoretical points of view.[244,248]

Semiempirical LCAO calculations for azoles, introducing σ-electrons, indicate that the charges are weak except for those on NH nitrogen atoms and that the σ-dipolar moments are close to those of lone pairs. It was therefore concluded that it is inappropriate to take σ-polarity into account in the approximations used in π-calculations for these heterocycles.[249] Applications of molecular orbital treatments to studies of protonation of imidazoles,[250,251] hydrogen bonding,[252,253] tautomerism,[251,253] isomerism, and conformation[254] have been made. In a comparison of the proton-donor and -acceptor properties of imidazole and pyrrole ab $initio$ SCF calculations have shown that the former is a stronger proton donor to water than pyrrole, and as proton acceptors the two are quite different. While imidazole forms a strong hydrogen bond with water through the N-σ lone pair of electrons, pyrrole forms a relatively weak and flexible π-complex.[252] Good agreement with experiment is found in theoretical studies of isomerism and conformations of azolides.[254]

Calculation of the heats of combustion for imidazoles suggest that, in substituent–nucleus tautomerism, the tautomer with the mobile proton on nitrogen should be more stable than that with it on carbon, and that the amino forms of amines, and the carbonyl forms of hydroxy compounds, are preferred.[255]

I. Hammett Studies

Attempts have been made with some success to apply Hammett treatments to reactions and reactivity of imidazoles. Complications due to the natures of the ring nitrogens and tautomerism, however, have created difficulties.

The use of σ_m and σ_o values of -0.34 and 0.42, respectively, for the ring NH group (replacement of a ring CH=CH), in conjunction with the Hammett relationship for pyridines, permits the prediction with reasonable

---

[248] Yu. B. Vysotskii and O. P. Shvaika, *Dopov. Akad. Nauk. Ukr. RSR, Ser. B* **9**, 804 (1975) [*CA* **84**, 30164 (1976)]
[249] M. Roche and L. Pujol, *Bull. Soc. Chim. Fr.*, 273 (1970).
[250] P. E. Grebow and T. N. Hooker, *Biopolymers* **14**, 871 (1975).
[251] S. Kang and D. Chou, *Chem. Phys. Lett.* **34**, 537 (1975).
[252] J. E. Del Bene and I. Cohen, *J. Am. Chem. Soc.* **100**, 5285 (1978).
[253] P. Mauret, J. P. Fayet, and M. Fabre, *Bull. Soc. Chim. Fr.*, 1675 (1975).
[254] H. Sauvaitre, J. Teysseyre, and J. Elguero, *Bull. Soc. Chim. Fr.*, 635 (1976).
[255] N. Bodor, M. J. S. Dewar, and A. J. Harget, *J. Am. Chem. Soc.* **92**, 2929 (1970).

accuracy of the pK_a values for imidazoles and pyrazoles.[256] Correlation of pK_a values for 1-, 2-, and 4-substituted imidazoles by the Hammett equation using σ_m and σ_I values has proved successful[184,185,257] (see Section III,B). Hammett treatments have been applied with some success to quaternization studies[168,187] and ring-proton exchange reactions[174] of imidazoles. Calculated σ constants for the imidazole ring in benzimidazoles[258] suggest that it is electron-releasing—a prediction which is borne out by experiment. Taft σ^* values for alkyl groups have been calculated from rate constants for N-acylimidazole hydrolysis and imidazole-catalyzed N-acylimidazole hydrolysis in water.[259] It was found that the steric effect was the same in both acid- and base-catalyzed reactions and that the range of σ^*_{rel} values (where $\sigma^*_{rel} = \sigma^*_{alkyl}/\sigma^*_{ethyl}$) for primary, secondary, and tertiary alkyl groups shows no dependence on the degree of branching. The corrected or "true" σ^*_{alkyl} values calculated (from the rate constants for C-substituted ester hydrolysis) show that in this case steric effects are not the same in the acid- and base-catalyzed reactions. Thus the Taft σ^* constants are probably a measure of a steric effect (the reactions differ in sensitivity to branching at the α- and β-carbon atoms of an alkyl group).[259] Applications of Hammett treatments to tautomerism studies are discussed in Section IV,A.

IV. Chemical Properties

A. Structure and Tautomerism

Studies of crystal structure have been made for imidazole[169,260,261]; the results from a low-temperature study[169] are illustrated in Fig. 1. In 4,5-di-t-butylimidazole the ring dimensions are essentially similar, although the 4,5-bond is slightly stretched.[169] Among derivatives of imidazole which have been examined are histidine hydrochloride,[170] 4-acetylamino-2-bromo-5-isopropyl-1-methylimidazole,[171] 4-acetyl-5-methyl-2-phenylimidazole,[82] imidazole-4-acetic acid hydrochloride,[172] imidazole imidazolium

---

[256] D. D. Perrin, *J. Chem. Soc.*, 5590 (1965).
[257] M. Charton, *J. Org. Chem.* **30**, 3346 (1965).
[258] L. I. Rudaya, I. Ya. Kvitko, and B. A. Porai-Koshits, *J. Org. Chem. USSR (Engl. Trans.)* **8**, 11 (1972).
[259] M. Charton, *J. Org. Chem.* **44**, 903 (1979).
[260] Yu. A. Omel'chenko and Yu. D. Kondrashev, *Kristallografiya* **16**, 115 (1971) [*CA* **74**, 147539 (1971)].
[261] G. Will, *Z. Krystallogr.* **129**, 211 (1969).

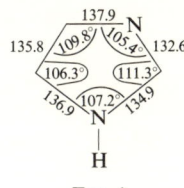

FIG. 1

perchlorate,[262] and histamine phosphate.[263,264] In the crystalline state histamine phosphate exists entirely as the *trans*-rotamer,[263,264] but in D_2O solution the univalent cation has approximately equal proportions of the *trans*- and *gauche*-rotamers.[265] The crystal structure and ^1H NMR spectrum of 1,3-dimethyl-2(3H)-imidazolethione has been interpreted[266] as showing partial double bond character in the N—C—N system, but no aromaticity; structure **87** is preferred to **88** or **89**. This suggestion should be treated with great caution.

Since 1970 there have been a number of studies of imidazole tautomerism, and aspects have been reviewed.[5,186,267] In fact imidazoles provide one of the best-studied examples of annular tautomerism of the type shown in Eq. (21), and K_T values have been calculated for a variety of 4(5)-substituted imidazoles and the results summarized.[186] Charton's application of the Hammett equation to heteroaromatic tautomerism[257,268] has

(21)

---

[262] A. Quick and D. J. Williams, *Can. J. Chem.* **54**, 2465 (1976).
[263] M. V. Veidis and G. J. Palenik, *Chem. Commun.*, 196 (1969).
[264] M. V. Veidis, G. J. Palenik, G. J. Schaffrin, and J. Trotter, *J. Chem. Soc. A*, 2659 (1969).
[265] A. F. Casy and R. R. Ison, *Chem. Commun.*, 1343 (1970).
[266] G. B. Ansell, D. M. Forkey, and D. W. Moore, *Chem. Commun.*, 56 (1970).
[267] B. I. Khristich, *Khim. Geterotsikl. Soedin.*, 1683 (1970).
[268] M. Charton, *J. Chem. Soc. B*, 1240 (1969).

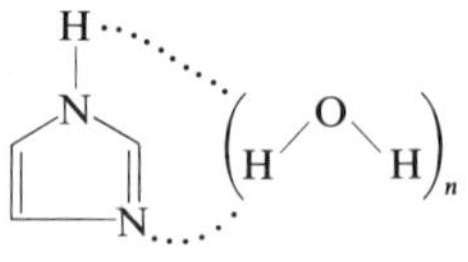

FIG. 2

suggested qualitatively that electron-donating groups ($\sigma_m < 0$) will favor the 5-substituted isomer, and electron-withdrawing groups ($\sigma_m > 0$) will favor the 4-substituted tautomer. It is possible to predict the position of the tautomeric equilibrium for any unsymmetrically substituted imidazole provided that $\Delta\rho$ is known) using the relationship log $K_T = pK_B - pK_A = (\rho_B - \rho_A)\sigma_m$ (where A and B refer to the two tautomers). More recently Blazevic[185] attempted to correlate the pK_a values of imidazoles with the sum of the Hammett coefficients, $\sum \sigma^*$; the σ values of the substituents are expressed as $\frac{1}{2}(\sigma_o + \sigma_m)$, but this treatment makes use of the inadmissable assumption that $K_T = 1$. The correlation of σ_m values with pK_a was used[269] to calculate K_T values for some 4-substituted histamine cations; while a methyl substituent makes little difference, electron-withdrawing groups cause the 1,4-tautomer to predominate.

The rate equation for the tautomerism of imidazole contains kinetic terms corresponding to catalysis by H^+, OH^-, and $C_3N_2H_5^+$. There is also an uncatalyzed term, and recent work has indicated that this is unexpectedly large; the suggested interpretation[270] involves the bifunctional participation of water as shown in Figure 2. Although prototropy between N-1 and N-3 of imidazole is normally very rapid, a decrease in the rate, such that the two forms can be demonstrated, is found with the benzimidazoles **90**. This retarding effect is ascribed to intramolecular hydrogen bonding occurring with the exocyclic carbonyl oxygen as indicated, rather than to ring–chain tautomerism, because of the low energy barrier to rotation and the strong carbonyl band in the infrared spectrum[271]. It has

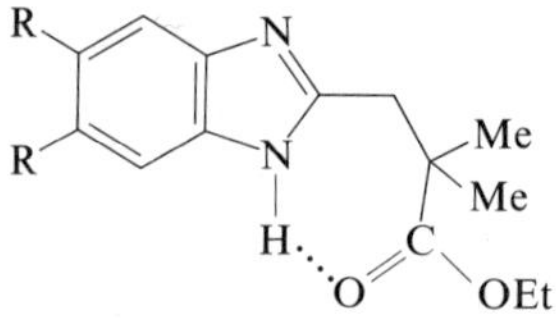

(90)

[269] C. R. Ganellin, *Jerusalem Symp. Quantum Chem. Biochem.* **7**, 43 (1974).
[270] K.-C. Chang and E. Grunwald, *J. Am. Chem. Soc.* **98**, 3737 (1976).
[271] J. Elguero, G. Llouquet, and C. Marzin, *Tetrahedron Lett.*, 4075 (1975).

been concluded that ^{13}C chemical shifts are of only limited value in ascertaining the positions of tautomeric equilibrium for rapidly interconverting tautomers.[222] Dipole moment studies have been used to determine the ratios of different forms in the tautomeric equilibrium mixtures for series of azoles and benzazoles.[253]

Tautomerism involving substituents will be discussed here rather than under the chemistry of functional groups.

Although it has been suggested[272] that some of the reactions of 2-aminoimidazole can best be explained by its existence as the imino tautomer in dimethylformamide, this result was based on incorrectly interpreted chemical evidence. In fact, a UV study[273] of 2-amino-1-methyl-4,5-diphenylimidazole provides convincing evidence for the amino form with $K_T = 3 \times 10^4$. Calculations[255] of heats of combustion show that in the 4-aminoimidazole $\rightleftharpoons$ 5-aminoimidazole equilibrium there is a slight predominance of the 4-amino form [a conclusion which is inconsistent with Charton's hypothesis since $\sigma_m(NH_2) = -0.161$]. The compounds, however, exist as amino rather than imino structures.[272,274]

The azido–tetrazole equilibrium found in other series applies in azidoimidazoles; under basic conditions 2-azidoimidazoles isomerize to the corresponding tetrazole anions (91) and in dimethyl sulfoxide the

(91)   (92)

N-acetyl derivatives of these azidoimidazoles give a 3:2 mixture of azido and tetrazole forms.[275] It has been suggested[276] that the azide form is much more prevalant in azoles than in azines and thiazoles. Substituents affect the equilibrium and a shift to the tetrazole form is mainly governed by resonance electron withdrawal.[277] Not only electron-withdrawing power, but the size of a substituent group can also affect the equilibrium, for in

---

[272] M. D. Coburn and P. N. Newman, *J. Heterocycl. Chem.* **7**, 1391 (1970).
[273] A. F. Pozharskii, I. S. Kashparov, Yu. P. Andreichikov, A. I. Buryak, A. A. Konstantinchenko, and A. M. Simonov, *Chem. Heterocycl. Compd. (Engl. Transl.)* **7**, 752 (1971).
[274] V. Sunjic, T. Fajdiga, M. Japelj, and P. Rems, *J. Heterocycl. Chem.* **6**, 53 (1969).
[275] E. Alcade and R. M. Claramunt, *Tetrahedron Lett.*, 1523 (1975).
[276] R. Granados, M. Rull, and J. Vilarrosa, *J. Heterocycl. Chem.* **13**, 281 (1976).
[277] M. Rull and J. Vilarrosa, *J. Heterocycl. Chem.* **14**, 33 (1977).

N-acyl-2-azidoimidazoles (**92**) when R = t-butyl the tetrazole form is more favored than when R = methyl.[278]

Many workers in recent years have been interested in the tautomeric equilibria pertaining to oxygenated imidazoles such as imidazolones,[86,279–284] N-hydroxyimidazoles, and imidazole N-oxides.[181,284–287] The earlier results have been critically summarized.[5,181,186,286] In the solid state 1,2-dimethyl-5-phenylimidazol-4-one exists in the OH form,[281] but insolubility prevented studies in solution. Theoretically 1-methylimidazol-5-ones can exist in the forms **93–96** shown in Eq. (22). French workers[281] have found that 1,2,4-trimethyl and 1,4-dimethyl-2-phenyl compounds both exist in the C-4H form **93** under a wide range of conditions. However, 4-phenyl analogs do not favor this form and may be mainly in the OH form (**94**) or zwitterionic NH form (**95**). The difficulty in preparing suitable model compounds and solubility problems have hindered firm conclusions. Studies of N-unsubstituted imidazol-5-ones have demonstrated that, although substituents affect the nature of the tautomer found, solvents have little effect.[281,284] Compounds stated to be 1-acylimidazol-4-ones have been formulated as the 4-oxo derivatives on infrared evidence.[282] The N-unsubstituted imidazol-4-ones can exist in the three possible forms (**97–99**) shown in Eq. (23). Infrared spectroscopy has demonstrated that in the solid state 2-methyl-5,5-diphenyl- and 2,5,5-triphenylimidazol-4-ones exist as

---

[278] M. L. Rull and J. Vilarrosa, *Tetrahedron Lett.*, 4175 (1976).
[279] J. T. Edward and I. Lantos, *J. Heterocycl. Chem.* **9**, 363 (1972).
[280] C. W. N. Cumper and G. D. Pickering, *J. C. S. Perkin II*, 2045 (1972).
[281] R. Jacquier, J.-M Lacombe, and G. Maury, *Bull. Soc. Chim. Fr.*, 1040 (1971).
[282] Y. Y. Usaevich, I. K. Feldman, and Y. I. Boksinev, *Khim. Geterotsikl. Soedin.*, 801 (1971).
[283] J. Nyitrai and K. Lempert, *Tetrahedron* **25**, 4265 (1969).
[284] O. Ceder and U. Stenhede, *Acta Chem. Scand.* **27**, 2221 (1973).
[285] V. S. Kobrin and L. B. Volodarskii, *Chem. Heterocycl. Compds.* (Engl. Transl.) **12**, 1280 (1976).
[286] H. Lettau, *Z. Chem.* **10**, 338 (1970).
[287] A. Maquestiau, Y. van Haverbeke, R. Flammang, S. O. Chua, M. J. Cook, and A. R. Katritzky, *Bull. Soc. Chim. Belg.* **83**, 105 (1974).

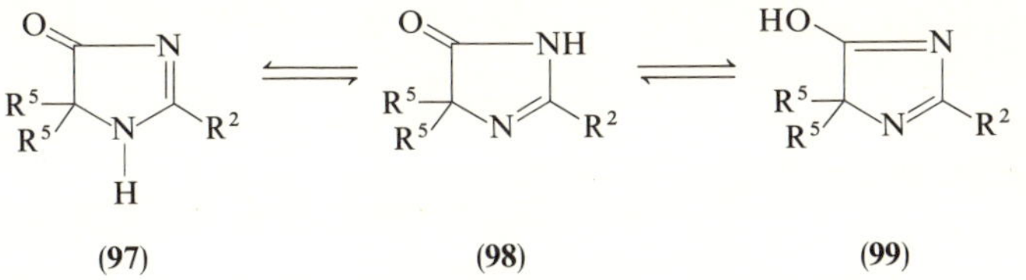

Sec. IV.A] ADVANCES IN IMIDAZOLE CHEMISTRY 287

(23)

(97)   (98)   (99)

a mixture of two of these forms, 97 and 98,[283] as does 2-phenyl-5,5-dimethylimidazol-4-one.[281] Recently, Edward[279] has shown that 5,5-diphenyl and 5-spirocyclohexyl derivatives exist in nonpolar solvents as the less conjugated and less polar tautomer 98, but as the solvent polarity increases the more polar tautomer 97 increases in importance.

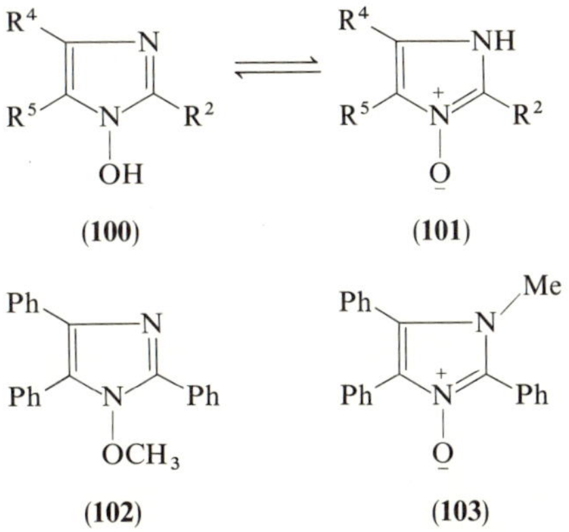

(100)   (101)   (24)

(102)   (103)

The equilibrium shown in Eq. (24) has K_T (101/100) ~ 3, but the ultraviolet spectrum for the compound in which $R^2 = R^4 = R^5$ = phenyl much more resembles the "hydroxy model" compound (102) than the "oxide model" (103).[181] The latter, however, is not a very suitable model because of steric interactions between the N-methyl and 2- and 4-phenyl groups. The authors[181] believe that in aqueous solution there are about equal quantities of the two tautomers, but in less polar solvents the hydroxy form (100) is preferred. On the basis of IR spectra in chloroform, acetonitrile, and dioxane it has been suggested[286] that 1-hydroxy-4,5-dimethylimidazole exists in these solvents as both possible tautomers. In the solid state mass spectrometry favors a polymorphism of a single tautomer,[287] but only a crystallographic study locating the hydrogen atoms involved in hydrogen bonding will provide a definite answer.

The tautomeric ratio, K_T, for the zwitterionic (104)/uncharged form (105) of 1-methylimidazole-2-carbaldehyde oxime is 5×10^{-3}.[177]

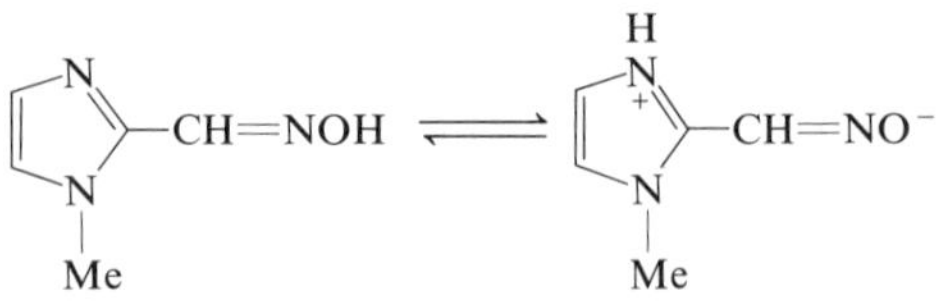

(104) (105)

Comparison of ^{13}C-NMR chemical shifts and pH profiles for histidine and its 1- and 3-methyl derivatives shows that in basic solution the preferred structure is **106**. This tautomeric form is maintained in a number of histidine

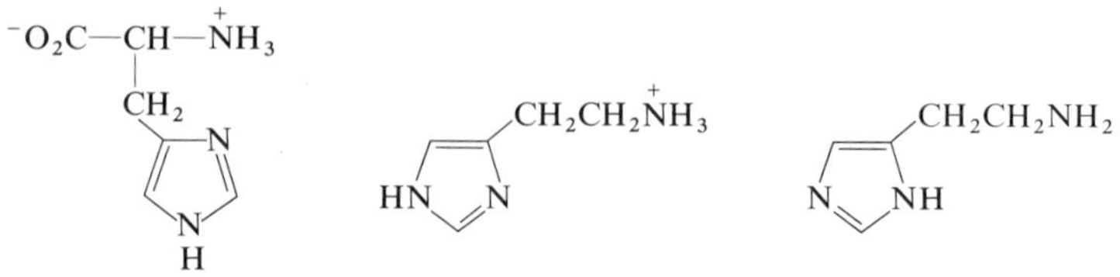

(106) (107) (108)

derivatives including the polypeptide bacitracin.[288] X-Ray crystallography has been used to determine that histamine is 5-(2-aminoethyl)imidazole[289] a result which contrasts with the structure determined for its cation.[183] From determinations of pK_a values of histamine and its two N-methyl derivatives a K_T value of 4 (4-:5-substituted isomer) has been found. In fact, in aqueous solution histamine is largely present as the protonated form **(107)**; on neutralization it reverts to the N-3 tautomer **(108)**. The stabilities of the two protomers can be correlated with results of *ab initio* and INDO molecular orbital calculations.[251] Electron-withdrawing groups in the imidazole ring of histamine cause the 4-aminoethyl tautomer to predominate.[269]

B. Electrophilic Substitution Reactions

1. *At Ring Nitrogen Atoms*

a. *Acylation and Related Reactions.* Because of the synthetic importance of the azolides, the part played by N-acylimidazoles in biological processes, the probable intermediacy of N-aroylimidazoles in 2-aroyl-

---

[288] W. F. Reynolds, I. R. Peat, M. H. Freeman, and J. R. Lyerla, *J. Am. Chem. Soc.* **95**, 328, 6510 (1973).

[289] J. J. Bonnet and J. A. Ibers, *J. Am. Soc.* **95**, 4829 (1973).

imidazole formation, and the value of N-acylimidazoles in the regiospecific alkylation of the heterocycles, there has been considerable study of this electrophilic substitution reaction.

Kinetic studies of the "acylation" of imidazoles with benzoyl halides[290-293] and sulfonyl halides[294,295] have been carried out. In acetonitrile, benzoyl fluoride reacts with imidazole in a mixed third- and fourth-order reaction; i.e., the rate-determining transition state contains the acid fluoride and two or three molecules of imidazole.[293] With benzoyl chloride in benzene a second-order equation is followed similar to the aroylation of anilines[290]; 2-alkyl- and 4,5-disubstituted imidazoles have steric hindrance to the reaction.[291]

Imidazoles have been "N-acylated" by isocyanates (at elevated temperatures)[296] (Scheme 21), acid halides,[297,298] and alkyl chlorocarbonates,[298] but 2-methylimidazole would not react with formamide and phosphoryl chloride.[299,300] Trifluoromethylsulfonation forms the imidazolide which is a convenient reagent for the introduction of the "triflate" group.[301] With highly basic 1-substituted imidazoles, acetyl halides form the 1-acetyl-3-substituted imidazolium salts,[292] while in the presence of antimony pentachloride, 1-acetylimidazole is quaternized by acetyl chloride.[302]

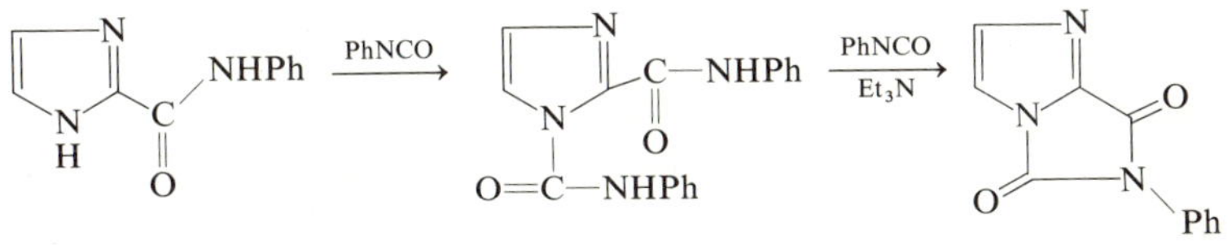

SCHEME 21

[290] Yu. S. Simanenko, L. M. Litvinenko, and V. A. Dadali, *J. Org. Chem. USSR (Engl. Transl.)* **10**, 1313 (1974).
[291] L. M. Litvinenko, V. A. Dadali, S. A. Lapshin, Yu. S. Simanenko, and V. I. Rybachenko, *J. Org. Chem. USSR (Engl. Transl.)* **11**, 245 (1975).
[292] V. L. Rybachenko, A. Yu. Chervinskii, L. M. Kapkan, R. G. Semenova, and E. V. Titov, *J. Org. Chem. USSR (Engl. Transl.)* **12**, 240 (1976).
[293] O. Rogne, *Chem. Commun.*, 25 (1975).
[294] O. Rogne, *J. C. S. Perkin II*, 1760 (1973).
[295] O. Rogne, *J. C. S. Perkin II*, 823 (1973).
[296] E. P. Papadopoulos and C. M. Schupbach, *J. Org. Chem.* **44**, 99 (1979).
[297] J. C. Cass, A. R. Katritzky, R. L. Harlow, and S. H. Simonsen, *Chem. Commun.*, 48 (1976).
[298] E. Guibe-Jampel, G. Bram, and M. Vilkas, *Bull. Soc. Chim. Fr.*, 1021 (1973).
[299] E. Guibe-Jampel, M. Wakselman, and M. Vilkas, *Bull. Soc. Chim. Fr.*, 1308 (1971).
[300] T. Hirota, T. Koyama, C. Basho, T. Nanba, K. Sasaki, and M. Yamato, *Chem. Pharm. Bull.* **25**, 3056 (1977).
[301] F. Effenberger and K. E. Mack, *Tetrahedron Lett.*, 3947 (1970).
[302] T. V. Stupnikova, A. I. Serdyuk, E. Portsel, and A. K. Sheinkman, *Dopov. Akad. Nauk Ukr. RSR., Ser. B*, 45 (1977) [*CA* **86**, 171332 (1977)].

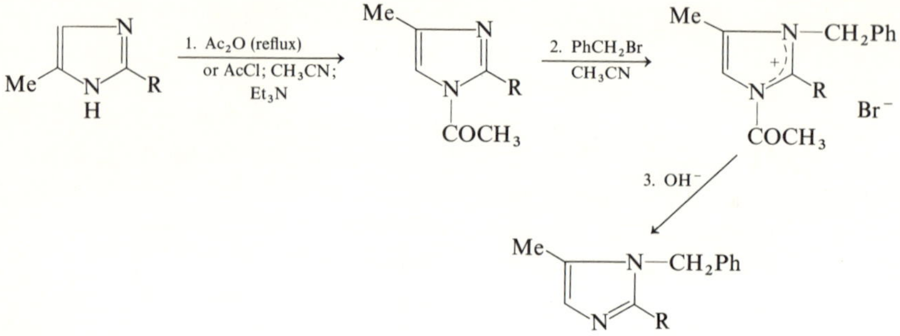

SCHEME 22

Acylation reactions leading ultimately to 2-acylimidazoles will be discussed in Section IV,B,2,a.

Since *N*-acylation is a reversible process, it has allowed the regiospecific alkylation of imidazoles to give the sterically less-favored derivative, i.e., the 1,5-disubstituted derivative (e.g., **109**; Scheme 22).[213,303-306] The sequence followed is: (1) acylation; (2) alkylation (often with oxonium reagents); and (3) deacylation with alcohol, water, or base. The *N*-acylation of 2-substituted imidazoles using ethyl chloroformate, triethylamine, and acetonitrile gives *N*-alkoxycarbonylimidazoles[307] which can lose carbon dioxide to give the *N*-alkyl derivatives. The reaction is of limited use in the synthesis of asymmetrically substituted imidazoles since, whereas 2-ethyl-4-methylimidazole gave >95% of 1-carbethoxy-2-ethyl-4-methylimidazole, the subsequent decarboxylation afforded a 3:1 mixture of 1,2-diethyl-4-methyl and 1,2-diethyl-5-methyl compounds.[307]

The acylation of imidazoles under Schotten–Baumann conditions has been studied.[308]

The imidazole-catalyzed hydrolysis of esters can be classified as an electrophilic attack by an acyl compound on the multiply bonded imidazole nitrogen, and there has been considerable effort expended in such studies.[2,6,309-320]

[303] H. C. Beyerman, L. Maat, and A. Van Zon, *Recl. Trav. Chim. Pays-Bas* **91**, 246 (1972).
[304] L. Maat, H. C. Beyerman, and A. Noordam, *Tetrahedron* **35**, 273 (1979).
[305] E. F. Godefroi and J. H. F. M. Mentjens, *Recl. Trav. Chim. Pays-Bas* **93**, 56 (1974).
[306] R. A. Olofson and R. V. Kendall, *J. Org. Chem.* **35**, 2246 (1970).
[307] H. J. J. Loozen, J. J. M. Drouen, and O. Piepers, *J. Org. Chem.* **40**, 3279 (1975).
[308] E. Babad and D. Ben-Ishai, *J. Heterocycl. Chem.* **6**, 235 (1969).
[309] J. G. Tillett, in "Annual Reports," Vol. 67B. Chemical Society, London, 1970.
[310] T. Kunitake and S. Shinkai, *J. Am. Chem. Soc.* **93**, 4247, 4256 (1971).
[311] C. A. Blythe and J. R. Knowles, *J. Am. Chem. Soc.* **93**, 3021 (1971).

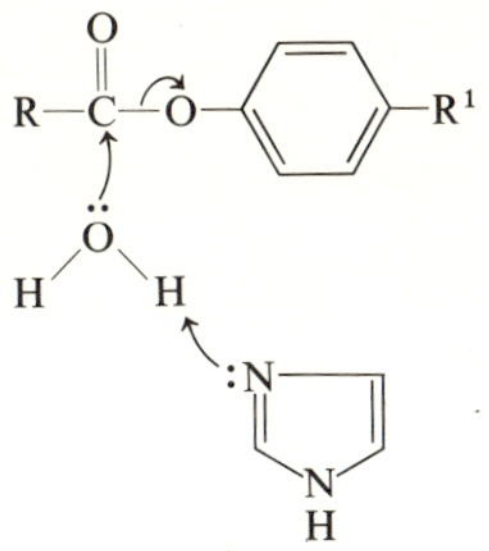

Sec. IV.B] ADVANCES IN IMIDAZOLE CHEMISTRY 291

FIG. 3

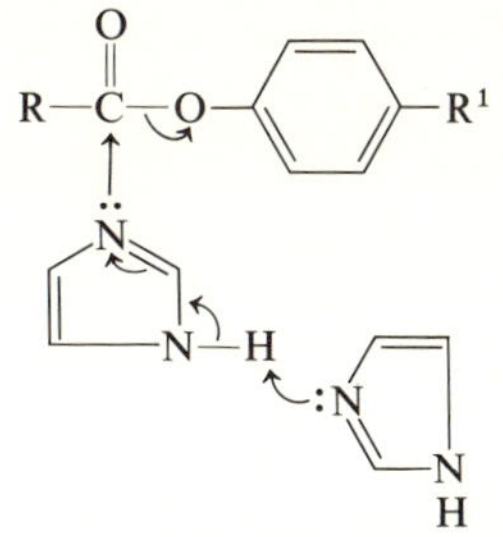

FIG. 4

1-Acetylimidazole can be made enzymically, probably by the following sequence:

acetyl phosphate + coenzyme A ⇌ acetyl coenzyme A + phosphate
acetyl coenzyme A + imidazole ⇌ 1-acetylimidazole + coenzyme A

Since the imidazolyl group of a histidine residue may be implicated in the mode of action of such hydrolytic enzymes as trypsin and chymotrypsin, the interest in imidazole catalysis has grown. Two pathways have been found to be involved: *general base catalysis* and *nucleophilic catalysis*. In the former the imidazole acts either as a base and activates a water molecule for attack at the carbonyl carbon of the ester (Fig. 3), and this is followed by the same steps as in simple OH^- ion hydrolysis, or the imidazole molecules may be involved directly, particularly at high imidazole concentrations (Fig. 4). The nucleophilic catalysis properties of imidazole are due to the intermediate formation of a 1-acylimidazole (Scheme 23). In the hydrolysis of esters with good leaving groups (e.g., *p*-nitrophenyl acetate) the imidazole neutral molecule is the effective catalyst, and the more basic the imidazole, the more effective it is. Where the esters have poorer leaving groups (e.g., *p*-cresol acetate) the imidazole anion is involved and general base catalysis predominates. Thus, for imidazoles with $pK_a > 4$ catalysis by the anion is the main reaction. Imidazole is a much more effective nucleophile than amines in this type of reaction since it is a tertiary amine with minimal steric

---

[312] F. D'Andrea and U. Tonellata, *Chem. Commun.*, 659 (1975).
[313] F. Schneider, *Angew. Chem., Int. Ed. Engl.* **17**, 583 (1978).
[314] W. Tagaki, S. Kobayashi, and D. Fukushima, *Chem. Commun.*, 29 (1977).
[315] C. G. Overberger and R. C. Glowaky, *J. Am. Chem. Soc.* **95**, 6014 (1973).
[316] P. Remuzon and M. Wakselman, *Tetrahedron* **33**, 3097 (1977).
[317] L. Anvardi and U. Tonellata, *Chem. Commun.*, 401 (1977).
[318] M. A. Wells and T. C. Bruice, *J. Am. Chem. Soc.* **99**, 5341 (1977).
[319] J. F. Patterson, W. P. Huskey, K. S. Venkatasubban, and J. L. Hogg, *J. Org. Chem.* **43**, 4935 (1978).
[320] J. L. Hogg, M. K. Phillips, and D. E. Jergens, *J. Org. Chem.* **42**, 2459 (1977).

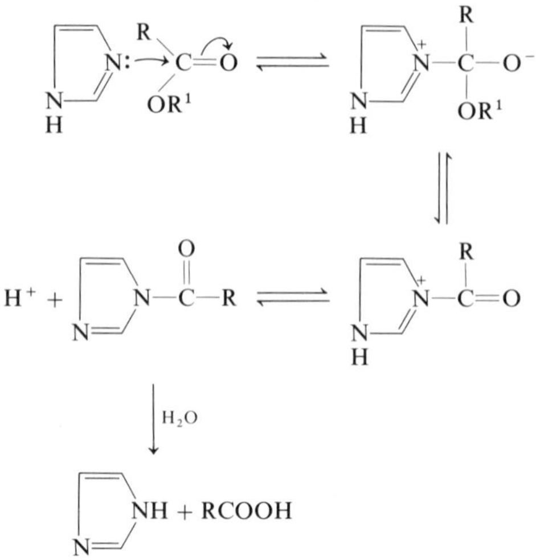

SCHEME 23

hindrance having the facility to delocalize the positive charge resulting from the nucleophilic addition to the carbonyl carbon. Furthermore, 1-acylimidazoles are much more highly reactive in aqueous solution than are other amides. This topic has been summarized.[2,5,6]

The imidazole-catalyzed isomerization of penicillin into penicillenic acids probably involves initial nucleophilic attack of imidazole on penicillin.[321] Imidazole-catalyzed N-acylimidazole hydrolysis was used by Charton[259] in obtaining σ^* values (see Section III,I).

b. *Alkylation.* A variety of processes have been used to introduce an alkyl group on to a ring nitrogen of an imidazole. Alkylation can be achieved by reaction with alkyl halides (in netural or basic medium), and this type of reaction has been adapted for the introduction of unsaturated groups. Thus, vinyl halides give the N-vinylimidazoles.[322] Other alkylation procedures include reaction of the imidazole with ω-bromoacetophenone (in the presence of base),[323] with halides and hydroxybenzyl alcohols,[324] Mannich bases,[325] triphenylmethyl chloride or fluoroborate,[326] with

[321] H. Bundegaard, *Tetrahedron Lett.*, 4613 (1971).
[322] G. G. Skvorstsova, E. S. Domnina, Yu. N. Ivlev, N. N. Chipanina, V. I. Skorobogatova, and Yu. I. Myachin, *J. Gen Chem. USSR* (*Engl. Transl.*) **42**, 592 (1972).
[323] S. V. Bogatkov, B. M. Kormanskaya, V. B. Mochalin, and E. M. Cherkasova, *Chem. Heterocycl. Compd.* (*Engl. Transl.*) **7**, 621 (1971).
[324] M. Wakselman, J. C. Robert, G. Decodts, and M. Vilkas, *Bull. Soc. Chim. Fr.*, 1179 (1973).
[325] F. B. Stocker, J. L. Kurtz, B. L. Gilman, and D. A. Forsyth, *J. Org. Chem.* **35**, 883 (1970).
[326] K. H. Buechel, W. Draber, E. Regel, and M. Plempel, *Arzneim.-Forsch.* **22**, 1260 (1972) [*CA* **77**, 152065 (1972)].

carbonyl compounds,[323,327] with alkyl esters of phosphonic and phosphinic acids,[328–330] and with a variety of unsaturated compounds.[331–335] A short communication has reported a novel Mannich reaction of 1,4-diisopropylimidazole which formed a dihydropyrazine.[336] 1-Tritylimidazoles can also be obtained using trityl chloride with either the imidazole Grignard reagent or 1-trimethylsilylimidazole.[326] Dimethyl sulfoxide reacts at elevated temperatures in a Pummerer reaction with 1-trimethylsilylimidazoles to give 1-(methylthio)methylimidazoles.[337] The sealed-tube reaction of imidazole and formaldehyde at 120–130°C gives a liquid (described as 1-hydroxymethylimidazole) which forms imidazole when boiled with water.[327] The N-alkylation reactivity of trialkylphosphates has been found to decrease as the alkyl groups increase in size.[330] In the alkylation of 2-amino-1-methylimidazole with α-bromocarbonyl compounds, the N,N-dialkylated iminoimidazole (110) is formed at low temperatures; above 66°C dehydration gives rise to a bicyclic product[338] (Scheme 24). Among the unsaturated reagents which are able to alkylate imidazoles are acetylene carboxylic esters,[335] acrylonitrile,[334] acetylene in the presence of potassium hydroxide,[332] and arylalkynes with lithium and hexamethylphosphorotriamide.[333] In addition to the last-named reagent,[333] styrene oxide can also give rise to 1-styrylimidazoles.[339,340] A two-phase alkylation procedure has been recommended.[341] When imidazoles are heated at 200–450°C with an alcohol or ether and an oxide or phosphate of calcium, aluminum, titanium, or thallium, the product is the 1-alkylimidazole.[342] Condensation of the mercury complex or silver salt of 4-aminoimidazole-5-carboxamide with protected carbohydrates provides a route to imidazole-1-glycosides.[343]

---

[327] P. W. Alley, *J. Org. Chem.* **40**, 1837 (1975).
[328] M. Hayashi, Y. Yamauchi, and M. Kinoshita, *Bull. Chem. Soc. Jpn.* **49**, 283 (1976).
[329] M. Oklobdzua, V. Sunjic, F. Kajfez, V. Caplar, and D. Kolbah, *Synthesis*, 596 (1975).
[330] K. Yamauchi and M. Kinoshita, *J. C. S. Perkin I*, 2506 (1973).
[331] F. Troxler, H. P. Weber, A. Jaunin, and H. R. Loosli, *Helv. Chim. Acta* **57**, 750 (1974).
[332] G. G. Skvorstsova, E. S. Domnina, N. P. Glazkova, and L. N. Makhno, *Khim. Atsetilena, Tr. Vses. Konf., 3rd, 1968.* 115 (1972) [*CA* **79**, 42414 (1973)].
[333] P. Bourgeois, J. Lucrece, and J. Dunogues, *J. Heterocycl. Chem.* **15**, 1543 (1978).
[334] M. Yamauchi and M. Masui, *Chem. Pharm Bull.* **24**, 1480 (1976).
[335] R. M. Acheson and N. F. Elmore, *Adv. Heterocycl. Chem.* **23**, 265 (1978).
[336] M. Masui, K. Suda, M. Yamauchi, and N. Yoshida, *Chem. Pharm. Bull.* **21**, 1387 (1973).
[337] A. F. Janzen, G. N. Lypka, and R. E. Wasylishen, *J. Heterocycl. Chem.* **16**, 415 (1979).
[338] B. A. Priimenko and P. M. Kochergin, *Chem. Heterocycl. Compd. (Engl. Transl.)* **7**, 1574 (1971).
[339] G. Cooper and W. J. Irwin, *J. C. S. Perkin I*, 545 (1976).
[340] G. Cooper and W. J. Irwin, *J. C. S. Perkin I*, 798 (1975).
[341] H. J. M. Dou and J. Metzger, *Bull. Soc. Chim. Fr.*, 1861 (1976).
[342] Badische Anilin- und Soda-Fabrik A.-G. French Patent 1,603,793 (1971) [*CA* **76**, 85820 (1972)].
[343] H. Gugliemi, *Justus Liebigs Ann. Chem.*, 1286 (1973).

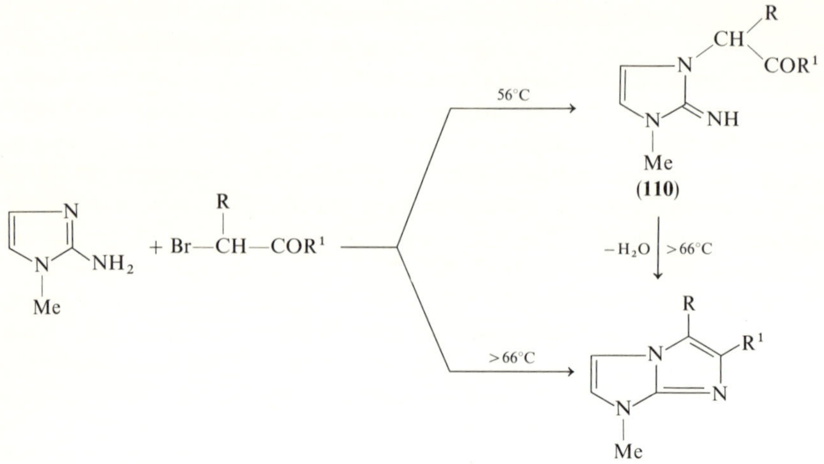

SCHEME 24

It is noteworthy that the majority of the most useful alkylating procedures involve the imidazole anion, either by using a metal salt of the azole, or by carrying out the reaction in strongly basic medium. Such conditions prevent quaternary salt and other salt formation.

The orientation of alkylation of unsymmetrical imidazoles depends on the reaction mechanism (S_E2cB or S_E2'), the influence of substituents, and steric factors; it seems too (on analogy with pyrazole methylation[344]) that the tautomeric nature of the imidazole cannot always be neglected. In basic medium the mechanism is S_E2cB (Scheme 25) when electron-withdrawing R groups should direct alkylation to give the 1,4-disubstituted compound (a $+I$ effect of R should give rise to mainly the 1,5-product). "Neutral" methylation (S_E2') can be affected by all of the factors listed above. Some recent results include the alkaline benzylation of 2,4-disubstituted imidazoles to give the sterically least hindered products (1,2,4-)[345]; the polyphosphoric alkyl ester methylation and ethylation of 2-methyl-4-nitroimidazole (S_E2')

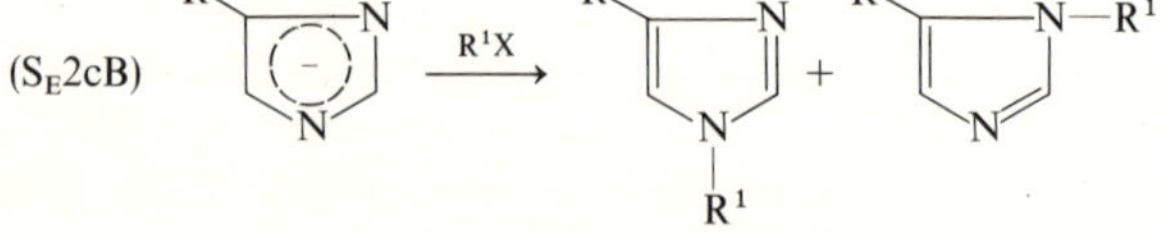

SCHEME 25

[344] M. R. Grimmett, K. H. R. Lim, and R. T. Weavers, *Aust. J. Chem.* **32**, 2203 (1979).
[345] M. Masui, K. Suda, M. Inoue, K. Izukura, and M. Yamauchi, *Chem. Pharm. Bull.* **21**, 2359 (1973).

gives mainly the 1,2-dialkyl-5-nitro compounds[329] (tautomerism effect; cf. methylation of 4-nitroimidazole[346,347]); with dimethyl sulfate in basic medium 2-chloro-4-nitroimidazole gives mainly 2-chloro-1-methyl-4-nitroimidazole[348,349] (S_E2cB mechanism); in another S_E2' process with dimethyl sulfate, 2-methyl-4-phenylimidazole produces 85% of the 4-phenyl (**112**) and 15% of the 5-phenyl (**111**) isomer.[206] In this last reaction (Scheme 26) the steric effect of the phenyl substituent must control the products which were identified by the use of lanthanide-shift reagents.

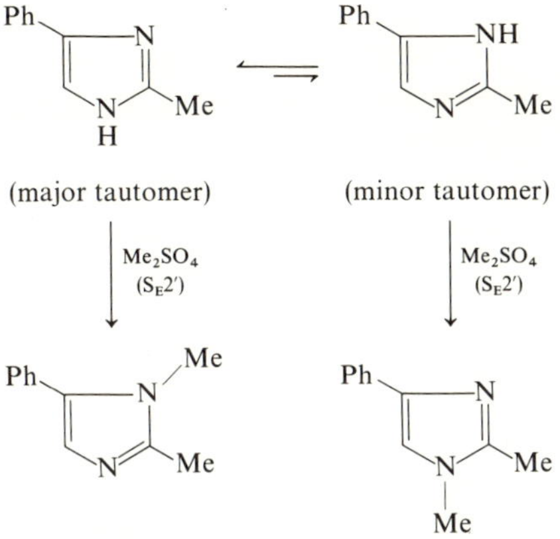

SCHEME 26

Quaternizing alkylation, too, has received some attention,[168,187,208,213,303–305,350,351] including a review.[187]

The effects of N-aryl substituents on rate constants for the quaternization of imidazoles can be correlated using a Hammett equation.[168] It has been shown[352] that the rates for quaternization of N-arylpyrazoles with dimethyl sulfate in sulfolane show greater steric and substituent effects than in the

---

[346] A. Grimison, J. H. Ridd, and B. V. Smith, *J. Chem. Soc.*, 1352, 1357 (1960).
[347] J. H. Ridd and B. V. Smith, *J. Chem. Soc.*, 1363 (1960).
[348] G. P. Sharnin, R. Kh. Fassakhov, T. A. Eneikina, and P. P. Orlov, *Chem. Heterocycl. Compd.* (*Engl. Transl.*) **13**, 529 (1977).
[349] G. P. Sharnin, R. Kh. Fassakhov, and T. A. Eneikina, *Chem. Heterocycl. Compd.* (*Engl. Transl.*) **13**, 1332 (1977).
[350] P. M. Kochergin, A. A. Druzhinina, and R. M. Palei, *Chem. Heterocycl. Compd.* (*Engl. Transl.*) **12**, 1274 (1976).
[351] L. W. Deady, *Aust. J. Chem.* **26**, 1949 (1973).
[352] L. W. Deady, R. G. McLoughlin, and M. R. Grimmett, *Aust. J. Chem.* **28**, 186 (1975).

ethylation of imidazoles. It is not possible to decide whether the value of ρ (greater sensitivity) of the pyrazoles (-1.10) over the imidazoles (-0.45) is primarily due to the presence of the quaternizing center adjacent to the site of attachment of the aryl group, or to steric effects which cause the aryl group to rotate out of the plane of the heterocyclic ring in the transition state.[352] The rate of quaternization of 1-methylimidazole with dimethyl sulfate or methyl iodide in dimethyl sulfoxide is controlled by the inductive effect of the other nitrogen and the basicity of the heterocycle.[351] In fact, in a series of azoles the ease of quaternization is affected by the heteroatom; i.e., the rate constants increase in the order of $O < S < NCH_3$, and azoles with heteroatoms 1,3 are more reactive than those with heteroatoms 1,2. A Brønsted correlation for the N-methylation of azoles and benzazoles shows a linear log–log correlation.[187]

As mentioned earlier (Section IV,B,1,a) N-acylimidazoles, although resistant to quaternization by virtue of the electron-withdrawing nature of the acyl group, can be methylated by reagents such as methyl fluorosulfonate[213] and trimethyloxonium fluoroborate.[303-305]

c. *Arylation.* There have been kinetic studies of the reactions of imidazoles with picryl chloride[353] and 2,4-dinitrofluorobenzene.[354,355] Imidazole acts as a nucleophile with 1-fluoro- or 1-chloro-2,4-dinitrobenzene.[355] The kinetics of these reactions are third-order overall and second-order with respect to imidazole at low imidazole concentrations. Some leveling-off of rates has been interpreted as indicating an addition–elimination reaction with decomposition into products being fast for chloro- and slow for fluorodinitrobenzene.

d. *Miscellaneous* N-*Substitutions.* Apart from the reactions surveyed above there are one or two other possible electrophilic substitutions on ring nitrogen of imidazole. Phosphino- and phosphinatoimidazoles can be prepared either by the reaction of imidazole with chlorophosphine in ether with triethylamine as HCl acceptor, or by the reaction of phosphorus trichloride with 1-trimethylsilylimidazole.[356] With phosphoryl chloride in alkaline medium imidazole can be phosphorylated.[299] The preparation of 1-trimethylsilylimidazoles has been reported again,[357] and although

---

[353] R. Minetti and A. Bruylants, *Bull. Cl. Sci., Acad. R. Belg.* **56**, 1047 (1970).
[354] R. H. de Rossi, A. B. Pierini, and R. A. Rossi, *J. Org. Chem.* **43**, 2982 (1978).
[355] F. Pietra and F. Del Cima, *J. C. S. Perkin II*, 1420 (1972).
[356] S. Fischer and L. K. Peterson, *Can. J. Chem.* **52**, 3981 (1974).
[357] V. Sheludyakov, N. A. Viktorov, G. V. Ryasin, and V. F. Mironov, *J. Gen. Chem. USSR* (*Engl. Transl.*) **42**, 354 (1972).

imidazole only forms the nitrate salt when treated with nitric acid in acetic anhydride, more weakly basic compounds such as 4-nitroimidazole can be N-nitrated under these condition.[358,359]

2. At Ring Carbon Atoms

In an endeavor to study the transmission of substituent effects in imidazoles, Noyce[360] examined the rates of solvolysis of a series of 1-(1-methylimidazolyl)ethyl p-nitrobenzoates (113–115; OPNB = p-nitrobenzoate). The

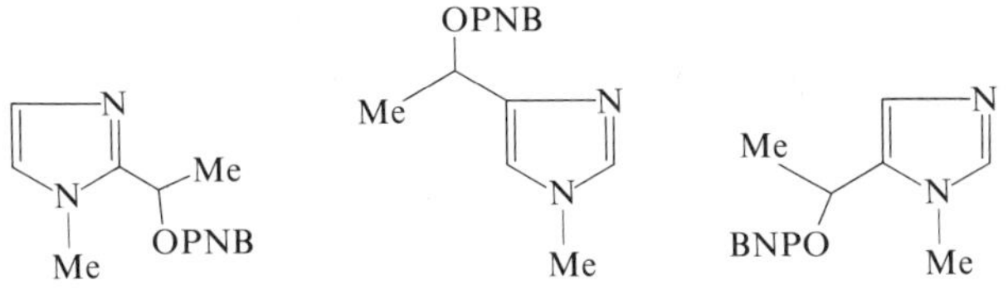

(113)   (114)   (115)

order of rates found (113:114:115 = 1:13:15) parallels the relative electron densities in 1-methylimidazole as deduced from chemical-shift data.[361] By comparison with other heteroarylethyl p-nitrobenzoates effective replacement constants, σ^+_{Ar}, were determined as $\sigma^+_{2\text{-Im}}$ −0.82, $\sigma^+_{4\text{-Im}}$ −1.01, $\sigma^+_{5\text{-Im}}$ −1.02. The effects on the 2-substituted compound of alkyl, aryl, and halogen substituents in the imidazole ring have also been examined, and although the rates for the 5-substituents were represented satisfactorily by σ^+_p, σ^+_m failed to account properly for the observed reactivities of the 4-substituents.[360] As mentioned earlier (Section III,H) various quantum-mechanical calculations of π-electron densities for the imidazole neutral molecule predict the order of substitution as 5 > 4 > 2, but the tautomeric equivalence of the 4- and 5-positions is not taken into account. In addition, there are probably few occasions on which electrophilic substitution takes place with the neutral molecule; the conjugate acid or conjugate base may be the reactive species.

a. *Carbonyl Reactions.* There have been some recent studies of the reaction of formaldehyde with imidazoles. A change in reaction conditions

---

[358] S. S. Novikov, L. I. Khmelnitskii, O. V. Lebedev, V. V. Sevastyanova, and L. V. Yepishina, *Chem. Heterocycl. Compd. (Engl. Transl.)* **6**, 503 (1970).

[359] M. R. Grimmett and T. H. Sio, unpublished; T. H. Sio, M.Sc. thesis, University of Otago, New Zealand, 1974.

[360] D. S. Noyce and G. T. Stowe, *J. Org. Chem.* **38**, 3762 (1973).

[361] G. B. Barlin and T. J. Batterham, *J. Chem. Soc. B*, 516 (1967).

from those which lead to a Mannich-type reaction with 2,4-diisopropylimidazole[336] allows hydroxymethylation.[345] The use of weakly acid media[362] (such as a formate buffer) with 1,2-dimethylimidazole gives 25–50% yields of a mixture of the 5- and 4,5-dihydroxymethylimidazoles. Sealed-tube reaction with formaldehyde also gives rise to low yields of the 5-hydroxymethyl compound.[345] The mechanistic possibilities include direct attack by formaldehyde at a ring carbon, or N-hydroxymethylation followed by rearrangement. The observation that the reaction of 1-methyl- or 1-benzylimidazoles also give the reaction does not support the second possibility unless an intermediate quaternary salt can be conceived. In sealed-tube reactions imidazole is hydroxymethylated at C-2[327,363] as well as at all three of the 2-, 4-, and 5-positions (in low yield).[327] 1,2,4-Trialkylimidazoles are substituted at C-5.[334]

In recent years there has been more than passing interest in the acylation of 1-substituted imidazoles at C-2 using an acid chloride in the presence of triethylamine,[364–369] or triethylamine–pyridine,[370] and in acetonitrile as solvent.[364,365,368,369] Under the same conditions (triethylamine–acetonitrile) imidazole gives triacylimidazolinylimidazoles,[371] but imidazole can be converted in a simple one-step synthesis into the 2-aroylimidazole **116** using an aroyl chloride and the triethylamine–pyridine medium.[370] Electron-withdrawing groups in the acid chloride assist the reaction,[365,368] and the mechanism has been demonstrated to involve an initial N-aroylation[368] followed by what is probably an intramolecular rearrangement (see Scheme 27). Benzimidazoles react similarly, but not pyrazoles.[368] When there is a methyl group at C-2 of the imidazole ring, the aroyl group gives an exocyclic C-aroyl product, but 1-benzyl-5-methylimidazole is aroylated normally at C-2.[364] At elevated temperatures imidazoles react with isocyanates to give 2-carboxamides (**117**) (Eq. 25) in a reaction which resembles the foregoing acylations.[296,372]

---

[362] E. F. Godefroi, H. J. J. Loozen, and J. T. J. Luderer-Platje, *Recl. Trav. Chim. Pays-Bas* **91**, 1383 (1972).
[363] C. Rufer, K. Schwarz, and E. Winterfeldt, *Justus Liebigs Ann. Chem.*, 1465 (1975).
[364] A. A. Macco, E. F. Godefroi, and J. M. Drouen, *J. Org. Chem.* **40**, 252 (1975).
[365] E. Regel and K. H. Buechel, *Justus Liebigs Ann. Chem.*, 145 (1977).
[366] E. Regel and K. H. Buechel, German Patent 1,926,206 (1971) [*CA* **74**, 31754 (1971)].
[367] E. Regel and K. H. Buechel, German Patent 1,956,711 (1971) [*CA* **75**, 49086 (1971)].
[368] M. R. Grimmett and K. H. R. Lim, unpublished; K. H. R. Lim, Ph.D. thesis, University of Otago, New Zealand, 1979.
[369] C. G. Begg, M. R. Grimmett, and Y. M. Lee, *Aust. J. Chem.* **26**, 415 (1973).
[370] L. A. M. Bastiaansen and E. F. Godefroi, *Synthesis*, 675 (1978).
[371] E. Regel, *Justus Liebigs Ann. Chem.*, 159 (1977).
[372] E. P. Papadopoulos, *J. Org. Chem.* **42**, 3925 (1977).

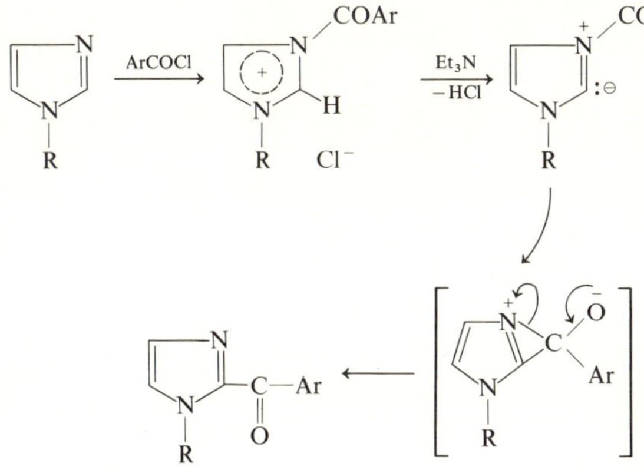

SCHEME 27

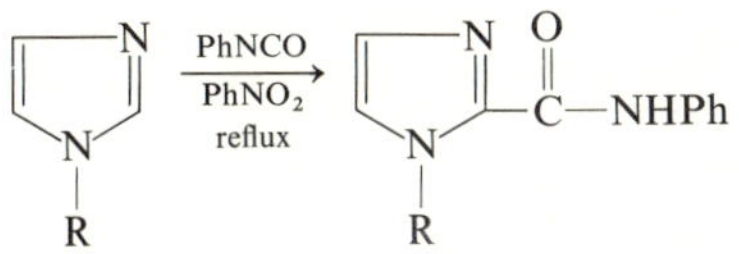

(116)

(117) (25)

Advantage can be taken of the enhanced polarity of a carbon–silicon bond over that of a carbon–hydrogen bond in the displacement of 2-trimethylsilyl groups by reaction with a number of carbonyl reagents to give imidazoles substituted at C-2 by secondary alcohol, acyl, aroyl, ester, and amide functions.[373]

b. *Hydrogen Exchange Reactions.* A significant amount of work has been devoted to the elucidation of the mechanisms of exchange of ring hydrogens on imidazoles.[174, 182, 374–380] Deuterium exchange takes place in neutral or basic conditions. While the proton attached to nitrogen exchanges rapidly in D_2O, that at C-2 is exchanged at 37°C with $t_{1/2} \sim 700$ min.[377]

---

[373] F. H. Pinkerton and S. F. Thames, *J. Heterocycl. Chem.* **9**, 67 (1972).
[374] J. H. Bradbury, B. E. Chapman, and F. A. Pellegrino, *J. Am. Chem. Soc.* **95**, 6139 (1973).
[375] J. L. Wong and J. H. Keck, *J. Org. Chem.* **39**, 2398 (1974).
[376] J. D. Vaughan, Z. Mughrabi, and E. Chung Wu, *J. Org. Chem.* **35**, 1141 (1970).
[377] H. A. Staab, H. Irngartinger, A. Mannschreck, and W. Mou-Thai, *Justus Liebigs Ann. Chem.* **695**, 55 (1966).
[378] C. G. Overberger, J. C. Salamone, and S. J. Yaroslavsky, *J. Org. Chem.* **30**, 3580 (1965).
[379] T. M. Harris and J. C. Randall, *Chem. Ind. (London)*, 1728 (1965).
[380] E. Chung-Wu and J. D. Vaughan, *J. Org. Chem.* **35**, 1146 (1970).

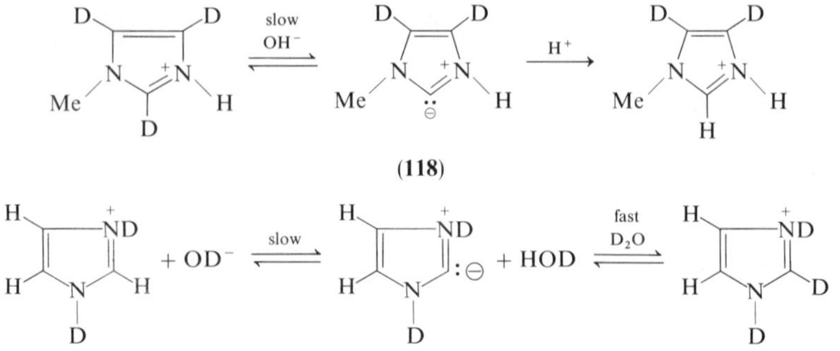

SCHEME 28

At 150°C the reaction is virtually complete in 2 hours with or without added base,[378] and in acid medium it does not occur.[377] 1-Methylimidazole can undergo 2-deuteration in D_2O with the rate of the reverse reaction being essentially independent of pH as long as the conditions do not become acid, when the rate falls rapidly to zero. A mechanism involving the ylide **118** derived from the conjugate acid is believed to be involved.[377,379] The relative rates of hydrogen–deuterium exchange at the 2-, 4-, and 5-positions of 1-methylimidazole are 54,500:1.6:1 (base-catalyzed exchange) via the ylide mechanism (Scheme 28), and 1:73:120 (acid-catalyzed exchange) via an S_EAr pathway involving the conjugate acid species.[375] Even under basic conditions the 1-methylimidazole reacts as the imidazolium ion (and consequently, after rate-determining loss of a proton attached to a ring carbon, through an ylide intermediate).

The deuteration of imidazole itself does not appear to involve the imidazole anion either, but rather depends on two parallel processes: rate-determining proton abstraction from the imidazolium cation by D_2O and by OD^- to give the ylide at C-2 followed by deuteration there (see Scheme 28). An additional path involving proton abstraction from the imidazole neutral molecule can account for the 4-substitution pD profile.[376] In strongly alkaline medium imidazoles with no N-substituent exchange more rapidly at C-4 or C-5 by a carbanion pathway involving proton abstraction from the neutral molecule.[182] In N-alkylimidazoles, however, only H-5 exchanges by this pathway and the resistance to C-4 exchange has been attributed[182] to the adjacent lone pair effect—a significant electrostatic repulsion between lone pairs of the coplanar sp^2 orbitals at N-3 and C-4.

There is a further mechanistic possibility at high pD values involving sigma intermediates. In the ylide mechanism the relative reactivities of the various species arising from the heterocycle are conjugate acid > neutral molecule > conjugate base (unreactive), while in any mechanism proceeding through a Wheland intermediate the sequence will be the opposite.

The ring protons of nitro- and fluoroimidazoles (and their N-methyl derivatives) also undergo base-catalyzed exchange in D_2O by a combination of ylide and carbanion pathways.[174] In the 4-substituted imidazoles carbanion exchange occurs more readily at C-5 than at C-2; e.g., for 1-methyl-4-nitroimidazole in 0.1 M NaOD at 50°C, $t_{1/2}$ = 3 min at C-5; $t_{1/2}$ = 2.7 h at C-2. In 1-methyl-5-nitro(or -fluoro)imidazoles exchange at C-4 occurs only via the ylide process since carbanion formation in the neutral species is again retarded by the adjacent lone-pair effect; this effect is considerably weaker, or nonexistent, for exchange at C-2; e.g., for 1-methyl-5-nitroimidazole in 0.1 M NaOD at 50°C, $t_{1/2}$ = >2 years at C-4; $t_{1/2}$ = 44 min at C-2.[174]

c. *Metalation.* Lithiation of 1-substituted imidazoles occurs readily at C-2.[360,381,382] Whereas Tertov[383] reported 5-metalation of 1,2-dimethylimidazole with butyllithium, it has been shown more recently[384] that certainly at low temperatures this does not occur to any extent; rather, lateral metalation on the 2-methyl group is noted.

d. *Diazo Coupling.* The rate of coupling of imidazole with phenyldiazonium cation is reported to be much greater in dimethyl sulfoxide than in water. Since the reaction involves the conjugate base of imidazole, results are complicated by the different acidities of the solvent systems.[385] Correlations with substituent effects have been found for diazo coupling of 2-alkylimidazoles.[386]

e. *Halogenation.* The halogenation of imidazoles has been reviewed to 1970.[5]

There has been considerable study of the bromination of imidazoles[387-391] and benzimidazoles.[243] Determination of the rate laws and isotope effects provide evidence that the bromination of imidazole (and other azoles)

[381] M. S. Shvartsberg, L. N. Bizhan, and I. L. Kotlyarevskii, *Izv. Akad. Nauk SSSR., Ser. Khim.*, 1534 (1971).
[382] R. J. Sundberg, *J. Heterocycl. Chem.* **14**, 517 (1977).
[383] B. Tertov, U. V. Burykin, and I. D. Sadekov, *Khim. Geterotsikl. Soedin.*, 520 (1969), [*CA* **71**, 124328 (1969)].
[384] D. S. Noyce, G. T. Stowe, and W. Wong, *J. Org. Chem.* **39**, 2301 (1974).
[385] Y. Hashida, F. Fujinuma, A. Katsuaki, and K. Matsui, *Nippon Kagaku Kaishi* **8**, 1369 (1975).
[386] O. Cartier, J. P. Paubel, and P. Niviere, *Hebd. Seances Acad. Sci., Ser. C* **284**, 355 (1977).
[387] B. E. Boulton and B. A. W. Coller, *Aust. J. Chem.* **27**, 2349 (1974).
[388] B. E. Boulton and B. A. W. Coller, *Aust. J. Chem.* **27**, 2331 (1974).
[389] B. E. Boulton and B. A. W. Coller, *Aust. J. Chem.* **27**, 2343 (1974).
[390] B. E. Boulton and B. A. W. Coller, *Aust. J. Chem.* **24**, 1413 (1971).
[391] T. Okuyama, K. Kunugiza, and T. Fueno, *Bull. Chem. Soc. Jpn.* **47**, 1267 (1974).

TABLE III

RATE CONSTANTS (k_{bi}^0) FOR NEUTRAL
MONOBROMINATION OF IMIDAZOLES[388]

	$k_{bi}^0 \times 10^{-6}/\text{dm}^3 \text{ mol}^{-1} \text{ sec}^{-1}$		
Substrate	C-5	C-4	C-2
Imidazole	0.5	0.4	0.2
1-Methylimidazole	2.3	1.7	0.8
2-Methylimidazole	90	—	—

involves a rate-determining bimolecular process leading to the formation of Wheland intermediates[243] between the neutral imidazole molecule and molecular bromine. The studies carried out by Coller's group employed halogenation by instalments with coulopotentiometric monitoring. Whereas in imidazoles all three ring hydrogens can be replaced with the order of reactivity 5 > 4 > 2 being followed (in agreement with CNDO/2 calculations), it is evident that the 2-position has unexpectedly high reactivity (see Table III), cf. Noyce.[360] It may be that bromination at C-2 involves an addition–elimination process as shown in Eq. (26). The relative rates of

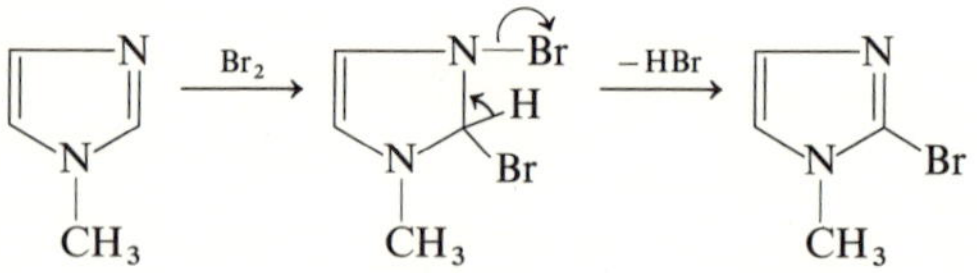

(26)

successive brominations are pH-dependent and conditions may be designed to obtain optimum yields of partially brominated derivatives,[388] a valuable finding in view of the difficulty of preparing monobromoimidazoles. The rather small effect of N-methylation (4-fold across the 1,2- or 1,5-bond) contrasts sharply with 2-methylation which has a 180-fold effect between C-2 and C-5. Such differences demonstrate the importance of bond-fixation effects in heterocyclic compounds. The fused benzene ring in benzimidazole deactivates the 2-position by a factor of greater than 5000.[243]

Direct monobromination of imidazole anions is possible using carbon tetrabromide[392] (monochlorination uses hexachloroethane), while the use of 2,4,6-tetrabromocyclohexa-2,5-dienone has been recommended for both imidazole and 1-methylimidazole.[393]

---

[392] M. Begtrup, *Chem. Commun.*, 334 (1975).
[393] V. Calo, F. Ciminale, L. Lopez, F. Naso, and P. E. Todesco, *J. C. S. Perkin I*, 2567 (1972).

From a study of base catalysis in the iodination of imidazole it is considered[394] that the rate-limiting step is a nucleophilic attack of the base on the hydrogen atom being abstracted from the sigma complex, and that the acceleration follows the nucleophilicity rather than the basicity of the catalyst. The iodinating agent is believed to be of the general form BI_2 where B is the base.[395] It may be that N-iodoimidazoles or imidazole complexes are implicated in the reaction; it has been shown that molecular iodine forms a 1:1 complex with 1-methylimidazole and the stability of this complex rises with solvent polarity.[396]

As with other halogenation processes, the monoiodination of imidazole has proved to be a difficult process. A number of recent approaches have shown some success. Thus, imidazoles can be monoiodinated at C-2 via the 2-lithio derivative or Grignard reagent.[381,397] With iodine and iodic acid 5-chloro-1-methylimidazole was converted to a mixture of 12.8% of the 2,4-diiodo compound and 27.5% of the 4-iodo product.[381] The previously undescribed 2-iodoimidazole (**119**) has been prepared in low yield by the sequence shown in Eq. (27).[382]

f. *Nitration.* The nitration of imidazoles has been reviewed in depth by Schofield.[5] In sulfuric acid medium ("mixed acids") the conjugate acid species is involved, and 1-phenylimidazole, for example, gives mainly 1-(*p*-nitrophenyl)imidazole via the cationic heterocycle.[398] When nitric acid in acetic anhydride is used, 1-phenylimidazole forms only the nitrate salt, a result which contrasts sharply with similar nitrations of 1-arylpyrazoles which give 4- (and sometimes 3-) nitropyrazoles.[398] That it is the more strongly basic nature of the imidazole (compared to pyrazoles) which is responsible for this behavior is evidenced by the failure of imidazole to be

---

[394] L. Schutte and E. Havinga, *Tetrahedron* **26**, 2297 (1970).
[395] W. J. Barry, P. Birkett, and I. L. Finar, *J. Chem. Soc. C*, 1328 (1969).
[396] V. N. Sheinker, A. D. Garnovskii, L. G. Tishchenko, and O. A. Osipov, *J. Gen. Chem. USSR (Engl. Transl.)* **44**, 2208 (1974).
[397] B. A. Tertov, V. V. Burykin, P. P. Onischenko, A. S. Morkovnik, and V. V. Bessonov, *Chem. Heterocycl. Compd. (Engl. Transl.)* **9**, 1025 (1973).
[398] M. R. Grimmett, S. R. Hartshorn, K. Schofield, and J. B. Weston, *J. C. S. Perkin II*, 1654 (1972).

N-nitrated in acetic anhydride, but the much more weakly basic 4-nitroimidazole forms good yields of an N-4-dinitro product.[359]

Both 1-methyl-4- and 1-methyl-5-chloroimidazoles can be nitrated in sulfuric acid.[399] Exhaustive nitration of imidazole in "mixed acids" yields in succession 4-mono-, 4,5-di-, and 2,4,5-trinitroimidazole.[358] With dinitrogen tetroxide in acetonitrile 4-substituted imidazoles with electron-withdrawing substituents yield a mixture of 5- and 2-nitro derivatives.[400] Nitrations at C-2 are most unexpected. The nitration of 4-(4′-alkoxyphenyl)imidazoles with 3–4 M nitric acid occurs *ortho* to the alkoxy group and in the imidazole 5-position. With concentrated nitric acid di- and trinitro compounds are formed.[401]

"Ipso" nitration in iodoimidazoles has been reported.[402] Even iodine atoms at C-2 can be replaced in nitric acid, while 4-iodo groups are displaced in nitric–sulfuric acid mixtures. With concentrated nitric acid the trinitroimidazole **120** is formed (Scheme 29), and this is unexpected in view of the

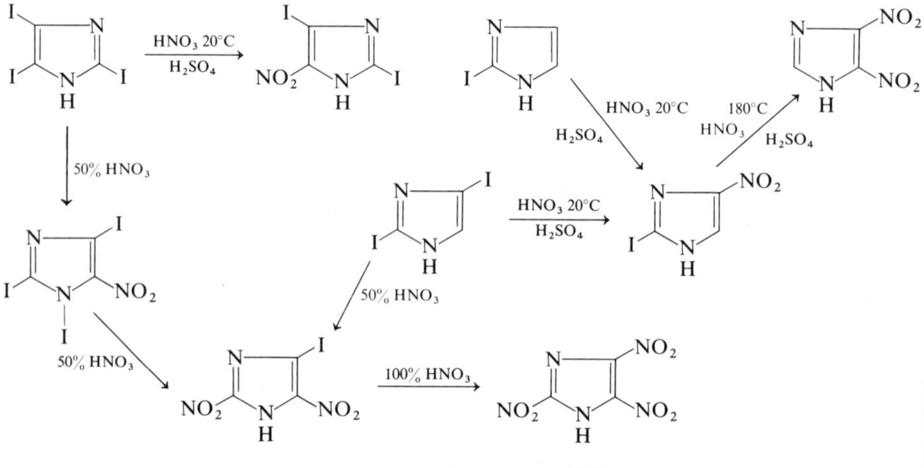

SCHEME 29

[399] K. Shimada, S. Kuriyama, T. Kanazawa, M. Satoh, and S. Toyoshima, *J. Pharm. Soc. Jpn.* **91**, 221 (1971).
[400] S. S. Novikov, L. U. Khmelnitskii, O. V. Lebedev, V. V. Sevastyanova, and L. V. Yepishina, *Chem. Heterocycl. Compd. (Engl. Transl.)* **6**, 669 (1970).
[401] M. A. Iradyan, A. G. Torosyan, R. G. Mirzoyan, and A. A. Aroyan, *Chem. Heterocycl. Compd. (Engl. Transl.)* **13**, 1110 (1977).
[402] S. S. Novikov, L. I. Khmelnitskii, O. V. Lebedev, L. V. Epishina, and V. V. Sevastyanova, *Chem. Heterocycl. Compd. (Engl. Transl.)* **6**, 614 (1970).

suggestion[403,404] that nitrodehalogenation should decrease in facility as acid concentration increases—water is believed to be necessary to remove Br^+ from the Wheland intermediate.

g. *Miscellaneous.* In a novel electrophilic substitution, trifluoromethylsulfenyl chloride gives 4-trifluoromethylthioimidazoles.[405,406] 2-Arylimidazoles can be chlorosulfonated at C-4 by chlorosulfonic acid.[407]

C. Nucleophilic Substitution Reactions

Molecular orbital calculations[248] predict that in unsubstituted imidazoles any nucleophilic attack should take place most easily at C-2. Electron-withdrawing substituents, however, are usually necessary before any such displacements are possible, and these may also modify the position of attack. Quaternization of the heterocycle makes it even more susceptible to such reactions.[408]

Many of the nucleophilic substitution reactions described are halogen displacements which take place most readily when the halogen is activated by an electron-withdrawing group in the molecule. Thus, attempts to prepare 1-methyl-4- and 1-methyl-5-piperidinoimidazole by nucleophilic displacement of halogen failed,[409] but 1-methyl-2-bromoimidazole could be converted into the 2-piperidino compound by heating with piperidine at 200°C for 60 hours. A much stronger nucleophile, lithium piperidide, reacts with 1-methyl-5-bromo- and 1-methyl-5-chloroimidazoles to give the results shown in Table IV. The formation of 4-halo compounds is an isomerization relatively common in basic medium, but the 2-piperidino compounds can only be explained in terms of the addition–elimination process shown in Scheme 30. Again, the reaction of 5-bromo-1-methylimidazole with potassium amide in liquid ammonia does not result in nucleophilic substitution, but rather in migration and elimination of halogen to give a mixture of

---

[403] R. B. Moodie, K. Schofield, and J. B. Weston, *J. C. S. Perkin II*, 1089 (1976).
[404] K. C. Chang, M. R. Grimmett, D. D. Ward, and R. T. Weavers, *Aust. J. Chem.* **32**, 1727 (1979).
[405] T. S. Croft and J. J. McBrady, *J. Heterocycl. Chem.* **12**, 845 (1975).
[406] D. M. Mulvey and H. Jones, *J. Heterocycl. Chem.* **12**, 597 (1975).
[407] J. J. Baldwin, P. K. Lumma, G. S. Ponticello, F. C. Novello, and J. M. Sprague, *J. Heterocycl Chem.* **14**, 889 (1977).
[408] M. Begtrup, *Angew. Chem., Int. Ed. Engl.* **13**, 347 (1974).
[409] D. A. de Bie, H. C. van der Plas, and G. Guertsen, *Recl. Trav. Chim. Pays-Bas* **90**, 594 (1971).

TABLE IV
PRODUCTS OF REACTION OF HALOIMIDAZOLES AND LITHIUM PIPERIDIDE[409]

Substrate	Products	Yield (%)
5-Bromo-1-methylimidazole	5-Bromo-1-methylimidazole	49
	4-Bromo-1-methylimidazole	3
	1-Methyl-5-piperidinoimidazole	25
	1-Methyl-2-piperidinoimidazole	16
5-Chloro-1-methylimidazole	5-Chloro-1-methylimidazole	70
	4-Chloro-1-methylimidazole	Trace
	1-Methyl-5-piperidinoimidazole	15–20
	1-Methyl-2-piperidinoimidazole	2–5

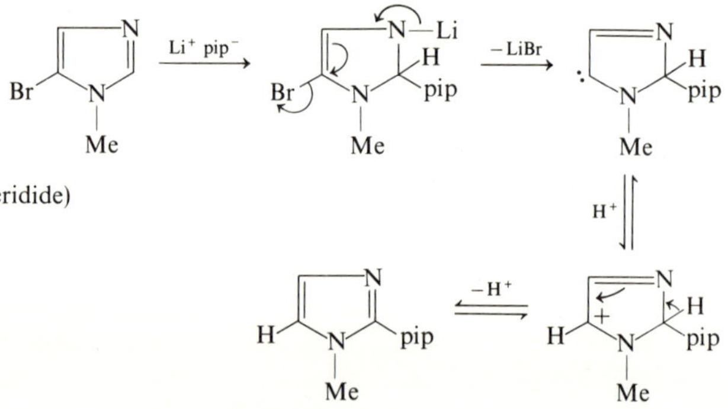

(pip⁻ = piperidide)

SCHEME 30

4-bromo-1-methyl- and 1-methylimidazole.[410] 2-Bromo-1,4-dimethyl-5-nitroimidazole reacts with much greater facility and forms the 2-amino compound when heated at 75°C in a sealed tube with ethanolic ammonia,[411] and at rather lower temperatures (20°C) 4-chloro-1-methyl-5-nitroimidazole reacts with cyclic secondary amines to give the 4-amino compounds. At 80°C some 4,5-diaminated product results.[412] A 2-bromo substituent may be replaced by amino even when activated only by phenyl groups at C-4 and C-5.[413,414] The behavior of 5-bromo-4-sulfonamidoimidazole is interesting;

[410] D. A. de Bie and H. C. van der Plas, *Recl. Trav. Chim. Pays-Bas* **88**, 1246 (1969).
[411] R. F. Miller and R. E. Bambury, *J. Med. Chem.* **14**, 1217 (1971).
[412] P. M. Kochergin, A. M. Tsyganova, V. S. Shlikhunova, and M. A. Klykov, *Chem. Heterocycl. Compd. (Engl. Transl.)* **7**, 648 (1971).
[413] B. A. Priimenko and P. M. Kochergin, *Khim. Geterotsikl. Soedin.* **7**, 1248 (1971) [*CA* **76**, 25174 (1972)]
[414] B. A. Priimenko, *Khim. Issled Farm.*, 55 (1970) [*CA* **75**, 129722 (1971)].

ammonia replaces the bromine by amino, while ethanolic ammonia at 180°C (or diethylamine) also displaces the sulfonamido group.[415,416]

Besides substitution by amino, a chloro or bromo substituent may also be replaced by iodo, hydroxy, and alkoxy,[412] though it is not possible to prepare 4-fluoroimidazoles by halogen exchange with metal fluorides on activated bromo-[417] or chloroimidazoles.[418] 4-Fluoroimidazoles are relatively resistant to nucleophiles, but the corresponding 2-fluoro compounds are subject to displacement (via an addition–elimination reaction) under mild conditions[180] (see also Section III,B). Undoubtedly, the intermediate formation of a stable symmetrical 2-fluoroimidazolium cation (120) is responsible for this ease of displacement. (Eq. 28). Iodine at all ring carbons

$$\underset{F}{\underset{\mathrm{N}}{\fbox{}}}\mathrm{NH} \xrightarrow{\mathrm{RSH}} \underset{F}{\underset{\mathrm{HN}}{\fbox{+}}}\mathrm{NH} \quad \mathrm{RS}^- \longrightarrow \underset{\mathrm{RS}\;\;F}{\mathrm{HN}\fbox{}\mathrm{NH}} \xrightarrow{-\mathrm{F}^-} \underset{\mathrm{SR}}{\underset{\mathrm{HN}}{\fbox{+}}}\mathrm{NH} \quad (28)$$

(120)

in 1-methylimidazole is catalytically displaced (as in the pyrazoles) by acetylenic groups.[419] The reactions of 4-halo- and 4-methylthioimidazolium salts with sulfide or methanethiolate anions are accompanied by (*i*) ring cleavage to give *N*-methylthioamides, (*ii*) replacement of halogen, or (*iii*) reduction. Subsequent thiation at C-2 can lead to 2-thiones.[420]

Nucleophilic displacements of imidazole diazonium salts have some synthetic utility, particularly for the introduction of groups at C-2. Thus 2-nitroimidazoles[421] and 2-azidoimidazoles[417,418] are available via the diazonium fluoroborates. The formation of halogen-substituted imidazoles by the photolytic (but not thermal) decomposition of diazonium fluoroborates has proved useful.[66,417,422] While 2- and 4-fluoro- and 2-chloroimidazoles can be prepared in this way, the reaction fails for iodo- and bromoimidazoles.[417,422]

---

[415] V. S. Mokrushin, V. I. Nifontov, Z. V. Pushkareva, and V. I. Ofitserov, *Khim. Geterotsikl. Soedin.* **7**, 1421 (1971) [*CA* **76**, 46137 (1972)].
[416] V. I. Nifontov, V. S. Mokrushin, Z. N. Pushkareva, and V. I. Ofitserov, *Chem. Heterocycl. Compd. (Engl. Transl.)* **7**, 262 (1971).
[417] K. L. Kirk and L. A. Cohen, *J. Am. Chem. Soc.* **95**, 4619 (1973).
[418] K. L. Kirk and L. A. Cohen, *J. Am. Chem. Soc.* **93**, 3060 (1971).
[419] M. S. Shvartsberg, L. N. Bizhan, E. E. Zaev, and I. L. Kotlyarevskii, *Bull. Acad. Sci. USSR* **21**, 426 (1972).
[420] M. Begtrup, *J. C. S. Perkin I*, 521 (1979).
[421] G. Lancini and E. Lazzari, British Patent 1,114,154 (1970) [*CA* **75**, 140848 (1971)].
[422] K. L. Kirk, *J. Org. Chem.* **43**, 4381 (1978).

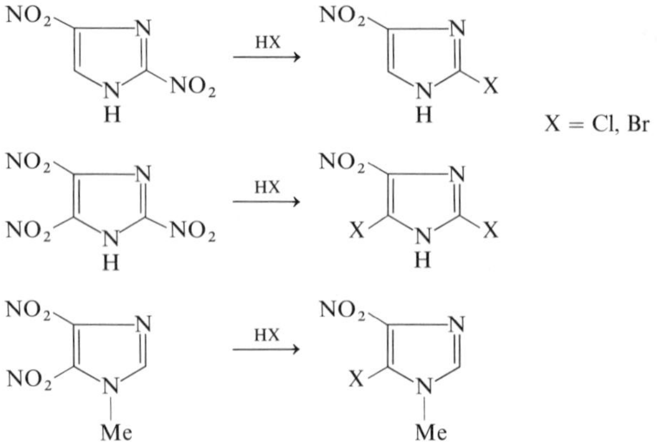

308   M. R. GRIMMETT   [Sec. IV.C

SCHEME 31

When there is more than one nitro group in the imidazole ring, halodenitrations are possible (see Scheme 31), the order of displacement being 2-nitro > 5-nitro > 4-nitro.[348,349]

Although strongly basic conditions tend to cleave the imidazole ring,[420] hydroxylation reactions are sometimes possible. The fusion of 1-substituted imidazoles at 270–320°C with powdered potassium hydroxide gives imidazolones (**121**)[423] (Scheme 32). It is possible to replace halogen groups by hydroxy in suitably substituted imidazoles,[412] but hydroxide ion (or thiolate anion) often causes decomposition. The cleavage of a 5-nitroimidazole derivative by hydroxide was ascribed to nucleophilic attack at C-4.[424] Takeuchi[174] has examined this instability of *N*-alkylnitroimidazoles and has demonstrated that the rate of decomposition increases sharply with base concentration and temperature. Under comparable conditions it has been shown that 1-methyl-4- and 1-methyl-2-nitroimidazoles are 50–150

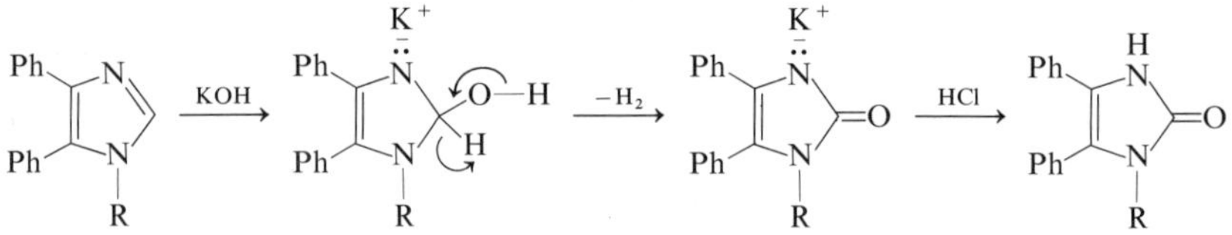

SCHEME 32

---

[423] I. S. Kashparov and A. F. Pozharskii, *Chem. Heterocycl. Compd.* (*Engl. Transl.*) **7**, 116 (1971).

[424] A. F. Cockerill, R. C. Harden, and D. N. B. Mallen, *J. C. S. Perkin II*, 1582 (1972).

times more stable than 1-methyl-5-nitroimidazole. Since the first step in the breakdown of the 4- and 5-nitro compounds is believed to be the β-addition of hydroxide ion to the 4,5-bond leading to **122** and **123**, respectively,[174]

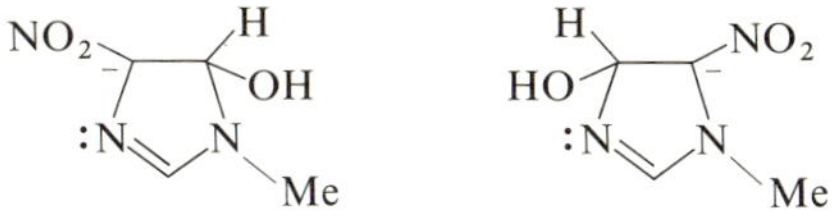

SCHEME 33

the greater stability of the 4-nitro compound may be a consequence of the fact that **122** cannot form so readily, being subject to an adjacent lone-pair effect not present in **123**.[174] The negative charge, however, might be expected to reside on the nitro group, and the foregoing argument may not be particularly sound.

Nucleophilic substitution in the not very common 1-substituted imidazole 3-oxides has been described.[77] Although the Reissert reaction failed to introduce a cyanide group into the ring, this was accomplished by converting the oxides into the quaternary salts, then treating these with potassium cyanide to generate 2-cyanoimidazoles. In the deoxygenation of the oxides with phosphorus trichloride some 2-chloroimidazoles were obtained.[77]

D. REACTIONS INVOLVING RADICALS: PHOTOCHEMICAL REACTIONS

This section covers radical substitution reactions, photochemical additions, cyclizations, and rearrangements, as well as a few other reactions that cannot be classified conveniently under other headings.

Radical substitution reactions include alkylations and arylations in the main. Nucleophilic radicals produced by the silver-catalyzed oxidative decarboxylation of carboxylic acids (by peroxydisulfate ion) attack protonated azoles at the most electron-deficient sites.[369,425-427] Thus, imidazole and 1-alkylimidazoles are methylated exclusively at C-2 in rather low yields.[369] The use of isopropyl and *t*-butyl radicals gives improved yields,[425] but benzyl and acyl radicals tend to dimerize rather than substitute the

---

[425] F. Bertini, R. Galli, F. Minisci, and O. Porta, *Chim. Ind.* (*Milan*) **54**, 223 (1972) [*CA* **77**, 5403 (1972)].
[426] R. Bernardi, F. Bertini, R. Galli, and M. Perchinummo, *Tetrahedron* **27**, 3575 (1971).
[427] F. Minisci, R. Galli, V. Malatesta, and T. Caronna, *Tetrahedron* **26**, 4083 (1970).

heterocycle.[359,369,428] The copper-catalyzed decomposition of benzoyl peroxide in acetic acid selectively phenylates imidazole at C-2.[429]

The photoaddition of acetone to 1-, 2-, and 1,2-dimethylimidazole gives α-hydroxyalkylimidazoles (but no oxetanes) which result from selective attack of excited carbonyl oxygen at C-5[430] (see Scheme 34). In the case of 2-methylimidazole the products are the 4-mono- (8%) and the 4,5-di- (14.5%) substituted compounds. Imidazole itself does not react, reflecting the requirement for more electron-rich substrates. The formation of mainly 4-substituted products is a consequence of the difference in reactivity of the 4- and 5-positions, and also the stabilities of the radical intermediates derived from carbonyl oxygen attack at C-4 or C-5. A suggested reaction mechanism is shown in Scheme 34. Simple Hückel calculations of radical

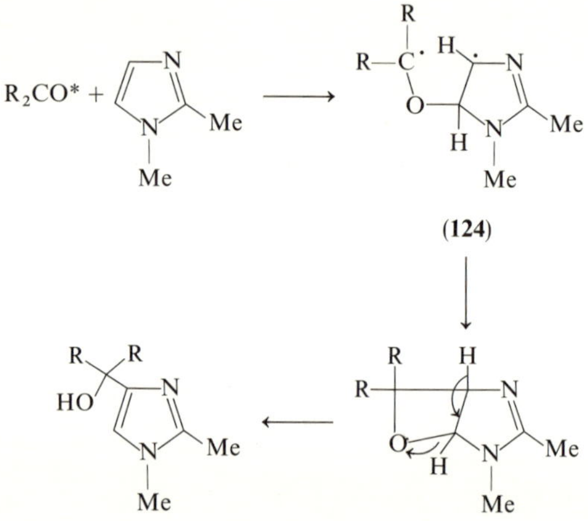

(124)

SCHEME 34

reactivity indices show that C-5 is more reactive, and that the biradical intermediate (124) is more stable than that substituted at C-4.[430] Benzophenone was found to form oxetanes (125) across the 4,5-bond in its photoaddition to 1-acylimidazoles.[431] However, with 1,2-dimethylimidazole (while acetone gives the secondary alcohol at C-4[430,431]) benzophenone adds at the 2-methyl group; with 1-benzylimidazole the methylene group is also involved[431] (see Scheme 35).

---

[428] A. Clerici, F. Minisci, and O. Porta, *Tetrahedron* **29**, 2775 (1973).
[429] H. J. M. Dou, G. Vernin, and J. Metzger, *Chim. Acta Turc.* **2**, 82 (1974).
[430] T. Matsuura, A. Banba, and K. Ogura, *Tetrahedron* **27**, 1211 (1971).
[431] T. Nakano, C. Rivas, C. Perez, and J. M. Larrauri, *J. Heterocycl. Chem.* **13**, 173 (1976).

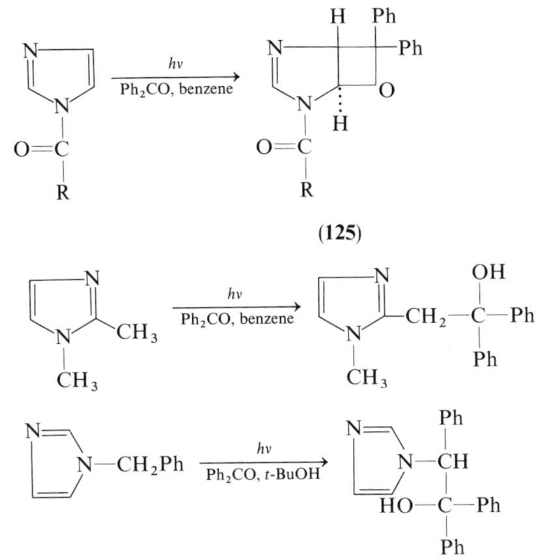

Sec. IV.D]   ADVANCES IN IMIDAZOLE CHEMISTRY

SCHEME 35

With acrylonitrile, 1-methyl-2,4,5-triphenylimidazole (**126**) also forms an addition product (**127**) across the 4,5-bond (similar to the oxetane formation)[432-434] (Eq. 29). That the reaction is remarkably sensitive to substituents,

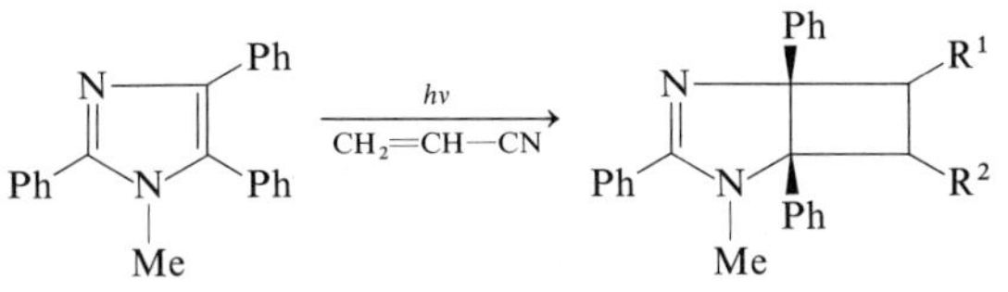

$$\begin{pmatrix} R^1 = H, R^2 = CN \\ R^1 = CN, R^2 = H \end{pmatrix}$$

(29)

nitriles, and solvents used is borne out by the observation[433] that 2,4,5-triphenylimidazole reacts with acrylonitrile in both ethanol and acetonitrile to give **128**, whereas 1-methyl-2,4,5-triphenylimidazole gives the [2 + 2]-cycloaddition product (**127**) in ethanol, and **129** with acetonitrile as solvent.

---

[432] M. J. Rance, C. W. Rees, P. Spagnolo, and R. C. Storr, *J. C. S., Chem. Commun.*, 658 (1974).
[433] Y. Ito and T. Matsuura, *J. Org. Chem.* **44**, 41 (1979).
[434] Y. Ito and T. Matsuura, *Tetrahedron Lett.*, 513 (1974).

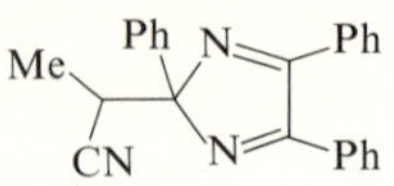

(128) 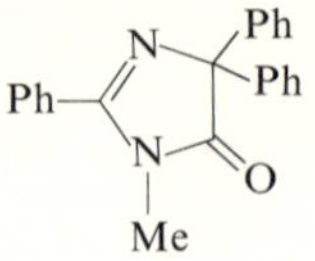   (129)

It has been concluded that encounter complexes (or exciplexes) and ion-pairs are the key intermediates in these reactions.[433] Addition of acrylonitrile to N-unsubstituted imidazoles generally gives α-cyanethylation while cycloaddition products arise with N-substituted compounds.[434]

The photochemical decomposition of imidazole diazonium fluoroborates to fluoroimidazoles has been discussed earlier (Section IV,C).

An example of radical cyclization onto an imidazole ring is provided by the reaction of the diazonium salt of **130** (Eq. 30) with H_3PO_2; again, attack is at C-2.[432] The photocyclization of 1-styrylimidazoles has been reported.[435]

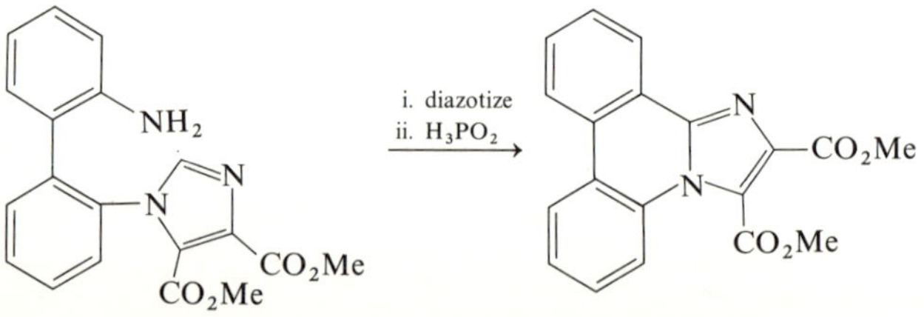

(30)

(130)

The photo-Fries rearrangement of 1-acylimidazoles could follow either a dissociative or intramolecular pathway as depicted in Scheme 36.[436,437] The rearrangement leads to 2- and 4-acylimidazoles. More complex acyl groups (e.g., stearoyl) also undergo cleavage in the side chain.[438,439] N-Acylimidazoles of dehydroabietic acid (**131**) did not undergo elimination reactions, but migrations, probably via a cyclobutanol intermediate (**132**), gave 2- and 4-acylimidazoles[193] (see Scheme 37).

The remarkable formation of 2-benzoyl-6-nitrobenzimidazole by irradiation of 1-(2,4-dinitrophenyl)-4-phenylimidazole has been reported.[440] Either a dipolar cycloaddition or a radical process could account for the formation

---

[435] G. Cooper and W. J. Irwin, *J. C. S. Perkin I*, 75 (1976).
[436] S. Iwasaki, *Helv. Chim. Acta* **59**, 2738 (1976).
[437] S. Iwasaki, *Helv. Chim. Acta* **59**, 2753 (1976).
[438] S. Iwasaki, *Helv. Chim. Acta* **61**, 2831 (1978).
[439] T. Yatsunami and S. Iwasaki, *Helv. Chim. Acta* **61**, 2823 (1978).
[440] P. Bouchet, C. Coquelet, J. Elguero, and R. Jacquier, *Bull. Soc. Chim. Fr.*, 192 (1976).

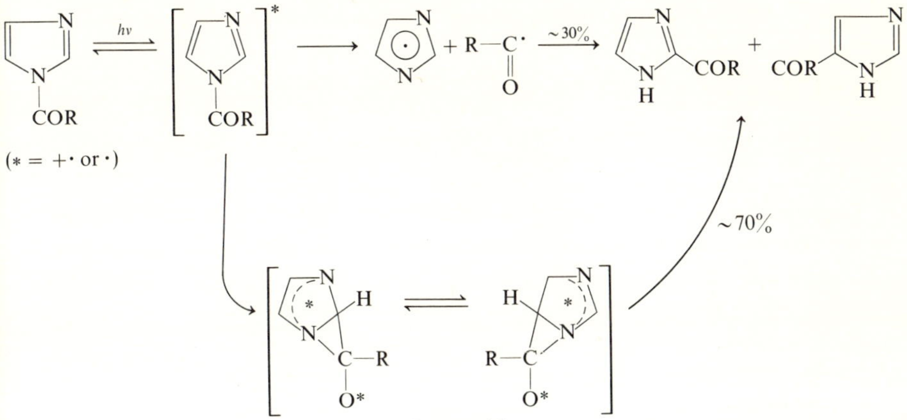

SCHEME 36

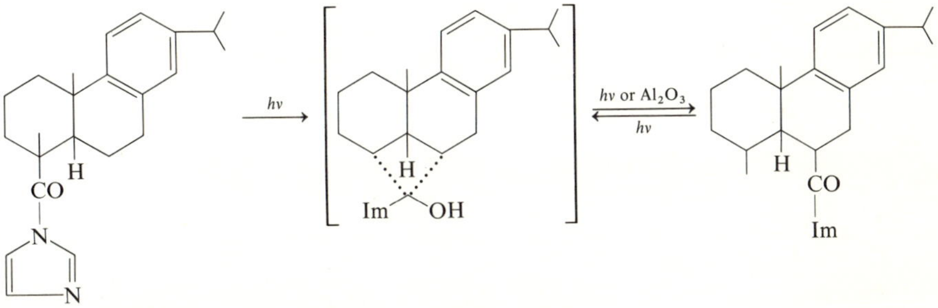

SCHEME 37

(Im = 2- or 4-imidazolyl)

of the *o*-nitrosoimine intermediate from the starting material. Intramolecular cyclization of this intermediate gives rise to the 2-substituted benzimidazole 1-oxide, an enamine which can be hydrolyzed to the 2-benzoyl-1-hydroxy compound. This is not isolated but is deoxygenated during the isolation procedure to give 2-benzoyl-6-nitrobenzimidazole (Eq. 31). The photolysis of 4-amino-5-cyanoimidazole gives 1,6-dihydroimidazo[4,5-*d*]imidazole.[441–443] When imidazole 3-oxides are photolyzed, the products are unsymmetrical benzil diimines.[444]

[441] J. P. Ferris and F. R. Antonucci, *J. Am. Chem. Soc.* **96**, 2010 (1974).
[442] J. P. Ferris and J. E. Kuder, *J. Am. Chem. Soc.* **92**, 2527 (1970).
[443] F. R. Antonucci and J. P. Ferris, *Chem. Commun.*, 126 (1972).
[444] G. J. Gainsford and A. D. Woolhouse, *J. C. S., Chem. Commun.*, 857 (1978).

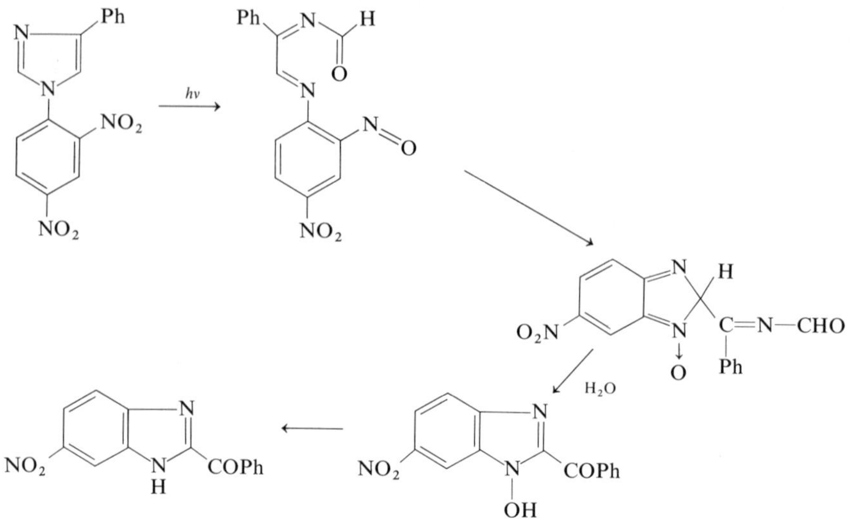

(31)

Although the photoisomerization of imidazoles to pyrazoles is unknown, there is a photoequilibrium between 1,4- and 1,2-dimethylimidazole.[445]

The luminescence of 1,2,5-triphenyl- and other arylimidazoles has continued to be of interest.[29,446,447]

The flash photolysis of hexaarylbiimidazole produces imidazolyl radicals[448] which have been shown[192] to be more nearly planar than the parent dimers; *ortho*-substituents in the aryl rings decrease the radical stability. The radicals oxidize electron-rich substrates by rapid electron abstraction from tertiary amines, iodide, and metal ions, and by hydrogen abstraction from phenols, mercaptans, secondary amines, and active methylene compounds.[449] Studies have been made of the photooxidation of *leuco*-triphenylmethane dyes by these radicals.[450,451]

---

[445] P. Beak and W. Messer, *Tetrahedron* **25**, 3287 (1969).
[446] M. I. Knyazhanskii, P. V. Gilyanovskii, and O. A. Osipov, *Chem. Heterocycl. Compd.* (*Engl. Transl.*), 1160 (1977).
[447] K. Maeda and T. Hayashi, *Bull. Chem. Soc. Jpn.* **44**, 533 (1971).
[448] R. H. Riem, A. MacLachlan, G. R. Coraor, and E. J. Urban, *J. Org. Chem.* **36**, 2272 (1971).
[449] L. A. Cescon, G. R. Coraor, R. Dessauer, A. S. Deutch, H. L. Jackson, A. MacLachlan, K. Marcali, E. M. Potrafke, R. E. Read, E. F. Silversmith, and E. J. Urban, *J. Org. Chem.* **36**, 2267 (1971).
[450] A. MacLachlan and R. H. Riem, *J. Org. Chem.* **36**, 2275 (1971).
[451] R. L. Cohen, *J. Org. Chem.* **36**, 2280 (1971).

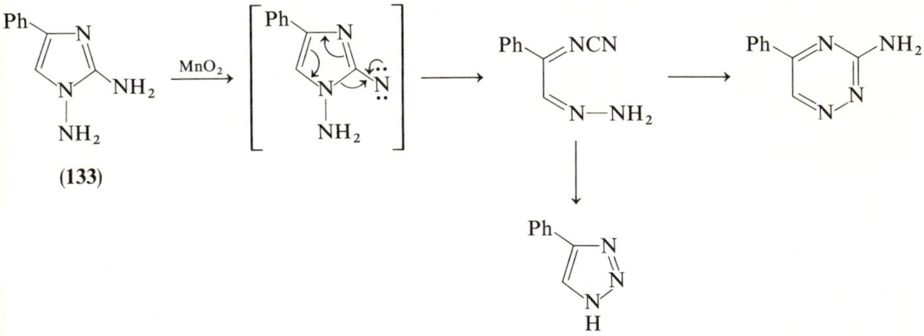

Sec. IV.E]   ADVANCES IN IMIDAZOLE CHEMISTRY   315

SCHEME 38

Oxidation of the diaminoimidazole **133** gives a mixture of triazine and triazole, possibly via the *C*-nitrene shown in Scheme 38.[452,453]

E. THERMAL REACTIONS

1. *Thermal Decomposition Reactions*

The pyrolysis of imidazolium halides substituted on the nitrogen atoms by alkyl or aryl groups leads to 1-substituted imidazoles. The likely mechanism is an S_N2 process in view of the behavior of differing alkyl groups and differing anions.[208] Thus, an ethyl group is cleaved less readily than the smaller methyl group (see Scheme 39). Electron-withdrawing groups at C-4 in the imidazole ring assist in cleavage of an adjacent N—C bond. The cleavage of an allyl group almost certainly involves an S_N2' process. Relative rates of cleavage of groups (based on methyl = 1) are: allyl (2.8), methyl (1.0), benzyl (0.56), ethyl (0.14), propyl or butyl (0.04 ~ 0.09), isopropyl (0.01 ~ 0.03), vinyl and phenyl (0).[208]

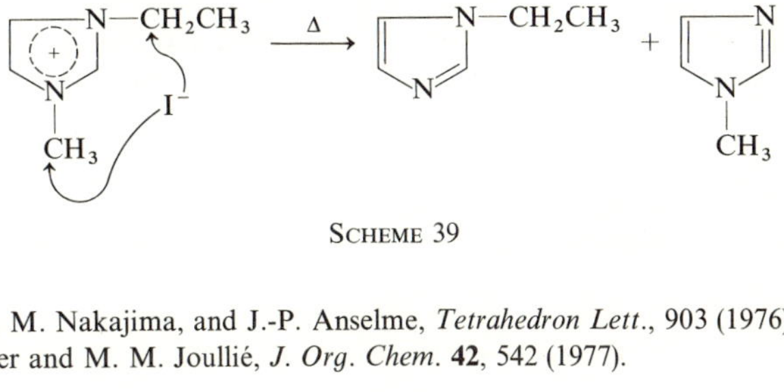

SCHEME 39

[452] R. Hisada, M. Nakajima, and J.-P. Anselme, *Tetrahedron Lett.*, 903 (1976).
[453] A. V. Zieger and M. M. Joullié, *J. Org. Chem.* **42**, 542 (1977).

2. Thermal Rearrangements

A detailed reexamination[233] of the thermal rearrangement of 1-methylimidazole originally reported by Wallach[454] has been extended to encompass a wide range of 1-substituted imidazoles.[233] The reaction appears to be of potential synthetic utility since it leads easily to 2-alkyl- and 2-arylimidazoles. It is an irreversible reaction, uncatalyzed, intramolecular, and does not involve radicals: [1,5]-sigmatropic shifts, as shown in Scheme 40, are probably implicated. The major product is the 2-substituted imidazole (**134**), but small amounts of the 4- (or 5-) isomer (**135**) are also formed. A 1-allyl substituent migrates with equal facility to both 2- and 4(or 5)-positions, suggesting that a Cope-type rearrangement may also be involved for this substituent.[233]

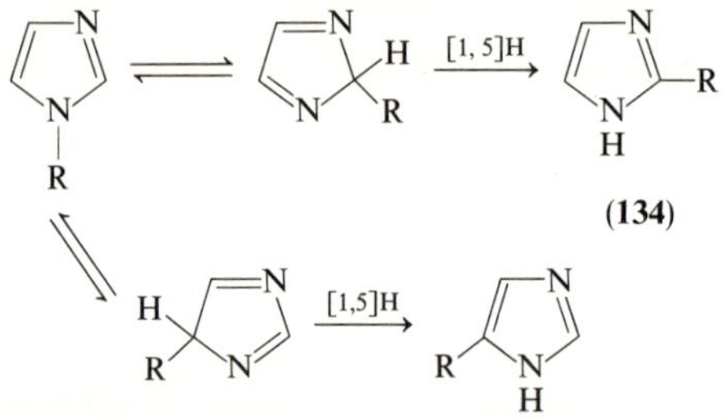

SCHEME 40

The migration of an N-nitro substituent to a ring carbon has also been noted: thus at 115–120°C in chlorobenzene[455] or anisole[359] 1,4(5)-dinitroimidazole gives 2,4-dinitroimidazole[359,455] and 4,5-dinitroimidazole.[359] There is also considerable denitration.[359] 2-Methyl-1,4(5)-dinitroimidazole also gives the 4,5-dinitro isomer on thermolysis, but 4-methyl-1,5-dinitroimidazole fails to rearrange.[359]

The thermal rearrangement of 2,2-dialkyl-2H-imidazoles to the 1,2-dialkyl isomers proceeds by a concerted mechanism.[456] In the absence of hydrogen sources the thermolysis of 2,4,4-triaryl-5-methylthio-4H-imidazoles gives 1,1′-biimidazolyl products by a free-radical process.[457]

---

[454] O. Wallach, *Ber. Dtsch. Chem. Ges.* **16**, 534 (1883).
[455] G. P. Sharnin, R. Kh. Fassakhov, and P. P. Orlov, USSR Patent 458,553 (1974) [*CA* **82**, 156316 (1975)].
[456] J. H. M. Hill, T. R. Fogg, and H. Guttman, *J. Org. Chem.* **40**, 2562 (1975).
[457] G. Domany, J. Nyitrai, K. Lempert, W. Voelter, and H. Horn, *Chem. Ber.* **111**, 1464 (1978).

F. REACTIONS OF SUBSTITUENT GROUPS

1. Acyl (and Aroyl) Groups

N-Acylimidazoles (azolides) have continued to provide a synthetic source of acid derivatives,[298,458–460] and the various "olysis" reactions of these compounds have been studied.[312,459,461,462] The reactions of acetylimidazolium ions with amines have been found to depend largely on amine basicity.[461,462] The thermal decarboxylation of N-alkoxycarbonylimidazoles provides a convenient synthetic route to N-alkylimidazoles.[307] At low temperatures the E-isomer of 1-acetylimidazole predominates.[463]

A facile synthesis of 2-formylimidazole (88% yield) from the "Regel intermediate" (136) has been described[464] (see Scheme 41).

Condensation reactions of imidazole C-aldehydes and -ketones with amino compounds, active methylene compounds, and other nucleophiles

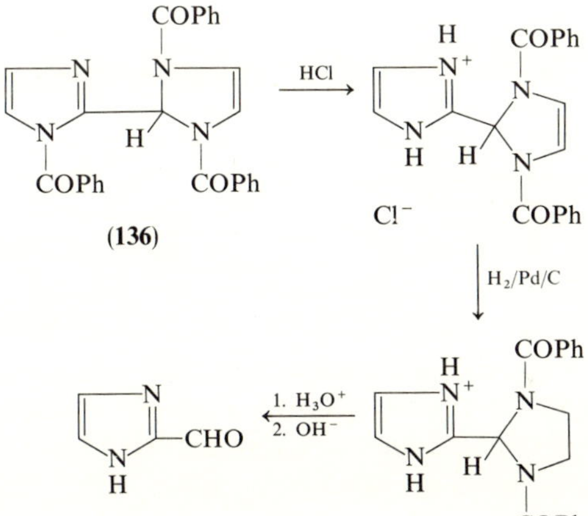

SCHEME 41

---

[458] M. J. Bourgeois, C. Filliatre, R. Lalande, B. Maillard, and J. J. Villenave, *Tetrahedron Lett.*, 3355 (1978).
[459] A. A. Zalikin, L. P. Nikitenova, and Yu. A. Strepikheev, *J. Gen. Chem. USSR (Engl. Transl.)* **43**, 1749 (1973).
[460] T. Katsuki, *Bull. Chem. Soc. Jpn.* **49**, 2019 (1976).
[461] D. G. Oakenfull and W. P. Jencks, *J. Am. Chem. Soc.* **93**, 178 (1971).
[462] D. G. Oakenfull, K. Salvesen, and W. P. Jencks, *J. Am. Chem. Soc.* **93**, 188 (1971).
[463] G. Yamamoto and M. Raban, *J. Org. Chem.* **41**, 3788 (1976).
[464] L. A. M. Bastiaansen and E. F. Godefroi, *J. Org. Chem.* **43**, 1603 (1978).

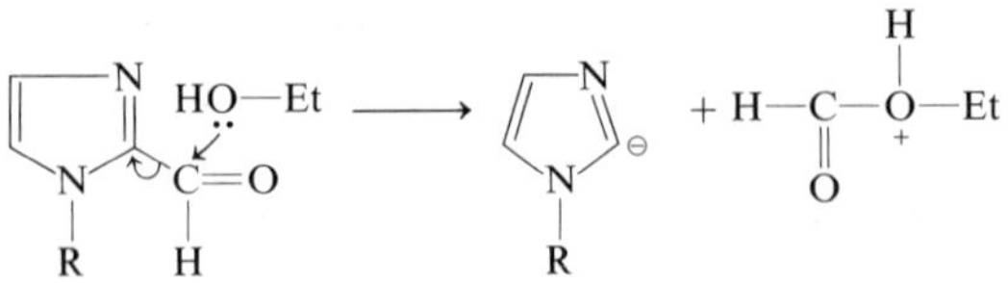

SCHEME 42

abound.[30,360,465-467] The decarbonylation of imidazole-2-aldehyde in ethanol involves nucleophilic attack by the alcohol on the carbonyl carbon (see Scheme 42) to give the imidazole and ethyl formate.[468] It was found that in the acid-catalyzed formation of acetals of imidazole-2-aldehydes the yield of acetal depended on the amount of catalyst. When the H^+:imidazole ratio is ≥ 1, good yields of the diethylacetal are obtained, but when the ratio is <1 decarbonylation occurs to some extent. When only ethanol is used (no acid) only decarbonylation takes place. Electron-withdrawing groups on imidazole assist the reaction, as does quaternization, and it does not occur with 4- and 5-aldehydes.[468]

2. Alkyl, Aryl, and Unsaturated Hydrocarbon Substituents

The thermal rearrangement of N-alkylimidazoles has already been discussed (Section IV,E,2).

Methyl groups, particularly at C-2, are active and can condense with carbonyl compounds to give unsaturated derivatives.[363,364,469,470] When heated with nitrosyl chloride or an N-oxide, 1,2-dimethyl-5-nitroimidazole (137) gives the corresponding 2-cyanoimidazole.[471] Although 137 reacts with dimethylformamide dicyclohexyl acetal to give the dimethylaminoalkene 138, and with pyridine and iodine to give the salt 139,[469] the less activated 2-methylimidazole failed to react with either formamide or trisformyl-

---

[465] L. Del Corona, G. G. Massaroli, G. Signorelli, and G. Musa, Boll. Chim. Farm. 110, 645 (1971) [CA 77, 5402 (1972)].
[466] I. G. Uryukina, I. I. Popov, A. M. Simonov, and L. M. Sitkina, Chem. Heterocycl. Compd. (Engl. Transl.) 9, 723 (1973).
[467] L. Tchissambou, M. Bénéchie, and F. Khuong-Huu, Tetrahedron Lett., 1801 (1978).
[468] I. Antonini, G. Cristalli, P. Franchetti, M. Grifantini, U. Gulini, and S. Martelli, J. Heterocycl. Chem. 15, 1201 (1978).
[469] J. D. Albright and R. G. Shepherd, J. Heterocycl. Chem. 10, 899 (1973).
[470] L. A. M. Bastiaansen, A. A. Macco, and E. F. Godefroi, J. C. S., Chem. Commun., 36 (1974).
[471] G. Berkelhammer and W. H. Gastrock, U.S. Patent 3,740,400 (1973) [CA 79, 42505 (1973)].

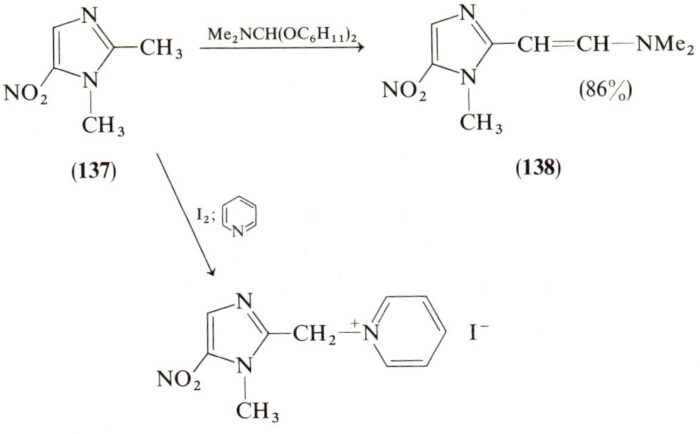

Sec. IV.F] ADVANCES IN IMIDAZOLE CHEMISTRY 319

aminomethane.[300] Some side-chain metalation of 1,2-dimethylimidazole occurs.[384]

Dipole moment studies of the conformations of arylimidazoles have shown that the aryl ring is not always coplanar with the heterocyclic ring.[166,168] ^{13}C-NMR spectroscopy of phenylimidazoles has been studied.[218] In mixed acids 1-phenylimidazole is nitrated as the conjugate acid to give mainly the *p*-nitrophenyl product; with nitric acid in acetic anhydride only the nitrate salt forms.[398] With 4-(*p*-alkoxyphenyl) imidazoles much of the nitration occurs ortho to the alkoxy group[401]; the alkoxynitroaryl substituents are readily oxidized to carboxyl by alkaline permanganate.[401]

Electron-donor substituents facilitate the polarographic oxidation of triarylimidazoles.[472] At 280°C iron(II) oxalate reduces 1-(*o*-nitrophenyl) imidazole mainly to the amino compound (140) with a trace of azo compound, but when there is a methyl group at C-2 or C-4 some of the condensed tricyclic product 141 is isolated.[473]

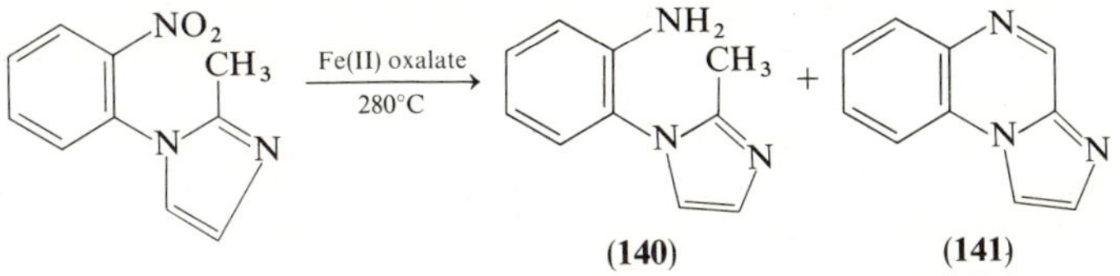

[472] V. N. Shishkin, B. S. Tanaseichuk, L. G. Tikhonova, and A. A. Bardina, *Chem. Heterocycl. Compd. (Engl. Transl.)* **9**, 358 (1978).
[473] R. G. R. Bacon and S. D. Hamilton, *J. C. S. Perkin I*, 1970 (1974).

Vinyl-[18,474–480] and styryl-[333,339,435,466] imidazoles have received considerable attention, particularly from Soviet researchers. Halogenation of 1-vinylimidazole is believed to occur through initial complex formation followed by some addition at the exocyclic double bond.[476] The vinyl function can be reduced by hydrogenation over Raney nickel.[475] Thiols add across the vinyl group and can be induced to add in an anti-Markovnikov mode under the influence of radical inhibitors.[478] At 530°C 1-vinylimidazole rearranges to an 82:12 ratio of 2- and 4-vinyl isomers.[233] There have been a few references to the chemistry of ethynylimidazoles.[246,481]

3. Amino and Diazo Groups

With α-diketones 1- and 2-amino- and 1,2- and 1,5-diaminoimidazoles condense to give fused-ring heterocycles,[482,483] with initial attack at the C-amino function.[483] An N-amino group can be removed by treatment with nitrous acid.[484] Oxidation of quaternary salts of 1-aminoimidazoles gives azoimidazolium salts,[56] while manganese dioxide converts 1,2-diaminoimidazoles into triazoles and triazines[452,453,485] (see Section IV,D).

2-Aminoimidazoles exist as the amino tautomers.[273] Carbonyl compounds condense normally with C-aminoimidazoles,[486,487] which can also be diazotized[213,417,488] although some amines are rather unstable.[213] The diazonium fluoroborates have been transformed into fluoroimid-

---

[474] M. S. Shostakovskii, N. P. Glazkova, E. S. Domnina, L. V. Belousova, and G. G. Skvortsova, *Chem. Heterocycl. Compd. (Engl. Transl.)* **7**, 894 (1971).

[475] M. S. Shostakovskii, G. G. Skvortsova, E. S. Domnina, and L. P. Makhno, *Khim. Geterotsikl. Soedin.*, 1289 (1970) [*CA* **74**, 141657 (1971)].

[476] G. G. Skvortsova, E. S. Domnina, N. P. Glazkova, N. N. Chipanina, and N. I. Chergina, *J. Gen. Chem. USSR (Engl. Transl.)* **41**, 620 (1971).

[477] G. G. Skvortsova, E. S. Domnina, L. P. Makhno, Yu. L. Frolov, V. K. Voronov, N. N. Chipanina, and N. I. Chergina, *Bull. Acad. Sci. USSR.*, 2570 (1971).

[478] G. G. Skvortsova, N. P. Glaskova, E. S. Domnina, and V. K. Voronov, *Chem. Heterocycl. Compd. (Engl. Transl.)* **6**, 153 (1970).

[479] L. B. Volodarskii, L. A. Fust, and V. S. Kobrin, *Chem. Heterocycl. Compd. (Engl. Transl.)* **8**, 994 (1972).

[480] G. Manecke and R. Schlegel, *Chem. Ber.* **107**, 892 (1974).

[481] A. M. Simonov and I. I. Popov, *Chem. Heterocycl. Compd. (Engl. Transl.)* **7**, 132 (1971).

[482] G. M. Golobushina, G. N. Poshtaruk, and V. A. Chuiguk, *Chem. Heterocycl. Compd. (Engl. Transl.)* **10**, 735 (1974).

[483] A. Bernardini, P. Viallefont, and R. Zniber, *J. Heterocycl. Chem.* **15**, 937 (1978).

[484] C. Yamazaki, *Bull. Chem. Soc. Jpn.* **51**, 1846 (1978).

[485] M. Nakajima, R. Hisada, and J. P. Anselme, *J. Org. Chem.* **43** 2693 (1978).

[486] G. Mackenzie and G. Shaw, *J. C. S. Perkin I*, 1381 (1978).

[487] A. Yamazuki and M. Okutsu, *J. Heterocycl. Chem.* **15**, 155 (1978).

[488] Y. F. Shealy and C. A. O'Dell, *J. Heterocycl. Chem.* **10**, 839 (1973).

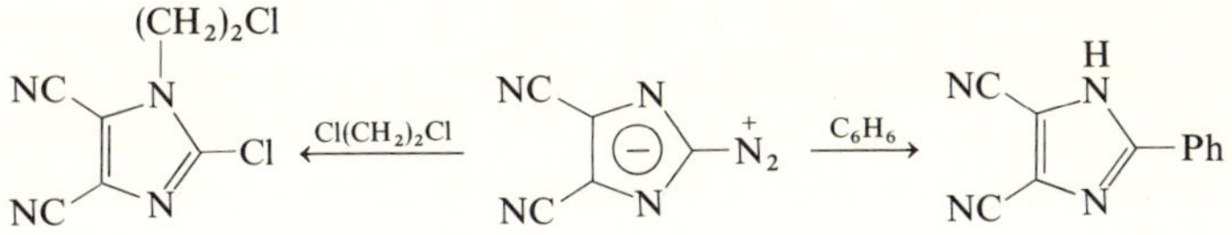

SCHEME 43

azoles.[213,417] With hydrazine, the 5-diazonium salt of imidazole-4-carboxamide gives the 5-azido product, probably via the tetrazene,[488] and other similar examples exist.[417,418] In solution at 80°C 2-diazo-4,5-dicyanoimidazole (142) eliminates nitrogen to form a highly reactive species which inserts 4,5-dicyanoimidazole into substrates which are prone to electrophilic attack[66] (see Scheme 43).

4. *Cyano and Cyanoalkyl Groups*

Alkaline hydrolysis of 2-cyano-1-hydroxyimidazole 3-oxide affords the 2-carboxamide; with barium hydroxide the corresponding carboxylic acid salt is formed.[489] Selective reaction at each of the two different cyano groups of 1,2-disubstituted 4,5-dicyanoimidazole allows the synthesis of purine derivatives of unequivocal structure.[490]

In a reaction specific for cyanoalkyl groups adjacent to an azole ring nitrogen atom, 1-methyl- and 1-benzyl-2-cyanomethylimidazoles are decyanated by hydrazine at 200°C.[491]

5. *Hydroxy and* N-*Oxide Functions*

N-Hydroxy functions can be removed by catalytic hydrogenation over Raney nickel,[489] and an unprecedented dehydroxylation occurs when 1-hydroxyl-2-methyl-5-phenylimidazole is treated in turn with butyl lithium and hexachlorodisilane.[492] Dehydrogenation of 1-hydroxy-2,4,5-triphenylimidazole and its 3-oxide produces the 1-oxide and 1,3-dioxide, respectively.[493] N-Hydroxyimidazole esters can act as acyl transfer reagents.[494]

---

[489] K. Hayes, *J. Heterocycl. Chem.* **11**, 615 (1974).
[490] T. Kojima and Y. Ohtsuka, *A. C. S. Congr. Abstr.* No. 167, (1979).
[491] H. C. van der Plas and H. Jongejan, *Recl. Trans. Chim. Pays-Bas* **91**, 133 (1972).
[492] A. G. Hortmann, J. Koo, and C. C. Yu, *J. Org. Chem.* **43**, 2289 (1978).
[493] K. Volkamer and H. Zimmermann, *Chem. Ber.* **103**, 296 (1970).
[494] D. G. McCarthy and A. F. Hegarty, *J. C. S. Perkin II*, 231 (1977).

Both O- and N-acylation are possible in 2,5-disubstituted 4-imidazolones.[282]

Although imidazole N-oxides cannot be prepared by direct oxidation of the heterocycles, a number have been prepared by cyclization methods (Section II,C and F) and their properties have been studied (Section IV,B and C).

The oxide function can be removed using phosphorus trichloride,[77,79] or phosphorus oxychloride,[77] by reduction with hydrogen and Raney nickel[489] with sodium borohydride,[25] and with hexachlorodisilane in chloroform.[492] The reduction of 4H-imidazole N-oxides with borohydride leads to the 1-hydroxyimidazoline or imidazolidine derivatives; under the same conditions 4H-imidazole N,N'-dioxides give 1,3-dihydroxyimidazolidines.[285]

The thermal rearrangement of 2,4,4-trimethyl-5-phenyl-4H-imidazole 1-oxide to 2,4,4-trimethyl-1-phenyl-2-imidazolin-5-one appears to involve a phenyl migration,[495] while alkaline cleavage gives oximes of α-acylaminoketones.[496]

6. Quaternary Salts

The thermal decomposition has been discussed (Section IV,E,1). Imidazole forms quaternary salts with methylmercury(II).[497] 1-Acetyl-3-alkylimidazolium salts can be dequaternized by hydroxide ion[305] or primary amines[298] to the 1-alkylimidazole.

7. Sulfur Derivatives

In acetic acid solution 30% hydrogen peroxide converts 2-imidazolyl methyl sulfides into sulfones[407,498] and sometimes sulfoxides.[407] In trifluoracetic acid the sulfoxides are formed preferentially[498]; periodate, too, can give the sulfoxides.[406] Oxidation of 4-mercaptoimidazoles under mild conditions gives bis(4-imidazolyl) disulfides which can be cleaved by hydrogen sulfide.[89] With 15% alkaline hydrogen peroxide at 90°C the sulfonic acid is the major product.[89] Imidazole-5-sulfonyl chlorides give sulfonamides

---

[495] L. B. Volodarskii and V. S. Kobrin, *Chem. Heterocycl. Compd. (Engl. Transl.)* **12**, 1179 (1976).

[496] L. B. Volodarskii, V. S. Kobrin, and Yu. G. Putsykin, *Chem. Heterocycl. Compd. (Engl. Transl.)* **8**, 1122 (1972).

[497] C. A. Evans, D. L. Robenstein, G. Geier, and I. W. Erni, *J. Am. Chem. Soc.* **99**, 8106 (1977).

[498] D. R. Hoff and C. S. Rooney, German Patent 2,124,103 (1972) [*CA* **76**, 72514 (1972)].

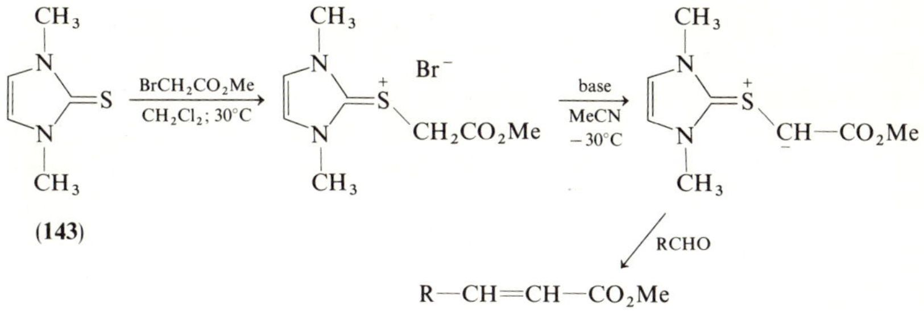

SCHEME 44

with ammonia and can be reduced to the thiols.[407,499] Mercaptoimidazoles readily form S-glycosides.[500] As shown in Scheme 44, 1,3-dimethylimidazole-2-thione (143) is able to act as a quasi-Wittig reagent.[501] With bromomethyldimethylchlorosilane 2-mercaptoimidazole gives the silylthioether.[502]

8. Miscellaneous Substituents

Reduction over platinum in acetic acid or acetic anhydride converts a nitroimidazole into the acetylamino compound.[503]

The decomposition (below 80°C) of 1-(dialkylphosphoryl)imidazoles to give 1-alkylimidazoles, trialkylphosphate, and 1,3-dialkylimidazolium polyphosphate appears to result from an initial nucleophilic attack by N-3 of one 1-(dialkylphosphoryl)imidazole on the α-alkyl carbon of another.[504]

V. Appendix

ADDITIONAL REFERENCES

Section II,A: Periodate oxidation of D-glucosamine, followed by reaction of the product with ammonium acetate and formaldehyde in the presence of Cu(II), has provided a route of L-*erythro*-β-hydroxyhistidine.[505] A

---

[499] V. I. Ofitserov, Z. S. Mokrushin, Z. V. Pushkareva, and N. V. Nikiforova, *Chem. Heterocycl. Compd.* (*Engl. Transl.*) **12**, 924 (1976).

[500] P. Nuhn and G. Wagner, *J. Prakt. Chem.* **312**, 90 (1970).

[501] E. M. Burgess and M. C. Pulcrano, *J. Am. Chem. Soc.* **100**, 6538 (1978).

[502] H. Alper and M. S. Wolin, *J. Org. Chem.* **40**, 437 (1975).

[503] R. Robinson and M. U. Zubair, *Bull. Chem. Soc. Jpn.* **50**, 561 (1977).

[504] N. Ranganathan and W. S. Briniger, *J. Org. Chem.* **43**, 4853 (1978).

[505] S. M. Hecht, K. M. Rupprecht, and P. M. Jacobs, *J. Am. Chem. Soc.* **101**, 3982 (1979).

synthetic method of this class has also been applied to the preparation of 1,4-dialkyl-5-bromo(or hydroxy)methylimidazoles.[506]

Section II,B: Nitration of symmetrical 1,3-dicarbonyl compounds using N_2O_4, followed by catalytic reduction of the product in acetic formic anhydride gives compounds of type MeCOC(NHCHO)=C(OH)Me which cyclize in formamide or formic acid to give 4-acetyl-5-methylimidazole.[507]

Section II,C: Cyclization of α-acylamino ketimines or α-acylamino hydrazones gives rise to 34–87% yields of imidazoles or 1-aminoimidazoles. The starting materials are readily prepared from α-amino acids.[508] Anhydromercaptoimidazolium hydroxides are available from N-formylsarcosine-N-methylthioamide.[509]

Section II,E: For the photochemical formation of an imidazole to proceed, enaminonitriles must contain an NH group.[510]

Section II,H: A patent[511] refers to the dehydrogenation of imidazolines.

Section III,H: A linear relationship between substituted imidazole pK_a values and CNDO lone-pair orbital energies was found to be limited to substituents which leave the n orbital electron density at nitrogen unchanged.[512]

Section IV,A: The structures of a number of iodonitroimidazoles[189,402] have been shown to be incorrect,[513] and doubt has also been cast on the assigned structures of some iodoimidazoles. Calculations, using the recently developed MNDO semiempirical SCF-MO method, for the cyclization of 2-azidoimidazole and its anion to the tetrazole forms show that the bicyclic anion is more stable than the neutral bicyclic molecule. Such a shift of the equilibrium toward the tetrazole form in the anion is thought to be due mainly to delocalization of the negative charge on the tetrazole moiety of the bicyclic system.[514]

Section IV,B,1,b: The alkylation of 2-substituted imidazoles using alcohols (e.g., dodecanol) in the presence of HCl at 260°C has been described.[515]

---

[506] H. G. Lennartz and W. Schunack, *Arch. Pharm.* (*Weinheim, Ger.*) **310**, 1019 (1977).
[507] E. P. Krebs and E. Bondi, *Helv. Chim. Acta* **62**, 497 (1979).
[508] N. Engel and W. Steglich, *Justus Liebigs Ann. Chem.*, 1916 (1978).
[509] M. Begtrup, *J. C. S. Perkin I*, 1132 (1979).
[510] J. P. Ferris, R. S. Narang, T. A. Newton, and V. R. Rao, *J. Org. Chem.* **44**, 1273 (1979).
[511] T. Dockner, A. Frank, and H. Krug, German Patent 2,733,466 (1978) [*CA* **90**, 204100 (1979)].
[512] B. G. Ramsey, *J. Org. Chem.* **44**, 2093 (1979).
[513] J. P. Dickens, R. L. Dyer, B. J. Hamill, and T. A. Harrow, *J. C. S., Chem. Commun.*, 523 (1979).
[514] S. Olivella and J. Vilarrasa, *J. Heterocycl. Chem.* **16**, 685 (1979).
[515] M. Takahashi, K. Yamada, and M. Nishimura, Japanese Patent 131,577 (1977) [*CA* **88**, 105340 (1978)].

Ethyl chloroacetate alkylates 4,5-disubstituted imidazoles.[516] While benzylation of 4-nitro-5-sulfonamidoimidazole occurred equally on both ring nitrogens,[517] a nitro substituent normally directs attack to the more remote nitrogen.[518] In the synthesis of N^π-alkylhistamines, bulky pivaloyloxymethyl and phthaloyl protecting groups on the N^τ-ring nitrogen and exocyclic amino groups, respectively, direct the incoming alkyl group into the desired N^π-position.[519] A number of 1-vinylimidazoles have been prepared by heating the unsubstituted imidazoles with acetylene in dioxane containing cadmium acetate in an autoclave at 210°C.[520]

Section IV,B,1,d: Phosphoimidazolium compounds have been prepared by the action of an N-alkylimidazole on a disubstituted acid chloride of tetracoordinated phosphorus in anhydrous conditions.[521] Phosphoryl imidazolides are of value for the conversion of aldoximes into nitriles.[522]

Section IV,B,2,e: NMR data indicate that, shortly after complexation of imidazole with iodine or iodine monobromide, the CH and NH ring protons are substituted by halogen. Although complex formation also occurs with 1-methylimidazole, there is no substitution.[523]

Section IV,C: Stable diazacyclopentadienylium salts are formed when nucleophilic displacement of tetrachloro-2H-imidazole takes place with diethyltrimethylisilylamine in the absence of air.[524]

Section IV,D: In methanol or acetone 2-unsubstituted imidazole 2-oxides are transformed photochemically to imidazol-2-ones.[525]

Section IV;E: The nature of the intermediate in the thermolysis of diazodicyanoimidazole has been discussed.[526]

Section IV,F1: An elimination–addition mechanism of acyl group transfer has been proposed for the decomposition of 1-(N-methylcarbamoyl)

---

[516] M. A. Iradyan, A. G. Torosyan, R. V. Agababyan, and A. Aroyan, *Arm. Khim. Zh.* **30**, 756 (1977) [*CA* **88**, 152498 (1978)].
[517] B. S. Huang, M. J. Lauzon, and J. C. Parham, *J. Heterocycl. Chem.* **16**, 811 (1979).
[518] M. A. Iradyan, N. S. Iradyan, and Sh. A. Avetyan, *Arm. Khim. Zh.* **31**, 435 (1978) [*CA* **89**, 197409 (1978)].
[519] J. C. Emmett, F. H. Holloway, and J. L. Turner, *J. C. S. Perkin I*, 1341 (1979).
[520] V. V. Kalmykov, I. B. Lopukh, and V. A. Ivanov, *Zh. Prikl. Khim.* **51**, 2377 (1978) [*CA* **90**, 38840 (1979)].
[521] P. E. Chabrier, H. P. Nguyen, T. T. Nguyen, and P. Chabrier, *C. R. Hebd. Seances Acad. Sci., Ser. C* **288**, 379 (1979).
[522] M. Konieczny and G. Sosnovsky, *Z. Naturforsch., Teil B* **33**, 1033 (1978).
[523] V. N. Sheinker, L. G. Tischenko, A. D. Garnovskii, and O. A. Osipov, *Zh. Org. Khim.* **13**, 2013 (1977).
[524] R. Gompper and K. Bichlmeyer, *Angew. Chem., Int. Ed. Engl.* **18**, 156 (1979).
[525] R. Bartnik and M. Grzegorz, *Rocz. Chem.* **51**, 1747 (1977) [*CA* **88**, 62332 (1978)].
[526] W. A. Sheppard, G. W. Gukel, O. W. Webster, K. Betterton, and J. W. Timberlake, *J. Org. Chem.* **44**, 1717 (1979).

imidazoles.[527] A Wittig reaction between 4-formylimidazole and carboethoxymethyltriphenylphosphonium bromide provides a route to ethyl urocanate.[528] Electrochemical reduction, with lead electrodes, of imidazole carboxylic esters gives hydroxymethyl and ethoxymethyl products.[529]

Section IV,F,8: The synthesis and SR $\rightleftharpoons$ NR rearrangement of thioethers in N- and S-substituted imidazoles has been studied, and an autocatalytic rearrangement process has been proposed.[530–532]

---

[527] H. Al-Rawi, A. Williams, and C. Dowrick, *J. C. S. Perkin II*, 1064 (1979).
[528] K. Honda and T. Yonetani, Japanese Patent 127,470 (1978) [*CA* **90**, 103958 (1979)].
[529] W. L. Mendelson, U.S. Patent 4,055,573 (1977) [*CA* **88**, 50861 (1978)].
[530] J. Kister, G. Assef, G. Mille, and J. Metzger, *Can. J. Chem.* **57**, 813 (1979).
[531] J. Kister, G. Assef, G. Mille, and J. Metzger, *Can. J. Chem.* **57**, 822 (1979).
[532] G. Assef, J. Kister, and J. Metzger, *Bull. Soc. Chim. Fr.*, 165 (1979).

Cumulative Index of Titles

A

Acetylenecarboxylic acids and esters, reactions with N-heterocyclic compounds, **1**, 125
Acetylenecarboxylic esters, reactions with nitrogen-containing heterocycles, **23**, 263
Acetylenic esters, synthesis of heterocycles through nucleophilic additions to, **19**, 297
Acid-catalyzed polymerization of pyrroles and indoles, **2**, 287
t-Amino effect, **14**, 211
Aminochromes, **5**, 205
Anils, olefin synthesis with, **23**, 171
Annulenes, N-bridged, cyclazines and, **22**, 321
Anthracen-1,4-imines, **16**, 87
Anthranils, **8**, 277
Applications of NMR spectroscopy to indole and its derivatives, **15**, 277
Applications of the Hammett equation to heterocyclic compounds, **3**, 209; **20**, 1
Aromatic azapentalenes, **22**, 183
Aromatic quinolizines, **5**, 291
Aromaticity of heterocycles, **17**, 255
Aza analogs of pyrimidine and purine bases, **1**, 189
7-Azabicyclo[2.2.1]hepta-2,5-dienes, **16**, 87
1-Azabicyclo[3.1.0]hexanes and analogs with further heteroatom substitution, **27**, 1
Azapentalenes, aromatic, chemistry of, **22**, 183
Azines, reactivity with nucleophiles, **4**, 145
Azines, theoretical studies of, physicochemical properties of reactivity of, **5**, 69
Azinoazines, reactivity with nucleophiles, **4**, 145
1-Azirines, synthesis and reactions of, **13**, 45

B

Base-catalyzed hydrogen exchange, **16**, 1
1-, 2-, and 3-Benzazepines, **17**, 45

Benzisothiazoles, **14**, 43
Benzisoxazoles, **8**, 277
Benzoazines, reactivity with nucleophiles, **4**, 145
Benzo[*c*]cinnolines, **24**, 151
1,5-Benzodiazepines, **17**, 27
Benzo[*b*]furan and derivatives, recent advances in chemistry of, Part I, occurrence and synthesis, **18**, 337
Benzo[*c*]furans, **26**, 135
Benzofuroxans, **10**, 1
2*H*-Benzopyrans (chrom-3-enes), **18**, 159
Benzo[*b*]thiophene chemistry, recent advances in, **11**, 177
Benzo[*c*]thiophenes, **14**, 331
1,2,3-(Benzo)triazines, **19**, 215
Biological pyrimidines, tautomerism and electronic structure of, **18**, 199

C

Carbenes, reactions with heterocyclic compounds, **3**, 57
Carbolines, **3**, 79
Cationic polar cycloaddition, **16**, 289 (**19**, xi)
Chemistry
  of aromatic azapentalenes, **22**, 183
  of benzo[*b*]furan, Part I, occurrence and synthesis, **18**, 337
  of benzo[*b*]thiophenes, **11**, 178
  of chrom-3-enes, **18**, 159
  of diazepines, **8**, 21
  of dibenzothiophenes, **16**, 181
  of 1,2-dioxetanes, **21**, 437
  of furans, **7**, 377
  of isatin, **18**, 1
  of isoxazolidines, **21**, 207
  of lactim ethers, **12**, 185
  of mononuclear isothiazoles, **14**, 1
  of 4-oxy- and 4-keto-1,2,3,4-tetrahydroisoquinolines, **15**, 99
  of phenanthridines, **13**, 315
  of phenothiazines, **9**, 321
  of 1-pyrindines, **15**, 197
  of tetrazoles, **21**, 323
  of 1,3,4-thiadiazoles, **9**, 165
  of thienothiophenes, **19**, 123
  of thiophenes, **1**, 1

Chrom-3-ene chemistry, advances in, **18**, 159
Claisen rearrangements, in nitrogen heterocyclic systems, **8**, 143
Complex metal hydrides, reduction of nitrogen heterocycles with, **6**, 45
Covalent hydration
  in heteroaromatic compounds, **4**, 1, 43
  in nitrogen heterocycles, **20**, 117
Current views on some physicochemical aspects of purines, **24**, 215
Cyclazines, and related N-bridged annulenes, **22**, 321
Cyclic enamines and imines, **6**, 147
Cyclic hydroxamic acids, **10**, 199
Cyclic peroxides, **8**, 165
Cycloaddition, cationic polar, **16**, 289 (**19**, xi)
(2 + 2)-Cycloaddition and (2 + 2)-cycloreversion reactions of heterocyclic compounds, **21**, 253

D

Developments in the chemistry
  of furans (1952–1963), **7**, 377
  of Reissert compounds (1968–1978), **24**, 187
2,4-Dialkoxypyrimidines, Hilbert–Johnson reaction of, **8**, 115
Diazepines, chemistry of, **8**, 21
1,4-Diazepines, 2,3-dihydro-, **17**, 1
Diazirines, diaziridines, **2**, 83; **24**, 63
Diazo compounds, heterocyclic, **8**, 1
Diazomethane, reactions with heterocyclic compounds, **2**, 245
Dibenzothiophenes, chemistry of, **16**, 181
2,3-Dihydro-1,4-diazepines, **17**, 1
1,2-Dihydroisoquinolines, **14**, 279
1,2-Dioxetanes, chemistry of, **21**, 437
Diquinolylmethane and its analogs, **7**, 153
1,2- and 1,3-Dithiolium ions, **7**, 39; **27**, 151

E

Electrolysis of N-heterocyclic compounds, **12**, 213

Electronic aspects of purine tautomerism, **13**, 77
Electronic structure of biological pyrimidines, tautomerism and, **18**, 199
Electronic structure of heterocyclic sulfur compounds, **5**, 1
Electrophilic substitutions of five-membered rings, **13**, 235
π-Excessive heteroannulenes, medium-large and large, **23**, 55

F

Ferrocenes, heterocyclic, **13**, 1
Five-membered rings, electrophilic substitutions of, **13**, 235
Free radical substitutions of heteroaromatic compounds, **2**, 131
Furans, development of the chemistry of (1952–1963), **7**, 377

G

Grignard reagents, indole, **10**, 43

H

Halogenation of heterocyclic compounds, **7**, 1
Hammett equation, applications to heterocyclic compounds, **3**, 209; **20**, 1
Hetarynes, **4**, 121
Heteroannulenes, medium-large and large π-excessive, **23**, 55
Heteroaromatic compounds
  free-radical substitutions of, **2**, 131
  homolytic substitution of, **16**, 123
  nitrogen, covalent hydration in, **4**, 1, 43
  prototropic tautomerism of, **1**, 311, 339; **2**, 1, 27; Suppl. 1
  quaternization of, **22**, 71
Heteroaromatic N-imines, **17**, 213
Heteroaromatic nitro compounds, ring synthesis of, **25**, 113
Heteroaromatic radicals, Part I, general properties; radicals with Group V ring heteroatoms, **25**, 205; Part II, radicals

with Group VI and Groups V and VI ring heteroatoms, 27, 31
Heteroaromatic substitution, nucleophilic, 3, 285
Heterocycles
 aromaticity of, 17, 255
 nomenclature of, 20, 175
 photochemistry of, 11, 1
 by ring closure of ortho-substituted t-anilines, 14, 211
 synthesis of, through nucleophilic additions to acetylenic esters, 19, 279
 thioureas in synthesis of, 18, 99
Heterocyclic betaine derivatives of alternant hydrocarbons, 26, 1
Heterocyclic chemistry, literature of, 7, 225; 25, 303
Heterocyclic compounds
 application of Hammett equation to, 3, 209; 20, 1
 (2 + 2)-cycloaddition and (2 + 2)-cycloreversion reactions of, 21, 253
 halogenation of, 7, 1
 isotopic hydrogen labeling of, 15, 137
 mass spectrometry of, 7, 301
 quaternization of, 3, 1; 22, 71
 reactions of, with carbenes, 3, 57
 reactions of diazomethane with, 2, 245
N-Heterocyclic compounds
 electrolysis of, 12, 213
 reaction of acetylenecarboxylic acids and esters with, 1, 125; 23, 263
Heterocyclic diazo compounds, 8, 1
Heterocyclic ferrocenes, 13, 1
Heterocyclic oligomers, 15, 1
Heterocyclic pseudobases, 1, 167; 25, 1
Heterocyclic sulphur compounds, electronic structure of, 5, 1
Heterocyclic synthesis, from nitrilium salts under acidic conditions, 6, 95
Hilbert–Johnson reaction of 2,4-dialkoxypyrimidines, 8, 115
Homolytic substitution of heteroaromatic compounds, 16, 123
Hydrogen exchange
 base-catalyzed, 16, 1
 one-step (labeling) methods, 15, 137
Hydroxamic acids, cyclic, 10, 199

I

Imidazole chemistry, advances in, 12, 103; 27, 241
N-Imines, heteroaromatic, 17, 213
Indole Grignard reagents, 10, 43
Indole(s)
 acid-catalyzed polymerization, 2, 287
 and derivatives, application of NMR spectroscopy to, 15, 277
Indolizine chemistry, advances in, 23, 103
Indolones, isatogens and, 22, 123
Indoxazenes, 8, 277
Isatin, chemistry of, 18, 1
Isatogens and indolones, 22, 123
Isoindoles, 10, 113
Isoquinolines
 1,2-dihydro-, 14, 279
 4-oxy- and 4-keto-1,2,3,4-tetrahydro-, 15, 99
Isothiazoles, 4, 107
 recent advances in the chemistry of monocyclic, 14, 1
Isotopic hydrogen labeling of heterocyclic compounds, one-step methods, 15, 137
Isoxazole chemistry, recent developments in, 2, 365; since 1963, 25, 147
Isoxazolidines, chemistry of, 21, 207

L

Lactim ethers, chemistry of, 12, 185
Literature of heterocyclic chemistry, 7, 225; 25, 303

M

Mass spectrometry of heterocyclic compounds, 7, 301
Medium-large and large π-excessive heteroannulenes, 23, 55
Meso-ionic compounds, 19, 1
Metal catalysts, action on pyridines, 2, 179
Monoazaindoles, 9, 27
Monocyclic pyrroles, oxidation of, 15, 67
Monocyclic sulfur-containing pyrones, 8, 219
Mononuclear isothiazoles, recent advances in chemistry of, 14, 1

N

Naphthalen-1,4-imines, **16,** 87
Naphthyridines, **11,** 124
Nitriles and nitrilium salts, heterocyclic syntheses involving, **6,** 95
Nitrogen-bridged six-membered ring systems, **16,** 87
Nitrogen heterocycles
    covalent hydration in, **20,** 117
    reactions of acetylenecarboxylic esters with, **23,** 263
    reduction of, with complex metal hydrides, **6,** 45
Nitrogen heterocyclic systems, Claisen rearrangements in, **8,** 143
Nomenclature of heterocycles, **20,** 175
Nuclear magnetic resonance spectroscopy, application to indoles, **15,** 277
Nucleophiles, reactivity of azine derivatives with, **4,** 145
Nucleophilic additions to acetylenic esters, synthesis of heterocycles through, **19,** 299
Nucleophilic heteroaromatic substitution, **3,** 285

O

Olefin synthesis with anils, **23,** 171
Oligomers, heterocyclic, **15,** 1
1,2,4-Oxadiazoles, **20,** 65
1,3,4-Oxadiazole chemistry, recent advances in, **7,** 183
1,3-Oxazine derivatives, **2,** 311; **23,** 1
Oxaziridines, **2,** 83; **24,** 63
Oxazole chemistry, advances in, **17,** 99
Oxazolone chemistry
    new developments in, **21,** 175
    recent advances in, **4,** 75
Oxidation of monocyclic pyrroles, **15,** 67
3-Oxo-2,3-dihydrobenz[d]isothiazole-1,1-dioxide (saccharin) and derivatives, **15,** 233
4-Oxy- and 4-keto-1,2,3,4-tetrahydroisoquinolines, chemistry of, **15,** 99

P

Pentazoles, **3,** 373
Peroxides, cyclic, **8,** 165 (see also 1,2-Dioxetanes)
Phenanthridine chemistry, recent developments in, **13,** 315
Phenanthrolines, **22,** 1
Phenothiazines, chemistry of, **9,** 321
Phenoxazines, **8,** 83
Photochemistry of heterocycles, **11,** 1
Physicochemical aspects of purines, **6,** 1; **24,** 215
Physicochemical properties
    of azines, **5,** 69
    of pyrroles, **11,** 383
3-Piperideines, **12,** 43
Polymerization of pyrroles and indoles, acid-catalyzed, **2,** 287
Prototropic tautomerism of heteroaromatic compounds, **1,** 311, 339; **2,** 1, 27; Suppl. 1
Pseudobases, heterocyclic, **1,** 167; **25,** 1
Purine bases, aza analogs of, **1,** 189
Purines
    physicochemical aspects of, **6,** 1; **24,** 215
    tautomerism, electronic aspects of, **13,** 77
Pyrazine chemistry, recent advances in, **14,** 99
Pyrazole chemistry, progress in, **6,** 347
Pyridazines, **9,** 211; **24,** 363
Pyridine(s)
    action of metal catalysts on, **2,** 179
    effect of substituents on substitution in, **6,** 229
    1,2,3,6-tetrahydro-, **12,** 43
Pyridoindoles (the carbolines), **3,** 79
Pyridopyrimidines, **10,** 149
Pyrimidine bases, aza analogs of, **1,** 189
Pyrimidines
    2,4-dialkoxy-, Hilbert–Johnson reaction of, **8,** 115
    tautomerism and electronic structure of biological, **18,** 199
1-Pyrindines, chemistry of, **15,** 197
Pyrones, monocyclic sulfur-containing, **8,** 219
Pyrroles
    acid-catalyzed polymerization of, **2,** 287
    oxidation of monocyclic, **15,** 67
    physicochemical properties of, **11,** 383
Pyrrolizidine chemistry, **5,** 315; **24,** 247
Pyrrolodiazines, with a bridgehead nitrogen, **21,** 1
Pyrrolopyridines, **9,** 27
Pyrylium salts, syntheses, **10,** 241

Q

Quaternization
  of heteroaromatic compounds, **22,** 71
  of heterocyclic compounds, **3,** 1
Quinazolines, **1,** 253; **24,** 1
Quinolizines, aromatic, **5,** 291
Quinoxaline chemistry
  developments 1963-1975, **22,** 367
  recent advances in, **2,** 203
Quinuclidine chemistry, **11,** 473

R

Reduction of nitrogen heterocycles with complex metal hydrides, **6,** 45
Reissert compounds, **9,** 1; **24,** 187
Ring closure of ortho-substituted t-anilines, for heterocycles, **14,** 211
Ring synthesis of heteroaromatic nitro compounds, **25,** 113

S

Saccharin and derivatives, **15,** 233
Selenazole chemistry, present state of, **2,** 343
Selenium-nitrogen heterocycles, **24,** 109
Selenophene chemistry, advances in, **12,** 1
Six-membered ring systems, nitrogen bridged, **16,** 87
Substitution(s),
  electrophilic, of five-membered rings, **13,** 235
  homolytic, of heteroaromatic compounds, **16,** 123
  nucleophilic heteroaromatic, **3,** 285
  in pyridines, effect of substituents, **6,** 229
Sulfur compounds, electronic structure of heterocyclic, **5,** 1
Synthesis and reactions of 1-azirines, **13,** 45
Synthesis of heterocycles through nucleophilic additions to acetylenic esters, **19,** 279

T

Tautomerism
  electronic aspects of purine, **13,** 77
  and electronic structure of biological pyrimidines, **18,** 199
  prototropic, of heteroaromatic compounds, **1,** 311, 339; **2,** 1, 27; Suppl. 1
Tellurophene and related compounds, **21,** 119
1,2,3,4-Tetrahydroisoquinolines, 4-oxy- and 4-keto-, **15,** 99
1,2,3,6-Tetrahydropyridines, **12,** 43
Tetrazole chemistry, recent advances in, **21,** 323
Theoretical studies of physicochemical properties and reactivity of azines, **5,** 69
1,2,4-Thiadiazoles, **5,** 119
1,2,5-Thiadiazoles, chemistry of, **9,** 107
1,3,4-Thiadiazoles, recent advances in the chemistry of, **9,** 165
Thiathiophthenes (1,6,6aS^{IV}-Trithiapentalenes), **13,** 161
1,2,3,4-Thiatriazoles, **3,** 263; **20,** 145
1,4-Thiazines and their dihydro-derivatives, **24,** 293
4-Thiazolidinones, **25,** 83
Thienopyridines, **21,** 65
Thienothiophenes and related systems, chemistry of, **19,** 123
Thiochromanones and related compounds, **18,** 59
Thiocoumarins, **26,** 115
Thiophenes, chemistry of, recent advances in, **1,** 1
Thiopyrones (monocyclic sulfur-containing pyrones), **8,** 219
Thioureas in synthesis of heterocycles, **18,** 99
Three-membered rings with two heteroatoms, **2,** 83; **24,** 63
1,3,5-, 1,3,6-, 1,3,7-, and 1,3,8-Triazanaphthalenes, **10,** 149
1,2,3-Triazines, **19,** 215
1,2,3-Triazoles, **16,** 33
1,6,6aS^{IV}-Trithiapentalenes, **13,** 161